AF871553

CLASSICAL GEOMETRY

CLASSICAL GEOMETRY

Euclidean, Transformational, Inversive, and Projective

I. E. Leonard, J. E. Lewis, A. C. F. Liu, G. W. Tokarsky
Department of Mathematical and Statistical Sciences
University of Alberta

WILEY

Copyright © 2014 by John Wiley & Sons, Inc. All rights reserved.

Published by John Wiley & Sons, Inc., Hoboken, New Jersey.
Published simultaneously in Canada.

No part of this publication may be reproduced, stored in a retrieval system or transmitted in any form or by any means, electronic, mechanical, photocopying, recording, scanning or otherwise, except as permitted under Section 107 or 108 of the 1976 United States Copyright Act, without either the prior written permission of the Publisher, or authorization through payment of the appropriate per-copy fee to the Copyright Clearance Center, Inc., 222 Rosewood Drive, Danvers, MA 01923, (978) 750-8400, fax (978) 750-4470, or on the web at www.copyright.com. Requests to the Publisher for permission should be addressed to the Permissions Department, John Wiley & Sons, Inc., 111 River Street, Hoboken, NJ 07030, (201) 748-6011, fax (201) 748-6008, or online at http://www.wiley.com/go/permission.

Limit of Liability/Disclaimer of Warranty: While the publisher and author have used their best efforts in preparing this book, they make no representation or warranties with respect to the accuracy or completeness of the contents of this book and specifically disclaim any implied warranties of merchantability or fitness for a particular purpose. No warranty may be created or extended by sales representatives or written sales materials. The advice and strategies contained herein may not be suitable for your situation. You should consult with a professional where appropriate. Neither the publisher nor author shall be liable for any loss of profit or any other commercial damages, including but not limited to special, incidental, consequential, or other damages.

For general information on our other products and services please contact our Customer Care Department within the United States at (800) 762-2974, outside the United States at (317) 572-3993 or fax (317) 572-4002.

Wiley also publishes its books in a variety of electronic formats. Some content that appears in print, however, may not be available in electronic formats. For more information about Wiley products, visit our web site at www.wiley.com.

Library of Congress Cataloging-in-Publication Data:

Leonard, I. Ed., 1938– author.
Classical geometry : Euclidean, transformational, inversive, and projective / I. E. Leonard, Department of Mathematical and Statistical Sciences, University of Alberta, Edmonton, Canada, J. E. Lewis, Department of Mathematical and Statistical Sciences, University of Alberta, Edmonton, Canada, A. C. F. Liu, Department of Mathematical and Statistical Sciences, University of Alberta, Edmonton, Canada, G. W. Tokarsky, Department of Mathematical and Statistical Sciences, University of Alberta, Edmonton, Canada.
pages cm
Includes bibliographical references and index.
ISBN 978-1-118-67919-7 (hardback)
1. Geometry. I. Lewis, J. E. (James Edward) author. II. Liu, A. C. F. (Andrew Chiang-Fung) author. III. Tokarsky, G. W., author. IV. Title.
QA445.L46 2014
516-dc23 2013042035

10 9 8 7 6 5 4 3 2 1

CONTENTS

PREFACE

It is sometimes said that geometry should be studied because it is a useful and valuable discipline, but in fact many people study it simply because geometry is a very enjoyable subject. It is filled with problems at every level that are entertaining and elegant, and this enjoyment is what we have attempted to bring to this textbook.

This text is based on class notes that we developed for a three-semester sequence of undergraduate geometry courses that has been taught at the University of Alberta for many years. It is appropriate for students from all disciplines who have previously studied high school algebra, geometry, and trigonometry.

When we first started teaching these courses, our main problem was finding a suitable method for teaching geometry to university students who have had minimal experience with geometry in high school. We experimented with material from high school texts but found it was not challenging enough. We also tried an axiomatic approach, but students often showed little enthusiasm for proving theorems, particularly since the early theorems seemed almost as self-evident as the axioms. We found the most success by starting early with problem solving, and this is the approach we have incorporated throughout the book.

The geometry in this text is synthetic rather than Cartesian or coordinate geometry. We remain close to classical themes in order to encourage the development of geometric intuition, and for the most part we avoid abstract algebra although we do demonstrate its use in the sections on transformational geometry.

Part I is about Euclidean geometry; that is, the study of the properties of points and lines that are invariant under isometries and similarities. As well as many of the usual topics, it includes material that many students will not have seen, for example, the theorems of Ceva and Menelaus and their applications. Part I is the basis for Parts II and III.

Part II discusses the properties of Euclidean transformations or isometries of the plane (translations, reflections, and rotations and their compositions). It also introduces groups and their use in studying transformations.

Part III introduces inversive and projective geometry. These subjects are presented as natural extensions of Euclidean geometry, with no abstract algebra involved.

We would like to acknowledge our late colleagues George Cree and Murray Klamkin, without whose inspiration and encouragement over the years this project would not have been possible.

Finally, we would like to thank our families for their patience and understanding in the preparation of the textbook. In particular, I. E. Leonard would like to thank Sarah for proofreading the manuscript numerous times.

ED, TED, ANDY, AND GEORGE

Edmonton, Alberta, Canada
January, 2014

PART I

EUCLIDEAN GEOMETRY

CHAPTER 1

CONGRUENCY

1.1 Introduction

Assumed Knowledge

This text assumes a bit of knowledge on the part of the reader. For example, it assumes that you know that the sum of the angles of a triangle in the plane is 180° ($x + y + z = 180^\circ$ in the figure below), and that in a right triangle with hypotenuse c and sides a and b, the Pythagorean relation holds: $c^2 = a^2 + b^2$.

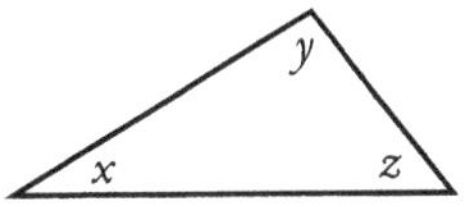

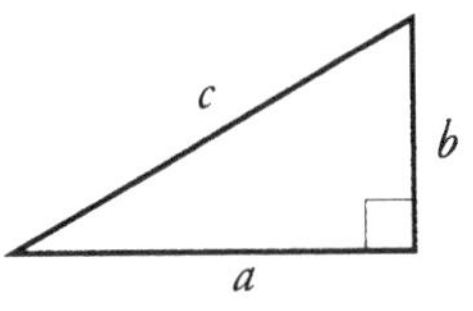

Classical Geometry: Euclidean, Transformational, Inversive and Projective, First Edition. By
I. E. Leonard, J. E. Lewis, A. C. F. Liu, G. W. Tokarsky
Copyright © 2014 John Wiley & Sons, Inc.

We use the word ***line*** to mean *straight line*, and we assume that you know that two lines either do not intersect, intersect at exactly one point, or completely coincide. Two lines that do not intersect are said to be ***parallel***.

We also assume certain knowledge about parallel lines, namely, that you have seen some form of the ***parallel axiom***:

Given a line l and a point P in the plane, there is exactly one line through P parallel to l.

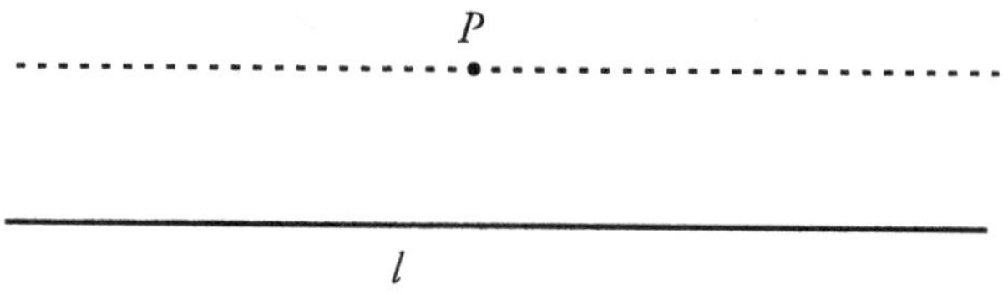

The preceding version of the parallel axiom is often called ***Playfair's Axiom***. You may even know something equivalent to it that is close to the original version of the ***parallel postulate***:

Given two lines l and m, and a third line t cutting both l and m and forming angles ϕ and θ on the same side of t, if $\phi + \theta < 180°$, then l and m meet at a point on the same side of t as the angles.

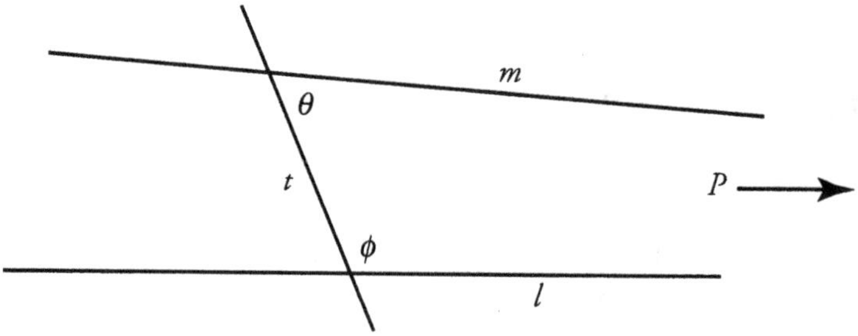

The subject of this part of the text is Euclidean geometry, and the above-mentioned parallel postulate characterizes Euclidean geometry. Although the postulate may seem to be obvious, there are perfectly good geometries in which it does not hold.

We also assume that you know certain facts about areas. A ***parallelogram*** is a quadrilateral (figure with four sides) such that the opposite sides are parallel.

The area of a parallelogram with base b and height h is $b \cdot h$, and the area of a triangle with base b and height h is $b \cdot h/2$.

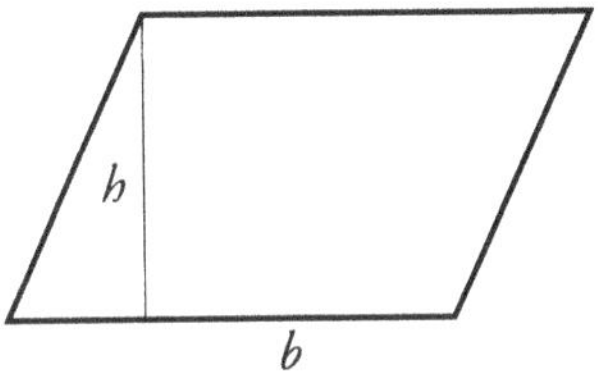

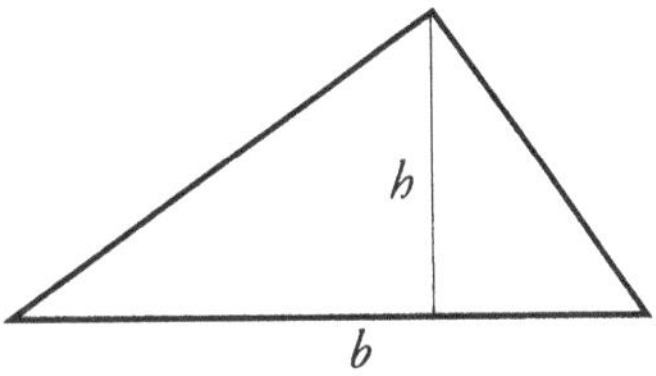

Notation and Terminology

Throughout this text, we use uppercase Latin letters to denote points and lowercase Latin letters to denote lines and rays. Given two points A and B, there is one and only one line through A and B. A ***ray*** is a half-line, and the notation $\overrightarrow{AB}$ denotes the ray starting at A and passing through B. It consists of the points A and B, all points between A and B, and all points X on the line such that B is between A and X.

Given rays $\overrightarrow{AB}$ and $\overrightarrow{AC}$, we denote by $\angle BAC$ the angle formed by the two rays (the shaded region in the following figure). When no confusion can arise, we sometimes use $\angle A$ instead of $\angle BAC$. We also use lowercase letters, either Greek or Latin, to denote angles.

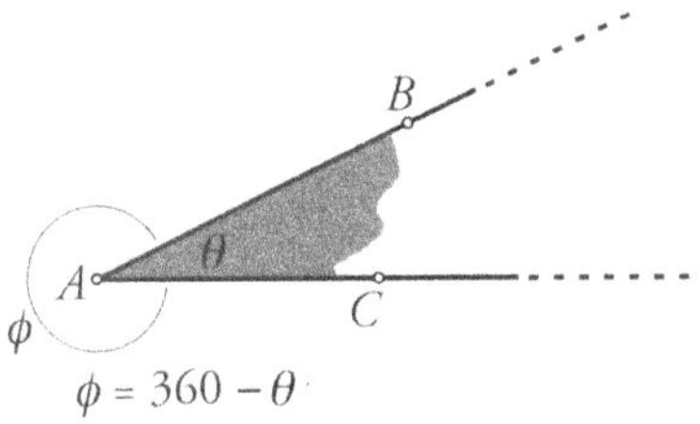

When two rays form an angle other than 180°, there are actually two angles to talk about: the smaller angle (sometimes called the ***interior angle***) and the larger angle (called the ***reflex angle***). When we refer to $\angle BAC$, we always mean the nonreflex angle.

Note. The angles that we are talking about here are *undirected angles*; that is, they do not have negative values, and so can range in magnitude from 0° to 360°. Some people prefer to use $m(\angle A)$ for the measure of the angle A; however, we will use the same notation for both the angle and the measure of the angle.

When we refer to a quadrilateral as $ABCD$ we mean one whose edges are AB, BC, CD, and DA. Thus, the quadrilateral $ABCD$ and the quadrilateral $ABDC$ are quite different.

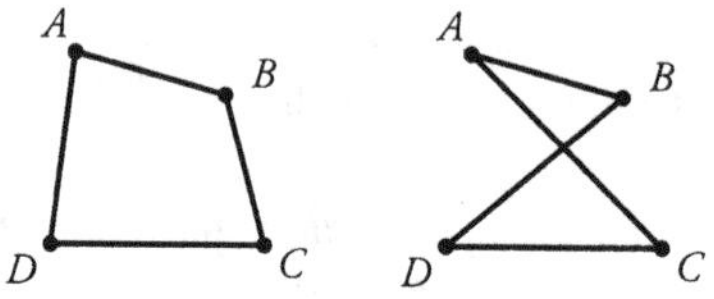

There are three classifications of quadrilaterals: convex, simple, and nonsimple, as shown in the following diagram.

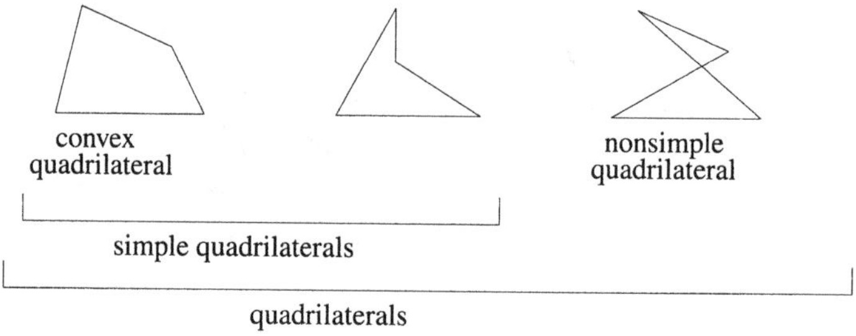

1.2 Congruent Figures

Two figures that have exactly the same shape and exactly the same size are said to be ***congruent***. More explicitly:

1. Two angles are ***congruent*** if they have the same measure.
2. Two line segments are ***congruent*** if they are the same length.
3. Two circles are ***congruent*** if they have the same radius.
4. Two triangles are ***congruent*** if corresponding sides and angles are the same size.
5. All rays are ***congruent***.
6. All lines are ***congruent***.

Theorem 1.2.1. *Vertically opposite angles are congruent.*

Proof. We want to show that $a = b$. We have

$$a + c = 180 \text{ and } b + c = 180,$$

and it follows from this that $a = b$.

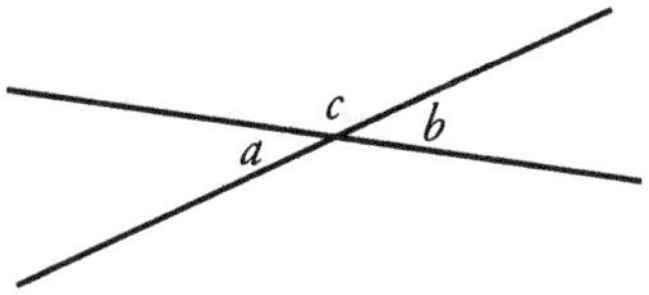

□

Notation. The symbol $\equiv$ denotes congruence. We use the notation $\triangle ABC$ to denote a triangle with vertices A, B, and C, and we use $\mathcal{C}(P, r)$ to denote a circle with center P and radius r.

Thus, $\mathcal{C}(P, r) \equiv \mathcal{C}(Q, s)$ if and only if $r = s$.

We will be mostly concerned with the notion of congruent triangles, and we mention that in the definition, $\triangle ABC \equiv \triangle DEF$ if and only if the following six conditions hold:

$$\begin{aligned}
\angle A &\equiv \angle D \\
\angle B &\equiv \angle E \\
\angle C &\equiv \angle F \\
AB &\equiv DE \\
BC &\equiv EF \\
AC &\equiv DF.
\end{aligned}$$

Note that the two statements $\triangle ABC \equiv \triangle DEF$ and $\triangle ABC \equiv \triangle EFD$ are not the same!

The Basic Congruency Conditions

According to the definition of congruency, two triangles are congruent if and only if six different parts of one are congruent to the six corresponding parts of the other. Do we really need to check all six items? The answer is no.

If you give three straight sticks to one person and three identical sticks to another and ask both to constuct a triangle with the sticks as the sides, you would expect the two triangles to be exactly the same. In other words, you would expect that it is possible to verify congruency by checking that the three corresponding sides are congruent. Indeed this is the case, and, in fact, there are several ways to verify congruency without checking all six conditions.

The three congruency conditions that are used most often are the Side-Angle-Side (**SAS**) condition, the Side-Side-Side (**SSS**) condition, and the Angle-Side-Angle (**ASA**) condition.

Axiom 1.2.2. *(SAS Congruency)*

Two triangles are congruent if two sides and the included angle of one are congruent to two sides and the included angle of the other.

Theorem 1.2.3. *(SSS Congruency)*

Two triangles are congruent if the three sides of one are congruent to the corresponding three sides of the other.

Theorem 1.2.4. *(ASA Congruency)*

Two triangles are congruent if two angles and the included side of one are congruent to two angles and the included side of the other.

You will note that the **SAS** condition is an axiom, and the other two are stated as theorems. We will not prove the theorems but will freely use all three conditions.

Any one of the three conditions could be used as an axiom with the other two then derived as theorems. In case you are wondering why the **SAS** condition is preferred as the basic axiom rather than the **SSS** condition, it is because *it is always possible to construct a triangle given two sides and the included angle*, whereas it is not always possible to construct a triangle given three sides (consider sides of length 3, 1, and 1).

Axiom 1.2.5. *(The Triangle Inequality)*

The sum of the lengths of two sides of a triangle is always greater than the length of the remaining side.

The congruency conditions are useful because they allow us to conclude that certain parts of two triangles are congruent by determining that certain other parts are congruent.

Here is how congruency may be used to prove two well-known theorems about isosceles triangles. (An ***isosceles*** triangle is one that has two equal sides.)

Theorem 1.2.6. *(The Isosceles Triangle Theorem)*

In an isosceles triangle, the angles opposite the equal sides are equal.

Proof. Let us suppose that the triangle is ABC with $AB = AC$.

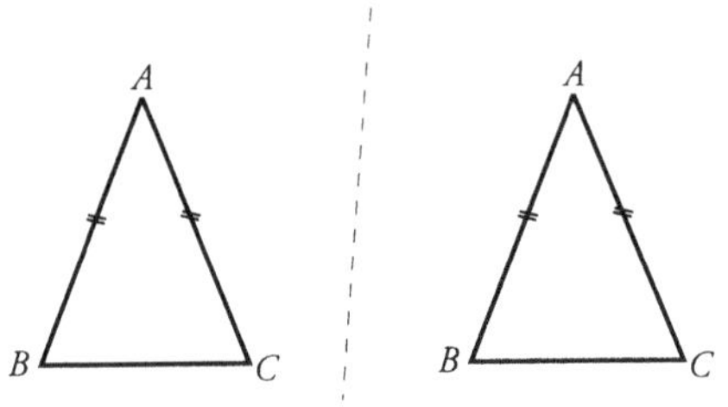

(same triangle, recopied for reference purposes)

In $\triangle ABC$ and $\triangle ACB$ we have

$$\begin{aligned} AB &= AC, \\ \angle BAC &= \angle CAB, \\ AC &= AB, \end{aligned}$$

so $\triangle ABC \equiv \triangle ACB$ by **SAS**.

Since the triangles are congruent, it follows that all corresponding parts are congruent, so $\angle B$ of $\triangle ABC$ must be congruent to $\angle C$ of $\triangle ACB$.

□

Theorem 1.2.7. *(Converse of the Isosceles Triangle Theorem)*

If in $\triangle ABC$ we have $\angle B = \angle C$, then $AB = AC$.

Proof. In $\triangle ABC$ and $\triangle ACB$ we have

$$\begin{aligned} \angle ABC &= \angle ACB, \\ BC &= CB, \\ \angle ACB &= \angle ABC, \end{aligned}$$

so $\triangle ABC \equiv \triangle ACB$ by **ASA**.

Since $\triangle ABC \equiv \triangle ACB$ it follows that $AB = AC$.

□

Perhaps now is a good time to explain what the converse of a statement is. Many statements in mathematics have the form

If $\mathcal{P}$, then $\mathcal{Q}$,

where $\mathcal{P}$ and $\mathcal{Q}$ are assertions of some sort.

For example:

If $ABCD$ is a square, then angles A, B, C, and D are all right angles.

Here, $\mathcal{P}$ is the assertion "$ABCD$ is a square," and $\mathcal{Q}$ is the assertion "angles A, B, C, and D are all right angles."

The ***converse*** of the statement "If $\mathcal{P}$, then $\mathcal{Q}$" is the statement

If $\mathcal{Q}$, then $\mathcal{P}$.

Thus, the converse of the statement "If $ABCD$ is a square, then angles A, B, C, and D are all right angles" is the statement

If angles A, B, C, and D are all right angles, then $ABCD$ is a square.

A common error in mathematics is to confuse a statement with its converse. Given a statement and its converse, if one of them is true, it does not automatically follow that the other is also true.

Exercise 1.2.8. *For each of the following statements, state the converse and determine whether it is true or false.*

1. *Given triangle ABC, if $\angle ABC$ is a right angle, then $AB^2 + BC^2 = AC^2$.*
2. *If $ABCD$ is a parallelogram, then $AB = CD$ and $AD = BC$.*
3. *If $ABCD$ is a convex quadrilateral, then $ABCD$ is a rectangle.*
4. *Given quadrilateral $ABCD$, if $AC \neq BD$, then $ABCD$ is not a rectangle.*

Solutions to the exercises are given at the end of the chapter.

The Isosceles Triangle Theorem and its converse raise questions about how sides are related to unequal angles, and there are useful theorems for this case.

Theorem 1.2.9. *(The Angle-Side Inequality)*
In $\triangle ABC$, if $\angle ABC > \angle ACB$, then $AC > AB$.

Proof. Draw a ray BX so that $\angle CBX \equiv \angle BCA$ with X to the same side of BC as A, as in the figure on the following page.

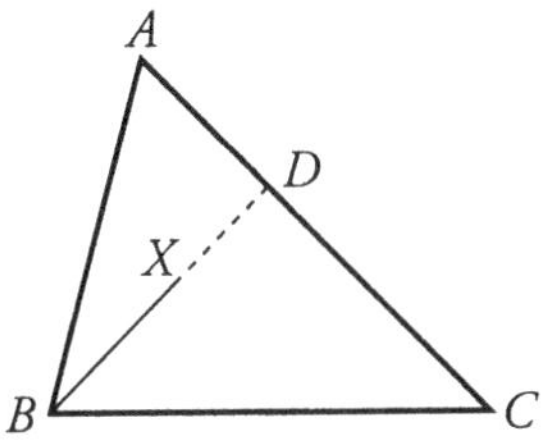

Since $\angle ABC > \angle CBX$, the point X is interior to $\angle ABC$ and so BX will cut side AC at a point D. Then we have

$$DB = DC$$

by the converse to the Isosceles Triangle Theorem.

By the Triangle Inequality, we have

$$AB < AD + DB,$$

and combining these gives us

$$AB < AD + DC = AC,$$

which is what we wanted to prove.

□

The converse of the Angle-Side Inequality is also true. Note that the proof of the converse uses the statement of the original theorem. This is something that frequently occurs when proving that the converse is true.

Theorem 1.2.10. *In $\triangle ABC$, if $AC > AB$, then $\angle ABC > \angle ACB$.*

Proof. There are three possible cases to consider:

(1) $\angle ABC = \angle ACB$.

(2) $\angle ABC < \angle ACB$.

(3) $\angle ABC > \angle ACB$.

If case (1) arises, then $AC = AB$ by the converse to the Isosceles Triangle Theorem, so case (1) cannot in fact arise. If case (2) arises, then $AC < AB$ by the Angle-Side Inequality, so (2) cannot arise. The only possibility is therefore case (3).

□

The preceding examples, as well as showing how congruency is used, are facts that are themselves very useful. They can be summarized very succinctly: in a triangle,

Equal angles are opposite equal sides.
The larger angle is opposite the larger side.

1.3 Parallel Lines

Two lines in the plane are ***parallel*** if

(a) they do not intersect or

(b) they are the same line.

Note that (b) means that a line is parallel to itself.

Notation. We use $l \parallel m$ to denote that the lines l and m are parallel and sometimes use $l \not\parallel m$ to denote that they are not parallel. If l and m are not parallel, they meet at precisely one point in the plane.

When a transversal crosses two other lines, various pairs of angles are endowed with special names:

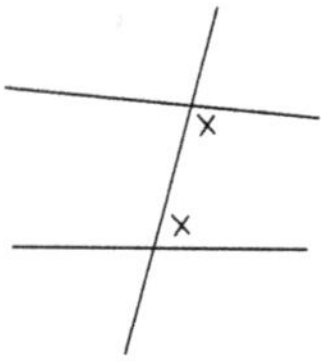

adjacent angles

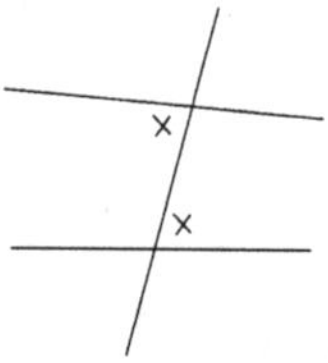

opposite or alternate angles

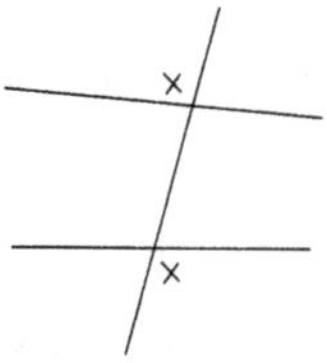

opposite or alternate exterior angles

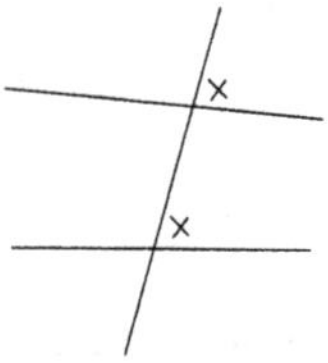

corresponding angles

The proofs of the next two theorems are omitted; however, we mention that the proof of Theorem 1.3.2 requires the parallel postulate, but the proof of Theorem 1.3.1 does not.

Theorem 1.3.1. *If a transversal cuts two lines and any one of the following four conditions holds, then the lines are parallel:*

(1) adjacent angles total 180°,

(2) alternate angles are equal,

(3) alternate exterior angles are equal,

(4) corresponding angles are equal.

Theorem 1.3.2. *If a transversal cuts two parallel lines, then all four statements of Theorem 1.3.1 hold.*

Remark. Theorem 1.3.1 can be proved using the External Angle Inequality, which is described below. The proof of the inequality itself ultimately depends on Theorem 1.3.1, but this would mean that we are using circular reasoning, which is not permitted. However, there is a proof of the External Angle Inequality which does not in any way depend upon Theorem 1.3.1, and so it is possible to avoid circular reasoning.

1.3.1 Angles in a Triangle

The parallel postulate is what distinguishes Euclidean geometry from other geometries, and as we see now, it is also what guarantees that the sum of the angles in a triangle is 180°.

Theorem 1.3.3. *The sum of the angles of a triangle is* 180°.

Proof. Given triangle ABC, draw the line XY through A parallel to BC, as shown. Consider AB as a transversal for the parallel lines XY and BC, then $x = u$ and similarly $y = v$. Consequently,

$$x + y + w = u + v + w = 180,$$

which is what we wanted to prove.

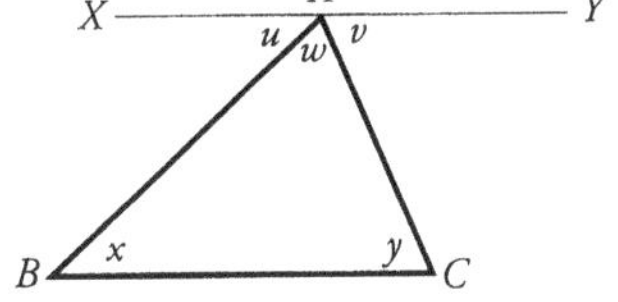

□

Given triangle ABC, extend the side BC beyond C to X. The angle ACX is called an ***exterior angle*** of $\triangle ABC$.

Theorem 1.3.4. *(The Exterior Angle Theorem)*

An exterior angle of a triangle is equal to the sum of the opposite interior angles.

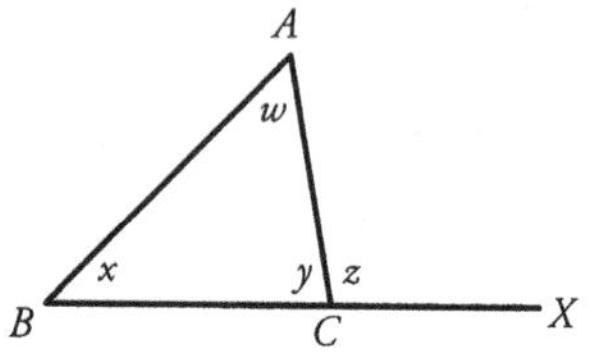

Proof. In the diagram above, we have $y + z = 180 = y + x + w$, so $z = x + w$.

□

The Exterior Angle Theorem has a useful corollary:

Corollary 1.3.5. *(The Exterior Angle Inequality)*

An exterior angle of a triangle is greater than either of the opposite interior angles.

Note. The proof of the Exterior Angle Inequality given above ultimately depends on the fact that the sum of the angles of a triangle is $180°$, which turns out to be equivalent to the parallel postulate. It is possible to prove the Exterior Angle Inequality without using any facts that follow from the parallel postulate, but we will omit that proof here.

1.3.2 Thales' Theorem

One of the most useful theorems about circles is credited to Thales, who is reported to have sacrificed two oxen after discovering the proof. (In truth, versions of the theorem were known to the Babylonians some one thousand years earlier.)

Theorem 1.3.6. *(Thales' Theorem)*

An angle inscribed in a circle is half the angle measure of the intercepted arc.

In the diagram, α is the measure of the inscribed angle, the arc CD is the intercepted arc, and β is the angle measure of the intercepted arc.

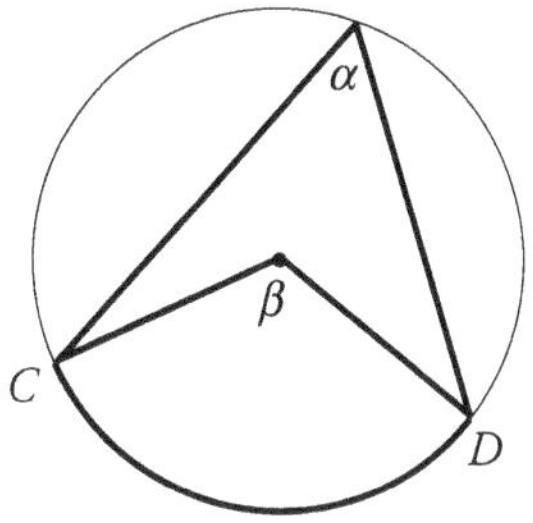

The following diagrams illustrate Thales' Theorem.

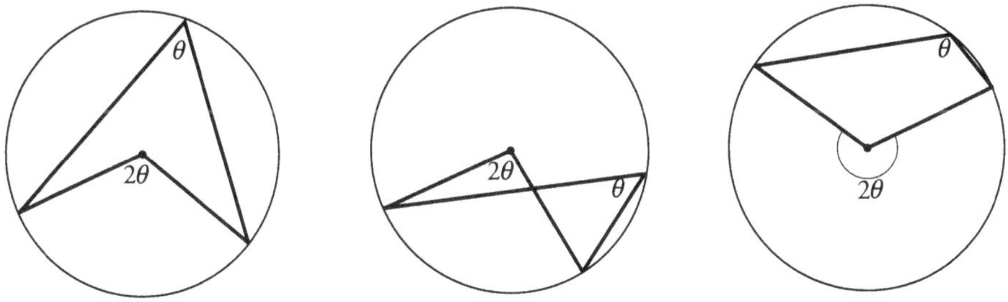

Proof. As the figures above indicate, there are several separate cases to consider. We will prove the first case and leave the others as exercises.

Referring to the diagram below, we have

$$\alpha = \nu + \omega \quad \text{and} \quad \beta = \mu + \eta.$$

But $\nu = \omega$ and $\mu = \eta$ (isosceles triangles). Consequently,

$$\angle BOC = \alpha + \beta = 2\omega + 2\eta = 2\angle BAC,$$

and the theorem follows.

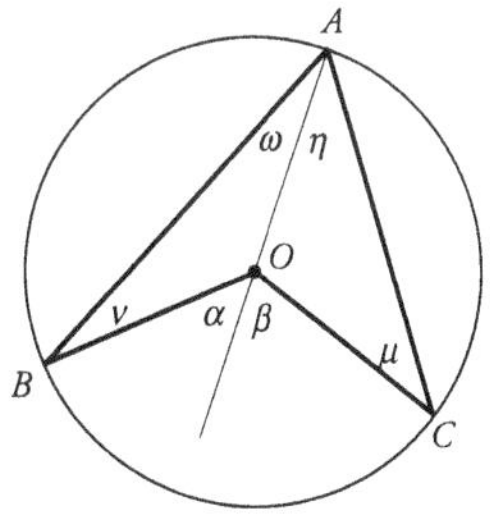

□

Thales' Theorem has several useful corollaries.

Corollary 1.3.7. *In a given circle:*

(1) All inscribed angles that intercept the same arc are equal in size.

(2) All inscribed angles that intercept congruent arcs are equal in size.

(3) The angle in a semicircle is a right angle.

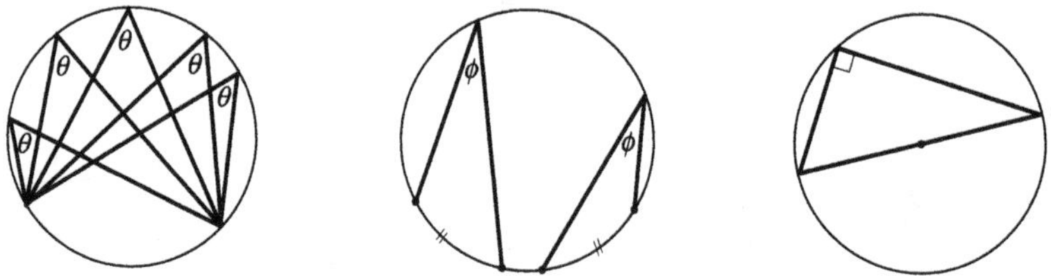

The converse of Thales' Theorem is also very useful.

Theorem 1.3.8. *(Converse of Thales' Theorem)*

Let $\mathcal{H}$ be a halfplane determined by a line PQ. The set of points in $\mathcal{H}$ that form a constant angle β with P and Q is an arc of a circle passing through P and Q.

Furthermore, every point of $\mathcal{H}$ inside the circle makes a larger angle with P and Q and every point of $\mathcal{H}$ outside the circle makes a smaller angle with P and Q.

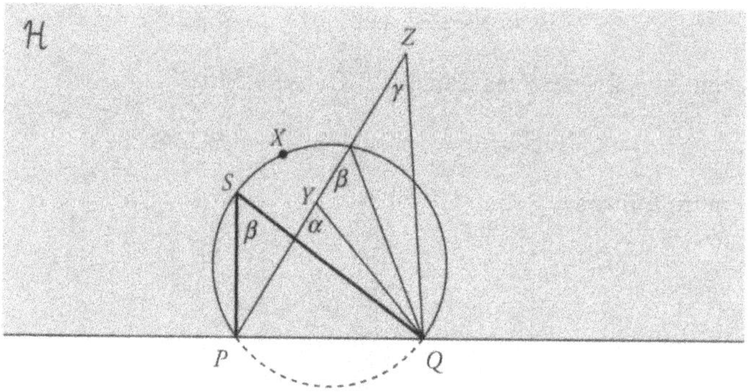

Proof. Let S be a point such that $\angle PSQ = \beta$ and let $\mathcal{C}$ be the circumcircle of $\triangle SPQ$. In the halfplane $\mathcal{H}$, all points X on $\mathcal{C}$ intercept the same arc of $\mathcal{C}$, so by Thales' Theorem, all angles PXQ have measure β.

From the Exterior Angle Inequality, in the figure on the previous page we have $\alpha > \beta > \gamma$. As a consequence, every point Z of $\mathcal{H}$ outside $\mathcal{C}$ must have $\angle PZQ < \beta$, and every point Y of $\mathcal{H}$ inside $\mathcal{C}$ must have $\angle PYQ > \beta$, and this completes the proof.

□

Exercise 1.3.9. *Calculate the size of* θ *in the following figure.*

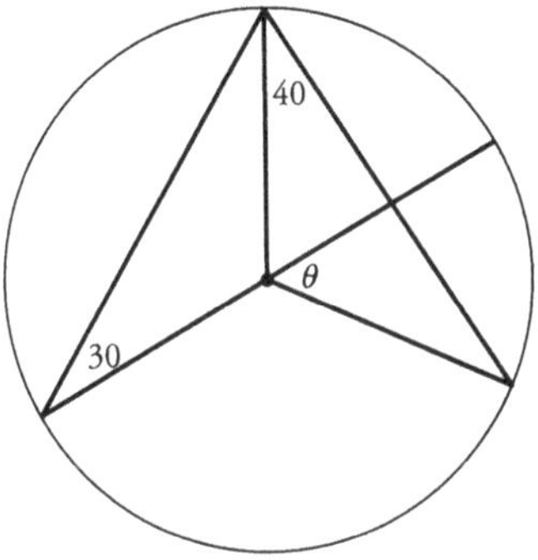

1.3.3 Quadrilaterals

The following theorem uses the fact that a simple quadrilateral always has at least one diagonal that is interior to the quadrilateral.

Theorem 1.3.10. *The sum of the interior angles of a simple quadrilateral is* $360°$.

Proof. Let the quadrilateral have vertices A, B, C, and D, with AC being an internal diagonal. Referring to the diagram, we have

$$\begin{aligned}\angle A + \angle B + \angle C + \angle D &= (\phi + \alpha) + \beta + (\gamma + \theta) + \delta \\ &= (\alpha + \beta + \gamma) + (\theta + \delta + \phi) \\ &= 180° + 180° \\ &= 360°.\end{aligned}$$

□

Note. This theorem is false if the quadrilateral is not simple, in which case the sum of the interior angles is less than $360°$.

Cyclic Quadrilaterals

A quadrilateral that is inscribed in a circle is called a ***cyclic quadrilateral*** or, equivalently, a ***concyclic quadrilateral***. The circle is called the ***circumcircle*** of the quadrilateral.

Theorem 1.3.11. *Let $ABCD$ be a simple cyclic quadrilateral. Then:*

(1) The opposite angles sum to $180°$.

(2) Each exterior angle is congruent to the opposite interior angle.

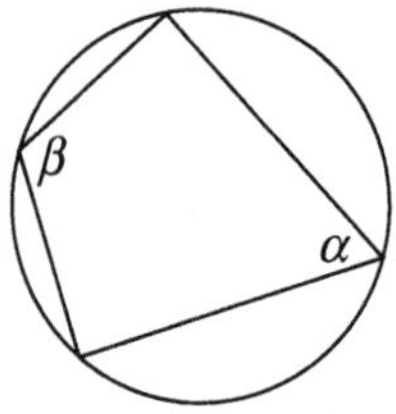

$\alpha + \beta = 180$

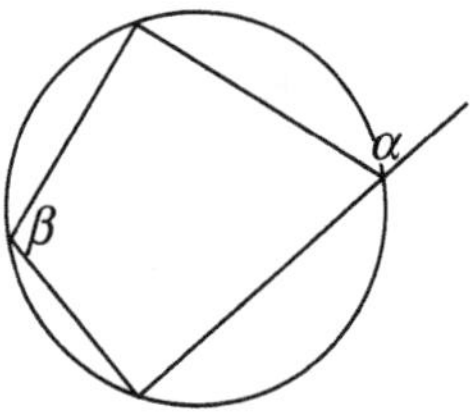

$\alpha = \beta$

Theorem 1.3.12. *Let $ABCD$ be a simple quadrilateral. If the opposite angles sum to $180°$, then $ABCD$ is a cyclic quadrilateral.*

We leave the proofs of Theorem 1.3.11 and Theorem 1.3.12 as exercises and give a similar result for nonsimple quadrilaterals.

Example 1.3.13. *(Cyclic Nonsimple Quadrilaterals)*

A nonsimple quadrilateral can be inscribed in a circle if and only if opposite angles are equal. For example, in the figure on the following page, the nonsimple quadrilateral $ABCD$ can be inscribed in a circle if and only if $\angle A = \angle C$ and $\angle B = \angle D$.

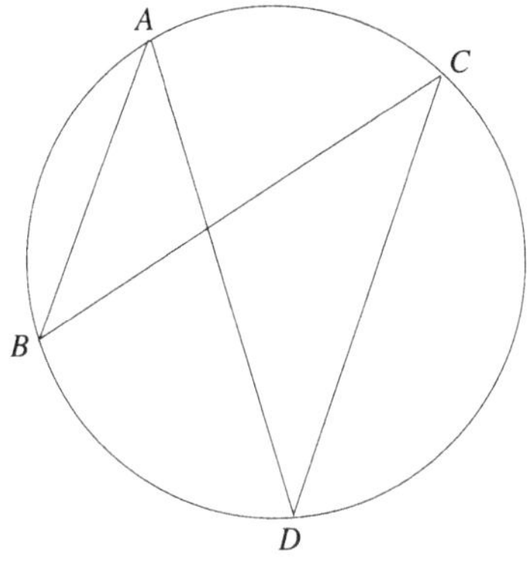

Solution. Suppose first that the quadrilateral $ABCD$ is cyclic. Then $\angle A = \angle C$ since they are both subtended by the chord BD, while $\angle B = \angle D$ since they are both subtended by the chord AC.

Conversely, suppose that $\angle A = \angle C$ and $\angle B = \angle D$, and let the circle below be the circumcircle of $\triangle ABC$.

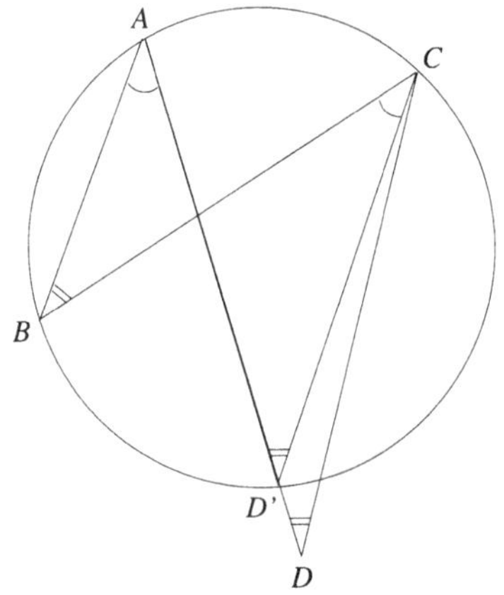

If the quadrilateral $ABCD$ is not cyclic, then the point D does not lie on this circumcircle. Assume that it lies outside the circle and let D' be the point where the line segment AD hits the circle. Since $ABCD'$ is a cyclic quadrilateral, then from the first part of the proof, $\angle B = \angle D'$ and therefore $\angle D = \angle D'$, which contradicts the External Angle Inequality in $\triangle CD'D$. Thus, if $\angle A = \angle C$ and $\angle B = \angle D$, then quadrilateral $ABCD$ is cyclic.

□

Exercise 1.3.14. *Show that a quadrilateral has an inscribed circle (that is, a circle tangent to each of its sides) if and only if the sums of the lengths of the two pairs of opposite sides are equal. For example, the quadrilateral $ABCD$ has an inscribed circle if and only if $AB + CD = AD + BC$.*

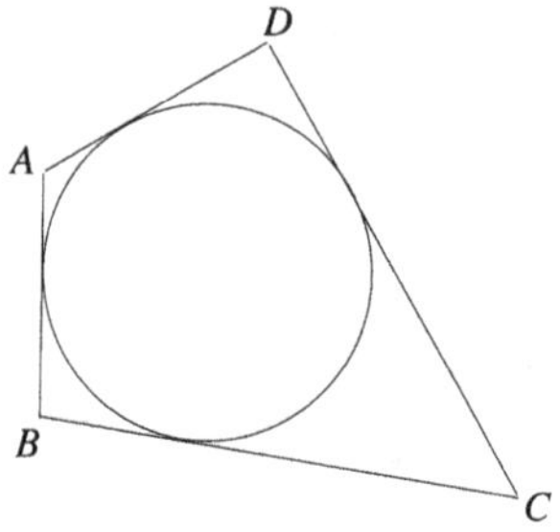

The following example is a result named after Robert Simson (1687–1768), whose *Elements of Euclid* was a standard textbook published in 24 editions from 1756 until 1834 and upon which many modern English versions of Euclid are based. However, in their book *Geometry Revisited*, Coxeter and Greitzer report that the result attributed to Simson was actually discovered later, in 1797, by William Wallace.

Example 1.3.15. *(Simson's Theorem)*

Given $\triangle ABC$ inscribed in a circle and a point P on its circumference, the perpendiculars dropped from P meet the sides of the triangle in three collinear points.

The line is called the **Simson line** *corresponding to P.*

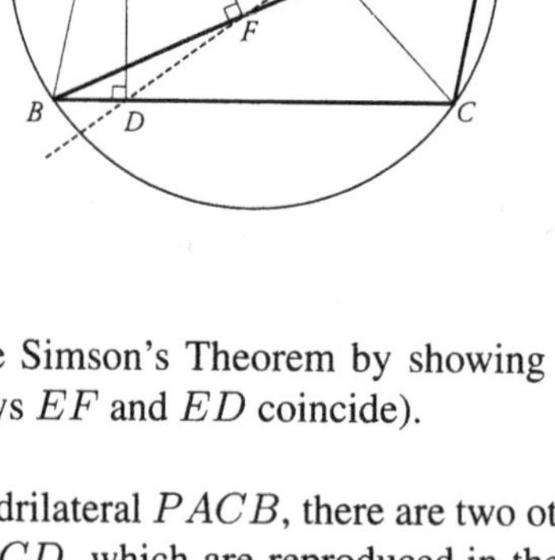

Solution. We will prove Simson's Theorem by showing that $\angle PEF = \angle PED$ (which means that the rays EF and ED coincide).

As well as the cyclic quadrilateral $PACB$, there are two other cyclic quadrilaterals, namely $PEAF$ and $PECD$, which are reproduced in the figure on the following page. (These are cyclic because in each case two of the opposite angles sum to $180°$).

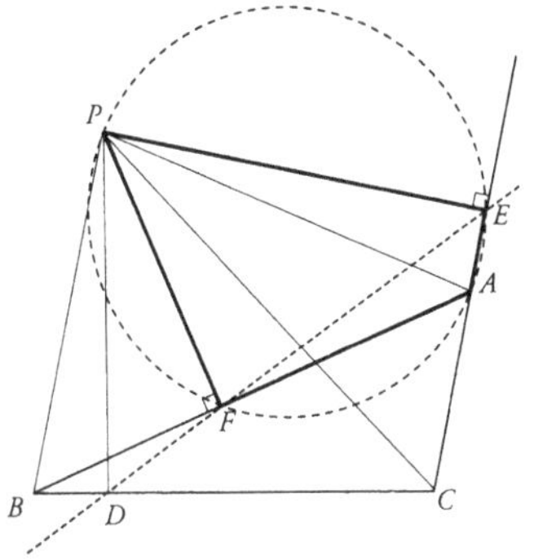

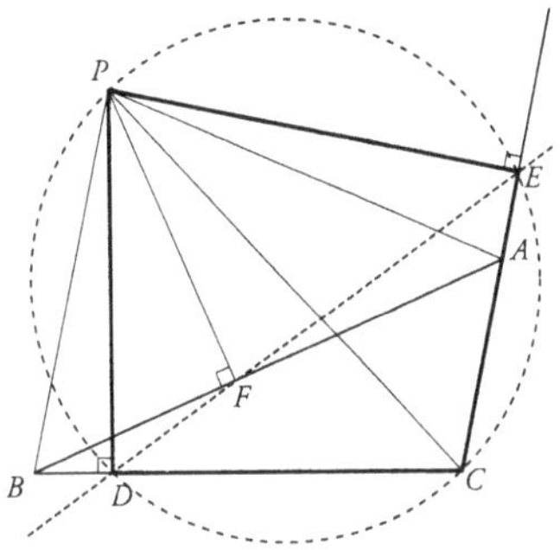

By Thales' Theorem applied to the circumcircle of $PEAF$, we get

$$\angle PEF = \angle PAF = \angle PAB.$$

By Thales' Theorem applied to the circumcircle of $PABC$, we get

$$\angle PAB = \angle PCB = \angle PCD.$$

By Thales' Theorem applied to the circumcircle of $PECD$, we get

$$\angle PCD = \angle PED.$$

Therefore, $\angle PEF = \angle PED$, which completes the proof. □

1.4 More About Congruency

The next theorem follows from **ASA** congruency together with the fact that the angle sum in a triangle is $180°$.

Theorem 1.4.1. *(***SAA*** Congruency)*

Two triangles are congruent if two angles and a side of one are congruent to two angles and the corresponding side of the other.

In the figure below we have noncongruent triangles ABC and DEF. In these triangles, $AB \equiv DE$, $AC \equiv DF$, and $\angle B \equiv \angle E$. This shows that, in general, **SSA** does not guarantee congruency.

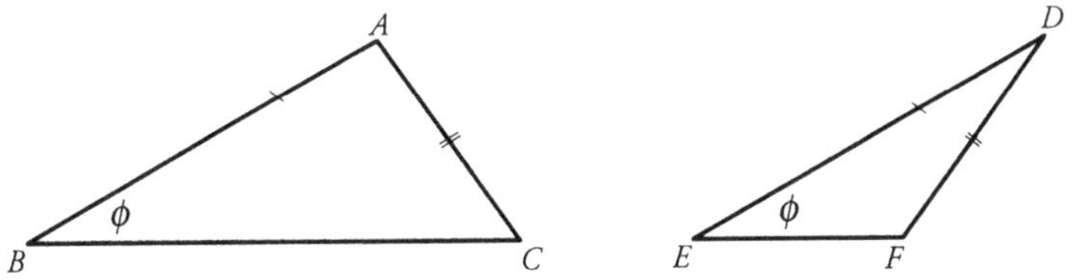

With further conditions we do get congruency:

Theorem 1.4.2. *(**SSA**$^+$ Congruency)*

***SSA** congruency is valid if the length of the side opposite the given angle is greater than or equal to the length of the other side.*

Proof. Suppose that in triangles ABC and DEF we have $AB = DE$, $AC = DF$, and $\angle ABC = \angle DEF$ and that the side opposite the given angle is the larger of the two sides.

We will prove the theorem by contradiction. Assume that the theorem is false, that is, assume that $BC \neq EF$; then we may assume that $BC < EF$. Let H be a point on EF so that $EH = BC$, as in the figure below.

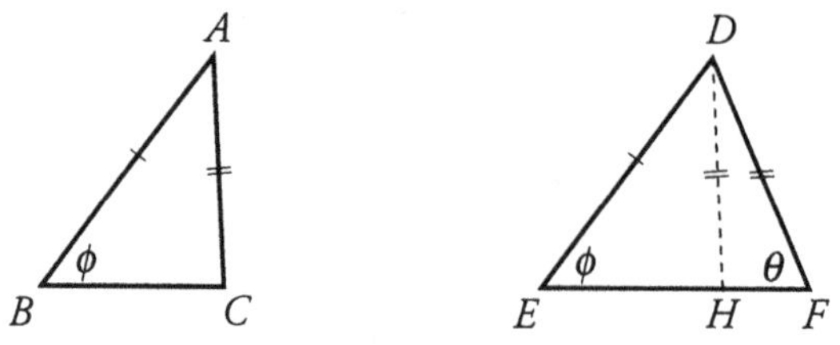

Then, $\triangle ABC \equiv \triangle DEH$ by **SSS** congruency. This means that $DH = DF$, so $\triangle DHF$ is isosceles. Then $\angle DFE = \angle DHF$.

Since we are given that $DF \geq DE$, the Angle-Side Inequality tells us that

$$\angle DEF \geq \angle DFE,$$

and so it follows that $\angle DEF \geq \angle DHF$. However, this contradicts the External Angle Inequality.

We must therefore conclude that the assumption that the theorem is false is incorrect, and so we can conclude that the theorem is true.

□

Since the hypotenuse of a right triangle is always the longest side, there is an immediate corollary:

Corollary 1.4.3. *(**HSR** Congruency)*

If the hypotenuse and one side of a right triangle are congruent to the hypotenuse and one side of another right triangle, the two triangles are congruent.

Counterexamples and Proof by Contradiction

If we were to say that

If ABCD is a rectangle, then $AB = BC$,

you would most likely show us that the statement is false by drawing a rectangle that is not a square. When you do something like this, you are providing what is called a ***counterexample***.

In the assertion "If $\mathcal{P}$, then $\mathcal{Q}$," the statement $\mathcal{P}$ is called the ***hypothesis*** and the statement $\mathcal{Q}$ is called the ***conclusion***. A counterexample to the assertion is any example in which the hypothesis is true and the conclusion is false.

To prove that an assertion is not true, all you need to do is find a single counterexample. (You do not have to show that it is never true, you only have to show that it is not always true!)

Exercise 1.4.4. *For each of the following statements, provide a diagram that is a counterexample to the statement.*

1. *Given triangle* ABC, *if* $\angle ABC = 60°$, *then* ABC *is isosceles.*

2. *Given that* ABC *is an isosceles triangle with* $AB = AC$ *and with* P *and* Q *on side* BC *as shown in the picture, if* $BP = PQ = QC$, *then* $\angle BAP = \angle PAQ = \angle QAC$.

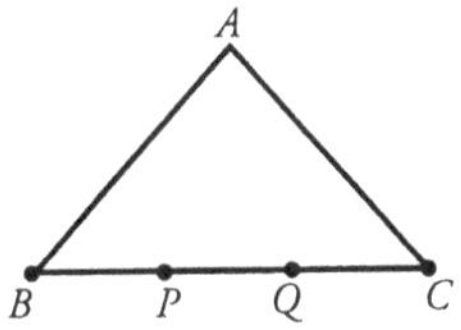

3. *In a quadrilateral* $ABCD$, *if* $AB = CD$ *and* $\angle BAD = \angle ADC = 90°$, *then* $ABCD$ *is a rectangle.*

A *proof by contradiction* to verify an assertion of the form "If $\mathcal{P}$, then $\mathcal{Q}$" consists of the following two steps:

(a) Assume that the assertion is false. This amounts to assuming that there is a counterexample to the assertion; that is, we assume that it is possible for the hypothesis $\mathcal{P}$ to be true while the conclusion $\mathcal{Q}$ is false. In other words, *assume that it is possible for the hypothesis* ***and*** *the negative of the conclusion to simultaneously be true.*

(b) Show that this leads to a contradiction of a fact that is known to be true. In such circumstances, somewhere along the way an error must have been made. Presuming that the reasoning is correct, the only possibility is that the assumption that the assertion is false must be in error. Thus, we must conclude that the assertion is true.

1.5 Perpendiculars and Angle Bisectors

Two lines that intersect each other at right angles are said to be ***perpendicular*** to each other. The ***right bisector*** or ***perpendicular bisector*** of a line segment AB is a line perpendicular to AB that passes through the midpoint M of AB.

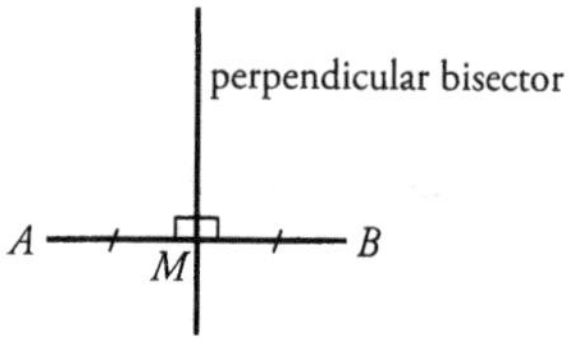

The following theorem is the characterization of the perpendicular bisector.

Theorem 1.5.1. *(Characterization of the Perpendicular Bisector)*

Given different points A and B, the perpendicular bisector of AB consists of all points P that are equidistant from A and B.

Proof. Let P be a point on the right bisector. Then in triangles PMA and PMB we have

$$PM = PM,$$
$$\angle PMA = 90° = \angle PMB,$$
$$MA = MB,$$

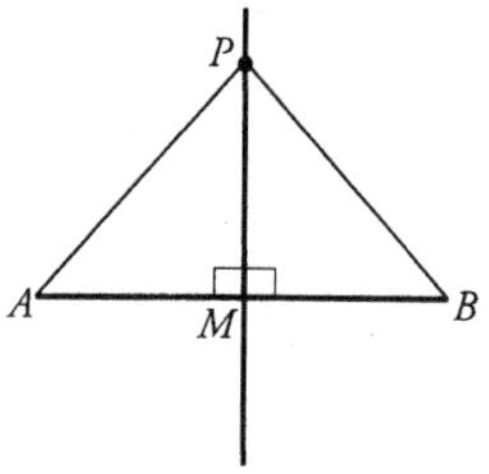

so triangles PMA and PMB are congruent by **SAS**. It follows that $PA = PB$.

Conversely, suppose that P is some point such that $PA = PB$. Then triangles PMA and PMB are congruent by **SSS**. It follows that $\angle PMA = \angle PMB$, and since the sum of the two angles is $180°$, we have $\angle PMA = 90°$. That is, P is on the right bisector of AB. □

Exercise 1.5.2. *If m is the perpendicular bisector of AB, then A and B are on opposite sides of m. Show that if P is on the same side of m as B, then P is closer to B than to A.*

The following exercise follows easily from Pythagoras' Theorem. Try to do it without using Pythagoras' Theorem.

Exercise 1.5.3. *Show that the hypotenuse of a right triangle is its longest side.*

Exercise 1.5.4. *Let l be a line and let P be a point not on l. Let Q be the foot of the perpendicular from P to l. Show that Q is the point on l that is closest to P.*

Exercise 1.5.5. *Let l be a line and let P be a point not on l. Show that there is at most one line through P perpendicular to l.*

Given a nonreflex angle $\angle ABC$, a ray BD such that $\angle ABD = \angle CBD$ is called an ***angle bisector*** of $\angle ABC$.

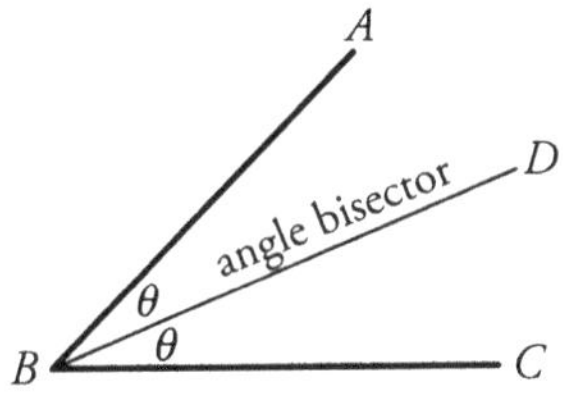

Given a line l and a point P not on l, the ***distance*** from P to l, denoted $d(P, l)$, is the length of the segment PQ where Q is the foot of the perpendicular from P to l.

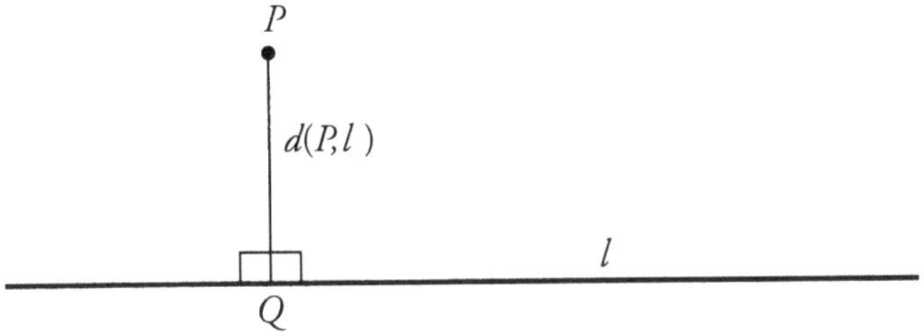

Theorem 1.5.6. *(Characterization of the Angle Bisector)*

The angle bisector of a nonreflex angle consists of all points interior to the angle that are equidistant from the arms of the angle.

Proof. Let P be a point on the angle bisector. Let Q and R be the feet of the perpendiculars from P to AB and CB, respectively. Triangles PQB and PRB have the side PB in common,

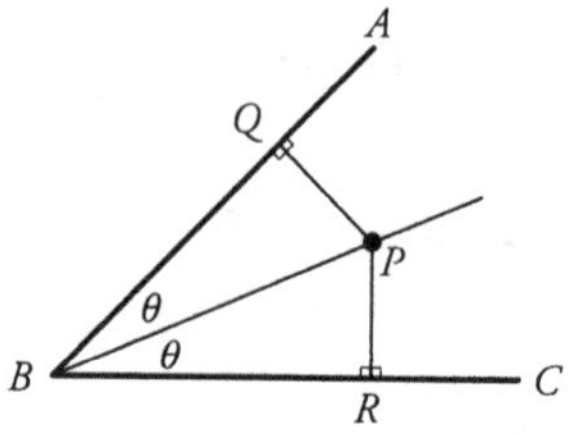

$$\angle PQB = \angle PRB \quad \text{and} \quad \angle PBQ = \angle PBR.$$

Thus, the triangles are congruent by **SAA**, hence $PQ = PR$. Therefore, P is equidistant from AB and CB.

Conversely, let P be a point that is equidistant from BA and BC. Let Q and R be the feet of the perpendiculars from P to AB and CB, respectively, so that $PQ = PR$. Thus, triangles PQB and PRB are congruent by **HSR**, and it follows that

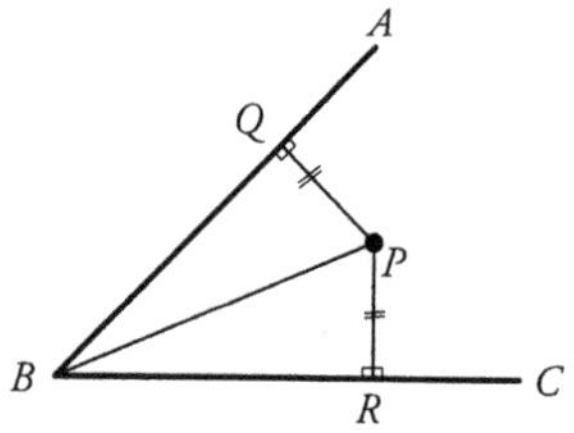

$$\angle PBA = \angle PBQ = \angle PBR = \angle PBC.$$

Hence, P is on the angle bisector of $\angle ABC$.

□

Inequalities in Proofs

Before turning to construction problems, we list the inequalities that we have used in proofs and add one more to the list.

1. Triangle Inequality
2. Exterior Angle Inequality
3. Angle-Side Inequality
4. Open Jaw Inequality

This last inequality is given in the following theorem.

Theorem 1.5.7. *(Open Jaw Inequality)*

Given two triangles $\triangle ABC$ and $\triangle DEF$ with $AB = DE$ and $BC = EF$. Then $\angle ABC < \angle DEF$ if and only if $AC < DF$, as in the figure.

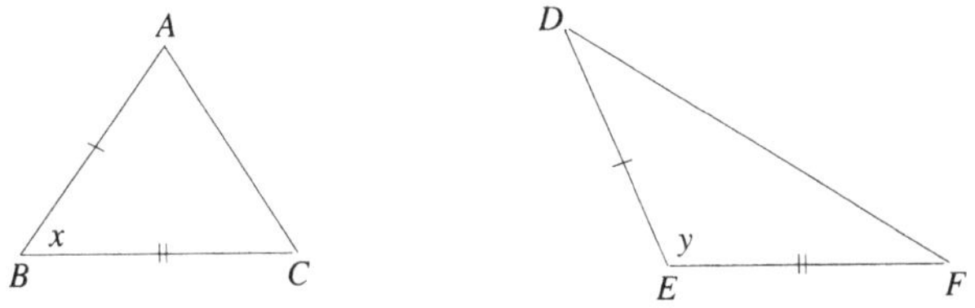

Proof. Suppose that $x < y$. Then we can build x in $\triangle DEF$ so that $EG = AB$. G can be inside or on the triangle. Here, we assume that G is outside $\triangle DEF$, as in the figure below.

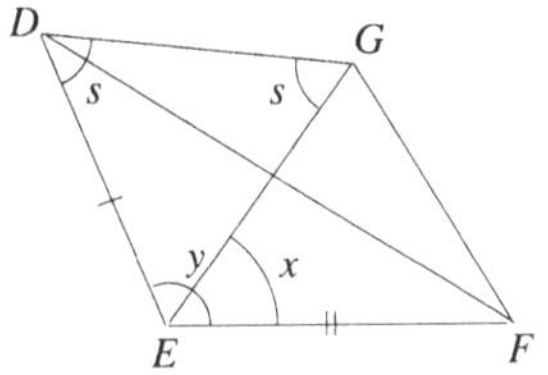

Now note that $\triangle ABC \equiv \triangle GEF$ by the **SAS** congruency theorem, and $\triangle EDG$ is isosceles with angles as shown. Also, $s < \angle DGF$, since GE is interior to $\angle G$, and $s > \angle GDF$, since DF is interior to $\angle D$. Therefore,

$$\angle DGF > s > \angle GDF,$$

and by the Angle-Side Inequality, this implies that

$$DF > GF = AC.$$

Now suppose that $AC < DF$. Then exactly one of the following is true:

$$x = y \qquad \text{or} \qquad x > y \qquad \text{or} \qquad x < y$$

(this is called the ***law of trichotomy*** for the real number system).

If $x = y$, then $\triangle ABC \equiv \triangle DEF$ by the **SAS** congruency theorem, which is a contradiction since we are assuming that $AC < DF$.

If $x > y$, then by the first part of the theorem we would have $AC > DF$, which is also a contradiction.

Therefore, the only possibility left is that $x < y$, and we are done.

□

1.6 Construction Problems

Although there are many ways to physically draw a straight line, the image that first comes to mind is a pencil sliding along a ruler. Likewise, the draftsman's compass comes to mind when one thinks of drawing a circle. To most people, the words *straightedge* and *compass* are synonymous with these physical instruments. In geometry, the same words are also used to describe idealized instruments. Unlike their physical counterparts, the geometric straightedge enables us to draw a line of arbitrary length, and the geometric compass allows us to draw arcs and circles of any radius we please. When doing geometry, you should regard the physical straightedge and compass as instruments that mimic the "true" or "idealized" instruments.

There is a reason for dealing with idealized instruments rather than physical ones. Mathematics is motivated by a desire to look at the basic essence of a problem, and to achieve this we have to jettison any unnecessary baggage. For example, we do not want to worry about the problem of the thickness of the pencil line, for this is a drafting problem rather than a geometry problem. However, as we strip away the unnecessary limitations of the draftsman's straightedge and compass, the effect is to create versions of the instruments that behave somewhat differently from their physical counterparts.

The idealized instruments are not "real," nor are the lines and circles that they draw. As a consequence, we cannot appeal to the properties of the physical instruments as verification for whatever we do in geometry. In order to work with idealized instruments, it is important to describe very clearly what they can do. The rules for the abstract instruments closely resemble the properties of the physical ones:

Straightedge Operations

> A straightedge can be used to draw a straight line that passes through two given points.

Compass Operations

> A compass can be used to draw an arc or circle centered at a given point with a given distance as radius. (The given distance is defined by two points.)

These two statements completely describe how the straightedge and compass operate, and there are no further restrictions, nor any additional properties. For example, the most common physical counterpart of the straightedge is a ruler, and it is a fairly easy matter to place a ruler so that the line to be drawn appears to be tangent to a given physical circle. With the true straightedge, this operation is forbidden. If you wish to draw a tangent line, you must first find two points on the line and then use the straightedge to draw the line through these two points.

A ruler has another property that the straightedge does not. It has a scale that can be used for measuring. A straightedge has no marks on it at all and so cannot be used as a measuring device.

We cannot justify our results by appealing to the physical properties of the instruments. Nevertheless, experimenting with the physical instruments sometimes leads to an understanding of the problem at hand, and if we restrict the physical instruments so that we only use the two operations described above, we are seldom led astray.

Useful Facts in Justifying Constructions

Recall that a ***rhombus*** is a parallelogram whose sides are all congruent, as in the figure on the right.

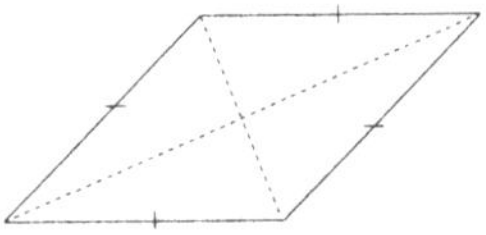

A ***kite*** is a convex quadrilateral with two pairs of adjacent sides congruent. Note that the diagonals intersect in the interior of the kite.

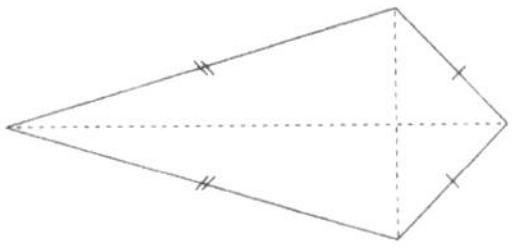

A ***dart*** is a nonconvex quadrilateral with two pairs of adjacent sides congruent. Note that the diagonals intersect in the exterior of the dart.

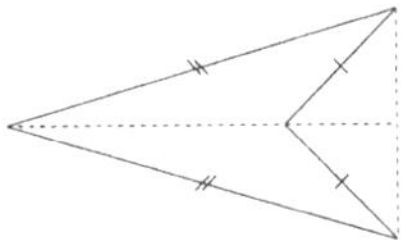

Theorem 1.6.1.

(1) The diagonals of a parallelogram bisect each other.

(2) The diagonals of a rhombus bisect each other at right angles.

(3) The diagonals (possibly extended) of a kite or a dart intersect at right angles.

Basic Constructions

The first three basic constructions are left as exercises.

Exercise 1.6.2. *To construct a triangle given two sides and the included angle.*

Exercise 1.6.3. *To construct a triangle given two angles and the included side.*

Exercise 1.6.4. *To construct a triangle given three sides.*

Example 1.6.5. *To copy an angle.*

Solution. Given $\angle A$ and a point D, we wish to construct a congruent angle FDE.

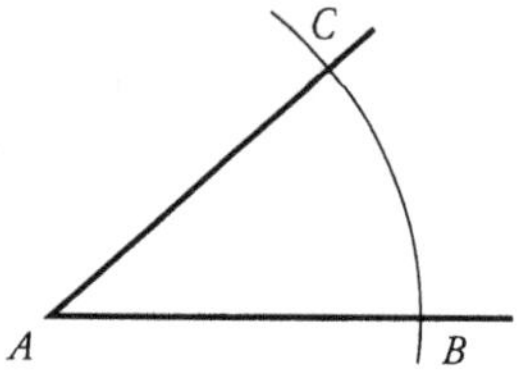

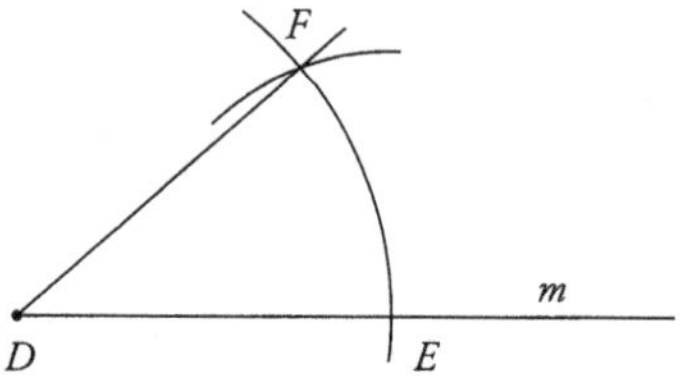

Draw a line m through the point D.
With center A, draw an arc cutting the arms of the given angle at B and C.
With center D, draw an arc of the same radius cutting m at E.
With center E and radius BC, draw an arc cutting the previous arc at F.
Then, $\angle FDE \equiv \angle A$.

Since triangles BAC and EDF are congruent by **SSS**, $\angle BAC \equiv \angle EDF$.

□

Example 1.6.6. *To construct the right bisector of a segment.*

Solution. Given points A and B, with centers A and B, draw two arcs of the same radius meeting at C and D. Then CD is the right bisector of AB.

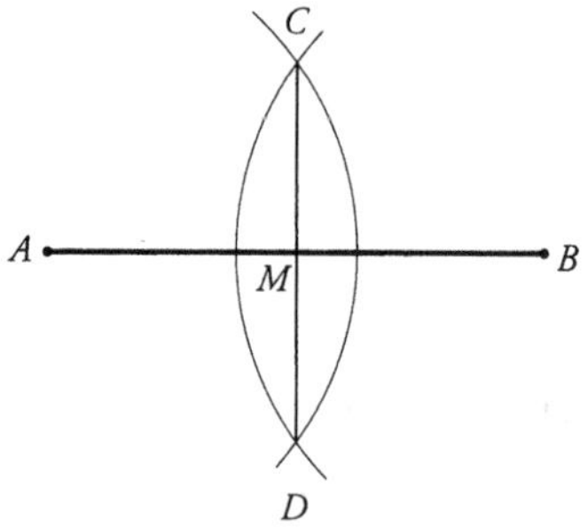

To see this, let M be the point where CD meets AB. First, we note that $\triangle ACD \equiv \triangle BCD$ by **SSS**, so $\angle ACD = \angle BCD$. Then in triangles ACM and BCM we have

$$AC = BC,$$
$$\angle ACM = \angle BCM,$$
$$CM \text{ is common,}$$

so $\triangle ACM \equiv \triangle BCM$ by **SAS**. Then

$$AM = BM \quad \text{and} \quad \angle AMC = \angle BMC = 90^\circ,$$

which means that CM is the right bisector of AB.

□

Example 1.6.7. *To construct a perpendicular to a line from a point not on the line.*

Solution. Let the point be P and the line be m.

With center P, draw an arc cutting m at A and B. With centers A and B, draw two arcs of the same radius meeting at Q, where $Q \neq P$. Then PQ is perpendicular to m.

Since by construction both P and Q are equidistant from A and B, both are on the right bisector of the segment AB, and hence PQ is perpendicular to m.

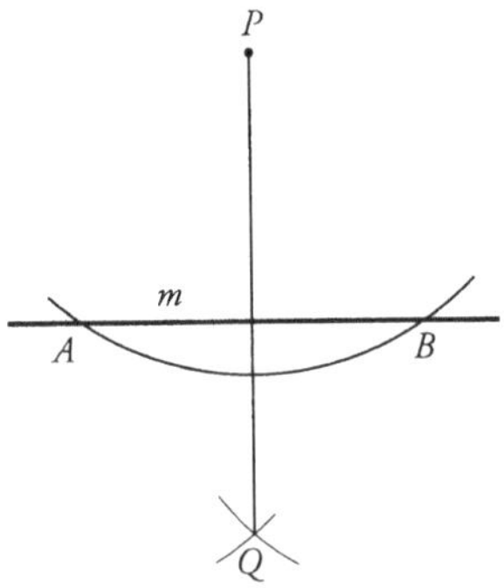

□

Example 1.6.8. *To construct the angle bisector of a given angle.*

Solution. Let P be the vertex of the given angle. With center P, draw an arc cutting the arms of the angle at S and T. With centers S and T, draw arcs of the same radius meeting at Q. Then PQ is the bisector of the given angle.

Since triangles SPQ and TPQ are congruent by **SSS**, $\angle SPQ = \angle TPQ$.

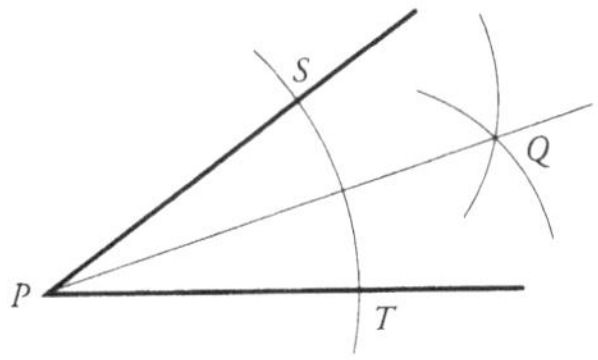

□

Exercise 1.6.9. *To construct a perpendicular to a line from a point on the line.*

1.6.1 The Method of Loci

The ***locus*** of a point that "moves" according to some condition is the traditional language used to describe the set of points that satisfy a given condition. For example, the locus of a point that is equidistant from two points A and B is the set of all points that are equidistant from A and B — in other words, the right bisector of AB.

The most basic method used to solve geometric construction problems is to locate important points by using the intersection of loci, which is usually referred to as the ***method of loci***. We illustrate with the following:

Example 1.6.10. *Given two intersecting lines l and m and a fixed radius r, construct a circle of radius r that is tangent to the two given lines.*

Solution. It is often useful to sketch the expected solution. We refer to this sketch as an ***analysis figure***. The more accurate the sketch, the more useful the figure. In the analysis figure you should attempt to include all possible solutions. The analysis figure for Example 1.6.10 is as follows, where l and m are the given lines intersecting at P.

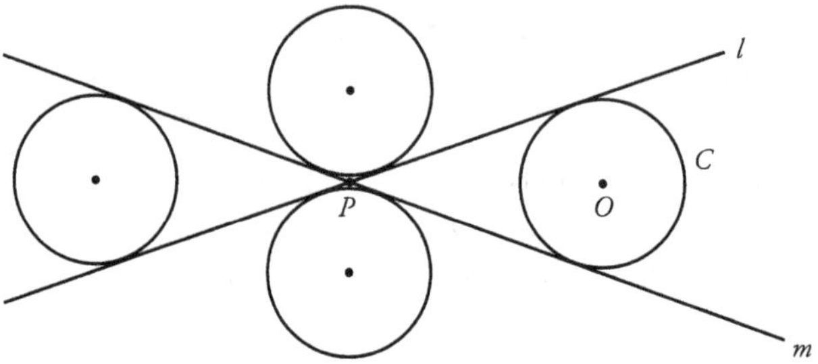

The analysis figure indicates that there are four solutions. The constructions of all four solutions are basically the same, so in this case it suffices to show how to construct one of the four circles.

Since we are given the radius of the circle, it is enough to construct O, the center of circle C. Since we only have a straightedge and a compass, there are only three ways to construct a point, namely, as the intersection of

- two lines,
- two circles, or
- a line and a circle.

The center O of circle C is equidistant from both l and m and therefore lies on the following constructible loci:

1. an angle bisector,
2. a line parallel to l at distance r from l,
3. a line parallel to m at distance r from m.

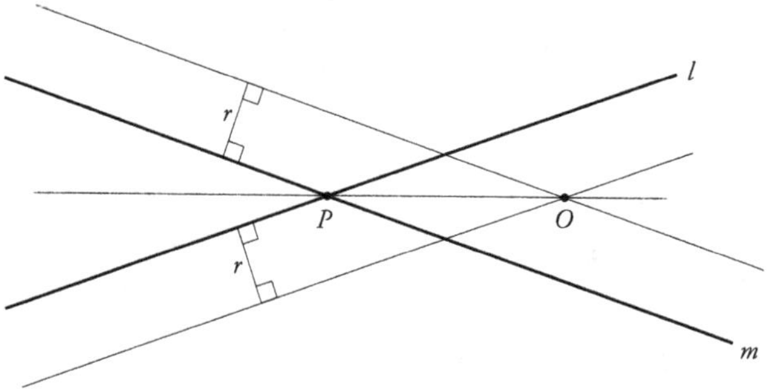

Any two of these loci determine the point O.

Having done the analysis, now write up the solution:

1. Construct line n parallel to l at distance r from l.
2. Construct line k parallel to m at distance r from m.
3. Let $O = n \cap k$.
4. With center O and radius r, draw the circle $\mathcal{C}(O, r)$.

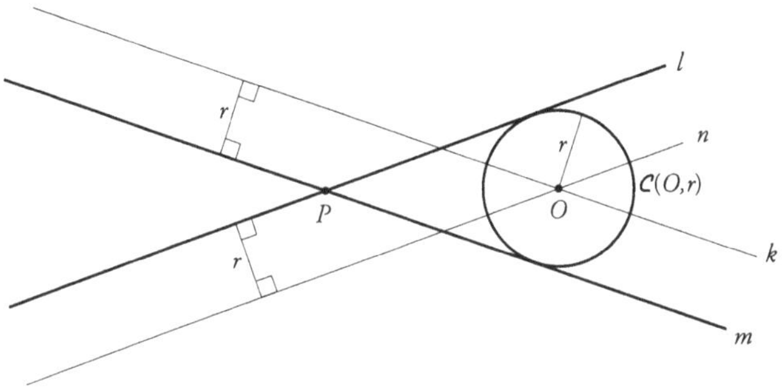

□

1.7 Solutions to Selected Exercises

Solution to Exercise 1.2.8

1. *Statement*: Given triangle ABC, if $\angle ABC$ is a right angle, then

$$AB^2 + BC^2 = AC^2.$$

Converse: Given triangle ABC, if

$$AB^2 + BC^2 = AC^2,$$

then $\angle ABC$ is a right angle.

Both the statement and its converse are true.

2. *Statement*: If $ABCD$ is a parallelogram, then $AB = CD$ and $AD = BC$.

   Converse: If $AB = CD$ and $AD = BC$, then $ABCD$ is a parallelogram.

   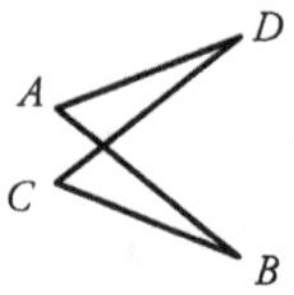

   The statement is true and the converse is false.

3. *Statement*: If $ABCD$ is a convex quadrilateral, then $ABCD$ is a rectangle.

   Converse: If $ABCD$ is a rectangle, then $ABCD$ is a convex quadrilateral.

   The statement is false and the converse is true.

4. *Statement*: Given quadrilateral $ABCD$, if $AC \neq BD$, then $ABCD$ is not a rectangle.

   Converse: Given quadrilateral $ABCD$, if $ABCD$ is not a rectangle, then $AC \neq BD$.

   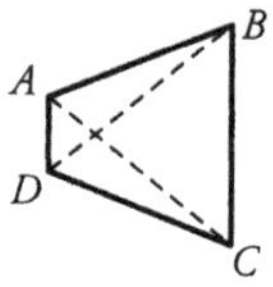

   The statement is true and its converse is false.

Solution to Exercise 1.3.9

In the figure below, we have $\alpha = 30$ (isosceles triangle).

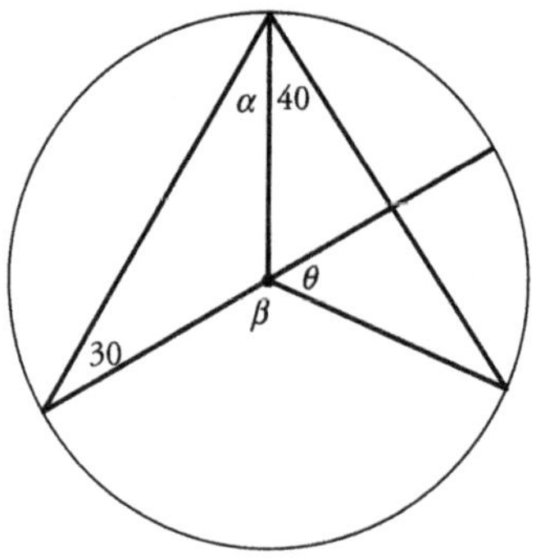

Thus, by Thales' Theorem,

$$\beta = 2(\alpha + 40) = 140,$$

so that $\theta = 180 - \beta = 40$.

Solution to Exercise 1.3.14

Suppose that the quadrilateral $ABCD$ has an inscribed circle that is tangent to the sides at points P, Q, R, and S, as shown in the figure.

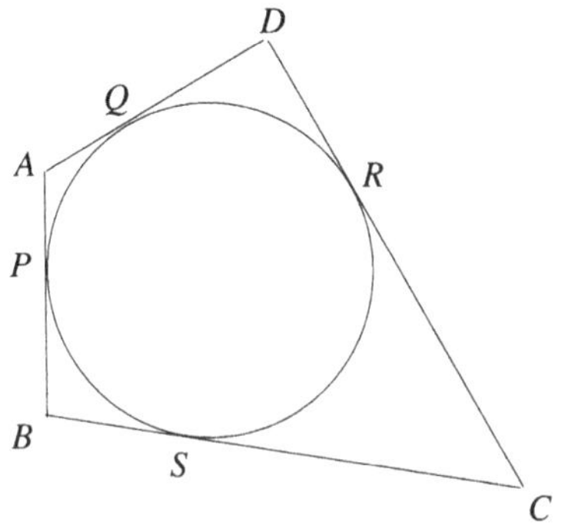

Since the tangents to the circle from an external point have the same length, then

$$AP = AQ, \quad PB = BS$$

and

$$SC = CR, \quad RD = QD,$$

so that

$$\begin{aligned} AB + CD &= AP + PB + CR + RD \\ &= AQ + BS + SC + QD \\ &= (AQ + QD) + (BS + SC) \\ &= AD + BC. \end{aligned}$$

Conversely, suppose that $AB + CD = AD + BC$, and suppose that the extended sides AD and BC meet at X. Introduce the incircle of $\triangle DXC$, that is, the circle internally tangent to each of the sides of the triangle, and suppose that it is not tangent to AB. Let E and F be on sides AD and BC, respectively, such that EF is parallel to AB and tangent to the incircle, as in the figure.

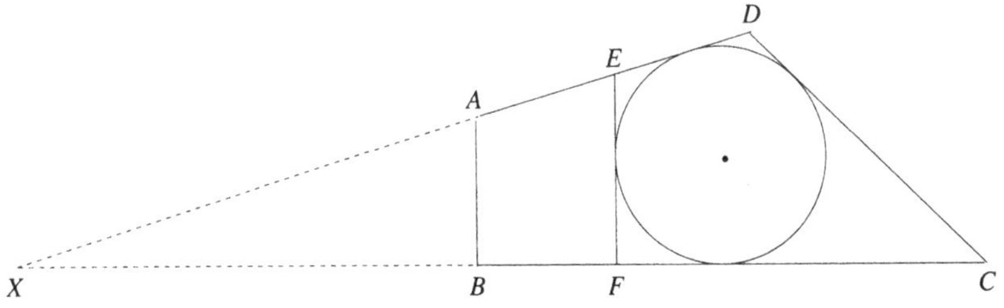

Note that since $\triangle XEF \sim \triangle XAB$ with proportionality constant $k > 1, EF > AB$, and since the quadrilateral $DEFC$ has an inscribed circle, then by the first part of the proof we must have

$$AB + CD < EF + CD = DE + CF < AD + BC,$$

which is a contradiction. When the side AB intersects the circle twice, a similar argument also leads to a contradiction. Therefore, if the condition

$$AB + CD = AD + BC$$

holds, then the incircle of $\triangle DXC$ must also be tangent to AB, and $ABCD$ has an inscribed circle.

The case when $ABCD$ is a parallelogram follows in the same way. First, we construct a circle that is tangent to three sides of the parallelogram. The center of the circle must lie on the line parallel to the sides AD and BC and midway between them. Let $2r$ be the perpendicular distance between AD and BC. Then the center must also lie on the line parallel to the side CD and at a perpendicular distance r from CD, as in the figure.

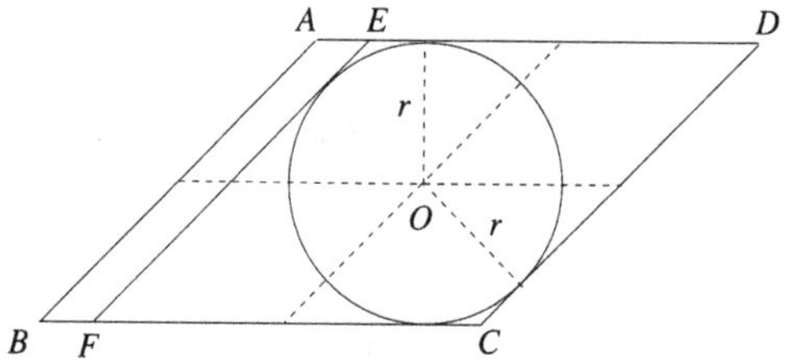

As before, suppose that the circle is not tangent to AB, and let E and F be on sides AD and BC, respectively, such that EF is parallel to AB and tangent to the circle, as in the figure above.

Now, $AB = EF$, and since the quadrilateral $DEFC$ has an inscribed circle, by the first part of the proof we must have

$$AB + CD = EF + CD = DE + CF < AD + BC,$$

which is a contradiction. When the side AB intersects the circle twice, a similar argument also leads to a contradiction. Therefore, if the condition

$$AB + CD = AD + BC$$

holds, then the circle must also be tangent to AB, and the parallelogram $ABCD$ has an inscribed circle.

Solution to Exercise 1.4.4

1.

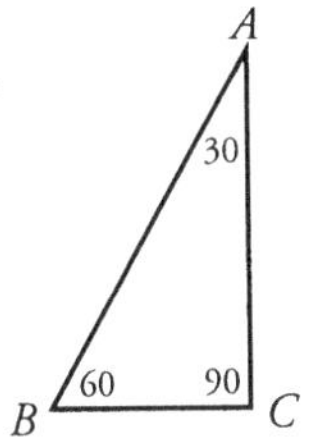

2. The assertion is always false.

3.

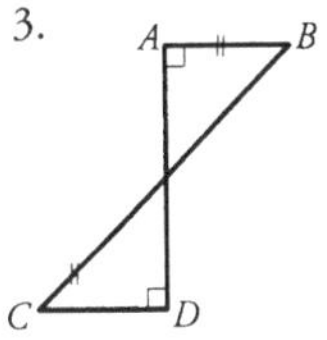

Solution to Exercise 1.5.2

In the figure below, we have

$$PB < PQ + QB = PQ + QA = PA.$$

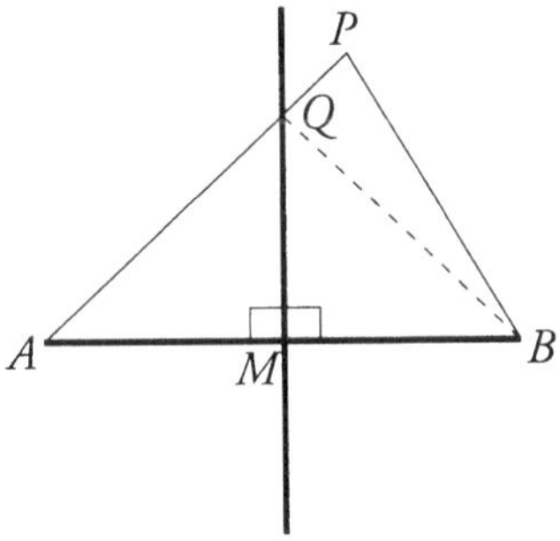

Solution to Exercise 1.5.3

Here are two different solutions.

1. The right angle is the largest angle in the triangle (otherwise the sum of the three angles of the triangle would be larger than $180°$). Since the hypotenuse is opposite the largest angle, it must be the longest side.

2. In the figure below, suppose that B is the right angle.

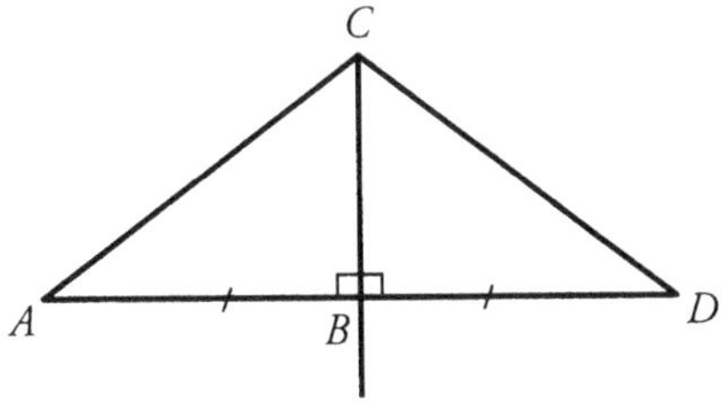

Extend AB beyond B to D so that $BD = AB$. Then C is on the right bisector of AD, so

$$AC = \frac{1}{2}(AC + CD) > \frac{1}{2}AD = AB.$$

This shows that $AC > AB$, and in a similar fashion it can be shown that $AC > CB$.

Solution to Exercise 1.5.4

Let Q be the foot of the perpendicular from P to the line l, and let R be any other point on l. Then $\triangle PQR$ is a right triangle, and by Exercise 1.5.3, $PR > PQ$.

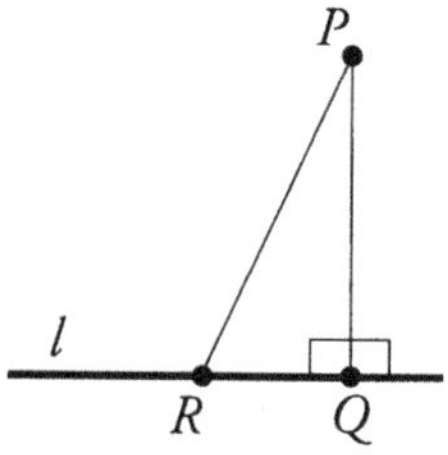

1.8 Problems

1. Prove that the internal and external bisectors of the angles of a triangle are perpendicular.

2. Let P be a point inside $\mathcal{C}(O, r)$ with $P \neq O$. Let Q be the point where the ray $\overrightarrow{OP}$ meets the circle. Use the Triangle Inequality to show that Q is the point on the circle that is closest to P.

3. Let P be a point inside $\triangle ABC$. Use the Triangle Inequality to prove that $AB + BC > AP + PC$.

4. Each of the following statements is true. State the converse of each statement, and if it is false, provide a figure as a counterexample.

   (a) If $\triangle ABC \equiv \triangle DEF$, then $\angle A = \angle D$ and $\angle B = \angle E$.

   (b) If $ABCD$ is a rectangle, then $\angle A = \angle C = 90°$.

   (c) If $ABCD$ is a rectangle, then $\angle A = \angle B = \angle C = 90°$.

5. Given the isosceles triangle ABC with $AB = AC$, let D be the foot of the perpendicular from A to BC. Prove that AD bisects $\angle BAC$.

6. Show that if the perpendicular from A to BC bisects $\angle BAC$, then $\triangle ABC$ is isosceles.

7. D is a point on BC such that AD is the bisector of $\angle A$. Show that

$$\angle ADC = 90 + \frac{\angle B - \angle C}{2}.$$

8. Construct an isosceles triangle ABC, given the unequal angle $\angle A$ and the length of the side BC.

9. Construct a right triangle given the hypotenuse and one side.

10. Calculate θ in the following figure.

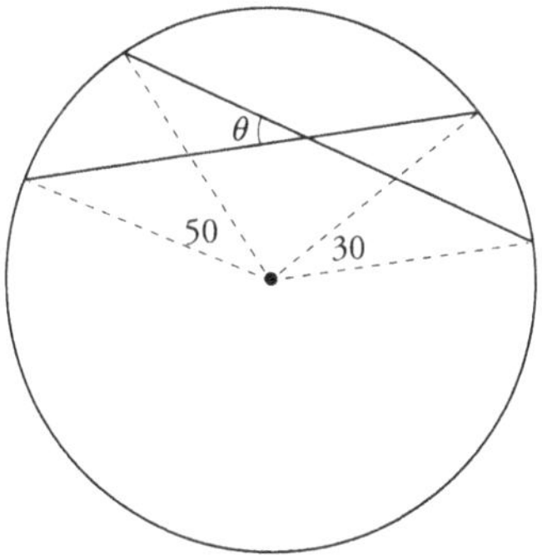

(not drawn to scale)

11. Let Q be the foot of the perpendicular from a point P to a line l. Show that Q is the point on l that is closest to P.

12. Let P be a point inside $C(O, r)$ with $P \neq O$. Let Q be the point where the ray $\overrightarrow{PO}$ meets the circle. Show that Q is the point of the circle that is farthest from P.

13. Let $ABCD$ be a simple quadrilateral. Show that $ABCD$ is cyclic if and only if the opposite angles sum to $180°$.

14. Draw the locus of a point whose sum of distances from two fixed perpendicular lines is constant.

15. Given a circle $\mathcal{C}(P, s)$, a line l disjoint from $\mathcal{C}(P, s)$, and a radius r ($r > s$), construct a circle of radius r tangent to both $\mathcal{C}(P, s)$ and l.

Note: The analysis figure indicates that there are four solutions.

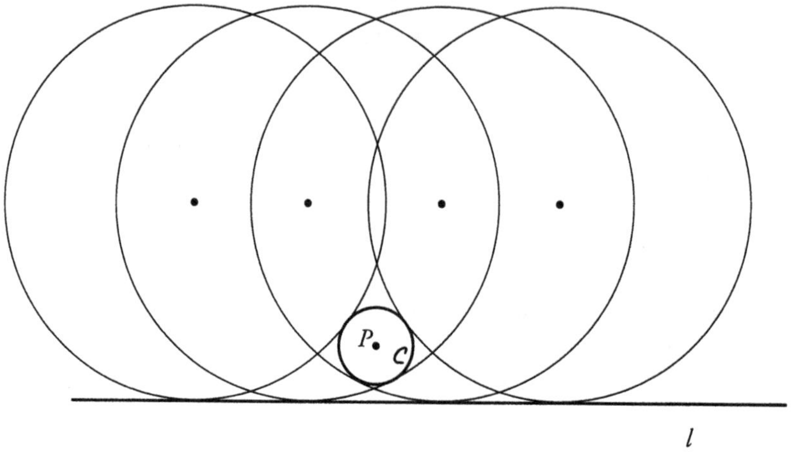

CHAPTER 2

CONCURRENCY

2.1 Perpendicular Bisectors

This chapter is concerned with concurrent lines associated with a triangle. A family of lines is ***concurrent*** at a point P if all members of the family pass through P.

In preparation, we need a few additional facts about parallel lines.

Theorem 2.1.1. *Let l_1 and l_2 be parallel lines, and suppose that m_1 and m_2 are lines with $m_1 \perp l_1$ and $m_2 \perp l_2$. Then m_1 and m_2 are parallel.*

Classical Geometry: Euclidean, Transformational, Inversive and Projective, First Edition. By
I. E. Leonard, J. E. Lewis, A. C. F. Liu, G. W. Tokarsky
Copyright © 2014 John Wiley & Sons, Inc.

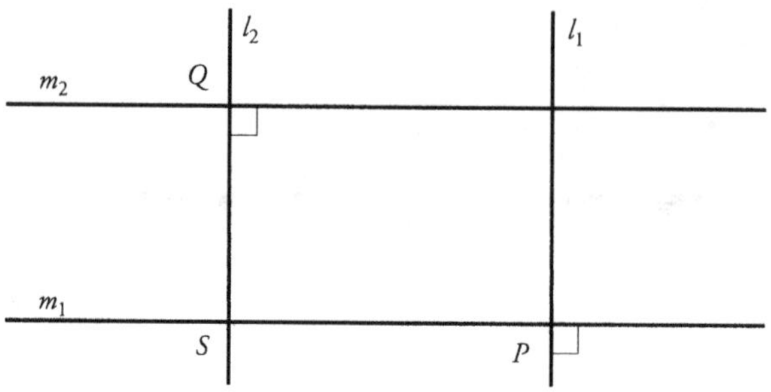

Proof. Let P be the point $l_1 \cap m_1$ and let Q be the point $l_2 \cap m_2$. By the parallel postulate, l_1 is the only line through P parallel to l_2, and so m_1 is not parallel to l_2 and consequently must meet l_2 at some point S. In other words, m_1 is a transversal for the parallel lines l_1 and l_2. Since the sum of the adjacent interior angles is $180°$, it follows that l_2 must be perpendicular to m_1.

But l_2 is also perpendicular to m_2 and so m_1 and m_2 must be parallel.

□

In the above proof, if we interchange l_1 and m_1 and l_2 and m_2, we have:

Corollary 2.1.2. *Suppose that $l_1 \perp m_1$ and $l_2 \perp m_2$. Then l_1 and l_2 are parallel if and only if m_1 and m_2 are parallel.*

One of the consequences of this corollary is that if two line segments intersect, then their perpendicular bisectors must also intersect. This fact is crucial in the following theorem, the proof of which also uses the fact that the right bisector of a segment can be characterized as being the set of all points that are equidistant from the endpoints of the segment.

Theorem 2.1.3. *The perpendicular bisectors of the sides of a triangle are concurrent.*

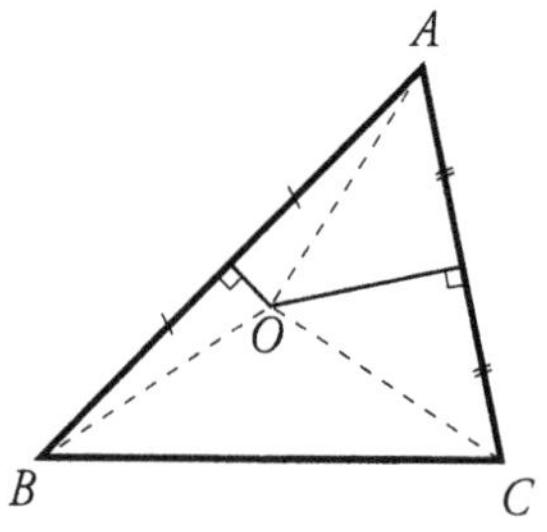

Proof. According to the comments preceding the theorem, the perpendicular bisectors of AB and AC meet at some point O, as shown in the figure.

It is enough to show that O lies on the perpendicular bisector of BC. We have

$$OB = OA \text{ (since } O \text{ is on the perpendicular bisector of } AB\text{)},$$
$$OA = OC \text{ (since } O \text{ is on the perpendicular bisector of } AC\text{)}.$$

Therefore $OB = OC$, which means that O is on the perpendicular bisector of BC.

□

The proof shows that the point O is equidistant from the three vertices, so with center O we can circumscribe a circle about the triangle. The circle is called the ***circumcircle***, and the point O is called the ***circumcenter*** of the triangle.

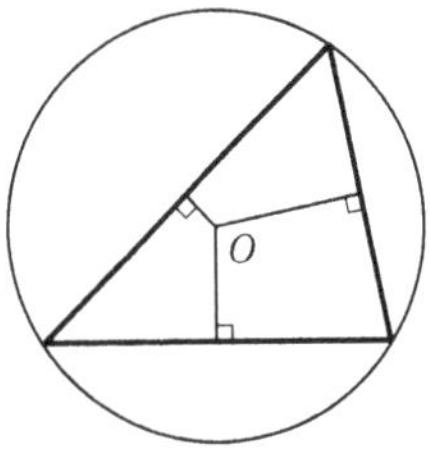

Exercise 2.1.4. *In the figure above, the circumcenter is interior to the triangle. In what sort of a triangle is the circumcenter*

1. *on one of the edges of the triangle?*
2. *outside the triangle?*

2.2 Angle Bisectors

Theorem 2.2.1. *For a circle $C(O, r)$:*

1. *The right bisector of a chord of $C(O, r)$ passes through O.*
2. *If a given chord is not a diameter, the line joining O to the midpoint of the chord is the right bisector of the chord.*
3. *The line from O that is perpendicular to a given chord is the right bisector of the chord.*
4. *A line is tangent to $C(O, r)$ at a point P if and only if it is perpendicular to the radius OP.*

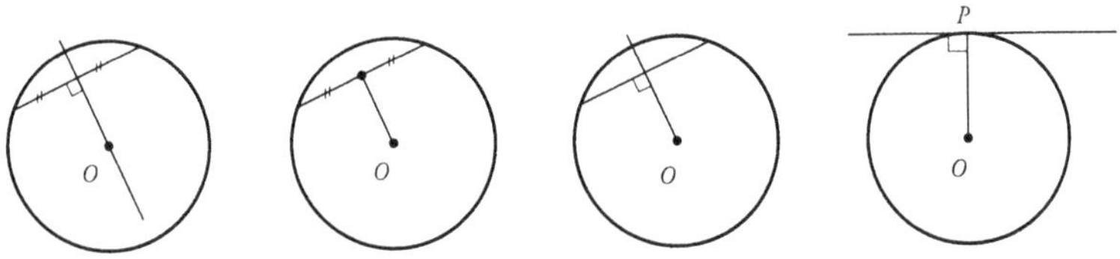

1. *2.* *3.* *4.*

Theorem 2.2.2. *The internal bisectors of the angles of a triangle are concurrent.*

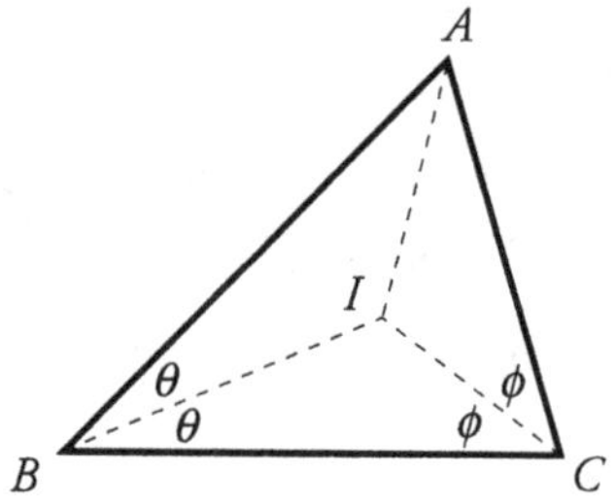

Proof. In $\triangle ABC$, let the internal bisectors of $\angle B$ and $\angle C$ meet at I, as shown above.

We will show that I lies on the internal bisector of $\angle A$. We have

$$d(I, AB) = d(I, BC),$$

since I lies on the internal bisector of $\angle B$, and

$$d(I, BC) = d(I, AC),$$

since I lies on the internal bisector of $\angle C$.

Hence, $d(I, AB) = d(I, AC)$, which means that I is on the internal bisector of $\angle A$.

□

The proof of Theorem 2.2.2 shows that if we drop the perpendiculars from I to the three sides of the triangle, the lengths of those perpendiculars will all be the same, say r. Then the circle $\mathcal{C}(I, r)$ will be tangent to all three sides by Theorem 2.2.1. That is, I is the center of a circle inscribed in the triangle. This inscribed circle is called the ***incircle*** of the triangle, and I is called the ***incenter*** of the triangle.

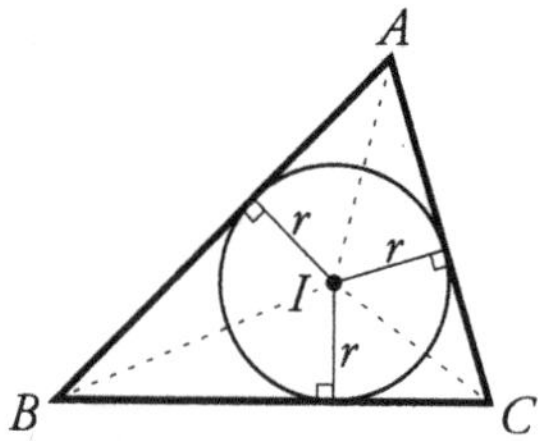

We will next prove a theorem about the external angle bisectors. First, we need to show that two external angle bisectors can never be parallel.

Proposition 2.2.3. *Each pair of external angle bisectors of a triangle intersect.*

Proof. Referring to the diagram,

$$2\theta = \alpha + \gamma,$$
$$2\phi = \beta + \gamma.$$

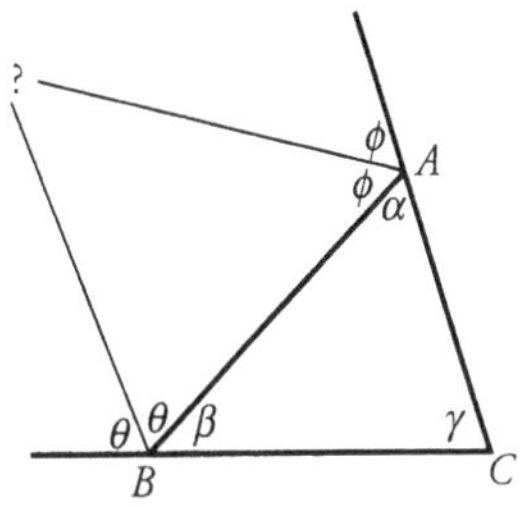

Adding and rearranging:

$$2(\theta + \phi) = 180 + \gamma.$$

Since $\gamma < 180$, it follows that $\theta + \phi < 180$, so the external angle bisectors cannot be parallel.

□

Theorem 2.2.4. *The external bisectors of two of the angles of a triangle and the internal bisector of the third angle are concurrent.*

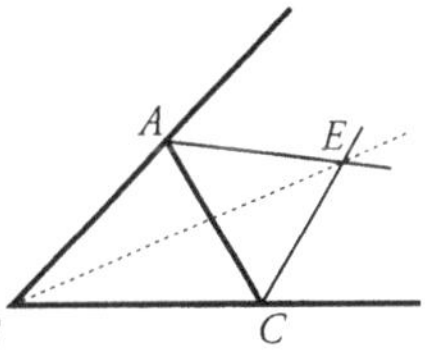

Proof. Let the two external bisectors meet at E, as shown in the figure. It is enough to show that E lies on the internal bisector of $\angle B$.

Now,

$$d(E, AB) = d(E, AC),$$

since E lies on the external bisector of $\angle A$, and

$$d(E, AC) = d(E, BC),$$

since E lies on the external bisector of $\angle C$. Hence, $d(E, AB) = d(E, BC)$, which means that E is on the internal bisector of $\angle B$.

□

The point of concurrency E is called an ***excenter*** of the triangle. Since E is equidistant from all three sides (some extended) of the triangle, we can draw an ***excircle*** tangent to these sides. Every triangle has three excenters and three excircles.

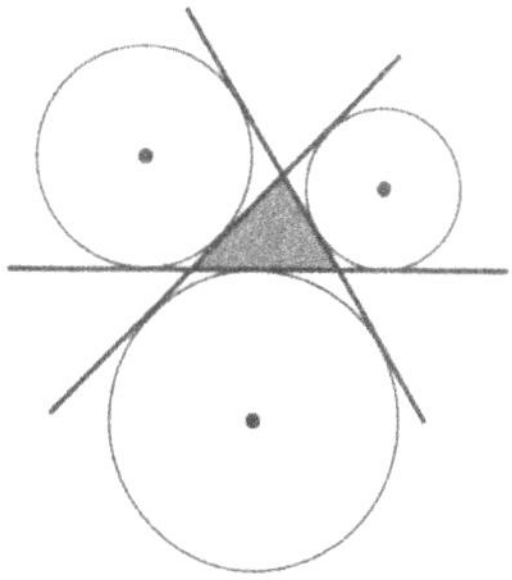

2.3 Altitudes

A line passing through a vertex of a triangle perpendicular to the opposite side is called an ***altitude*** of the triangle.

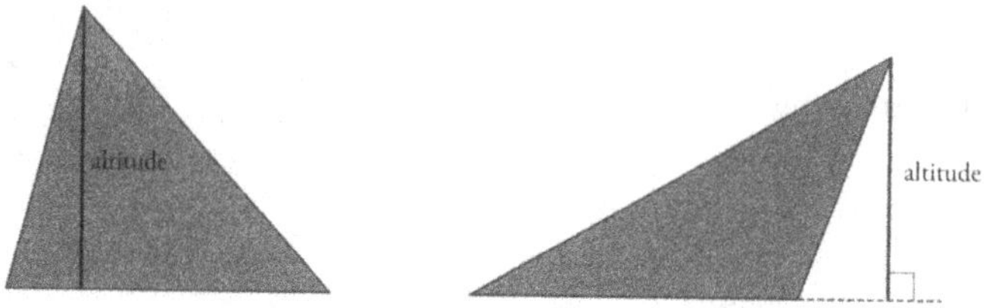

To prove that the altitudes of a triangle are concurrent, we need some facts about parallelograms.

A ***parallelogram*** is a quadrilateral whose opposite sides are parallel. A parallelogram whose sides are equal in length is called a ***rhombus***. Squares and rectangles are special types of parallelograms.

Theorem 2.3.1. *In a parallelogram:*

(1) Opposite sides are congruent.

(2) Opposite angles are congruent.

(3) The diagonals bisect each other.

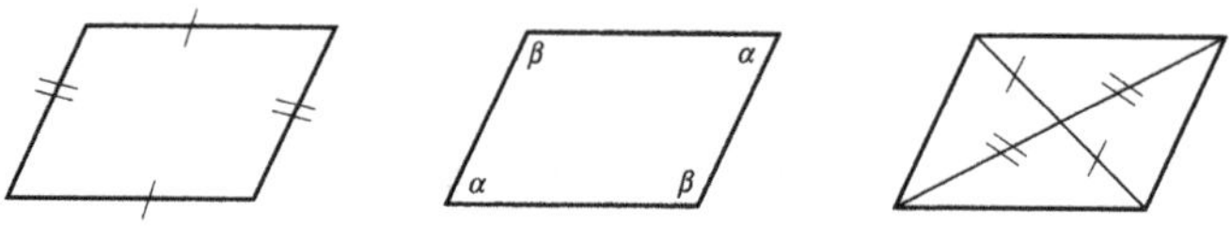

Proof. We will prove (1) and (2). Given parallelogram $ABCD$, as in the figure below, we will show that $\triangle ABC \equiv \triangle CDA$.

Diagonal AC is a transversal for parallel lines AB and CD, and so $\angle BAC$ and $\angle DCA$ are opposite interior angles for the parallel lines. So we have

$$\angle BAC = \angle DCA.$$

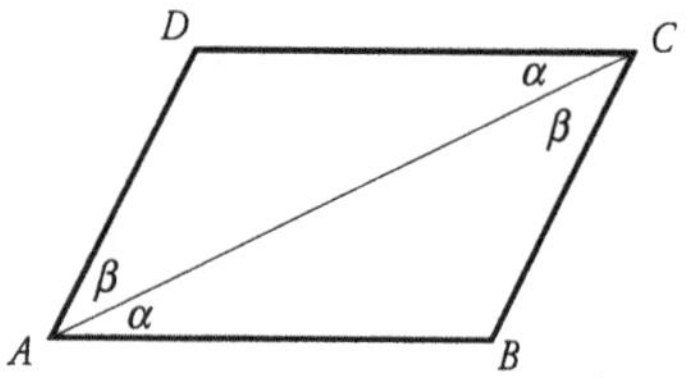

Treating AC as a transversal for AD and CB, we have

$$\angle BCA = \angle DAC.$$

Since AC is common to $\triangle BAC$ and $\triangle DCA$, **ASA** implies that they are congruent. Consequently, $AB = CD$ and $BC = DA$, proving (1).

Also, $\angle BAD = \alpha + \beta = \angle BCD$ and $\angle ADC = \angle ABC$ because triangles ADC and ABC are congruent, proving (2).

□

Exercise 2.3.2. *Prove statement (3) of Theorem 2.3.1.*

Theorem 2.3.3. *If the diagonals of a quadrilateral bisect each other, then the quadrilateral is a parallelogram.*

Exercise 2.3.4. *Prove Theorem 2.3.3.*

Theorem 2.3.3 tells us that statement (3) of Theorem 2.3.1 will guarantee that the quadrilateral is a parallelogram. However, neither statement (1) nor statement (2) of Theorem 2.3.1 is by itself enough to guarantee that a quadrilateral is a parallelogram. For example, a nonsimple quadrilateral whose opposite sides are congruent is not a parallelogram. However, if the quadrilateral is a simple polygon, then either statement (1) or (2) is sufficient. As well, there is another useful condition that can help determine if a simple quadrilateral is a parallelogram:

Theorem 2.3.5. *A simple quadrilateral is a parallelogram if any of the following statements are true:*

(1) Opposite sides are congruent.

(2) Opposite angles are congruent.

(3) One pair of opposite sides is congruent and parallel.

Proof. We will justify case (1), leaving the others to the reader.

Suppose that $ABCD$ is the quadrilateral, as in the figure on the right. Since $ABCD$ is simple, we may suppose that the diagonal AC is interior to the quadrilateral. Since the opposite sides of the quadrilateral are congruent, the **SSS** congruency condition implies that $\triangle ABC \equiv \triangle CDA$.

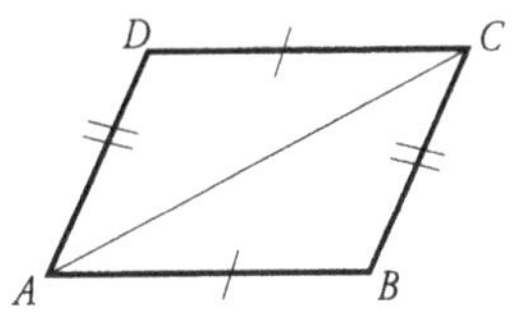

This in turn implies that the alternate interior angles $\angle BAC$ and $\angle DCA$ are congruent and also that the alternate interior angles $\angle BCA$ and $\angle DAC$ are congruent. The fact that the edges are parallel now follows from the well-known facts about parallel lines.

□

The next theorem uses a clever trick. It embeds the given triangle in a larger one in such a way that the altitudes of the given triangle are right bisectors of the sides of the larger one.

Theorem 2.3.6. *The altitudes of a triangle are concurrent.*

Proof. Given $\triangle ABC$, we embed it in a larger triangle $\triangle DEF$ by drawing lines through the vertices of $\triangle ABC$ that are parallel to the opposite sides so that $ABCE$, $ACBF$, and $CABD$ are parallelograms, as in the diagram on the right. Clearly, A, B, and C are midpoints of the sides of $\triangle DEF$, and an altitude of $\triangle ABC$ is a perpendicular bisector of a side of $\triangle DEF$. However, since the perpendicular bisectors of $\triangle DEF$ are concurrent, so are the altitudes of $\triangle ABC$.

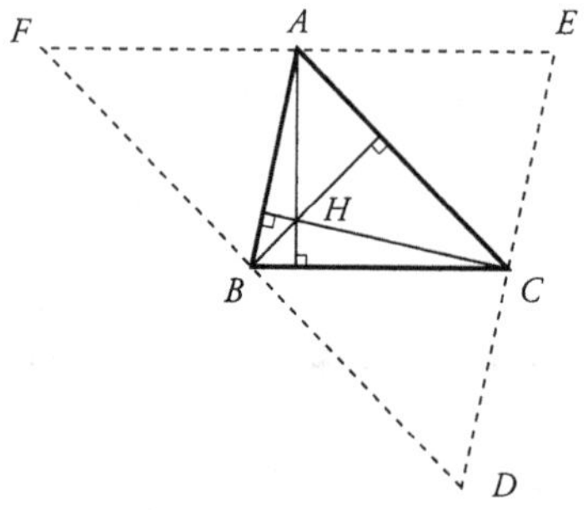

□

The point of concurrency of the altitudes is called the ***orthocenter*** of the triangle, and it is usually denoted by the letter H. In the proof, the orthocenter of $\triangle ABC$ is the circumcenter of $\triangle DEF$.

Note that the orthocenter can lie outside the triangle (for obtuse-angled triangles) or on the triangle (for right-angled triangles). The same proof also works for these types of triangles.

Exercise 2.3.7. *The figure above shows a triangle whose orthocenter is interior to the triangle. Give examples of triangles where:*

1. *The orthocenter is on a side of the triangle.*
2. *The orthocenter is exterior to the triangle.*

2.4 Medians

A ***median*** of a triangle is a line passing through a vertex and the midpoint of the opposite side.

Exercise 2.4.1. *Show that in an equilateral triangle ABC the following are all the same:*

1. *The perpendicular bisector of BC.*
2. *The bisector of $\angle A$.*
3. *The altitude from vertex A.*
4. *The median passing through vertex A.*

The next theorem, which is useful on many occasions, is also proved by using the properties of a parallelogram.

Theorem 2.4.2. *(The Midline Theorem)*

If P and Q are the respective midpoints of sides AB and AC of triangle ABC, then PQ is parallel to BC and $PQ = BC/2$.

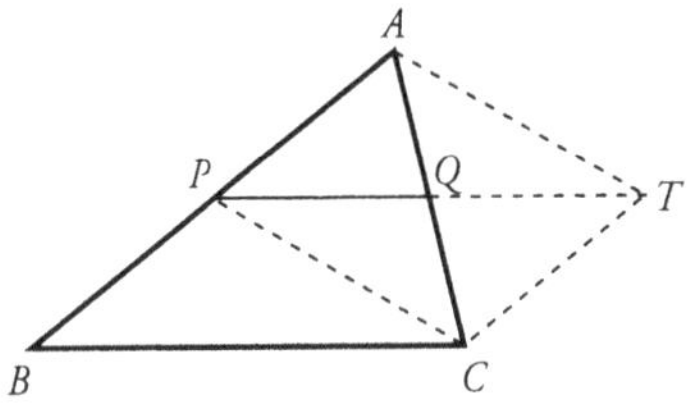

Proof. First, extend PQ to T so that $PQ = QT$, as in the figure. Then $ATCP$ is a parallelogram, because the diagonals bisect each other. This means that TC is parallel to and congruent to AP. Since P is the midpoint of AB, it follows that TC is parallel to and congruent to BP, and so $TCBP$ is also a parallelogram. From this we can conclude that PT is parallel to and congruent to BC; that is, PQ is parallel to BC and half the length of BC.

□

Lemma 2.4.3. *Any two medians of a triangle trisect each other at their point of intersection.*

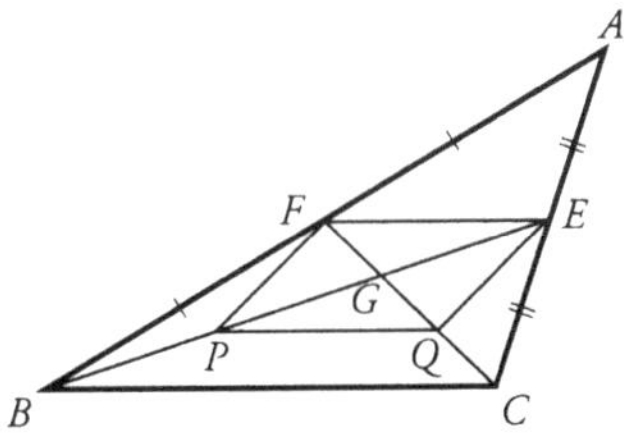

Proof. Let the medians BE and CF intersect at G, as shown in the figure. Draw FE. By the Midline Theorem, FE is parallel to BC and half its length.

Let P be the midpoint of GB and let Q be the midpoint of GC. Then, again by the Midline Theorem, PQ is parallel to BC and half its length. Since FE and PQ are parallel and equal in length, $EFPQ$ is a parallelogram, and so the diagonals of $EFPQ$ bisect each other. It follows that $FG = GQ = QC$ and $EG = GP = PB$, which proves the lemma.

□

There are two different points that trisect a given line segment. The point of intersection of the two medians is the trisection point of each that is farthest from the vertex.

Theorem 2.4.4. *The medians of a triangle are concurrent.*

Proof. Let the medians be AD, BE, and CF. Then BE and CF meet at a point G for which $EG/EB = 1/3$. Also, AD and BE meet at a point G' for which $EG'/EB = 1/3$. Since both G and G' are between B and E, we must have $G = G'$, which means that the three medians are concurrent.

□

The point of concurrency of the three medians is called the ***centroid***. The centroid always lies inside the triangle. A thin triangular plate can be balanced at its centroid on the point of a needle, so physically the centroid corresponds to the center of gravity.

The partial converses of the Midline Theorem are useful:

Theorem 2.4.5. *Let P be the midpoint of side AB of triangle ABC, and let Q be a point on AC such that PQ is parallel to BC. Then Q is the midpoint of AC.*

Proof. Let Q' be the midpoint of AC, then $PQ' \parallel BC$. Since there is only one line through P parallel to BC, the lines PQ and PQ' must be the same, so the points Q and Q' are also the same.

□

2.5 Construction Problems

The two basic construction problems associated with parallel lines are constructing a parallelogram given two of its adjacent edges (that is, completing a parallelogram) and constructing a line through a given point parallel to a given line. Once we have solved the first problem, the second one is straightforward.

Example 2.5.1. *To construct a parallelogram given two adjacent edges.*

Solution. Given edges AB and AC, with center B and radius AC, draw an arc. With center C and radius AB, draw a second arc cutting the first at D on the same side of AC as B. Then $ABDC$ is the desired parallelogram.

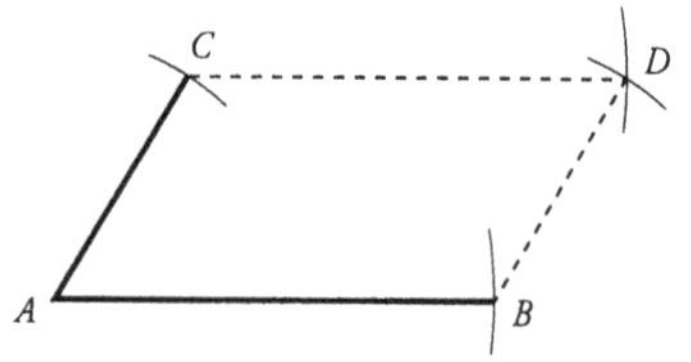

□

Example 2.5.2. *To construct a line parallel to a given line through a point not on the line.*

Solution. With center P, draw an arc cutting the given line at Q and S. With the same radius and center S, draw a second arc. With center P and radius QS, draw a third arc cutting the second at T. Then PT is the desired line, because $PQST$ is a parallelogram.

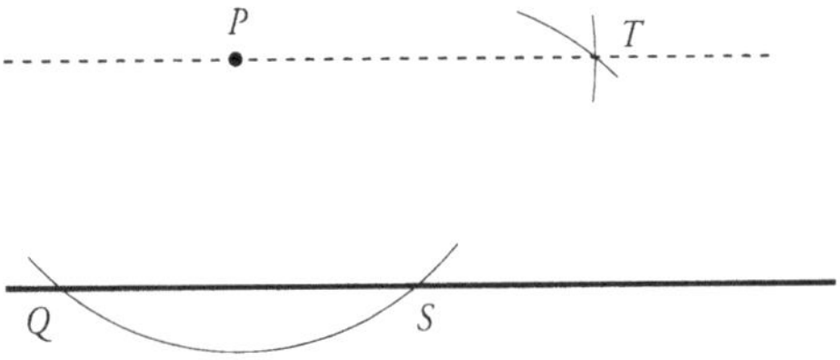

□

Example 2.5.3. *Given a circle with center O and a point P outside the circle, construct the lines through P tangent to the circle.*

Solution. Draw the line OP and construct the right bisector of OP, obtaining the midpoint M of OP. With center M and radius MP, draw a circle cutting the given circle at S and T. Then PS and PT are the desired tangent lines.

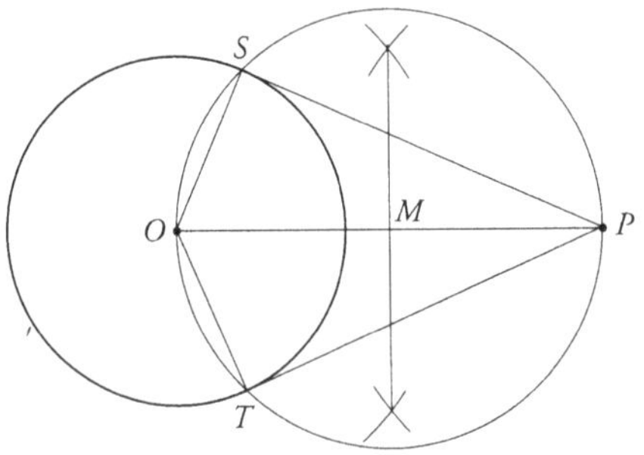

Note that angles OSP and OTP are angles in a semicircle, so by Thales' Theorem both are right angles. Thus, PS and PT are perpendicular to radii OS and OT, respectively, and so must be tangent lines by Theorem 2.2.1. □

Thales' Locus

Given a segment AB and an angle θ, the set of all points P such that $\angle APB = \theta$ forms the union of two arcs of a circle, which we shall call ***Thales' Locus***, shown in the figure on the following page.

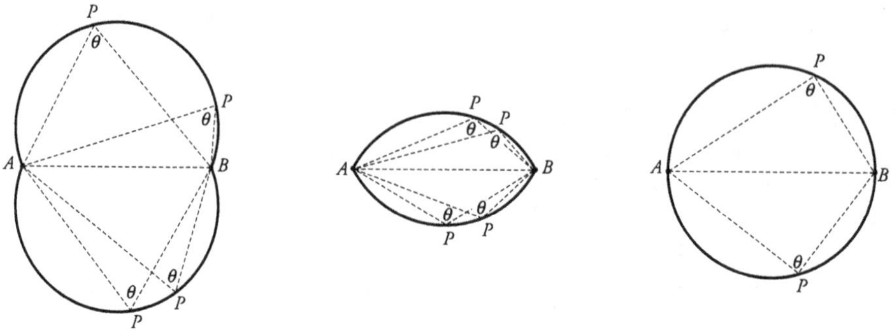

θ acute　　θ obtuse　　θ = 90

To help us construct Thales' Locus, we need the following useful theorem.

Theorem 2.5.4. *Let AB be a chord of the circle $C(O, r)$ and let P be a point on the line tangent to the circle, as shown in the diagram below. Then $\angle AOB = 2\angle PAB$.*

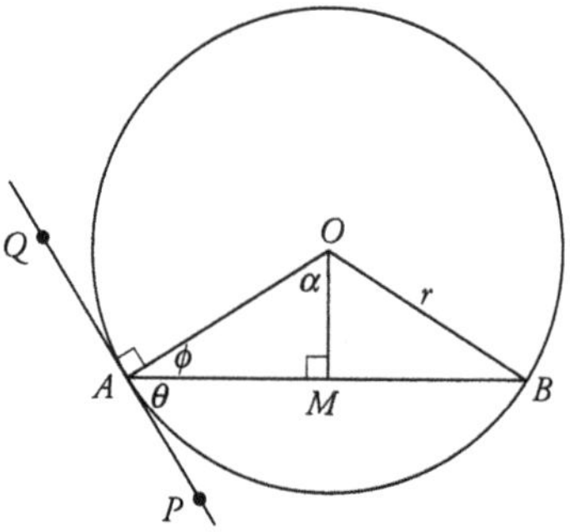

Proof. Referring to the diagram, drop the perpendicular OM from O to AB. Then $\angle AOB = 2\angle AOM = 2\alpha$. Now, $\theta = 90 - \phi = \alpha$, and it follows that $\angle AOB = 2\angle PAB$.

□

Note. In a similar way, it can be shown that the reflex angle AOB is twice the size of $\angle QAB$.

Example 2.5.5. *Given a segment AB and an angle $\theta < 90°$, construct Thales' Locus for the given data:*

Data:

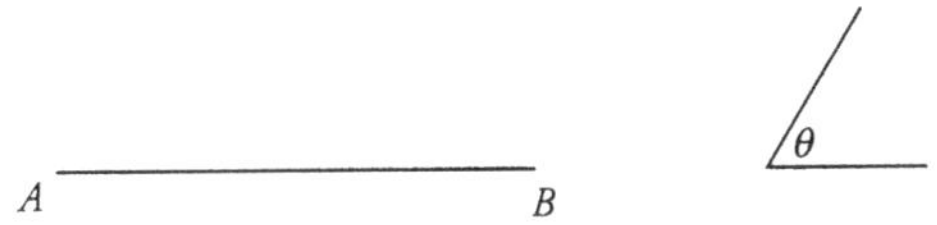

Solution. Here is one method of doing this construction.

1. Copy angle θ to point A so that AB is one arm of the angle, and l is the line containing the other arm.
2. Construct the right bisector m of AB.
3. Construct the line t through A perpendicular to l. Let O be the point $t \cap m$.
4. Draw $\mathcal{C}(O, OA)$. Then one of the arcs determined by $\mathcal{C}(O, OA)$ is part of Thales' Locus for the given data.

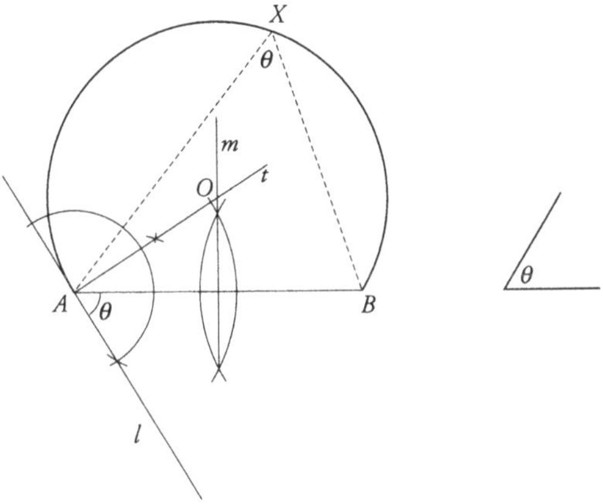

By Theorem 2.5.4, $\angle AOB = 2\theta$, so $\angle AXB$ at the circumference is of size θ.

□

Example 2.5.6. *Construct a triangle ABC given the size θ of $\angle A$, the vertices B and C, and the length of the altitude h from A.*

Solution. We give a quick outline of the solution.

1. Construct Thales' Locus for BC and θ.
2. Construct the perpendicular BX to BC so that $BX = h$.
3. Through X, draw the line l parallel to BC cutting the locus at a point A.

Then ABC is the desired triangle.

□

2.6 Solutions to the Exercises

Solution to Exercise 2.1.4

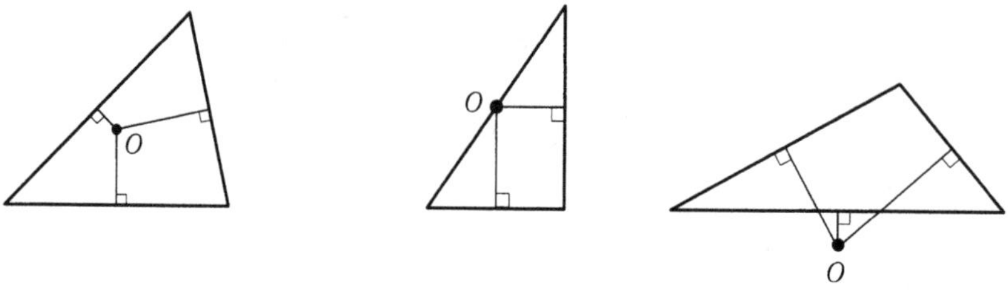

In an acute-angled triangle, the circumcenter is always interior to the triangle.

1. In a right triangle, the circumcenter is always on the hypotenuse.
2. In an obtuse-angled triangle, the circumcenter is always outside the triangle.

Solution to Exercise 2.3.2

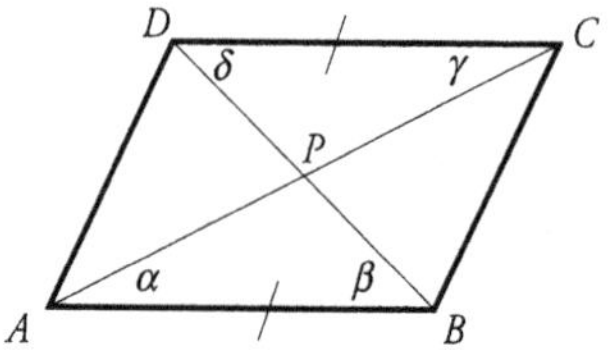

In triangles ABP and CDP we have

$$\alpha = \gamma,$$
$$\beta = \delta,$$

because $AB \parallel CD$, and

$$AB = CD,$$

by statement (1) of Theorem 2.3.1.

So, $\triangle ABP \equiv \triangle CDP$ by **ASA**, and it follows that $AP = CP$ and $BP = DP$.

Solution to Exercise 2.3.4

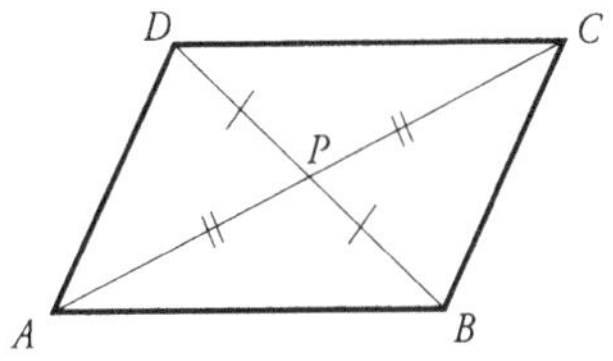

In triangles ABP and CDP we have

$$AP = CP,$$
$$\angle APB = \angle CPD$$

since they are vertically opposite angles, and

$$BP = DP.$$

So triangles ABP and CDP are congruent, and thus $\angle PAB = \angle PCD$, which implies that $AB \parallel CD$.

Similarly, $AD \parallel CB$, which shows that $ABCD$ is a parallelogram.

Solution to Exercise 2.3.7

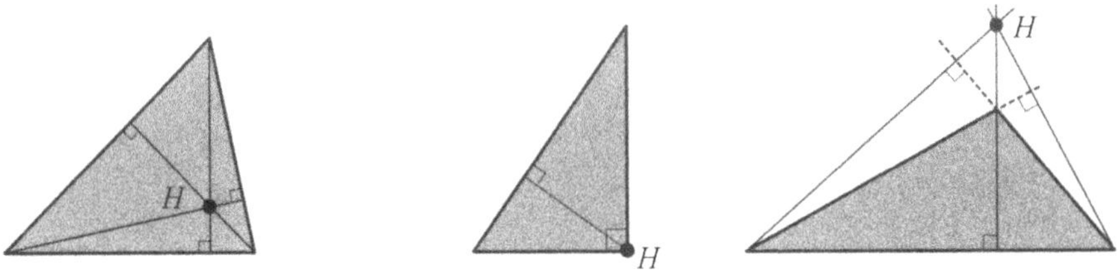

In an acute-angled triangle, the orthocenter is always interior to the triangle.

1. In a right triangle, the orthocenter is always the vertex of the right angle.
2. In an obtuse-angled triangle, the orthocenter is always outside the triangle.

Solution to Exercise 2.4.1

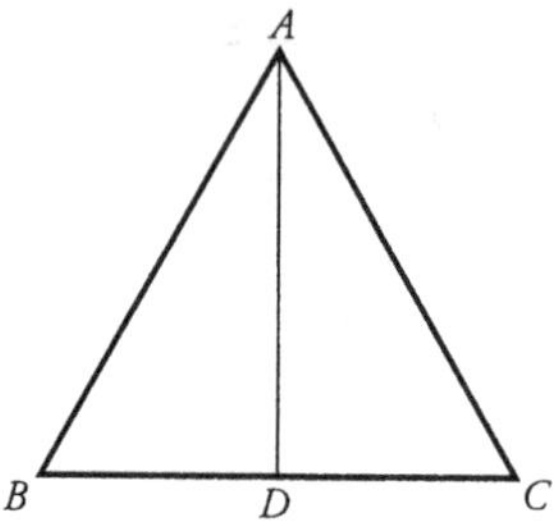

We will show that the line through the vertex A and the midpoint D of BC is simultaneously the perpendicular bisector of BC, the bisector of $\angle A$, the altitude from A, and the median from A.

1. Since $AB = AC$, the point A is equidistant from B and C and so A is on the right bisector of BC. It follows that AD is the right bisector of BC.
2. Triangles ADB and ADC are congruent by **SSS**, so $\angle DAB = \angle DAC$; that is, AD is the bisector of $\angle A$.
3. Since $AD \perp BC$, by statement 1 above, AD is an altitude of $\triangle ABC$.
4. Since D is the midpoint of BC, AD is a median of $\triangle ABC$.

2.7 Problems

1. BE and CF are altitudes of $\triangle ABC$, and M is the midpoint of BC. Show that $ME \equiv MF$.
2. BE and CF are altitudes of $\triangle ABC$, and EF is parallel to BC. Prove that $\triangle ABC$ is isosceles.
3. The perpendicular bisector of side BC of $\triangle ABC$ meets the circumcircle at D on the opposite side of BC from A. Prove that AD bisects $\angle BAC$.
4. Given $\triangle ABC$ with incenter I, prove that

$$\angle BIC = 90 + \tfrac{1}{2}\angle BAC.$$

Note. This is an important property of the incenter that will prove useful later.

5. In the given figure, calculate the sizes of the angles marked α and β.

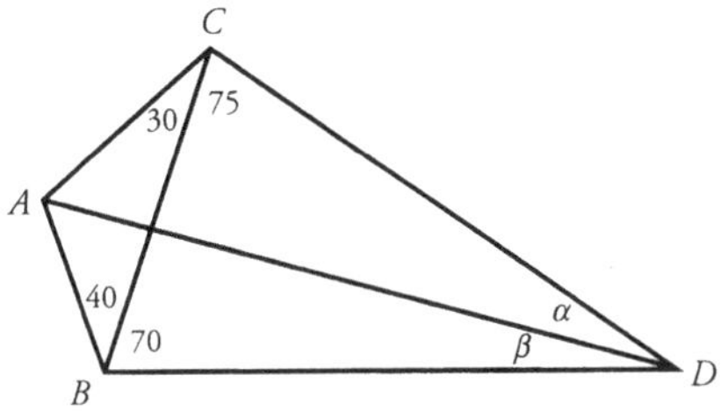

6. In $\triangle ABC$, $\angle BAC = 100°$ and $\angle ABC = 50°$. AD is an altitude and BE is a median. Find $\angle CDE$.

7. Segments PS and PT are tangent to the circle at S and T. Show that

   (a) $PS \equiv PT$ and

   (b) $ST \perp OP$.

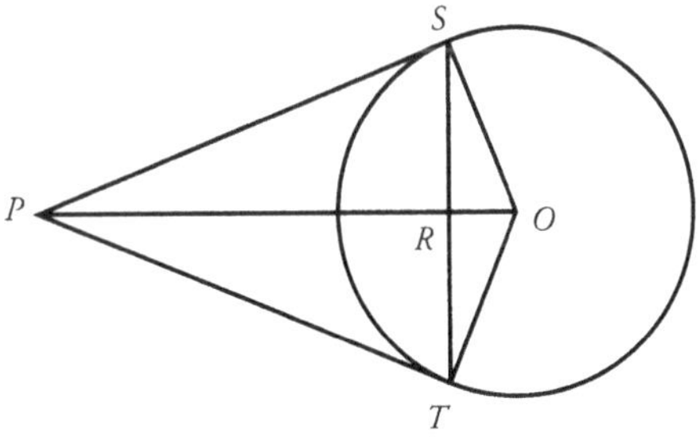

8. Construct $\triangle ABC$ given the side BC and the lengths h_b and h_c of the altitudes from B and C, respectively.

9. Construct $\triangle ABC$ given the side BC, the length h_b of the altitude from B, and the length m_a of the median from A.

10. Construct $\triangle ABC$ given the length of the altitude h from A and the length of the sides b and c. Here, b is the side opposite $\angle B$, and c is the side opposite $\angle C$.

11. Construct triangle ABC given BC, an angle β congruent to $\angle B$, and the length t of the median from B.

12. Given a point P inside angle ABC as shown below, construct a segment XY with endpoints in AB and CB such that P is the midpoint of XY.

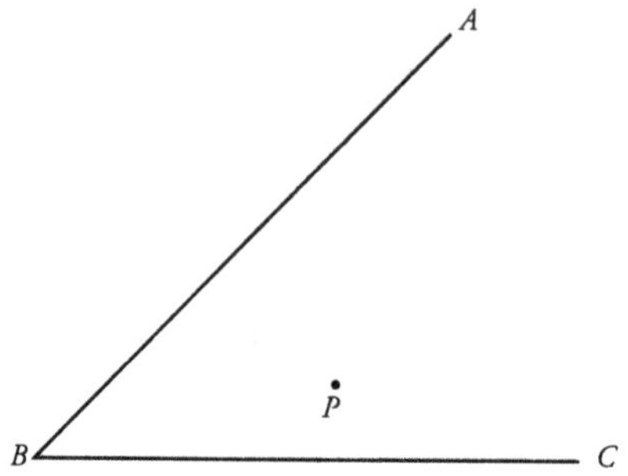

13. Given segments AB and CD, which meet at a point P off the page, construct the bisector of $\angle P$. All constructions must take place within the page.

14. Let AB be a diameter of a circle. Show that the points where the right bisector of AB meet the circle are the points of the circle that are farthest from AB.

15. M is the midpoint of the chord AB of a circle $\mathcal{C}(O, r)$. Show that if a different chord CD contains M, then $AB < CD$. (You may use Pythagoras' Theorem.)

16. $ABCD$ is a nonsimple quadrilateral. P, Q, R, and S are the midpoints of AB, BC, CD, and DA, respectively. Show that $PQRS$ is a parallelogram.

17. A ***regular polygon*** is onc in which all sides are equal and all angles are equal. Show that the vertices of a regular convex polygon lie on a common circle.

CHAPTER 3

SIMILARITY

3.1 Similar Triangles

The word *similar* is used in geometry to describe two figures that have identical shapes but are not necessarily the same size. A working definition of similarity can be obtained in terms of angles and ratios of distances.

Two polygons are ***similar*** if corresponding angles are congruent and the ratios of corresponding sides are equal.

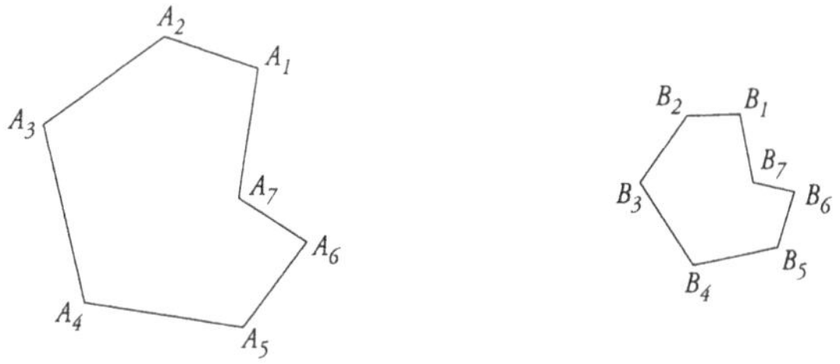

Classical Geometry: Euclidean, Transformational, Inversive and Projective, First Edition. By
I. E. Leonard, J. E. Lewis, A. C. F. Liu, G. W. Tokarsky
Copyright © 2014 John Wiley & Sons, Inc.

Notation. We use the symbol $\sim$ to denote similarity. Thus, we write

$$ABCDE \sim QRSTU$$

to denote that the polygons $ABCDE$ and $QRSTU$ are similar. As with congruency, the order of the letters is important.

To say that $\triangle ABC \sim \triangle DEF$ means that

$$\angle A \equiv \angle D, \qquad \angle B \equiv \angle E, \qquad \angle C \equiv \angle F,$$

and

$$\frac{AB}{DE} = \frac{BC}{EF} = \frac{AC}{DF} = k\,,$$

where k is a positive real number.

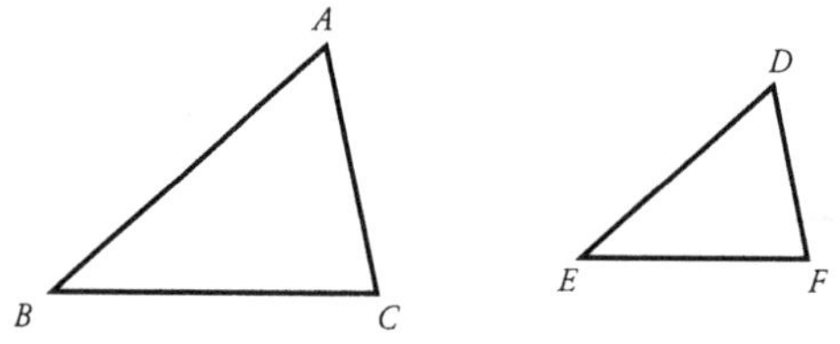

The constant k is called the ***proportionality constant*** or the ***magnification factor***. If $k > 1$, triangle ABC is larger than triangle DEF; if $0 < k < 1$, triangle ABC is smaller than triangle DEF; and if $k = 1$, the triangles are congruent.

Note that congruent figures are necessarily similar, but similar figures do not need to be congruent.

3.2 Parallel Lines and Similarity

There is a close relationship between parallel lines and similarity.

Lemma 3.2.1. *In $\triangle ABC$, suppose that D and E are points of AB and AC, respectively, and that DE is parallel to BC. Then*

$$\frac{AD}{DB} = \frac{AE}{EC}\,.$$

Figures (i), (ii), and (iii) on the following page illustrate the three different possibilities that can occur.

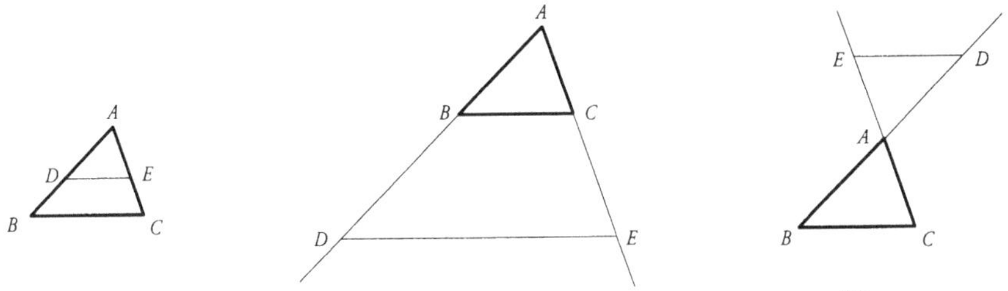

(i) (ii) (iii)

The proof of this lemma uses some simple facts about areas:

Theorem 3.2.2. *Given parallel lines l and m and two triangles each with its base on one line and its remaining vertex on the other, the ratio of the areas of the triangles is the ratio of the lengths of their bases.*

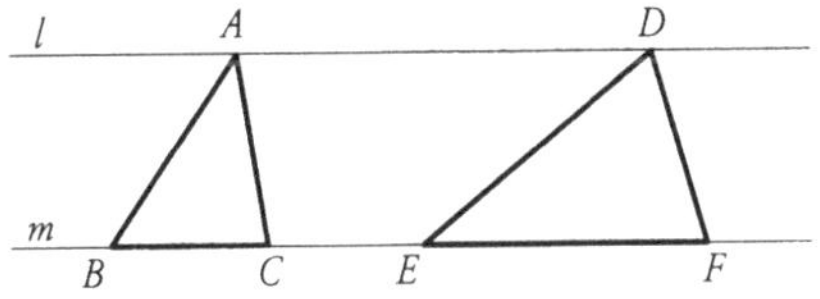

For the diagram above, the theorem says that

$$\frac{[ABC]}{[DEF]} = \frac{BC}{EF}.$$

Note that we use square brackets to denote area; that is, $[XYZ]$ is the area of $\triangle XYZ$.

There are two special cases worth mentioning, and these are illustrated by the figures below:

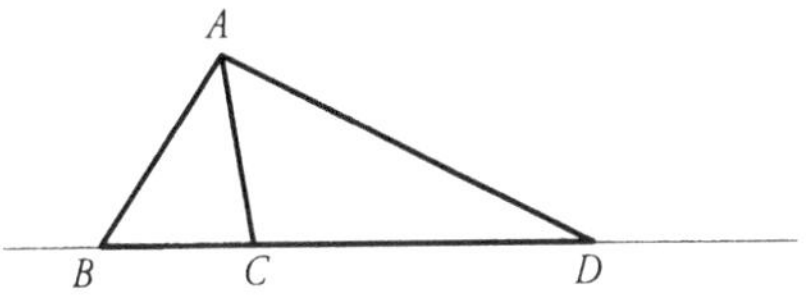

$$\frac{[ABC]}{[ACD]} = \frac{BC}{CD}$$

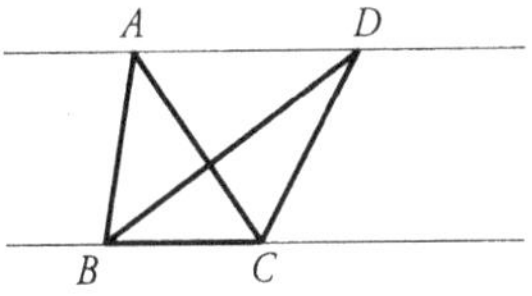

$$[ABC] = [DBC]$$

We sketch the proof of Lemma 3.2.1 for case (i). The proof of the other cases is very much the same.

Proof. In case (i), the line DE enters the triangle ABC through side AB and so must exit either through vertex C or through one of the other two sides.[1] Since DE is parallel to BC, the line DE cannot pass through vertex C or any other point on side BC. It follows that DE must exit the triangle through side AC.

Insert segments BE and CD and use the previously cited facts about areas of triangles to get

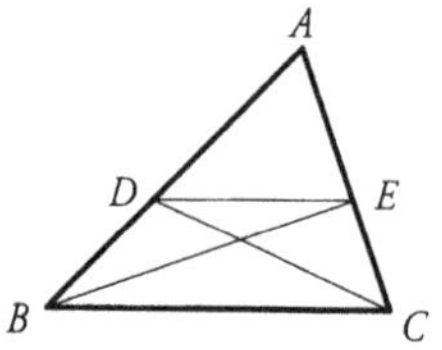

$$\frac{[ADE]}{[BDE]} = \frac{AD}{BD} \quad \text{and} \quad \frac{[ADE]}{[CDE]} = \frac{AE}{CE}.$$

Since $[BDE] = [CDE]$, we must have

$$\frac{AD}{DB} = \frac{AE}{EC}.$$

□

A useful extension of the lemma is the following:

Theorem 3.2.3. *Parallel projections preserve ratios. Suppose that l, m, and n are parallel lines that are met by transversals t and t' at points A, B, C and A', B', C', respectively. Then*

$$\frac{AB}{BC} = \frac{A'B'}{B'C'}.$$

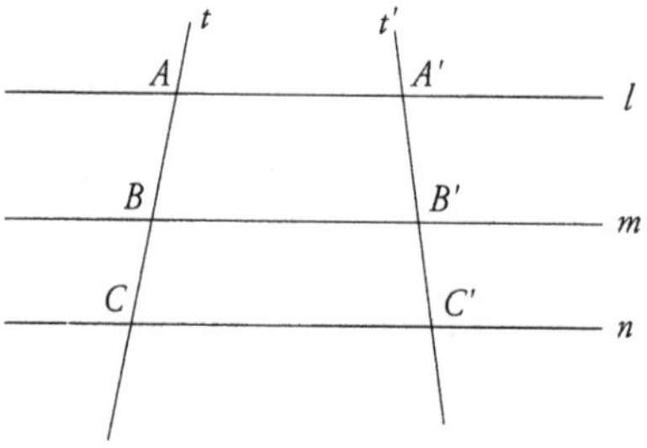

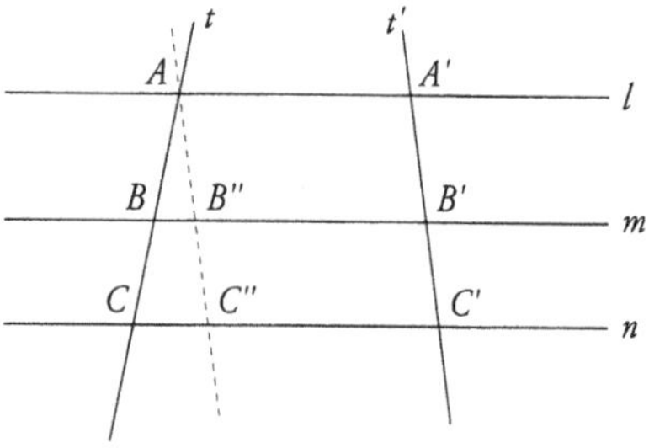

Proof. Draw a line through A parallel to t' meeting m at B'' and n at C''. Then $AA'B'B''$ and $B''B'C'C''$ are parallelograms, so $AB'' = A'B'$ and $B''C'' = B'C'$. Applying Lemma 3.2.1 to triangle ACC'', we have

$$\frac{AB}{BC} = \frac{AB''}{B''C''},$$

and the theorem follows. □

[1] This statement is known as Pasch's Axiom.

Next is the basic similarity theorem for triangles:

Theorem 3.2.4. *In $\triangle ABC$, suppose that DE is parallel to BC. If D and E are points of AB and AC, with D being neither A nor B nor C, then*

$$\triangle ABC \sim \triangle ADE.$$

Proof. As in the case of Lemma 3.2.1, there are three cases to consider:

(1) D is between A and B.

(2) D is on the ray $\overrightarrow{AB}$ beyond B.

(3) D is on the ray $\overrightarrow{BA}$ beyond A.

We will prove the theorem for case (1). The proofs for the other cases are almost identical. Because of the properties of parallel lines, the corresponding angles of ABC and ADE are equal, so it remains to show that the ratios of the corresponding sides are equal.

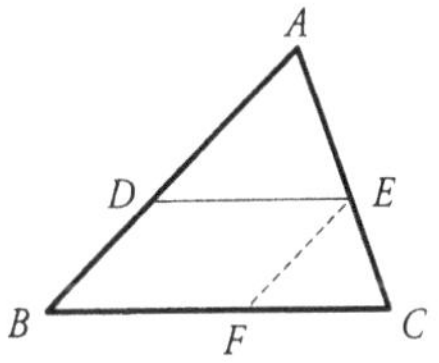

By Lemma 3.2.1, we have

$$\frac{AD}{DB} = \frac{AE}{EC},$$

which implies that

$$\frac{DB}{AD} = \frac{EC}{AE},$$

and this in turn implies that

$$\frac{DB}{AD} + 1 = \frac{EC}{AE} + 1,$$

so that

$$\frac{AD + DB}{AD} = \frac{AE + EC}{AE},$$

and therefore

$$\frac{AB}{AD} = \frac{AC}{AE}.$$

Now, through E, draw EF parallel to AB, with F on BC. Using an argument similar to the one above, we get

$$\frac{AC}{AE} = \frac{BC}{BF}.$$

Since $DE = BF$ (because $BDEF$ is a parallelogram), we get

$$\frac{AC}{AE} = \frac{BC}{DE}.$$

Thus,

$$\frac{AB}{AD} = \frac{AC}{AE} = \frac{BC}{DE},$$

which completes the proof that triangles ABC and ADE are similar.

□

There are other conditions that allow us to conclude that two triangles are similar, but there are many occasions where Theorem 3.2.4 is the most appropriate one to use.

Congruency versus Similarity

Congruency and similarity are both ***equivalence relations*** — both relations are reflexive, symmetric, and transitive.

Both are ***reflexive***:
$\triangle ABC \equiv \triangle ABC$.
$\triangle ABC \sim \triangle ABC$.

Both are ***symmetric***:
If $\triangle ABC \equiv \triangle DEF$, then $\triangle DEF \equiv \triangle ABC$.
If $\triangle ABC \sim \triangle DEF$, then $\triangle DEF \sim \triangle ABC$.

Both are ***transitive***:
If $\triangle ABC \equiv \triangle DEF$ and $\triangle DEF \equiv \triangle GHI$, then $\triangle ABC \equiv \triangle GHI$.
If $\triangle ABC \sim \triangle DEF$ and $\triangle DEF \sim \triangle GHI$, then $\triangle ABC \sim \triangle GHI$.

3.3 Other Conditions Implying Similarity

According to the definition, two triangles are similar if and only if the three angles are congruent and the three ratios of the corresponding sides are equal. As with congruent triangles, it is not necessary to check all six items. Here are some of the conditions that will allow us to verify that triangles are similar without checking all six.

Theorem 3.3.1. *(**AAA** or Angle-Angle-Angle Similarity)*

Two triangles are similar if and only if all three corresponding angles are congruent.

Exercise 3.3.2. *Show that two quadrilaterals need not be similar even if all their corresponding angles are congruent.*

Theorem 3.3.3. *(**sAs** or side-Angle-side Similarity)*

If in $\triangle ABC$ and $\triangle DEF$ we have

$$\frac{AB}{DE} = \frac{AC}{DF} \quad \text{and} \quad \angle A \equiv \angle D,$$

then $\triangle ABC$ and $\triangle DEF$ are similar.

Theorem 3.3.4. *(**sss** or side-side-side Similarity)*

Triangles ABC and DEF are similar if and only if

$$\frac{AB}{DE} = \frac{BC}{EF} = \frac{AC}{DF}.$$

The lowercase letter **s** in **sAs** and **sss** is to remind us that the sides need only be *in proportion* rather than congruent, while the uppercase letter **A** is to remind us that the angles must be *congruent.*

It is worth mentioning that since there are $180°$ in a triangle, the **AAA** similarity condition is equivalent to:

Theorem 3.3.5. *(**AA** Similarity)*

Two triangles are similar if and only if two of the three corresponding angles are congruent.

Proof of the **AAA** *similarity condition.* Make a congruent copy of one triangle so that it shares an angle with the other, that is, so that two sides of the one triangle fall upon two sides of the other. The congruency of the angles then guarantees that the third sides are parallel, and the proof is completed by applying Theorem 3.2.4.

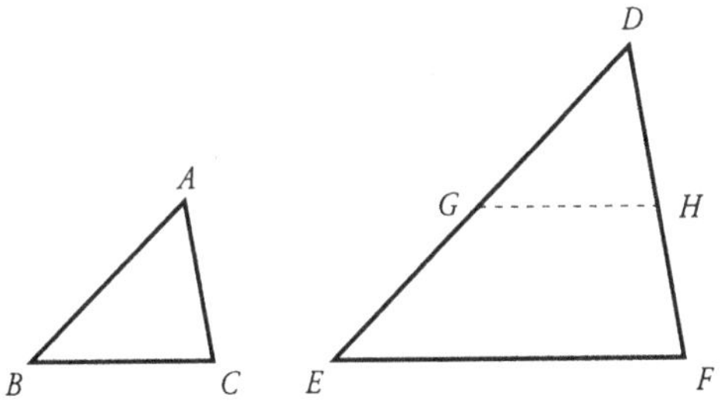

□

The proof for the **sAs** similarity condition uses the converse of Lemma 3.2.1.

Lemma 3.3.6. *Let P and Q be points on AB and AC with P between A and B and Q between A and C. If*

$$\frac{AP}{PB} = \frac{AQ}{QC},$$

then PQ is parallel to BC.

Proof. Through P, draw a line PR parallel to BC with R on AC. Then, since PR enters triangle ABC through side AB, it must exit the triangle through side AC or side BC. Since PR is parallel to BC, it follows that R is between A and C.

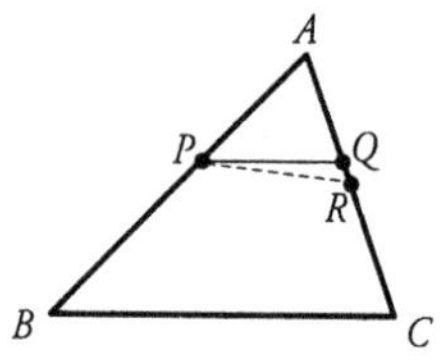

By Lemma 3.2.1,

$$\frac{AP}{PB} = \frac{AR}{RC},$$

so it follows that

$$\frac{AR}{RC} = \frac{AQ}{QC} = k\,.$$

However, given the positive number k, there is only one point X between A and C such that

$$\frac{AX}{XC} = k,$$

and so it follows that $R = Q$, and so $PQ = PR$, showing that PQ is parallel to BC.

□

Proof of the **sAs** *similarity condition.*

Suppose that in triangles ABC and DEF we have $\angle A \equiv \angle D$ and that

$$\frac{AB}{DE} = \frac{AC}{DF} = k\,.$$

We may assume that $k > 1$, that is, that $AB > DE$ and $AC > DF$. Cut off a segment AP on AB so that $AP = DE$. Cut off a segment AQ on AC so that $AQ = DF$.

Then it follows that

$$\frac{AB}{AP} = \frac{AC}{AQ},$$

and subtracting 1 from both sides of the equation gives us

$$\frac{PB}{AP} = \frac{QC}{AQ}.$$

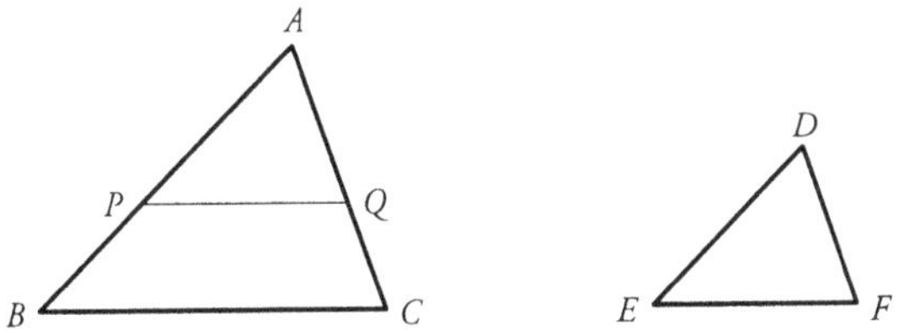

It now follows from Lemma 3.3.6 that PQ is parallel to BC, and Theorem 3.2.4 implies that $\triangle ABC \sim \triangle APQ$. Since $\triangle APQ \equiv \triangle DEF$ (by the **SAS** congruency condition), it follows that $\triangle ABC \sim \triangle DEF$.

□

Exercise 3.3.7. *Prove that two triangles are similar if they satisfy the* **sss** *similarity condition.*

3.4 Examples

Pythagoras' Theorem

Theorem 3.4.1. *(Pythagoras' Theorem)*

If two sides of a right triangle have lengths a and b and the hypotenuse has length c, then $a^2 + b^2 = c^2$.

Proof. In the figure below, drop the perpendicular CD to the hypotenuse, and let $AD = p$ and $BD = q$.

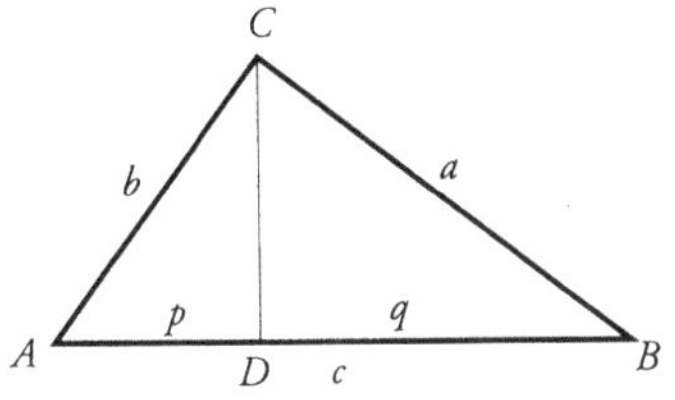

From the **AA** similarity condition,

$$\triangle CBD \sim \triangle ABC \qquad \text{and} \qquad \triangle ACD \sim \triangle ABC,$$

implying that

$$\frac{a}{q} = \frac{c}{a} \qquad \text{and} \qquad \frac{b}{p} = \frac{c}{b},$$

or, equivalently, that

$$a^2 = cq \qquad \text{and} \qquad b^2 = cp\,,$$

from which it follows that

$$a^2 + b^2 = cq + cp = c(q + p)\,,$$

and since $q + p = c$ we have $a^2 + b^2 = c^2$. □

Theorem 3.4.2. *(Converse of Pythagoras' Theorem)*

Let ABC be a triangle. If

$$AB^2 = BC^2 + CA^2,$$

then $\angle C$ is a right angle.

Exercise 3.4.3. *Prove the converse of Pythagoras' Theorem.*

Apollonius' Theorem

As an application of Pythagoras' Theorem, we prove the following result.

Theorem 3.4.4. *(Apollonius' Theorem)*

Let M be the midpoint of the side BC of triangle ABC. Then

$$AB^2 + AC^2 = 2AM^2 + 2BM^2.$$

Proof. Let AD be the altitude on the base BC, and assume that D lies between M and C.

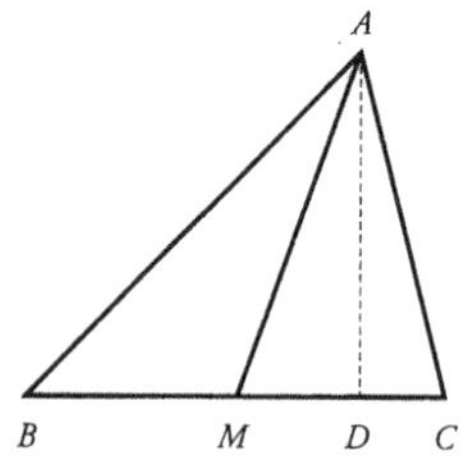

By Pythagoras' Theorem,

$$\begin{aligned} AB^2 + AC^2 &= BD^2 + 2AD^2 + CD^2 \\ &= (BM + MD)^2 + (BM - MD)^2 + 2AD^2 \\ &= 2BM^2 + 2MD^2 + 2AD^2 \\ &= 2BM^2 + 2AM^2. \end{aligned}$$

For other positions of D, the argument is essentially the same.

□

Example 3.4.5. *Use Apollonius' Theorem to prove that in $\triangle ABC$, if $m_a = AM$ is the median from the vertex A, then*

$$m_a = \frac{1}{2}\sqrt{2(b^2 + c^2) - a^2}$$

where $a = BC$, $b = AC$, and $c = AB$, as in the figure.

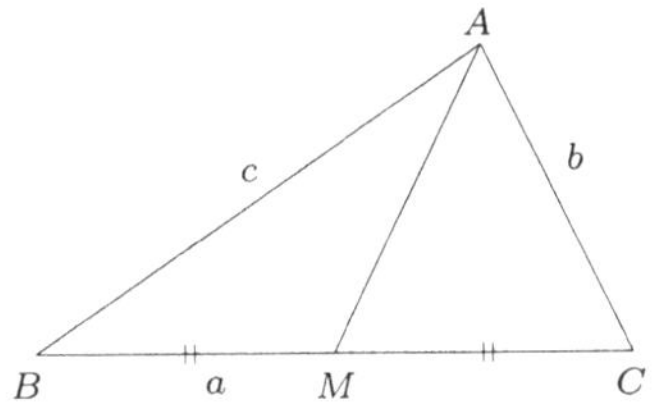

Solution. From Apollonius' Theorem, we have

$$AB^2 + AC^2 = 2BM^2 + 2AM^2,$$

that is,

$$c^2 + b^2 = 2(a/2)^2 + 2m_a^2,$$

so that

$$4m_a^2 = 2(b^2 + c^2) - a^2.$$

Therefore,

$$m_a = \frac{1}{2}\sqrt{2(b^2 + c^2) - a^2}.$$

□

Stewart's Theorem

A related theorem is the following, interesting in its own right, which allows us to express the lengths of the internal angle bisectors of a triangle in terms of the lengths of the sides.

Theorem 3.4.6. *(Stewart's Theorem)*

In $\triangle ABC$, if D is any point internal to the segment $\overline{BC}$, $d = AD$, $m = BD$, $n = DC$, $b = AC$, $c = AB$, and $a = BC$, as in the figure, then

$$c^2 \cdot n + b^2 \cdot m = a \cdot (d^2 + m \cdot n).$$

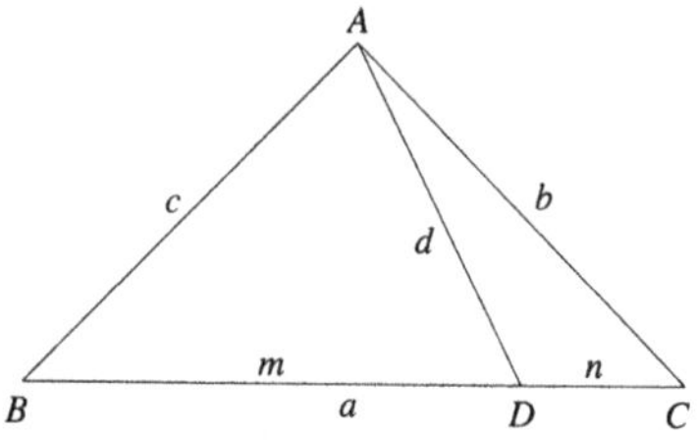

Proof. Drop a perpendicular from A to BC, hitting BC at E, and let $h = AE$ and $k = DE$, as in the figure below.

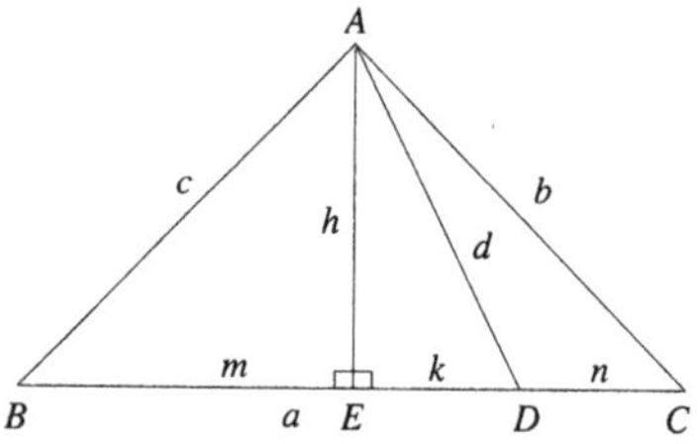

From Pythagoras' Theorem, we have

$$h^2 + k^2 = d^2,$$

so that

$$c^2 = (m - k)^2 + h^2 = m^2 - 2mk + k^2 + h^2 = m^2 - 2mk + d^2$$

and

$$b^2 = (n + k)^2 + h^2 = n^2 + 2nk + k^2 + h^2 = n^2 + 2nk + d^2.$$

Multiplying the equation for c^2 by n and the equation for b^2 by m, we have

$$n \cdot c^2 = n \cdot m^2 - 2mnk + n \cdot d^2,$$
$$m \cdot b^2 = m \cdot n^2 + 2mnk + m \cdot d^2,$$

so that

$$n \cdot c^2 + m \cdot b^2 = m \cdot n(m + n) + (m + n) \cdot d^2,$$

and since $m + n = a$, then

$$c^2 \cdot n + b^2 \cdot m = a \cdot (d^2 + m \cdot n).$$

□

Angle Bisector Theorem

Theorem 3.4.7. *(The Angle Bisector Theorem)*

Let D be a point on side BC of triangle ABC.

(1) If AD is the internal bisector of $\angle BAC$, then $AB/AC = DB/DC$.

(2) If AD is the external bisector of $\angle BAC$, then $AB/AC = DB/DC$.

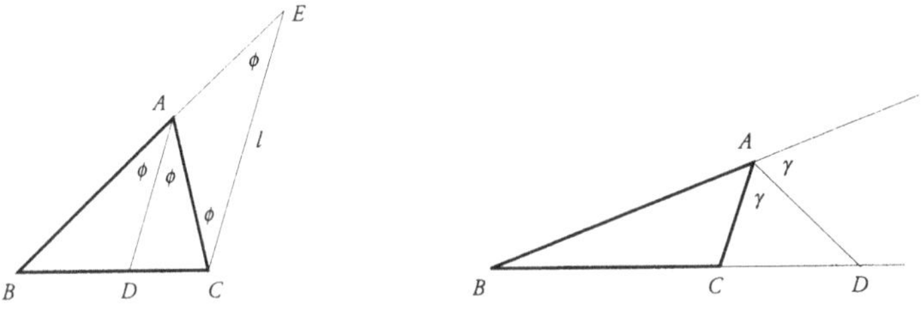

(i) (ii)

Proof. (1) Let l be a line parallel to AD through C meeting AB at E. Then

$$\angle BAD \equiv \angle BEC$$

and

$$\angle CAD \equiv \angle ACE,$$

and so $\angle AEC \equiv \angle ACE$; that is, $\triangle ACE$ is isosceles. Thus,

$$AC = AE,$$

and, again using the fact that $CE \parallel DA$,

$$\triangle ABD \sim \triangle EBC.$$

Hence,

$$\frac{DB}{DC} = \frac{AB}{AE} = \frac{AB}{AC}.$$

The proof of (2) using similarity is left as an exercise.

□

Example 3.4.8. *Use Stewart's Theorem to prove that in $\triangle ABC$, if $AD = f_A$ is the internal angle bisector at $\angle A$, then*

$$f_A^2 = bc\left[1 - \left(\frac{a}{b+c}\right)^2\right]$$

where $a = BC$, $b = AC$, and $c = AB$, as in the figure.

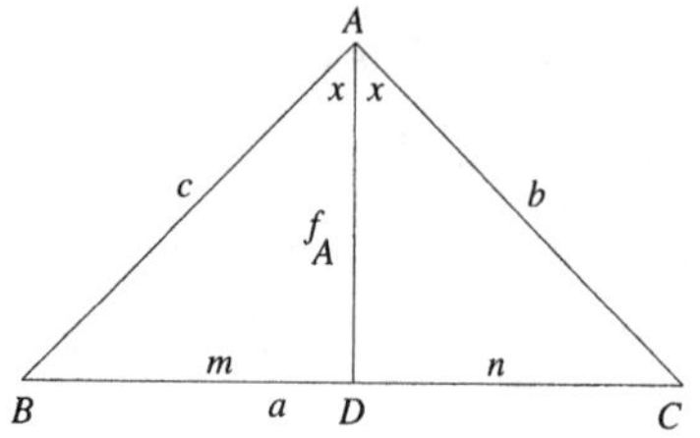

Solution. From the Internal Angle Bisector Theorem, we have

$$\frac{m}{n} = \frac{DB}{DC} = \frac{AB}{AC} = \frac{c}{b},$$

so that

$$m = \frac{c \cdot n}{b},$$

and since $a = m + n$, then

$$a = m + n = \frac{c \cdot n}{b} + n$$

and

$$n\left(1 + \frac{c}{b}\right) = a.$$

That is,

$$n = \frac{ab}{b+c} \qquad \text{and} \qquad m = \frac{ac}{b+c}.$$

From Stewart's Theorem, we have

$$\frac{c^2ab}{b+c} + \frac{b^2ac}{b+c} = a\left(f_A^2 + \frac{a^2bc}{(b+c)^2}\right);$$

that is,

$$abc\left(\frac{c}{b+c} + \frac{b}{b+c}\right) = a\left(f_A^2 + \frac{a^2bc}{(b+c)^2}\right).$$

Therefore,

$$bc = f_A^2 + \frac{a^2bc}{(b+c)^2};$$

that is,

$$f_A^2 = bc\left[1 - \left(\frac{a}{b+c}\right)^2\right].$$

□

Corollary 3.4.9. *In $\triangle ABC$, if f_A, f_B, and f_C denote the lengths of the internal angle bisectors at $\angle A$, $\angle B$, and $\angle C$, respectively, then*

$$f_A^2 = bc\left[1 - \left(\frac{a}{b+c}\right)^2\right],$$

$$f_B^2 = ac\left[1 - \left(\frac{b}{a+c}\right)^2\right],$$

$$f_C^2 = ab\left[1 - \left(\frac{c}{a+b}\right)^2\right].$$

Exercise 3.4.10. *In $\triangle ABC$, let AD, BE, and CF be the angle bisectors of $\angle BAC$, $\angle ABC$, and $\angle ACB$, respectively. Show that if*

$$\angle BAC < \angle ABC < \angle ACB,$$

then

$$|AD| > |BE| > |CF|.$$

Medians

In Chapter 2, we used area to show that the medians of a triangle trisect each other and are concurrent. Here is how similar triangles can be used to prove the same thing:

Example 3.4.11. *Let BE and CF be two medians of a triangle meeting each other at G. Show that*

$$\frac{EG}{GB} = \frac{FG}{GC} = \frac{1}{2}$$

and deduce that all three medians are concurrent at G.

Solution. In the figure below,

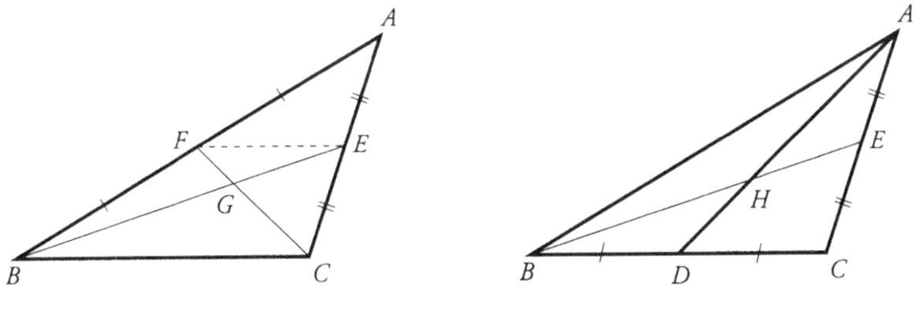

by the **sAs** similarity condition we have $\triangle ABC \sim \triangle AFE$, with proportionality constant 1/2. This means that $\angle ABC \equiv \angle AFE$, from which it follows that FE is parallel to BC with $FE = BC/2$. This means that $\triangle BCG \sim \triangle EFG$ with a proportionality constant of $1/2$. Hence,

$$GE = \frac{1}{2}GB \quad \text{and} \quad GF = \frac{1}{2}GC,$$

which means that $EG/GB = FG/GC = 1/2$.

Applying the same reasoning, medians BE and AD intersect at a point H for which $EH/HB = HD/AH = 1/2$. But this means that

$$\frac{EG}{GB} = \frac{EH}{HB} = \frac{1}{2},$$

and since both G and H are between B and E, we must have $H = G$. This shows that the three medians are concurrent at G.

□

We will end this section with a theorem that is sometimes useful in construction problems.

Theorem 3.4.12. *In $\triangle ABC$, suppose that D and E are points on AB and AC, respectively, such that $DE \parallel BC$. Let P be a point on BC between B and C, and let Q be a point on DE between D and E. Then A, P, and Q are collinear if and only if*

$$\frac{BP}{PC} = \frac{DQ}{QE}.$$

The figures below illustrate the three cases that can arise.

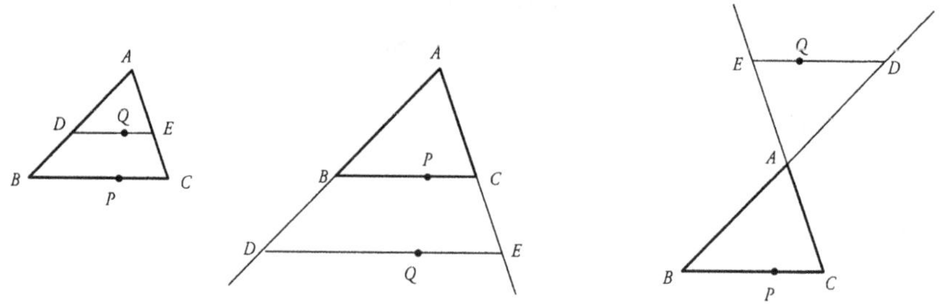

Proof.
(i) Suppose A, P, and Q are collinear. Since DE is parallel to BC, we have

$$\triangle APB \sim \triangle AQD \quad \text{and} \quad \triangle APC \sim \triangle AQE,$$

and so

$$\frac{BP}{DQ} = \frac{AP}{AQ} \quad \text{and} \quad \frac{AP}{AQ} = \frac{PC}{QE},$$

from which it follows that

$$\frac{BP}{DQ} = \frac{PC}{QE},$$

or, equivalently, that

$$\frac{BP}{PC} = \frac{DQ}{QE}.$$

(ii) Conversely, suppose that

$$\frac{BP}{PC} = \frac{DQ}{QE}.$$

Draw the line AP meeting DE at Q'. We will show that $Q = Q'$.

Since A, P, and Q' are collinear, the first part of the proof shows that

$$\frac{BP}{PC} = \frac{DQ'}{Q'E}.$$

We are given that

$$\frac{BP}{PC} = \frac{DQ}{QE},$$

so that

$$\frac{DQ'}{Q'E} = \frac{DQ}{QE},$$

from which it follows that $Q' = Q$, and this completes the proof.

□

3.5 Construction Problems

If you were asked to bisect a given line segment, you would probably construct the right bisector. How would you solve the following problem?

Example 3.5.1. *To divide a given line segment into three equal parts.*

Many construction problems involve similarity. The idea is to create a figure similar to the desired one and then, by means of parallel lines or some other device, transform the similar figure into the desired figure.

Creating a similar figure effectively removes size restrictions, and this can make a difficult problem seem almost trivial. Without size restrictions, Example 3.5.1 becomes:

To construct any line segment that is divided into three equal parts.

This is an easy task: using any line, fix the compass at any radius, and strike off three segments of equal length AB, BC, and CD. Then AD is a line segment that has been divided into three equal parts. What follows shows how this can be used to solve the original problem.

Solution. We are given the line segment PQ, which is to be divided into three equal parts. First, draw a ray from P making an angle with PQ. With the compass set at a convenient radius, strike off congruent segments PR', $R'S'$, and $S'Q'$ along the ray.

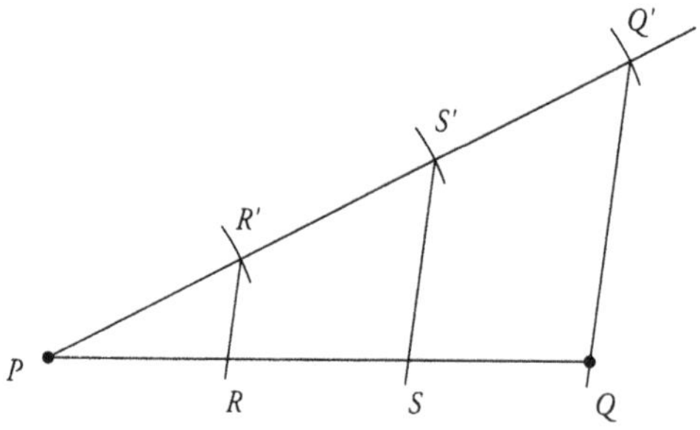

Join Q' and Q. Through R' and S', draw lines parallel to $Q'Q$ that meet PQ at R and S. Since parallel projections preserve ratios, we have

$$\frac{RS}{SQ} = \frac{R'S'}{S'Q'} = 1,$$

and so $RS = SQ$. Similarly, $PR = RS$.

□

The preceding construction can be modified to divide a line segment into given proportions. That is, it can be used to solve the following: given AB and line segments of length p and q, construct the point C between A and B so that

$$\frac{AC}{CB} = \frac{p}{q}.$$

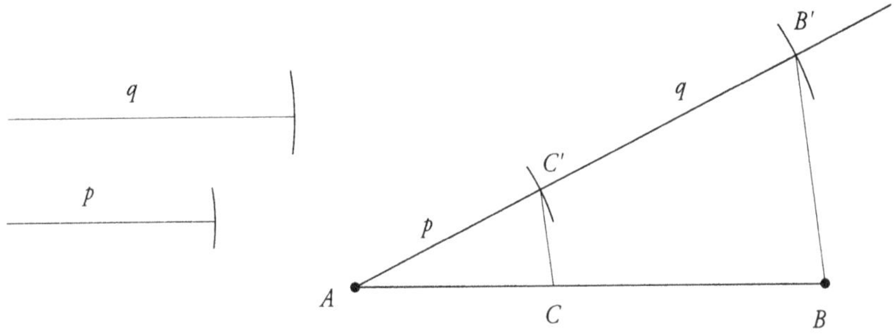

There is frequently more than one way to use similarity to solve a construction problem.

Example 3.5.2. *Given two parallel lines l and m and given points A on l, B on m, and P between l and m, construct a line n through P that meets l at C and m at D so that*

$$AC = \frac{1}{2}BD,$$

as in figure (a) below.

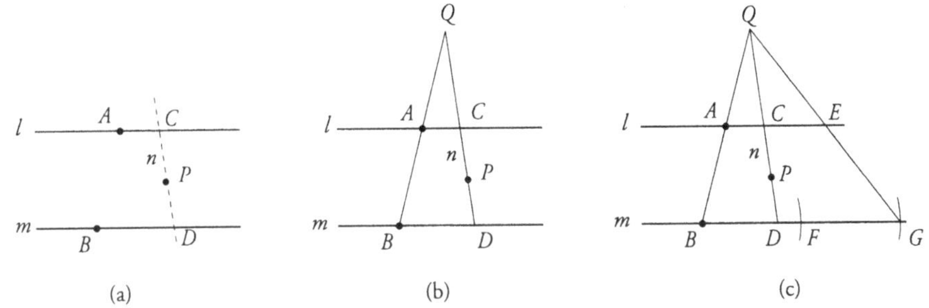

Solution.

(1) As in figure (b), draw the segment AB and extend it beyond A to Q so that $BA = AQ$. Draw the line PQ, meeting l at C and m at D, and then by similar triangles, $AC = BD/2$.

(2) As in figure (c), on line l, strike off a segment AE. On line m, strike off segments BF and FG of the same length as AE. Let Q be the point where the lines AB and EG meet. Draw the line PQ, meeting l at C and m at D, and then by similar triangles, $AC = BD/2$.

□

Example 3.5.3. *Given a point P inside an angle A, construct the two circles passing through P that are tangent to the arms of the angle.*

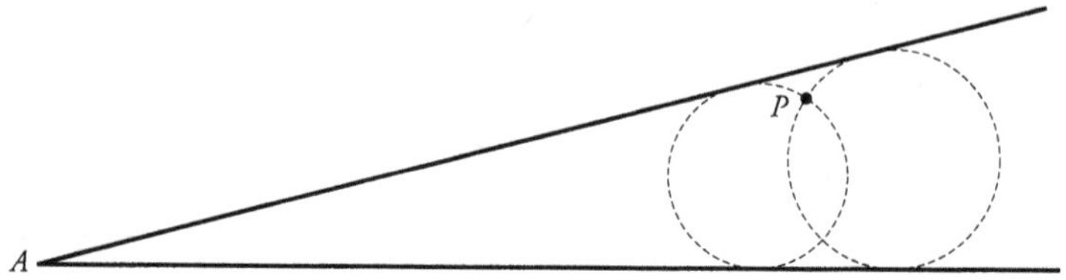

Solution. In the diagram, several construction arcs have been omitted for clarity.

Analysis Figure:

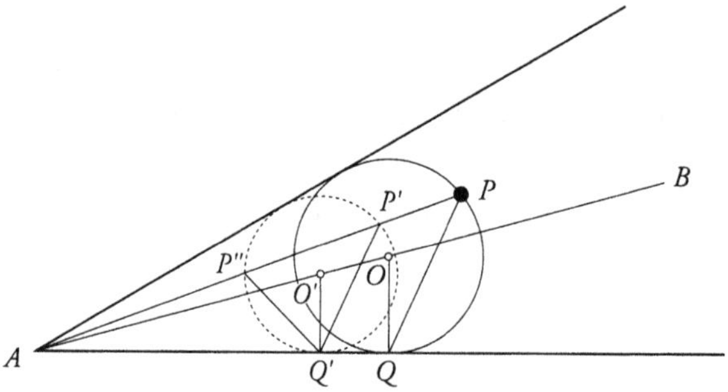

Construction:

(1) Construct the angle bisector AB of $\angle A$.

(2) Choose a point O' on AB, and drop the perpendicular $O'Q'$ to one of the arms of the angle.

(3) Draw the circle $\mathcal{C}(O', O'Q')$. This is the dotted circle in the diagram and it is tangent to both arms of the angle because O' is on the angle bisector.

(4) Draw the ray AP cutting the circle at P' and P''.

(5) Through P, construct the line PQ parallel to $P'Q'$, with Q on AQ'.

(6) Through Q, construct the line perpendicular to AQ meeting AO' at O.

(7) Draw the circle $\mathcal{C}(O, OQ)$. This is one of the desired circles.

(8) Through P, construct the line PQ'' parallel to $P''Q'$, with Q'' on AQ' (not shown).

(9) Through Q'', construct the line perpendicular to AQ'' meeting AO' at O'' (not shown).

(10) Draw the circle $\mathcal{C}(O'', O''Q'')$ (not shown). This is the second desired circle.

Justification:

To show that $\mathcal{C}(O, OQ)$ is one of the desired circles, we know that it is tangent to both arms of the angle because O is on the angle bisector and the radius OQ is perpendicular to an arm of the angle. It remains to show that the circle passes through P; that is, that $OP = OQ$, the radius of the circle.

Since $O'Q' \parallel OQ$, it follows that

$$\frac{OQ}{O'Q'} = \frac{AQ}{AQ'}.$$

Since $Q'P' \parallel QP$, it follows that

$$\frac{QP}{Q'P'} = \frac{AQ}{AQ'}$$

and, consequently, that

$$\frac{OQ}{O'Q'} = \frac{QP}{Q'P'}.$$

Again, since $Q'P' \parallel QP$, it follows that $\angle AQ'P' = \angle AQP$, and so

$$\angle O'Q'P' = \angle OQP.$$

Then, by the **sAs** criteria, $\triangle O'Q'P' \sim \triangle OQP$, and so

$$\frac{OP}{O'P'} = \frac{OQ}{O'Q'} = \frac{OQ}{O'P'}.$$

Thus $OP = OQ$, and this completes the proof.

□

Remark. Example 3.5.3 uses the idea that given a point A and a figure $\mathcal{F}$, we can construct a similar figure $\mathcal{G}$ by constructing all points Q such that $AQ = k \cdot AP$ for points P in $\mathcal{F}$, where k is a fixed nonzero constant. In the example, we constructed a circle centered at a point O' and then constructed a magnified version of this circle centered at point O.

The transformation that maps each point P to the corresponding point Q so that $AQ = k \cdot AP$ is called a ***homothety*** and is denoted $H(A, k)$. The number k is the magnification constant. The figure on the following page illustrates the effect of $H(A, 3)$.

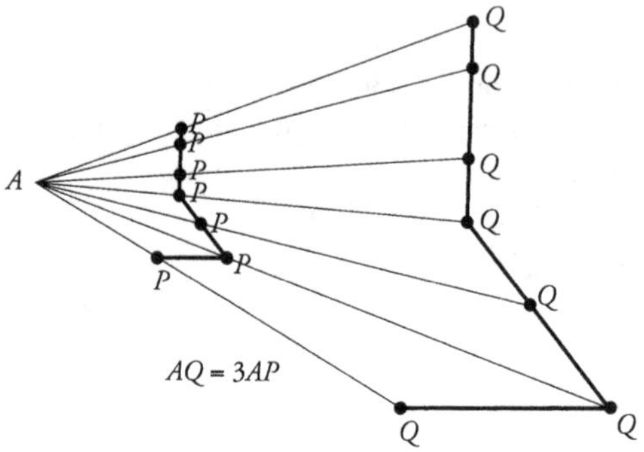

Here are two problems that are solved by translations rather than a homothety. The strategy is very similar: we construct a circle that fulfills part of the criteria and then use a translation to move it to the correct place.

Example 3.5.4. *Given a line l and points P and Q such that P and Q are on the same side of l and such that $PQ \perp l$, construct the circle tangent to l that passes through P and Q.*

Solution. In the figure, we have omitted construction lines for the standard constructions (like dropping a perpendicular from a point to a line).

Analysis Figure:

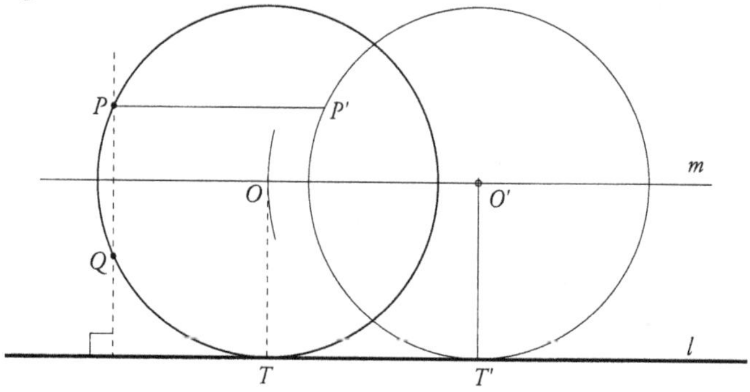

Construction:

(1) Construct the right bisector m of PQ.

(2) Choose any point O' on m and drop the perpendicular $O'T'$ from O' to l.

(3) Construct $\mathcal{C}(O', r)$ where $r = O'T'$. Note that $\mathcal{C}(O', r)$ is tangent to l.

(4) Construct a line through P perpendicular to PQ that meets $\mathcal{C}(O', r)$ at P'.

(5) With center O' and radius PP', draw an arc cutting m at O.

(6) Draw $\mathcal{C}(O, r)$. This is the desired circle.

Justification:

$PP'O'O$ is a parallelogram because PP' is congruent to and parallel to OO'. Then $OP = O'P' = r$, showing that P is on the circle $\mathcal{C}(O, r)$. Since O is on the right bisector of PQ, it follows that $OQ = OP = r$, showing that Q is on $\mathcal{C}(O, r)$.

Finally, since m and l are parallel, the perpendicular distance from O to l is the same as the perpendicular distance from O' to l, and it follows that $\mathcal{C}(O, r)$ is tangent to l.

□

Example 3.5.5. *Given two disjoint circles of radii R and r, construct the four tangents to the circle.*

The diagram on the right illustrates that two nonoverlapping circles have four common tangent lines. The problem is to construct those lines. We will show how to construct the "external" tangents and leave the construction of the "internal" ones as an exercise.

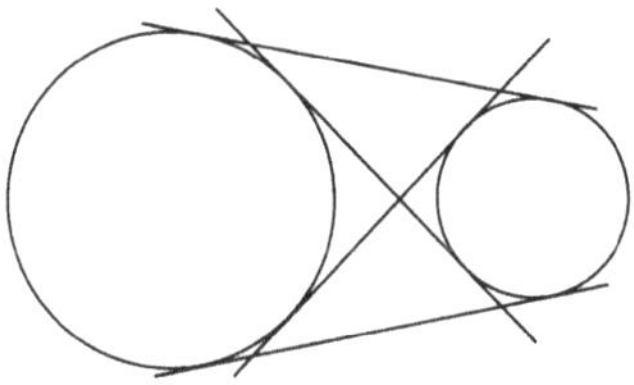

Solution.

Analysis Figure:

Let us suppose that the radii of the circles are R and r, as shown on the right. If we draw a line m through the center Q of the smaller circle parallel to a tangent line l as shown, then its distance from the center O of the larger circle will be $R - r$, and it will be tangent to $\mathcal{C}(O, R - r)$.

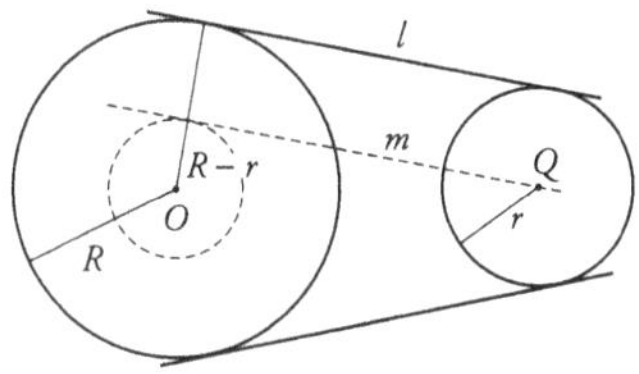

Construction:

Here is the step-by-step construction.

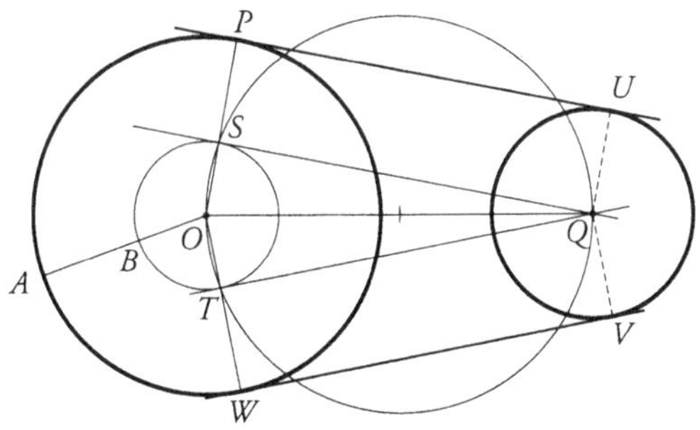

(1) Draw a radius OA of the larger circle. With the compass, cut off a point B so that $AB = r$. Then $OB = R - r$. Draw the circle $\mathcal{C}(O, OB)$.

(2) Through the point Q, draw the tangents QS and QT to $\mathcal{C}(O, OB)$.

(3) Draw the rays OS and OT, cutting the large circle at P and W, respectively.

(4) Through P, draw a line parallel to QS. Through W, draw a line parallel to QT. These are the desired tangent lines.

Justification:

Let QU be parallel to SP so that $QSPU$ is a rectangle. Then $QU = PS = r$ where r is the radius of the smaller circle. Since $PU \perp QU$, then PU is tangent to the smaller circle.

Since OP is a radius of the larger circle and since $PU \perp OP$, then PU is tangent to the larger circle.

Let QV be parallel to TW so that $QTWV$ is a rectangle. Reasoning as before, WV is tangent to both circles.

□

3.6 The Power of a Point

The following problem is taken from *AHA! Insight*, a delightful book written by Martin Gardner. Gardner attributes the problem to Henry Wadsworth Longfellow.

A lily pad floats on the surface of a pond as far as possible from where its root is attached to the bottom. If it is pulled out of the water vertically, until its stem is taut, it can be lifted 10 cm out of the water. The stem enters the water at a point 50 cm from where the lily pad was originally floating. What is the depth of the pond ?

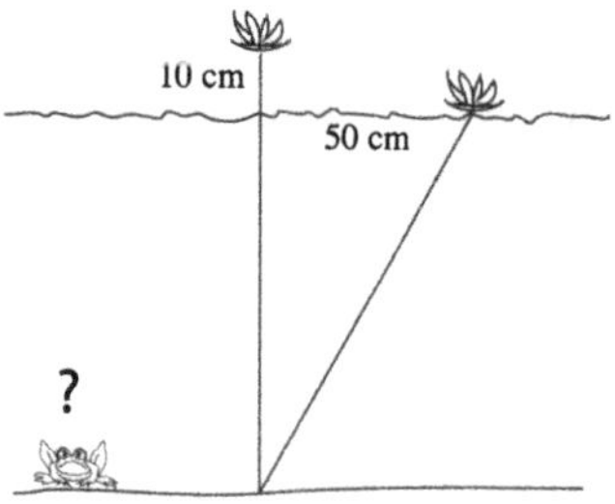

Most people attack this problem by using Pythagoras' Theorem together with the fact that a chord is perpendicular to the radius that bisects it. At the end of this section we will see that there is a much more elegant approach. We begin with a valuable fact about intersecting secants and chords.

Theorem 3.6.1. *Let ω be a circle, let P be any point in the plane, and let l and m be two lines through P meeting the circle at A and B and C and D, respectively. Then*

$$PA \cdot PB = PC \cdot PD.$$

Proof. If P is outside the circle, suppose that A is between P and B and that C is between P and D. Insert the line segments AD and BC, and consider triangles PAD and PCB.

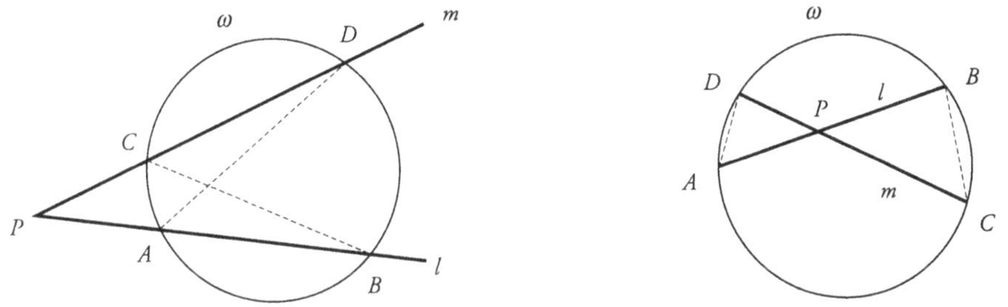

By Thales' Theorem,

$$\angle PDA \equiv \angle PBC \qquad \text{and} \qquad \angle P \text{ is common,}$$

so by the **AA** similarity criteria, $\triangle PAD \sim \triangle PCB$.

Consequently,

$$\frac{PA}{PC} = \frac{PD}{PB},$$

and so

$$PA \cdot PB = PC \cdot PD.$$

If P is inside the circle, again insert line segments AD and BC. Then, triangles PDA and PBC are similar, and so we again have $PA \cdot PB = PC \cdot PD$.

□

When P is outside the circle, an interesting thing happens if we swing the secant line PCD to a tangent position, as in the figure on the following page. In this case, the points C and D approach each other and coalesce at the point T, and so the product $PC \cdot PD$ approaches PT^2.

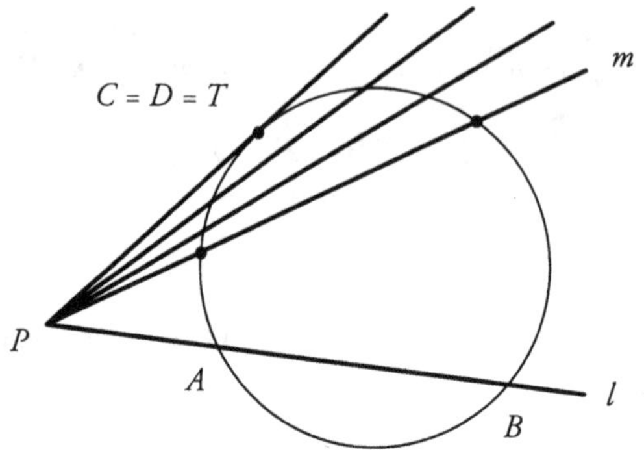

As a consequence:

Theorem 3.6.2. *Let P be a point in the plane outside a circle ω. Let PT be tangent to the circle at T, and let l be a line through P meeting the circle at A and B. Then*

$$PT^2 = PA \cdot PB.$$

Proof. Here is a proof that does not involve limits.

We may assume that A is between P and B. Insert the segments TA and TB, and we have the figure below.

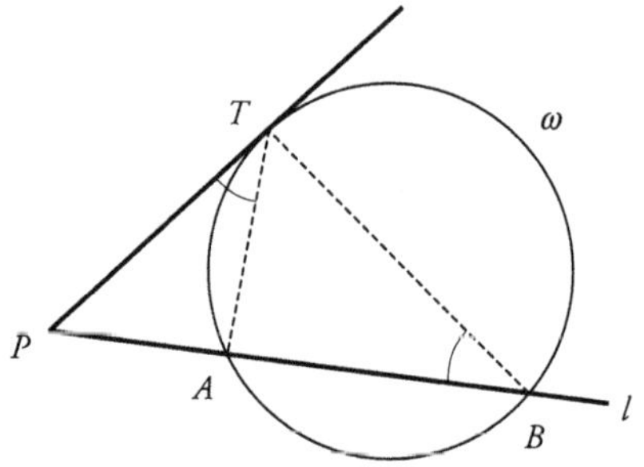

A consequence of Thales' Theorem is that $\angle PTA$ and $\angle PBT$ are equal in size. It follows that $\triangle PTA \sim \triangle PBT$, and so

$$PT/PB = PA/PT.$$

It follows immediately that $PT^2 = PA \cdot PB$.

□

If P is inside the circle, then, of course, no tangent to the circle passes through P. However, there is a result that looks somewhat like the previous theorem when P is inside the circle:

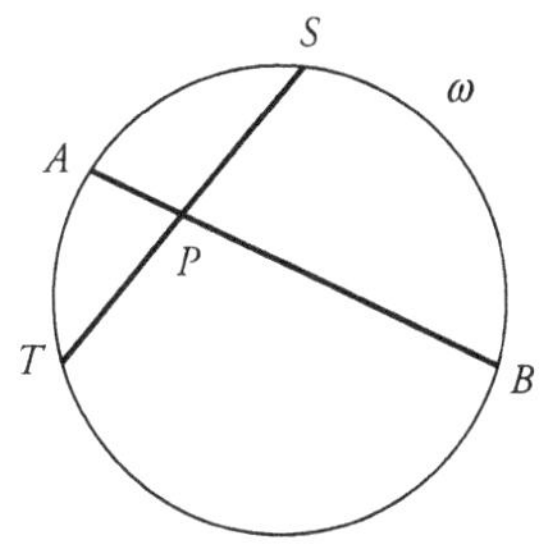

Theorem 3.6.3. *Let P be a point in the plane inside a circle ω. Let TS be a chord whose midpoint is P and let AB be any other chord containing P. Then*

$$PT^2 = PA \cdot PB.$$

Now let l be a line, and assign a direction to the line. For two points A and B on the line, with $A \neq B$, let AB be the distance between A and B. The ***directed distance*** or ***signed distance*** from A to B, denoted $\overline{AB}$, is defined as follows:

$$\overline{AB} = \begin{cases} AB & \text{if } A \text{ is before } B \text{ in the direction along } l, \\ 0 & \text{if } A = B, \\ -AB & \text{if } B \text{ is before } A \text{ in the direction along } l. \end{cases}$$

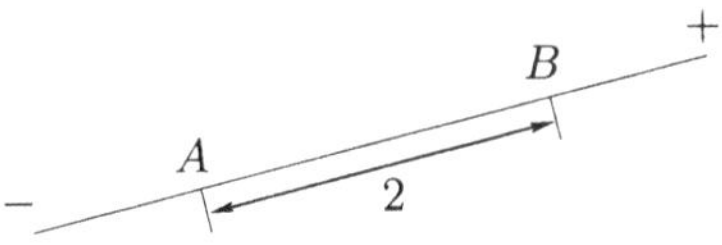

By using directed distances, the theorems above can be combined and extended as follows:

Theorem 3.6.4. *Let P be any point in the plane, let ω be a given circle, and let l be a line through P meeting the circle at A and B. Then the value of the product $\overline{PA} \cdot \overline{PB}$ is independent of the line l, and*

(1) $\overline{PA} \cdot \overline{PB} > 0$ if P is outside the circle,

(2) $\overline{PA} \cdot \overline{PB} = 0$ if P is on the circle,

(3) $\overline{PA} \cdot \overline{PB} < 0$ if P is inside the circle.

The value $\overline{PA} \cdot \overline{PB}$ is called the ***power*** of the point P with respect to the circle ω.

Suppose we are given a circle ω with center O and radius r, and suppose that we are given a point P and that we know the distance OP. It is fairly obvious how we could experimentally determine the power of P with respect to ω: draw any line through P meeting the circle at A and B, and measure PA and PB. Of course, if one chooses a line passing through O, then PA and PB are readily determined without recourse to measurement (see the figure on the following page).

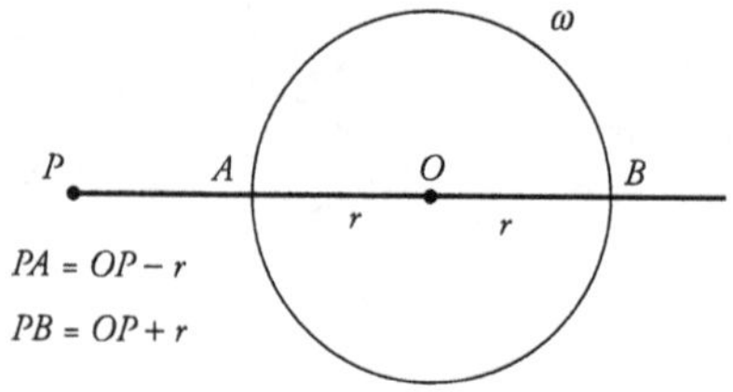

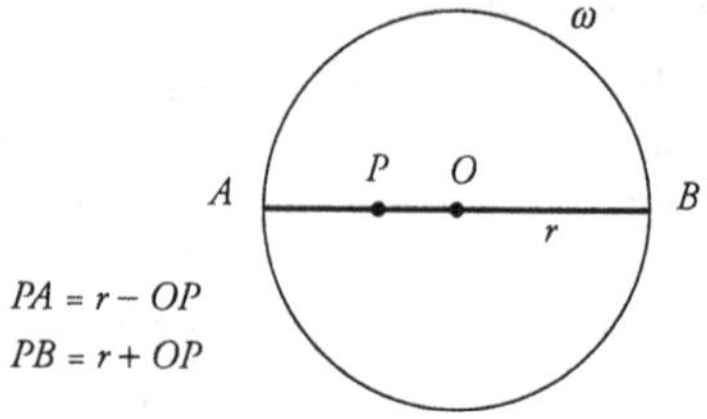

By doing this, we find:

Corollary 3.6.5. *The power of a point P with respect to a circle with center O and radius r is $OP^2 - r^2$.*

The Lily Pad Problem

The figure below illustrates the geometry of the lily pad problem. The depth of the pond is $OP = x$, and the length of the stem of the lily pad is the radius of a circle centered at O.

We know now that $PA \cdot PB = PT^2$, so referring to the diagram,

$$(2x + 10)(10) = 50^2,$$

and therefore the pond is 120 cm deep.

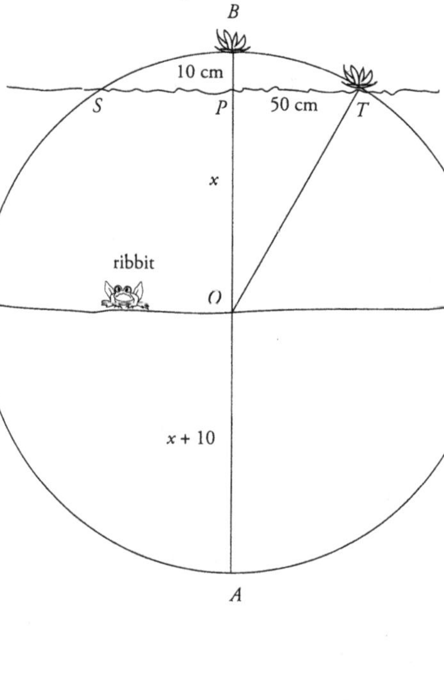

Example 3.6.6. *A person whose eyes are exactly 1 km above sea level looks towards the horizon at sea. The point on the horizon is 113 km from the observer's eyes. Use this to estimate the diameter of the earth.*

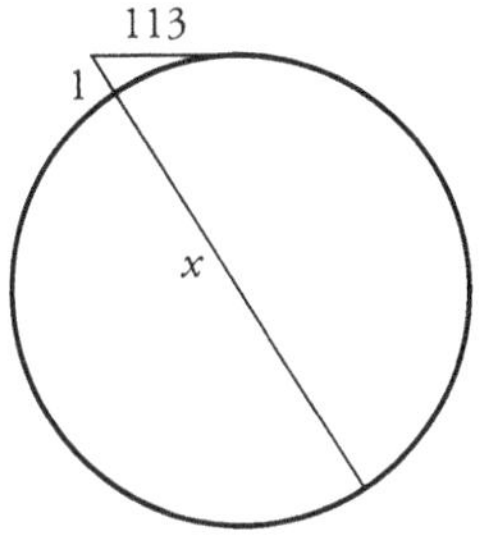

Solution. The line of sight to the horizon is tangent to the earth. If x is the diameter of the earth, we have

$$(x+1)1 = 113^2,$$

so

$$x = 113^2 - 1 = 12768 \text{ km}.$$

□

3.7 Solutions to the Exercises

Solution to Exercise 3.3.2

There are many examples, as the following figures indicate:

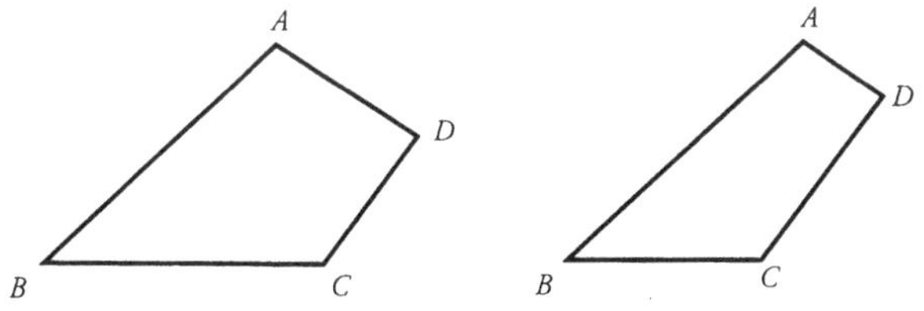

Solution to Exercise 3.3.7

We are given that

$$\frac{AB}{DE} = \frac{BC}{EF} = \frac{AC}{DF} = k,$$

and so we have to prove that the angles are equal.

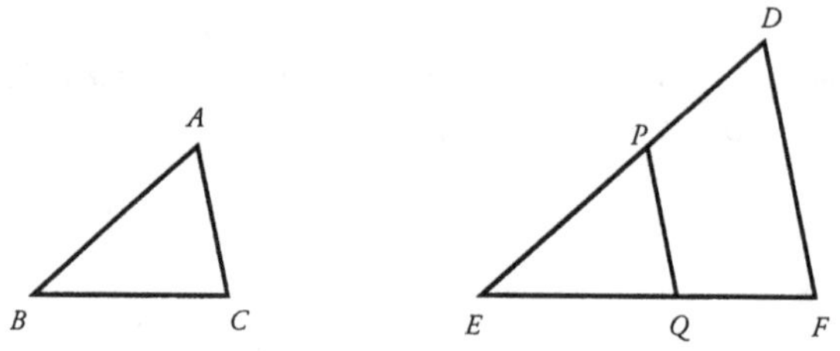

We may assume that $k < 1$, that is, that $AB < DE$.

Let P be the point on ED such that $EP = BA$. Draw PQ parallel to DF with Q on EF. By Theorem 3.2.4, $\triangle EPQ \sim \triangle EDF$, and so

$$\frac{EQ}{EF} = \frac{EP}{ED} = \frac{AB}{ED} = k = \frac{BC}{EF}.$$

This implies that $EQ = BC$.

Similarly, $PQ = AC$, so $\triangle EPQ \equiv \triangle BAC$, and it follows that

$$\angle DEF = \angle PEQ = \angle ABC.$$

Since $PQ \parallel DF$ we also have

$$\angle EDF = \angle EPQ = \angle BAC \quad \text{and} \quad \angle EFD = \angle EQP = \angle BCA,$$

and this completes the proof.

Solution to Exercise 3.4.3

Given triangle ABC with $AB^2 = BC^2 + CA^2$, let PQR be a triangle with a right angle at R, such that $QR = BC$ and $RP = CA$. Then

$$PQ^2 = QR^2 + RP^2 = BC^2 + CA^2 = AB^2.$$

Hence, $PQ = AB$, so ABC and PQR are congruent triangles by **SSS**. It follows that $\angle C$ is a right angle.

Solution to Exercise 3.4.10

Let $\beta = \angle CBE$ and let $\gamma = \angle BCF$. First, we note that if $\alpha = \beta$, then $\triangle ABC$ is isosceles, so that $BE = CF$. Thus, we may assume, without loss of generality, that $\beta < \gamma$.

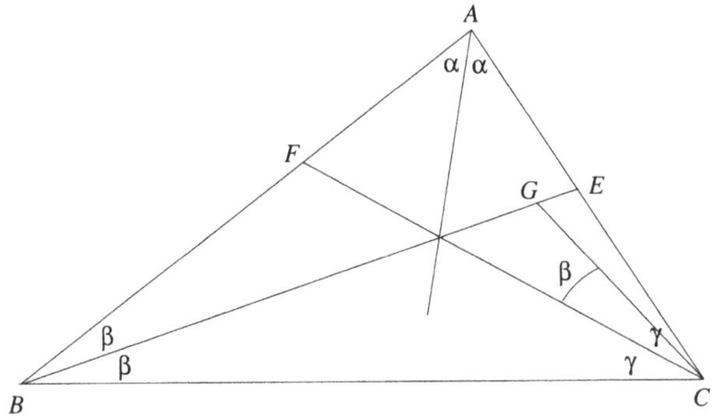

Let G be the point on BE between B and E such that $\angle FCG = \beta$. Then

$$\angle FBG = \angle FCG = \beta,$$

so by Thales' Theorem, $F,\ G,\ B,$ and C lie on a circle.

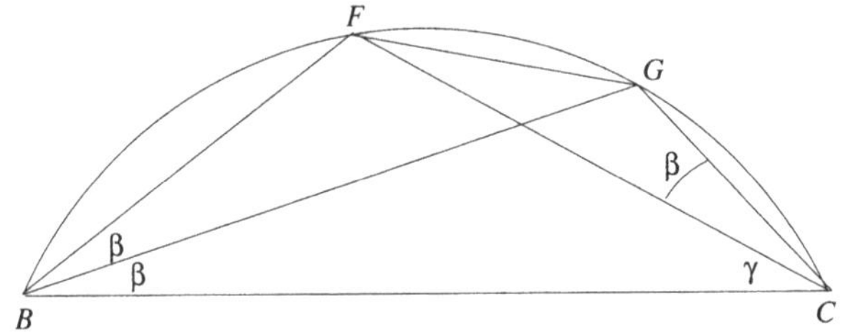

Now,

$$\angle B = 2\beta < \beta + \gamma = \frac{1}{2}(2\beta + 2\gamma) < \frac{1}{2}(2\alpha + 2\beta + 2\gamma) = 90,$$

so that $\angle CBF < \angle BCG < 90$, and therefore $CF < BG < BE$.

These last inequalities follow from the fact that given two chords in a circle, one is smaller if and only if it is further from the center, and this is true if and only if it subtends a smaller central angle (the **SAS** inequality and its converse) and therefore a smaller acute angle at the circumference.

So, if a triangle has two different angles, the smaller angle has the longer internal bisector, which is what we wanted to show.

3.8 Problems

1. P and Q are points on the side BC of $\triangle ABC$ with $BP = PQ = QC$. The line through P parallel to AC meets AB at X, and the line through Q parallel to AB meets AC at Y. Show that $\triangle ABC \sim \triangle AXY$.

2. BE and CF are altitudes of triangle ABC that meet at H. Prove that

$$BH \cdot HE = CH \cdot HF.$$

3. In the figure on the right, $\triangle ABC$ is an isosceles triangle with altitude AD,

$$AB = \tfrac{3}{2}BC, \quad AF = 4FD, \quad \text{and} \quad FD = 1.$$

FE is a perpendicular from F to AC. Find the length of FE.

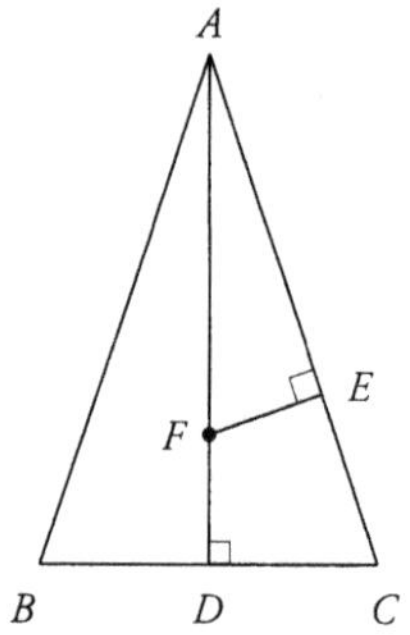

4. Triangle ABC is isosceles with $AB = AC$, and D is the midpoint of BC. The point K is on the line through C parallel to AB and $K \neq C$. The line KD intersects the sides AB and AC (or the extensions of the sides) at P and Q, respectively. Show that $AP \cdot QK = BP \cdot QP$.

5. The bisector of $\angle BAC$ meets BC at D. The circle with center C passing through D meets AD at X. Prove that $AB \cdot AX = AC \cdot AD$.

6. (*The Eyeball Theorem*) In the diagram on the following page, P and Q are centers of circles, and tangent lines are drawn from the centers to the other circle, intersecting the circles at A, B, C, and D, as shown. Prove that the chords AB and CD are equal in length.

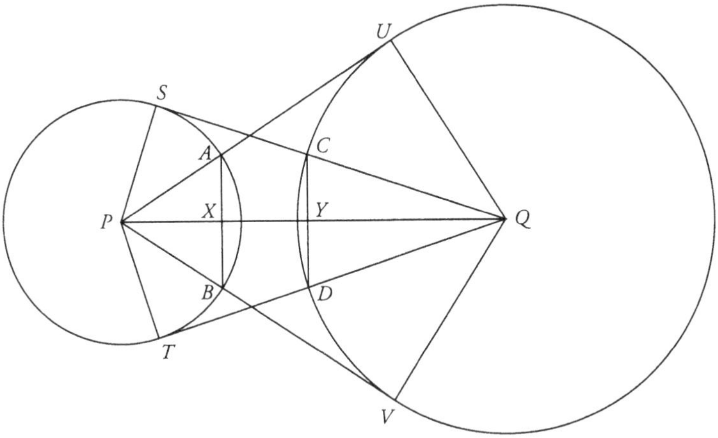

7. Given a triangle ABC with acute angles B and C, construct a square $PQRS$ with PQ in BC and vertices R and S in AB and AC, respectively.

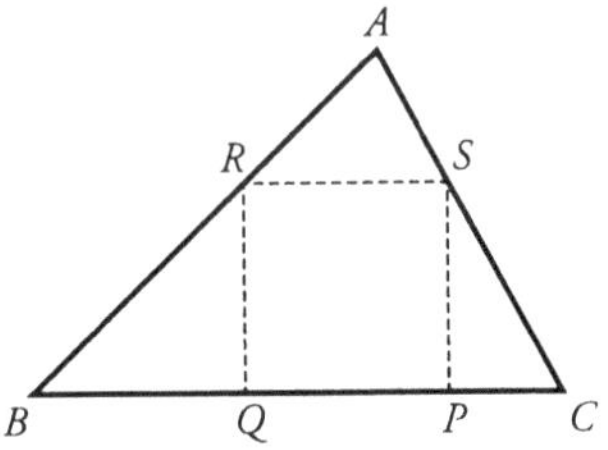

8. Given a line l and two points P and Q on the same side of l but with PQ not parallel to l, construct a circle tangent to l passing through P and Q.

9. Given two disjoint circles $\mathcal{C}(O, R)$ and $\mathcal{C}(Q, r)$, with $R > r$, construct the two "internal" tangents:

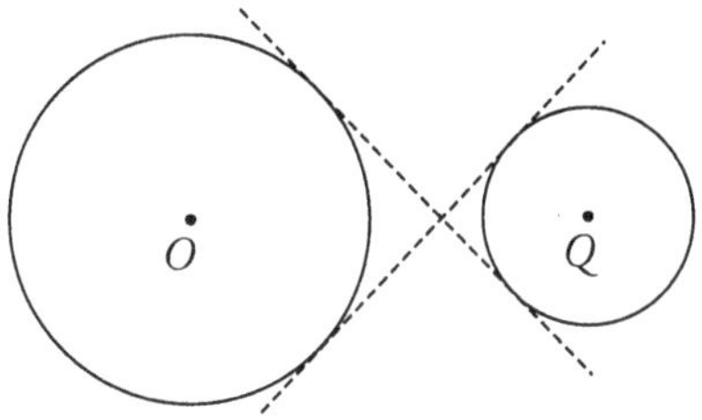

10. Construct a triangle ABC given the length of BC and the lengths of the medians m_B and m_C from B and C, respectively.

11. In the diagram, $PA = 3$, $BC = 4$, and $PC = 6$. Find the length of the segment AB.

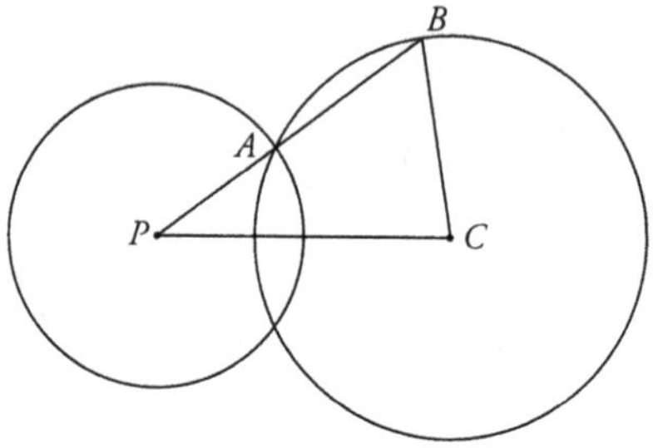

12. Given a circle of radius 1, find all points P such that the power of P with respect to the circle is 3.

13. In the following diagram, the segment AB is of length 3. Construct all points on the line AB whose power with respect to ω is 4.

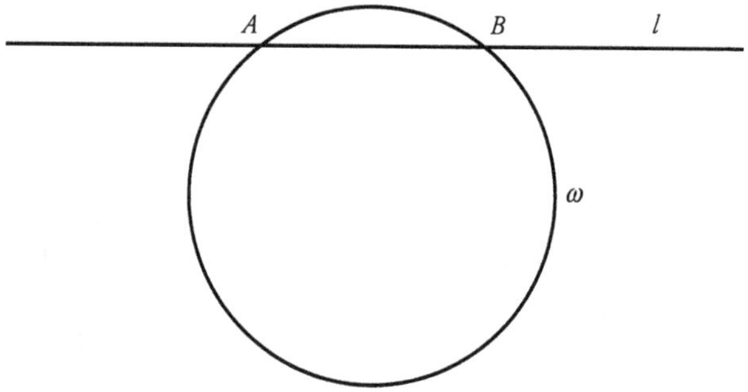

14. Prove the following partial converse of Theorem 3.6.4. Note the use of directed distances.

Let l and m be two different lines intersecting at P. Let A and B be points on l. Let C and D be points on m. If

$$\overline{PA} \cdot \overline{PB} = \overline{PC} \cdot \overline{PD},$$

then $ABCD$ is a cyclic quadrilateral.

15. Two different circles intersect at two points A and B. Find all points P such that the power of P is the same with respect to both circles.

16. M is the midpoint of the chord PQ of a circle. AB and CD are two other chords through M. PQ meets AD at X and BC at Y. Prove that $MX = MY$. This result is called the ***Butterfly Theorem***.

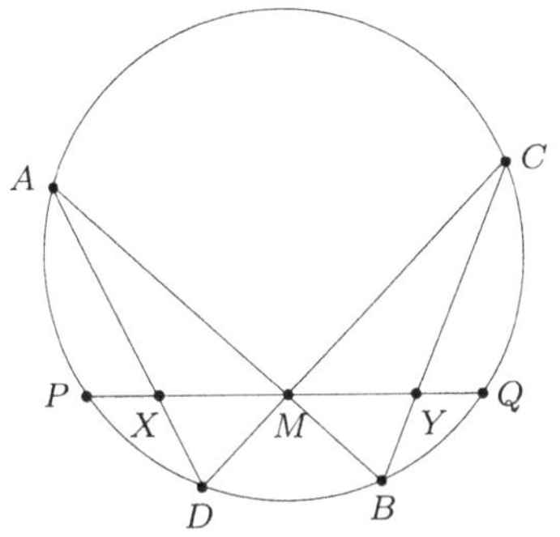

Hint. Drop perpendiculars from X and Y to AB and CD. Find four pairs of similar triangles.

CHAPTER 4

THEOREMS OF CEVA AND MENELAUS

4.1 Directed Distances, Directed Ratios

This chapter provides a more algebraic approach to concurrency and collinearity through the theorems of Ceva and Menelaus. The theorems are best understood using the notions of directed distances and directed ratios. We repeat the definition given in the previous chapter.

Let l be a line and assign a direction to the line. For two points A and B on the line, with $A \neq B$, let AB be the distance between A and B. The ***directed distance*** or ***signed distance*** from A to B, denoted $\overline{AB}$, is defined as follows:

$$\overline{AB} = \begin{cases} AB & \text{if } A \text{ is before } B \text{ in the direction along } l, \\ 0 & \text{if } A = B, \\ -AB & \text{if } B \text{ is before } A \text{ in the direction along } l. \end{cases}$$

Classical Geometry: Euclidean, Transformational, Inversive and Projective, First Edition. By I. E. Leonard, J. E. Lewis, A. C. F. Liu, G. W. Tokarsky
Copyright © 2014 John Wiley & Sons, Inc.

The directed distances $\overline{AB}$ and $\overline{BA}$ for the given direction along l are shown in the figure below, where $AB = 2$.

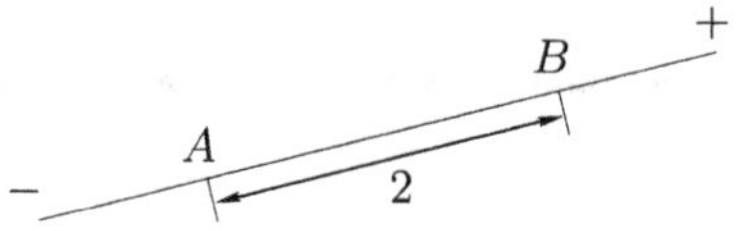

$$AB = 2$$
$$\overline{AB} = 2$$
$$\overline{BA} = -2$$

Properties of Directed Distance

(1) $\overline{AB} = -\overline{BA}$.

(2) For points A, B, and C on a line, $\overline{AB} + \overline{BC} = \overline{AC}$.

(3) If $\overline{AB} = \overline{AC}$, then $B = C$.

Sometimes property (2) is stated as $\overline{AB} + \overline{BC} + \overline{CA} = 0$. Note that properties (2) and (3) do not hold for unsigned distances.

Ratios

In the theorems of Ceva and Menelaus, given three points A, B, and C on a line, the following ***directed ratio*** occurs frequently:

$$\frac{\overline{AC}}{\overline{CB}}.$$

Note that if C is between A and B, then

$$\frac{\overline{AC}}{\overline{CB}} = +\frac{AC}{CB},$$

while if C is external to the segment AB, then

$$\frac{\overline{AC}}{\overline{CB}} = -\frac{AC}{CB}.$$

4.2 The Theorems

The two theorems are considered as companions of each other, although their discoveries were separated by many centuries. Menelaus proved his theorem around the year 100 CE. It languished in obscurity until 1678, when it was uncovered by Giovanni Ceva, who published it along with the theorem that bears his name. The two theorems are strikingly similar, and it is surprising that there was such a time span between the two discoveries.

Theorem 4.2.1. *(Ceva's Theorem)*

Let AD, BE, and CF be lines from the vertices A, B, and C of a triangle to nonvertex points D, E, and F of the opposite sides. The lines are either concurrent or parallel if and only if

$$\frac{\overline{AF}}{\overline{FB}} \cdot \frac{\overline{BD}}{\overline{DC}} \cdot \frac{\overline{CE}}{\overline{EA}} = +1\,.$$

Given a triangle, any line that passes through a vertex of the triangle and also through a nonvertex point of the opposite side is a called a ***cevian line*** or ***cevian*** of the triangle.[2] Medians, angle bisectors, and altitudes are cevians, but the right bisector of a side of a nonisosceles triangle is not. Also, a line through a vertex parallel to the opposite side of a triangle is not a cevian, although we will later incorporate this when we discuss the extended Euclidean plane.

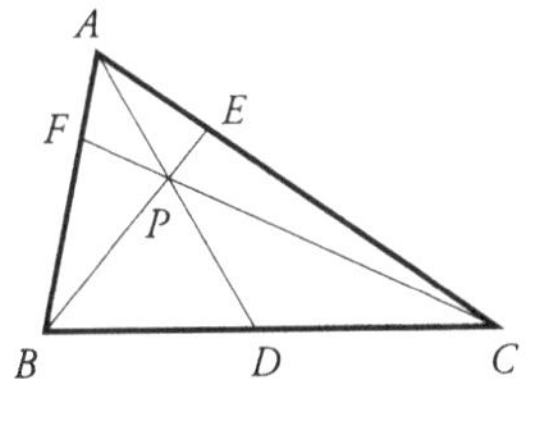

concurrent cevians

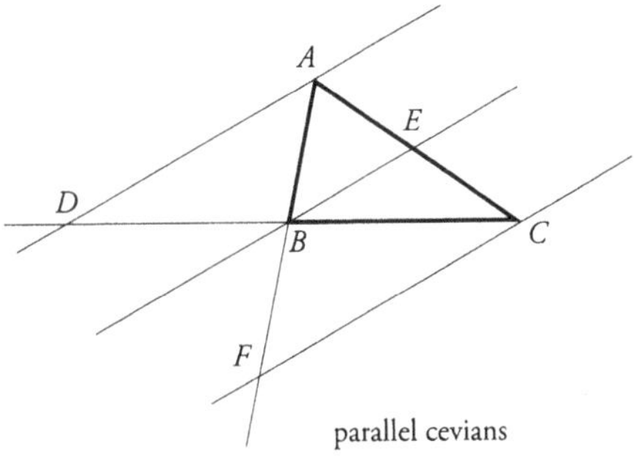

parallel cevians

Given a triangle, a nonvertex point in a side or in an extended side of the triangle is called a ***menelaus point***. A line that passes through each of the three edges of a triangle, but not through any of the vertices, is called a ***transversal line*** or a ***transversal***. Menelaus' Theorem tells us when three menelaus points lie on a transversal.

[2]The word "cevian" is a combination of Ceva's name and the word "median."

Theorem 4.2.2. *(Menelaus' Theorem)*

Let D, E, and F be menelaus points on the (extended) sides BC, CA, and AB of a triangle ABC. The points are collinear if and only if

$$\frac{\overline{AF}}{\overline{FB}} \cdot \frac{\overline{BD}}{\overline{DC}} \cdot \frac{\overline{CE}}{\overline{EA}} = -1\,.$$

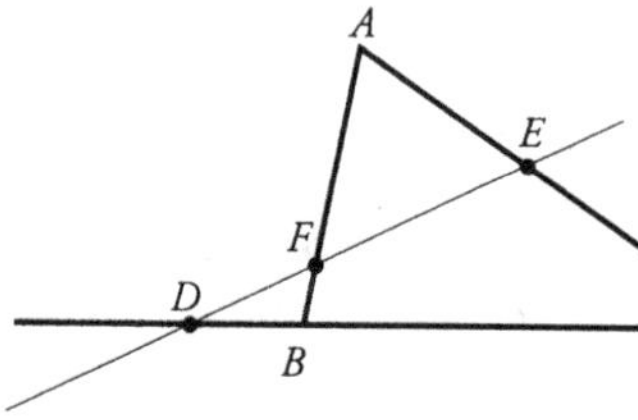

To use the two theorems, simply assign a direction to each side of the triangle, and proceed around the triangle in either a clockwise or counterclockwise direction. The product of the ratios in the theorems is sometimes called the ***cevian product***.

Note that the cevian product

$$\frac{\overline{AF}}{\overline{FB}} \cdot \frac{\overline{BD}}{\overline{DC}} \cdot \frac{\overline{CE}}{\overline{EA}}$$

involves the directed ratios

$\dfrac{\overline{AF}}{\overline{FB}}$ (from A to F to B along side AB),

$\dfrac{\overline{BD}}{\overline{DC}}$ (from B to D to C along side BC),

$\dfrac{\overline{CE}}{\overline{EA}}$ (from C to E to A along side CA).

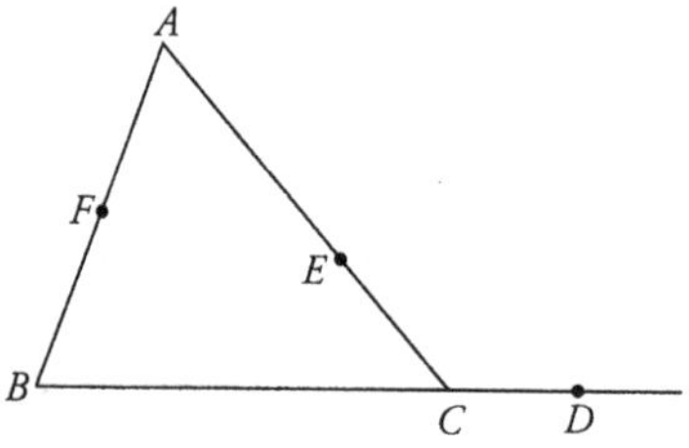

These ratios are meaningful because points F, D, and E are in the (perhaps extended) sides AB, BC, and CA of the triangle, respectively.

Note also that changing the direction assigned to the sides of the triangle does not affect the sign of the directed ratio. It should also be noted that if you compute the cevian product by proceeding clockwise rather than counterclockwise, the two cevian products will be reciprocals of each other. This does not alter the theorems because $+1$ and -1 are the only two real numbers that are their own reciprocals!

The original versions of the theorems used undirected distances, and, although the theorems are often stated that way, the modern versions are easier to use.

Ceva's Theorem will be the object of study in the next section, and Menelaus' Theorem will be discussed in the section following that.

4.3 Applications of Ceva's Theorem

The theorems of Ceva and Menelaus involve directed ratios of the form

$$\frac{\overline{AX}}{\overline{XB}}$$

where X is a point on the line AB other than A or B. Consequently, either X divides AB *internally* (meaning that X is between A and B) or else X divides AB *externally* (X is not between A and B). It is useful to recall that in these cases,

$$\frac{\overline{AX}}{\overline{XB}} = \begin{cases} +\dfrac{AX}{XB} & \text{if } X \text{ divides } AB \text{ internally,} \\[2ex] -\dfrac{AX}{XB} & \text{if } X \text{ divides } AB \text{ externally.} \end{cases}$$

As well, it is convenient to define directed ratios for segments that belong to *different but parallel* lines. Thus,

$$\frac{\overline{AB}}{\overline{CD}} = \begin{cases} +\dfrac{AB}{CD} & \text{if } AB \parallel CD \text{ and } ABDC \text{ is convex,} \\[2ex] -\dfrac{AB}{CD} & \text{if } AB \parallel CD \text{ and } ABDC \text{ is nonconvex.} \end{cases}$$

This is illustrated in the figure on the following page.

$ABDC$ is convex:

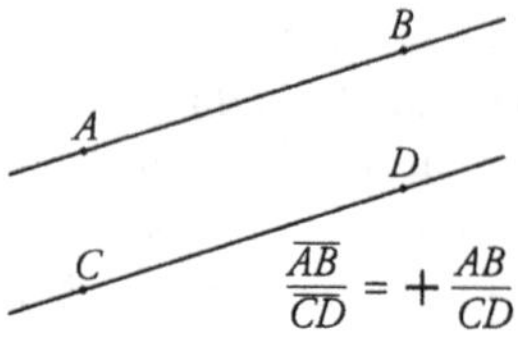

$ABDC$ is nonconvex:

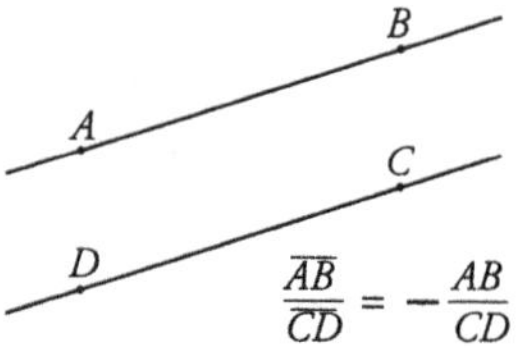

Before proving Ceva's Theorem, we will illustrate how it can be used to show that the most familar cevians are concurrent. Some of the results use the so-called ***Crossbar Theorem***, which we state without proof:

Theorem 4.3.1. *(Crossbar Theorem)*

If P is an interior point of triangle ABC, then the ray $\overrightarrow{AP}$ meets side B at some point between B and C.

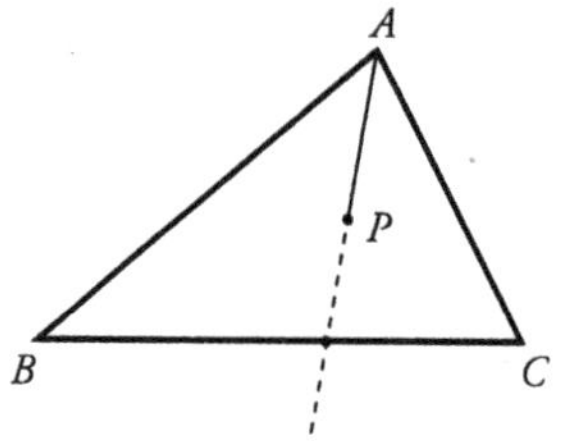

A frequently used consequence, that we again state without proof, is:

Theorem 4.3.2. *Two cevians that pass through the interior of a triangle are not parallel.*

Here is how Ceva's Theorem can be used to prove that medians, angle bisectors, and altitudes are concurrent.

Example 4.3.3. *The medians of a triangle are concurrent.*

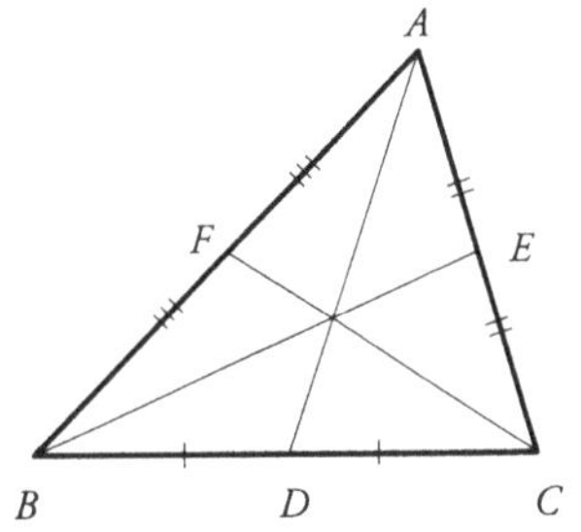

Solution. We have

$$\frac{\overline{AF}}{\overline{FB}} \cdot \frac{\overline{BD}}{\overline{DC}} \cdot \frac{\overline{CE}}{\overline{EA}} = 1 \cdot 1 \cdot 1 = +1,$$

since $\overline{AF} = \overline{FB}$, $\overline{BD} = \overline{DC}$, and $\overline{CE} = \overline{EA}$.

By Ceva's Theorem, the medians are either concurrent or parallel. Since they cannot be parallel by Theorem 4.3.2, they must be concurrent.

□

Example 4.3.4. *The internal bisectors of the angles of a triangle are concurrent.*

Solution.

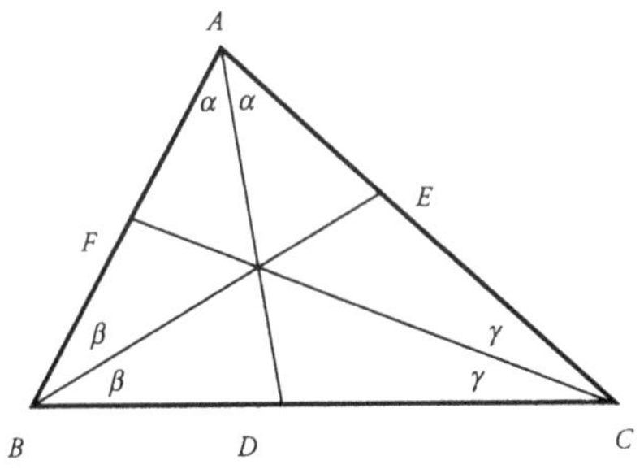

Recall from the Angle Bisector Theorem that if the interior angle bisector of $\angle A$ meets side BC at D, then $BD/DC = AB/AC$. Therefore,

$$\frac{\overline{AF}}{\overline{FB}} \cdot \frac{\overline{BD}}{\overline{DC}} \cdot \frac{\overline{CE}}{\overline{EA}} = \frac{CA}{CB} \cdot \frac{AB}{AC} \cdot \frac{BC}{BA} = +1.$$

By Theorem 4.3.2, the cevians in this case cannot be parallel, so by Ceva's Theorem, they are concurrent. □

Example 4.3.5. *The altitudes of a triangle are concurrent.*

Solution.

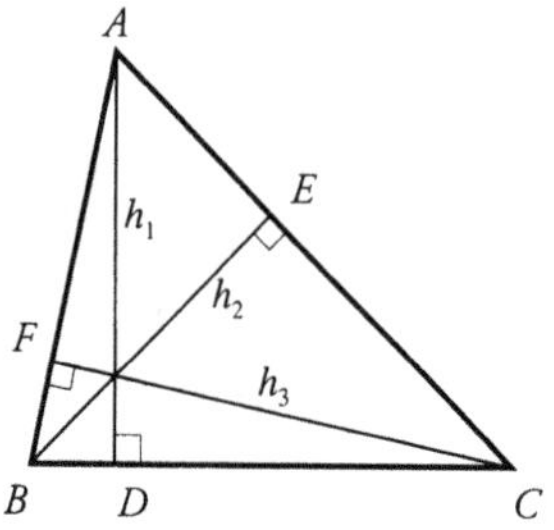

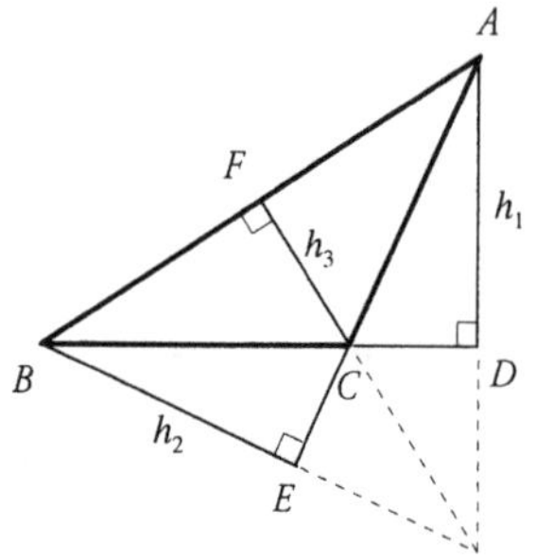

For a right triangle, the three altitudes are concurrent at the vertex of the right angle. For an acute-angled triangle, the three altitudes divide the sides internally, and for an obtuse-angled triangle, the altitudes divide exactly two of the sides externally. (A proof of this may be obtained from the External Angle Inequality and is left as an exercise.)

Therefore, for a triangle ABC with altitudes AD, BE, and CF, we have

$$\frac{\overline{AF}}{\overline{FB}} \cdot \frac{\overline{BD}}{\overline{DC}} \cdot \frac{\overline{CE}}{\overline{EA}} = \frac{AF}{FB} \cdot \frac{BD}{DC} \cdot \frac{CE}{EA},$$

since either all ratios are internal or else exactly two are external.

Using similar triangles to relate the ratios to the length of altitudes,

$$\triangle BEC \sim \triangle ADC, \qquad \text{so} \quad \frac{CE}{DC} = \frac{h_2}{h_1},$$

$$\triangle ABE \sim \triangle ACF, \qquad \text{so} \quad \frac{AF}{AE} = \frac{h_3}{h_2},$$

$$\triangle ABD \sim \triangle CBF, \qquad \text{so} \quad \frac{BD}{BF} = \frac{h_1}{h_3}.$$

Hence,

$$\frac{AF}{FB} \cdot \frac{BD}{DC} \cdot \frac{CE}{EA} = \frac{h_2}{h_1} \cdot \frac{h_3}{h_2} \cdot \frac{h_1}{h_3} = 1,$$

and Ceva's Theorem shows that the altitudes are concurrent.

□

The next example illustrates how Ceva's Theorem can be used to compute a ratio or a distance rather than to conclude that certain cevians are concurrent.

Example 4.3.6. *In the figure, $AB = 4$, $BC = 5$, and $AC = 6$. AD is an angle bisector, and BE is a median. Find the length of AF.*

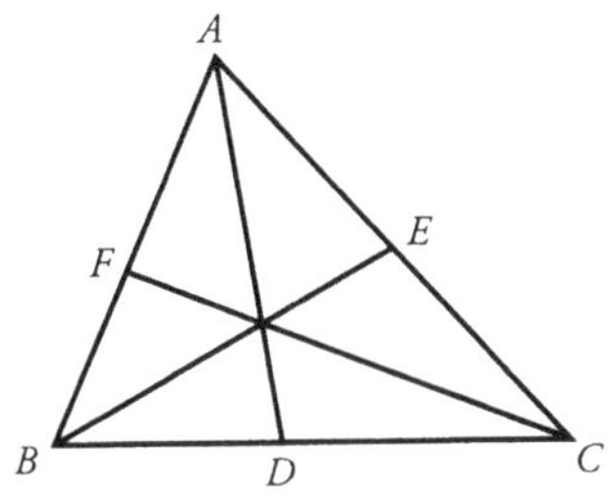

Solution. By Ceva's Theorem,

$$\frac{AF}{FB} \cdot \frac{BD}{DC} \cdot \frac{CE}{EA} = 1.$$

By the Angle Bisector Theorem, $BD/DC = 4/6$, and since BE is a median, $CE/EA = 1$, so that

$$\frac{AF}{FB} \cdot \frac{BD}{DC} \cdot \frac{CE}{EA} = \frac{AF}{4 - AF} \cdot \frac{4}{6} \cdot 1 = 1,$$

from which we get $AF = 12/5$.

□

4.4 Applications of Menelaus' Theorem

Angle Bisectors

There are three internal bisectors and three external bisectors of the angles of a triangle. If we take one bisector, either internal or external, from each vertex, then there are four possible combinations:

(i) All three are internal bisectors.

(ii) Two are external bisectors and one is an internal bisector.

(iii) One is an external bisector and two are internal bisectors.

(iv) All three are external bisectors.

For (i) and (ii), the bisectors are concurrent, and so it is probably no surprise to the reader that there is something significant about (iii) and (iv).

Example 4.4.1. *The internal bisectors of two angles of a triangle and the external bisector of the third meet the opposite sides in three collinear points.*

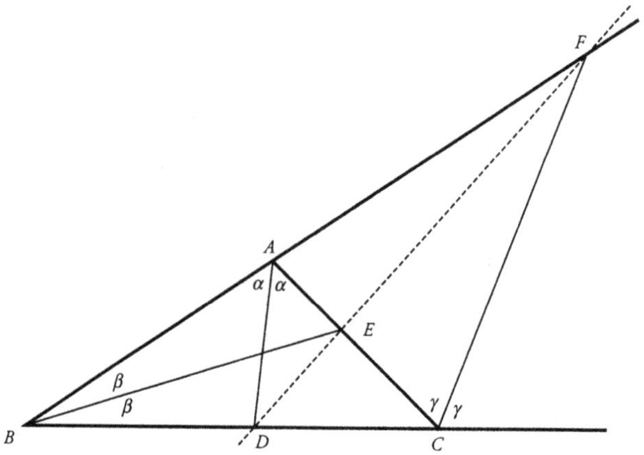

Solution. Let us consider the internal bisectors of $\angle A$ and $\angle B$ and the external bisector of $\angle C$. Using the Internal and External Bisector Theorems,

$$\frac{\overline{AF}}{\overline{FB}} \cdot \frac{\overline{BD}}{\overline{DC}} \cdot \frac{\overline{CE}}{\overline{EA}} = \left(-\frac{CA}{CB}\right) \cdot \frac{AB}{AC} \cdot \frac{BC}{BA} = -1\,,$$

and so by Menelaus' Theorem, the points are collinear.

□

Simson's Theorem

The following proof of Simson's Theorem uses Menelaus' Theorem.

Example 4.4.2. *Given $\triangle ABC$ and a point P on its circumcircle, the perpendiculars dropped from P meet the sides of the triangle in three collinear points.*

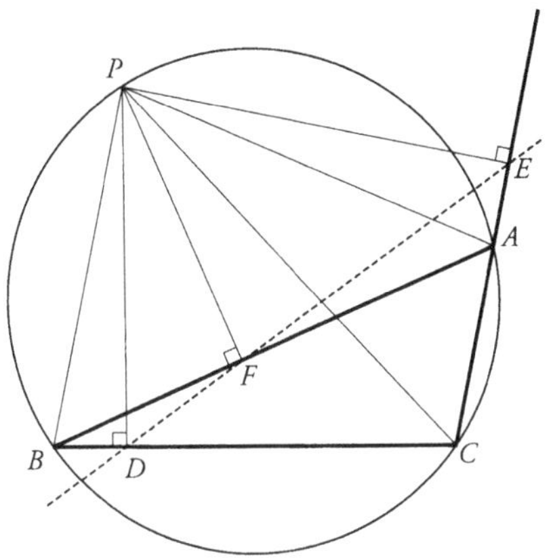

Solution. Let D, E, and F be the feet of the perpendiculars on BC, CA, and AB, respectively. Since D, E, and F are three menelaus points of $\triangle ABC$, it is enough to show that the cevian product is -1.

Introducing PA, PB, and PC, Thales' Theorem reveals that $\angle PAF \equiv \angle PCD$. Since both $\triangle PAF$ and $\triangle PCD$ are right triangles, they are similar. In fact, we find that

$$\triangle PAF \sim \triangle PCD, \quad \triangle PCE \sim \triangle PBF, \quad \text{and} \quad \triangle PBD \sim \triangle PAE,$$

with the last similarity following from Theorem 1.3.11. It follows that

$$\frac{AF}{CD} = \frac{PF}{PD},$$
$$\frac{EC}{FB} = \frac{PE}{PF},$$
$$\frac{BD}{AE} = \frac{PD}{PE},$$

and the result is that

$$\frac{\overline{AF}}{\overline{FB}} \cdot \frac{\overline{BD}}{\overline{DC}} \cdot \frac{\overline{CE}}{\overline{EA}} = \pm 1.$$

To check that the sign is actually negative, we have to show that an odd number of the points D, E, and F divide the sides externally. This may be accomplished by constructing a hexagon $AQBRCS$ whose sides are perpendicular to the sides of $\triangle ABC$, as in figure (a) on the following page.

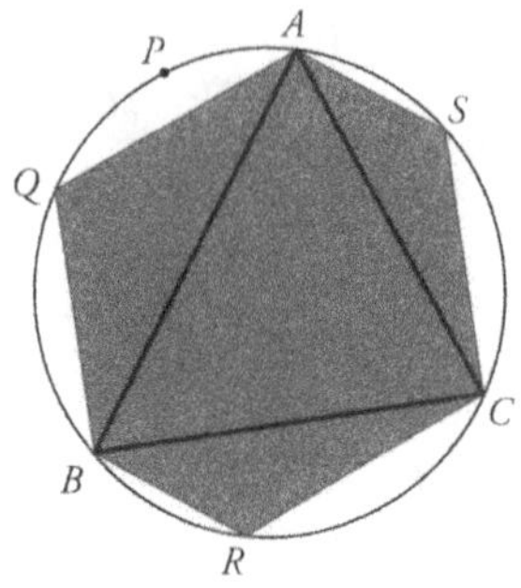

(a)

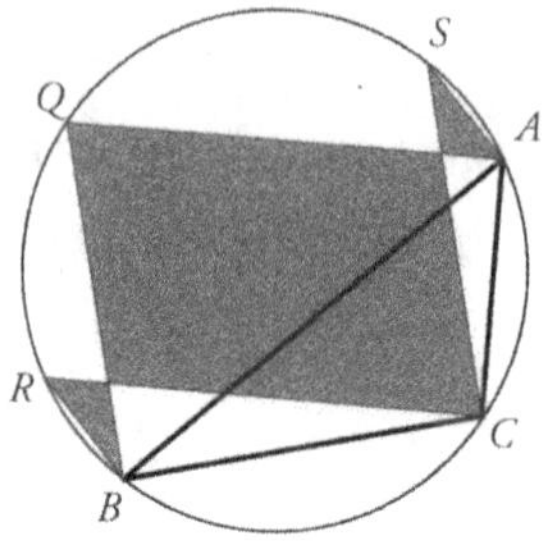

(b)

Then, the vertices Q, R, and S are on the circumcircle of $\triangle ABC$, since the quadrilaterals $AQBC$, $BRCA$, and $CSAB$ are cyclic.

Referring to figure (a), it is not difficult to see that if P is strictly between A and Q on the small arc AQ, then the perpendicular PE from P to AC divides AC externally. (The reason is that in order to divide AC internally, the point P would have to be between the lines AQ and CR.) On the other hand, the perpendiculars from P to AB and BC divide those sides internally.

Similar reasoning applies when P is one of the other six arcs of the circumcircle and also to the case where the hexagon is nonconvex, as in figure (b) above. □

Desargues' Two Triangle Theorem

In this section we will prove Desargues' Two Triangle Theorem, but first the definitions of copolar and coaxial triangles.

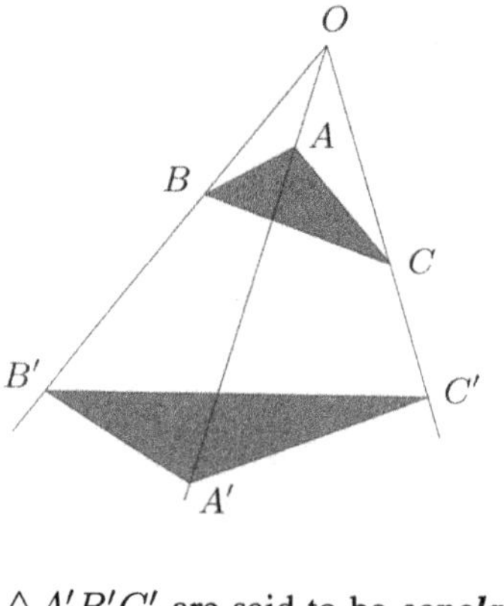

Two triangles $\triangle ABC$ and $\triangle A'B'C'$ are said to be ***copolar*** from a point O if and only if the joins of corresponding vertices are concurrent at the point O; that is, if and only if the lines AA', BB', and CC' are concurrent at O, as in the figure above. The point O is called the ***pole***.

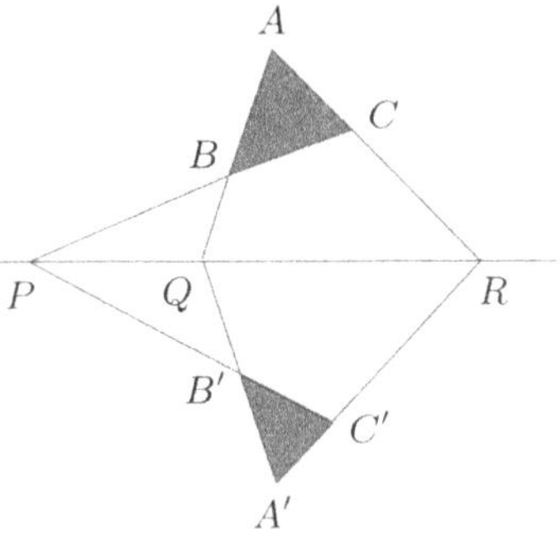

Two triangles $\triangle ABC$ and $\triangle A'B'C'$ are ***coaxial*** from a line l if and only if the points of intersection of corresponding sides

$$P = \overline{BC} \cap \overline{B'C'}, \qquad Q = \overline{AB} \cap \overline{A'B'}, \qquad R = \overline{AC} \cap \overline{A'C'}$$

are collinear, as in the figure above.

Theorem 4.4.3. *(Desargues' Two Triangle Theorem)*

Two triangles in the plane are copolar if and only if they are coaxial, as in the figure below.

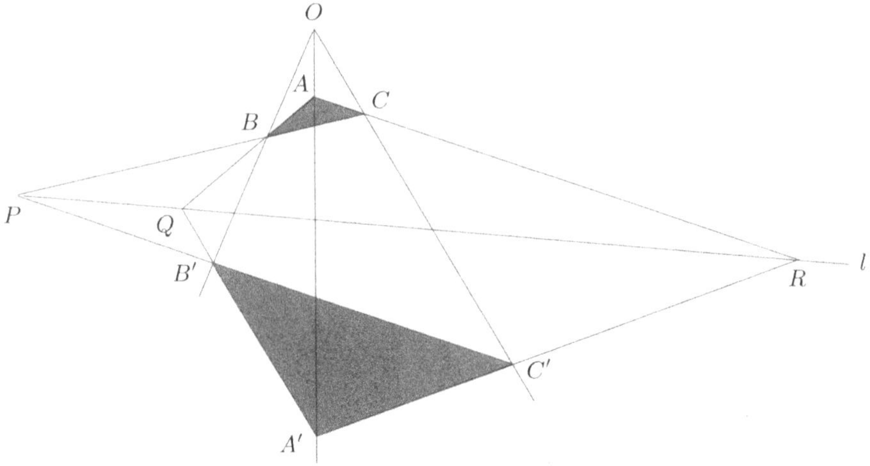

The line l is called the ***Desargues line***.

Proof. *Copolar implies Coaxial.*

Given that $\triangle ABC$ and $\triangle A'B'C'$ are copolar from O, we need to show that

$$P = \overline{BC} \cap \overline{B'C'}, \qquad Q = \overline{AB} \cap \overline{A'B'}, \qquad R = \overline{AC} \cap \overline{A'C'}$$

are collinear, and from Menelaus' Theorem we only have to show that

$$\frac{\overline{AQ}}{\overline{QB}} \cdot \frac{\overline{BP}}{\overline{PC}} \cdot \frac{\overline{CR}}{\overline{RA}} = -1.$$

We will use a decomposition type argument. We decompose $\triangle BOC$ into three triangles $\triangle AOB$, $\triangle AOC$, and $\triangle ABC$ and apply Menelaus' Theorem to three transversals as follows:

For $\triangle AOB$ with transversal $A'B'Q$, we have

$$\frac{\overline{AQ}}{\overline{QB}} \cdot \frac{\overline{BB'}}{\overline{B'O}} \cdot \frac{\overline{OA'}}{\overline{A'A}} = -1.$$

For $\triangle BOC$ with transversal $PB'C'$, we have

$$\frac{\overline{BP}}{\overline{PC}} \cdot \frac{\overline{CC'}}{\overline{C'O}} \cdot \frac{\overline{OB'}}{\overline{B'B}} = -1.$$

For $\triangle AOC$ with transversal $A'C'R$, we have

$$\frac{\overline{CR}}{\overline{RA}} \cdot \frac{\overline{AA'}}{\overline{A'O}} \cdot \frac{\overline{OC'}}{\overline{C'C}} = -1.$$

Multiplying these three expressions together, we have

$$\frac{\overline{AQ}}{\overline{QB}} \cdot \frac{\overline{BP}}{\overline{PC}} \cdot \frac{\overline{CR}}{\overline{RA}} \cdot X = (-1)^3 = -1,$$

where

$$X = \frac{\overline{BB'}}{\overline{B'O}} \cdot \frac{\overline{OA'}}{\overline{A'A}} \cdot \frac{\overline{CC'}}{\overline{C'O}} \cdot \frac{\overline{OB'}}{\overline{B'B}} \cdot \frac{\overline{AA'}}{\overline{A'O}} \cdot \frac{\overline{OC'}}{\overline{C'C}}.$$

Thus, $X = 1$, so that

$$\frac{\overline{AQ}}{\overline{QB}} \cdot \frac{\overline{BP}}{\overline{PC}} \cdot \frac{\overline{CR}}{\overline{RA}} = -1,$$

and P, Q, and R are collinear.

Coaxial implies Copolar.

Given a pair of coaxial triangles $\triangle ABC$ and $\triangle A'B'C'$, as in the figure on the following page, we want to show that they are copolar; that is, we want to show that AA', BB', and CC' are concurrent.

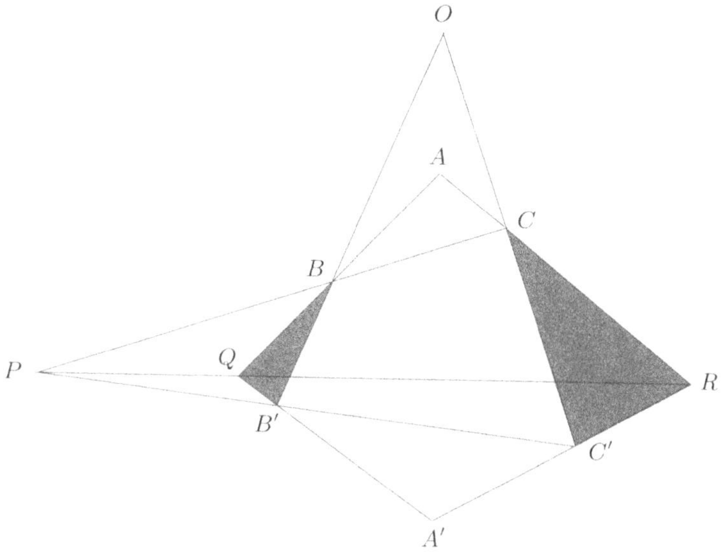

Let O be the intersection of BB' and CC'. Then we have to show that AA' also goes through O.

Observe that $\triangle QBB'$ and $\triangle RCC'$ are copolar from P. From the first half of the proof, we know that $\triangle QBB'$ and $\triangle RCC'$ are also coaxial, so that

$$QB \cap RC = A, \qquad QB' \cap RC' = A', \qquad \text{and} \qquad BB' \cap CC' = O$$

are collinear, and so AA' passes through O. Therefore, $\triangle ABC$ and $\triangle A'B'C'$ are copolar from O.

□

Pascal's Mystic Hexagon Theorem

In this section we will prove Pascal's Mystic Hexagon Theorem, but first we give a convention about labeling the sides of a hexagon, simple or nonsimple.

Given a hexagon $ABCDEF$, simple or nonsimple, inscribed in a circle, we label the sides with the positive integers so that

$$1 \leftrightarrow AB, \quad 2 \leftrightarrow BC, \quad 3 \leftrightarrow CD, \quad 4 \leftrightarrow DE, \quad 5 \leftrightarrow EF, \quad \text{and} \quad 6 \leftrightarrow FA,$$

as in the figure on the following page.

We say that sides 1 and 4, sides 2 and 5, and sides 3 and 6 are ***opposite sides***, whether the hexagon is simple or nonsimple.

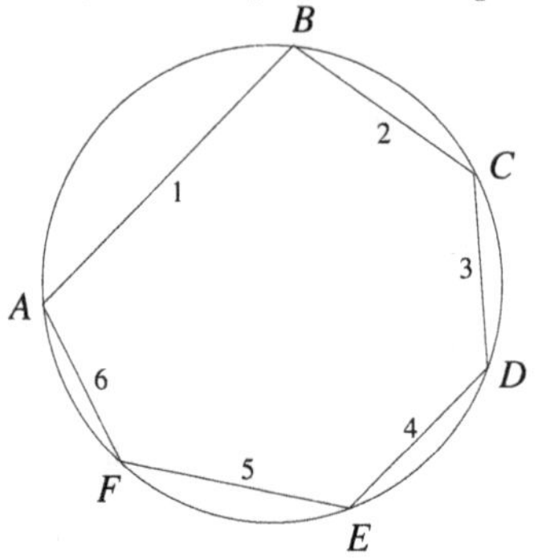

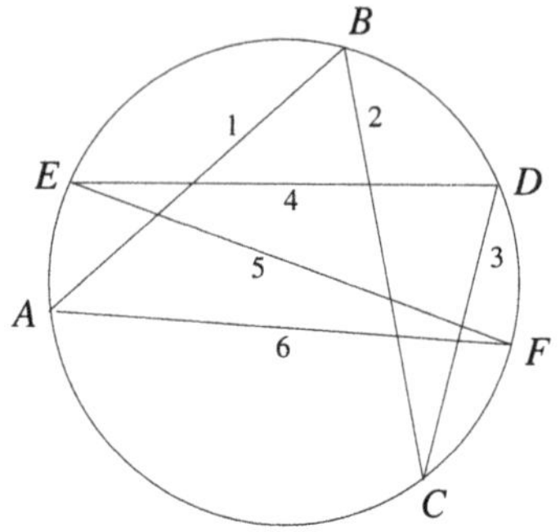

Theorem 4.4.4. *(Pascal's Mystic Hexagon Theorem)*

Given a hexagon (simple or nonsimple) inscribed in a circle, the points of intersection of opposite sides are collinear and form the ***Pascal line****.*

Proof. Given an inscribed hexagon as shown, we want to show that P, Q, and R are collinear.

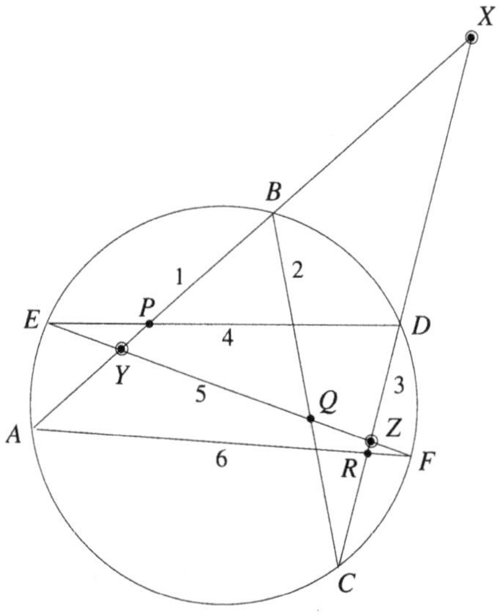

Create $\triangle XYZ$ by taking every second side of the hexagon so that

$$X = \text{side } 1 \cap \text{ side } 3, \quad Y = \text{side } 1 \cap \text{ side } 5, \quad \text{and} \quad Z = \text{side } 3 \cap \text{ side } 5.$$

Note that

$$P \text{ is on side } 1, \quad Q \text{ is on side } 5, \quad \text{and} \quad R \text{ is on side 3 (extended)},$$

so that P, Q, and R are menelaus points of $\triangle XYZ$.

In order to show that P, Q, and R are collinear, by Menelaus' Theorem we need only show that

$$\frac{\overline{XP}}{\overline{PY}} \cdot \frac{\overline{YQ}}{\overline{QZ}} \cdot \frac{\overline{ZR}}{\overline{RX}} = -1.$$

Applying Menelaus' Theorem to the following *labeled* transversals of $\triangle XYZ$, we have

$$\overleftrightarrow{EPD}: \qquad \frac{\overline{XP}}{\overline{PY}} \cdot \frac{\overline{YE}}{\overline{EZ}} \cdot \frac{\overline{ZD}}{\overline{DX}} = -1,$$

$$\overleftrightarrow{BQC}: \qquad \frac{\overline{XB}}{\overline{BY}} \cdot \frac{\overline{YQ}}{\overline{QZ}} \cdot \frac{\overline{ZC}}{\overline{CX}} = -1,$$

$$\overleftrightarrow{ARF}: \qquad \frac{\overline{XA}}{\overline{AY}} \cdot \frac{\overline{YF}}{\overline{FZ}} \cdot \frac{\overline{ZR}}{\overline{RX}} = -1,$$

and multiplying these together we get

$$\begin{aligned}
&\frac{\overline{XP}}{\overline{PY}} \cdot \frac{\overline{YQ}}{\overline{QZ}} \cdot \frac{\overline{ZR}}{\overline{RX}} \cdot \frac{\overline{YE}}{\overline{EZ}} \cdot \frac{\overline{ZD}}{\overline{DX}} \cdot \frac{\overline{XB}}{\overline{BY}} \cdot \frac{\overline{ZC}}{\overline{CX}} \cdot \frac{\overline{XA}}{\overline{AY}} \cdot \frac{\overline{YF}}{\overline{FZ}} \\
&\quad = \frac{\overline{XP}}{\overline{PY}} \cdot \frac{\overline{YQ}}{\overline{QZ}} \cdot \frac{\overline{ZR}}{\overline{RX}} \cdot \left(\frac{\overline{XA} \cdot \overline{XB}}{\overline{XC} \cdot \overline{XD}}\right) \cdot \left(\frac{\overline{YE} \cdot \overline{YF}}{\overline{YA} \cdot \overline{YB}}\right) \cdot \left(\frac{\overline{ZC} \cdot \overline{ZD}}{\overline{ZE} \cdot \overline{ZF}}\right) \\
&\quad = (-1)^3 = -1.
\end{aligned}$$

Now, the power of the point X with respect to the circle is

$$\overline{XA} \cdot \overline{XB} = \overline{XC} \cdot \overline{XD},$$

the power of the point Y with respect to the circle is

$$\overline{YA} \cdot \overline{YB} = \overline{YE} \cdot \overline{YF},$$

while the power of the point Z with respect to the circle is

$$\overline{ZC} \cdot \overline{ZD} = \overline{ZE} \cdot \overline{ZF},$$

and therefore,

$$\left(\frac{\overline{XA} \cdot \overline{XB}}{\overline{XC} \cdot \overline{XD}}\right) \cdot \left(\frac{\overline{YE} \cdot \overline{YF}}{\overline{YA} \cdot \overline{YB}}\right) \cdot \left(\frac{\overline{ZC} \cdot \overline{ZD}}{\overline{ZE} \cdot \overline{ZF}}\right) = 1 \cdot 1 \cdot 1 = 1,$$

so that

$$\frac{\overline{XP}}{\overline{PY}} \cdot \frac{\overline{YQ}}{\overline{QZ}} \cdot \frac{\overline{ZR}}{\overline{RX}} = -1,$$

and the points P, Q, and R are collinear.

□

Pappus' Theorem

In this section, we will prove Pappus' Theorem. Although it belongs to the realm of projective geometry, we can give an elementary proof using Euclidean geometry via Menelaus' Theorem.

Theorem 4.4.5. *(Pappus' Theorem)*

Given points A, B, and C on a line l and points A', B', and C' on a line l', the points of intersection

$$P = AB' \cap A'B, \qquad Q = AC' \cap A'C, \qquad R = BC' \cap B'C$$

are collinear, as in the figure.

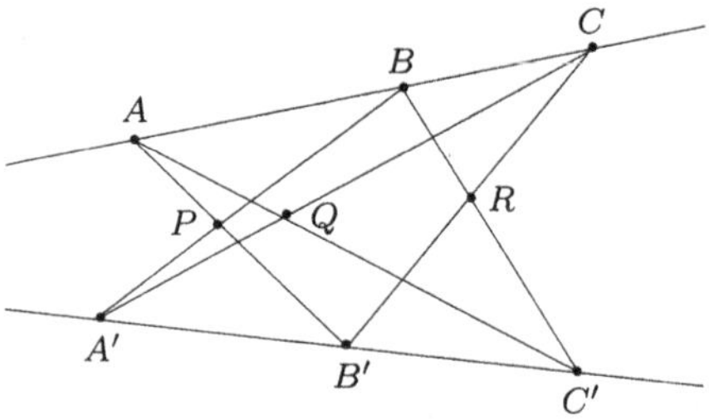

Proof. Extending the sides of the triangle formed by the sides $A'B$, $B'C$, and $C'A$, we have the figure on the following page.

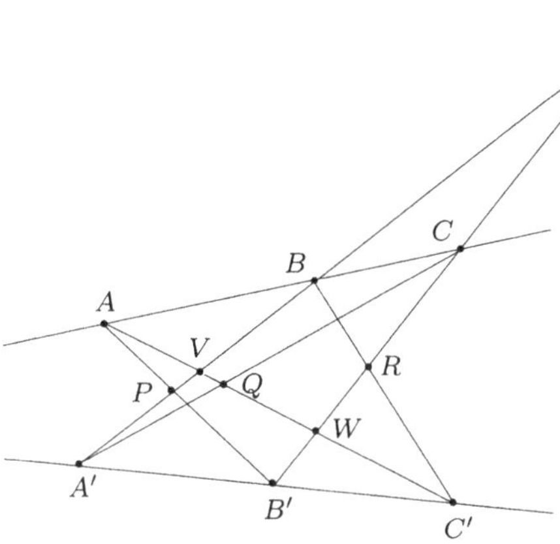

We will apply Menelaus' Theorem to each of the five transversals of the triangle $\triangle UVW$

$$APB', \qquad CQA', \qquad BRC', \qquad ABC, \qquad A'B'C'$$

in turn.

For the transversal APB', we have

$$\frac{\overline{UP}}{\overline{PV}} \cdot \frac{\overline{VA}}{\overline{AW}} \cdot \frac{\overline{WB'}}{\overline{B'U}} = -1$$

by Menelaus' Theorem.

For the transversal CQA', we have

$$\frac{\overline{UA'}}{\overline{A'V}} \cdot \frac{\overline{VQ}}{\overline{QW}} \cdot \frac{\overline{WC}}{\overline{CU}} = -1$$

by Menelaus' Theorem.

For the transversal BRC', we have

$$\frac{\overline{UB}}{\overline{BV}} \cdot \frac{\overline{VC'}}{\overline{C'W}} \cdot \frac{\overline{WR}}{\overline{RU}} = -1$$

by Menelaus' Theorem.

For the transversal ABC, we have

$$\frac{\overline{UB}}{\overline{BV}} \cdot \frac{\overline{VA}}{\overline{AW}} \cdot \frac{\overline{WC}}{\overline{CU}} = -1$$

by Menelaus' Theorem.

For the transversal $A'B'C'$, we have

$$\frac{\overline{UA'}}{\overline{A'V}} \cdot \frac{\overline{VC'}}{\overline{C'W}} \cdot \frac{\overline{WB'}}{\overline{B'U}} = -1$$

by Menelaus' Theorem.

Multiplying the first three expressions together, we get

$$\frac{\overline{UP}}{\overline{PV}} \cdot \frac{\overline{VQ}}{\overline{QW}} \cdot \frac{\overline{WR}}{\overline{RU}} \cdot X = -1,$$

where

$$X = \frac{\overline{VA}}{\overline{AW}} \cdot \frac{\overline{WB'}}{\overline{B'U}} \cdot \frac{\overline{UA'}}{\overline{A'V}} \cdot \frac{\overline{WC}}{\overline{CU}} \cdot \frac{\overline{UB}}{\overline{BV}} \cdot \frac{\overline{VC'}}{\overline{C'W}}.$$

Rearranging the terms in X, we get

$$X = \left(\frac{\overline{UB}}{\overline{BV}} \cdot \frac{\overline{VA}}{\overline{AW}} \cdot \frac{\overline{WC}}{\overline{CU}}\right) \cdot \left(\frac{\overline{UA'}}{\overline{A'V}} \cdot \frac{\overline{VC'}}{\overline{C'W}} \cdot \frac{\overline{WB'}}{\overline{B'U}}\right) = (-1)(-1) = 1$$

from the last two expressions, and therefore,

$$\frac{\overline{UP}}{\overline{PV}} \cdot \frac{\overline{VQ}}{\overline{QW}} \cdot \frac{\overline{WR}}{\overline{RU}} = -1,$$

so that P, Q, and R are collinear by Menelaus' Theorem. □

Numerical Applications

In addition to using Menelaus' Theorem to determine when menelaus points are collinear, we can also use it to calculate distances when three points are known to be collinear.

Such problems often present us with several different ways of viewing the same diagram.

In the figure on the right, there are several "menelaus patterns":

- AB is a transversal for $\triangle PEC$.
- AC is a transversal for $\triangle PDB$.
- BE is a transversal for $\triangle ADC$.
- CD is a transversal for $\triangle ABE$.

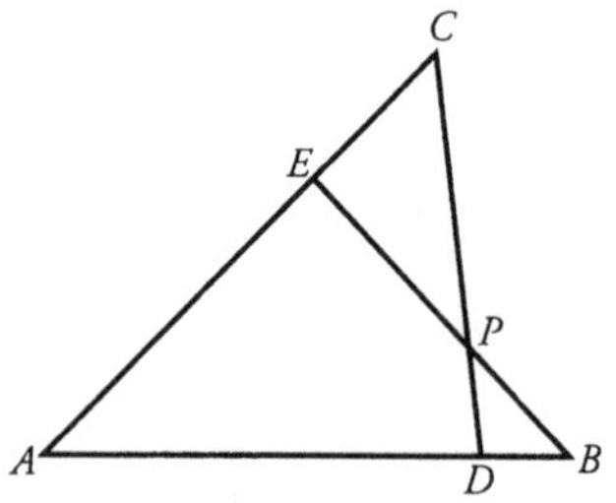

In solving a problem, it is sometimes a matter of trying various combinations until we find one that works.

Example 4.4.6. *BD is a median of $\triangle ABC$, and E is the midpoint of BD. Line CE extended meets AB at F. If $AB = 6$, find the length of FB.*

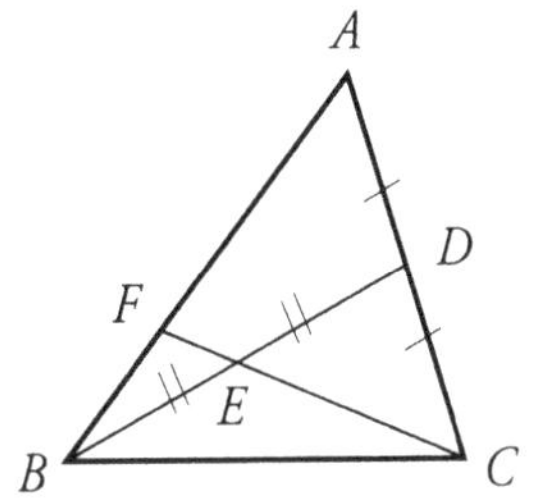

Solution. We wish to find a menelaus pattern whose triangle has two of its edges divided into known ratios. Referring to the figure, note that if we consider CF as a transversal for $\triangle ABD$, then CF divides the sides BD and AD into known ratios. So, let $FB = x$ and apply Menelaus' Theorem to $\triangle ABD$ with transversal CF:

$$\frac{\overline{AF}}{\overline{FB}} \cdot \frac{\overline{BE}}{\overline{ED}} \cdot \frac{\overline{DC}}{\overline{CA}} = -1,$$

which implies that

$$\frac{AF}{x} \cdot 1 \cdot \left(-\frac{1}{2}\right) = -1,$$

which in turn implies that

$$\frac{AF}{x} = 2.$$

Thus, $AF = 2x$, so $AB = 3x$, and it follows that $x = AB/3 = 2$.

□

4.5 Proofs of the Theorems

Both Ceva's and Menelaus' Theorems are of the "if and only if" variety, and so in each case there are two things to prove.

Proof of Ceva's Theorem

We will first prove the necessity; that is, we will prove the following:

Theorem 4.5.1. *(Ceva's Theorem: Necessity)*
Let AD, BE, and CF be cevians of triangle ABC, with D on BC, E on CA, and F on AB. If AD, BE, and CF are concurrent or parallel, then

$$\frac{\overline{AF}}{\overline{FB}} \cdot \frac{\overline{BD}}{\overline{DC}} \cdot \frac{\overline{CE}}{\overline{EA}} = 1.$$

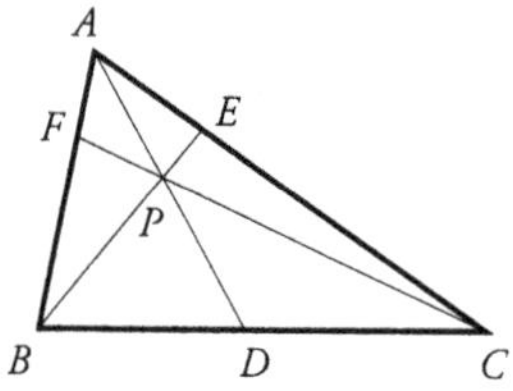

concurrent cevians

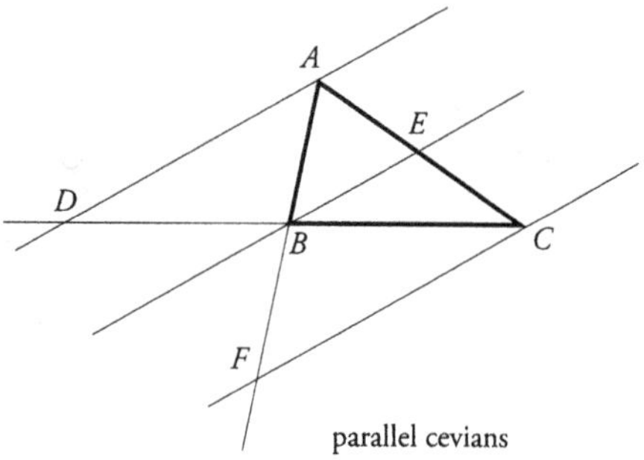

parallel cevians

Proof. We will prove the theorem for the case where AD, BE, and CF are concurrent. The case where the cevians are parallel is left as an exercise. Suppose the cevians AD, BE, and CF are concurrent at a point P. Let l be a line through A parallel to BC, let S be the point where BE meets l, and let R be the point where CF meets l.

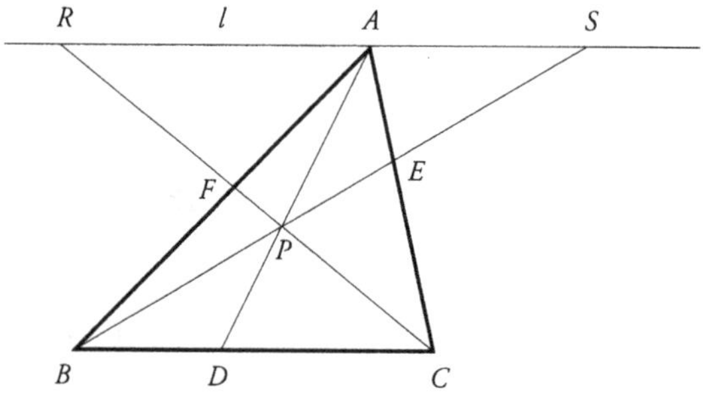

Then

$$\frac{\overline{AF}}{\overline{FB}} = \frac{\overline{AR}}{\overline{BC}} \quad \text{(because } \triangle RAF \sim \triangle CBF\text{)},$$
$$\frac{\overline{CE}}{\overline{EA}} = \frac{\overline{BC}}{\overline{AS}} \quad \text{(because } \triangle ASE \sim \triangle CBE\text{)},$$
$$\frac{\overline{BD}}{\overline{DC}} = \frac{\overline{AS}}{\overline{AR}},$$

with the last equality arising from the fact that

$$\frac{\overline{BD}}{\overline{AS}} = \frac{\overline{DP}}{\overline{PA}} = \frac{\overline{DC}}{\overline{AR}}$$

because $\triangle BPD \sim \triangle SPA$ and $\triangle DPC \sim \triangle APR$. It follows that

$$\frac{\overline{AF}}{\overline{FB}} \cdot \frac{\overline{BD}}{\overline{DC}} \cdot \frac{\overline{CE}}{\overline{EA}} = \frac{\overline{AR}}{\overline{BC}} \cdot \frac{\overline{AS}}{\overline{AR}} \cdot \frac{\overline{BC}}{\overline{AS}} = 1,$$

which completes the proof of necessity.

□

Next, we wish to prove the sufficiency, that is:

Theorem 4.5.2. *(Ceva's Theorem: Sufficiency)*
Let AD, BE, and CF be cevians of triangle ABC, with D on BC, E on CA, and F on AB. If

$$\frac{\overline{AF}}{\overline{FB}} \cdot \frac{\overline{BD}}{\overline{DC}} \cdot \frac{\overline{CE}}{\overline{EA}} = 1,$$

then AD, BE, and CF are either concurrent or parallel.

Proof. There are two cases to consider: either (i) BE and CF are parallel or else (ii) BE and CF meet at a single point P.

Case (i). Let l be a line through A parallel to BE. Since BE is not parallel to BC (by the definition of a cevian), it follows that l is not parallel to BC, and so l must meet the line BC at some point G. Then, the cevians AG, BE, and CF are parallel, so by the "necessary" part of the theorem we must have

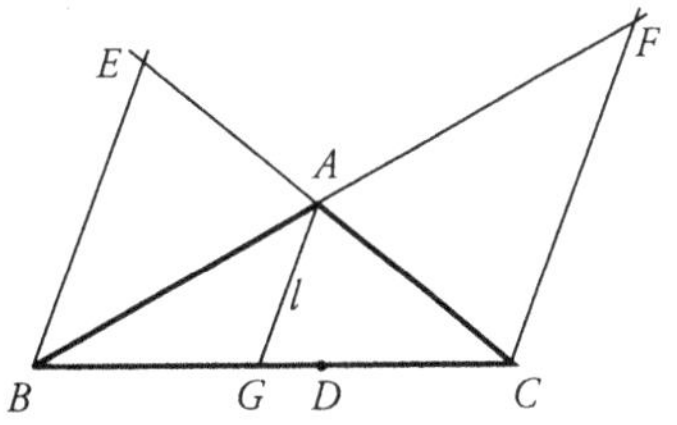

$$\frac{\overline{AF}}{\overline{FB}} \cdot \frac{\overline{BG}}{\overline{GC}} \cdot \frac{\overline{CE}}{\overline{EA}} = 1.$$

By hypothesis, we also have

$$\frac{\overline{AF}}{\overline{FB}} \cdot \frac{\overline{BD}}{\overline{DC}} \cdot \frac{\overline{CE}}{\overline{EA}} = 1,$$

so that

$$\frac{\overline{AF}}{\overline{FB}} \cdot \frac{\overline{BG}}{\overline{GC}} \cdot \frac{\overline{CE}}{\overline{EA}} = \frac{\overline{AF}}{\overline{FB}} \cdot \frac{\overline{BD}}{\overline{DC}} \cdot \frac{\overline{CE}}{\overline{EA}}.$$

Thus,

$$\frac{\overline{BG}}{\overline{GC}} = \frac{\overline{BD}}{\overline{DC}},$$

and this implies that $G = D$. This shows that the cevians AD, BE, and CF are parallel.

Case (ii). Let l be a line through A and P. We will first show that if l is not parallel to BC, then l is actually the cevian AD. To see why, let us suppose that l meets BC at G. Then the cevians AG, BE, and CF are concurrent at P, so by the first part of the theorem we must have

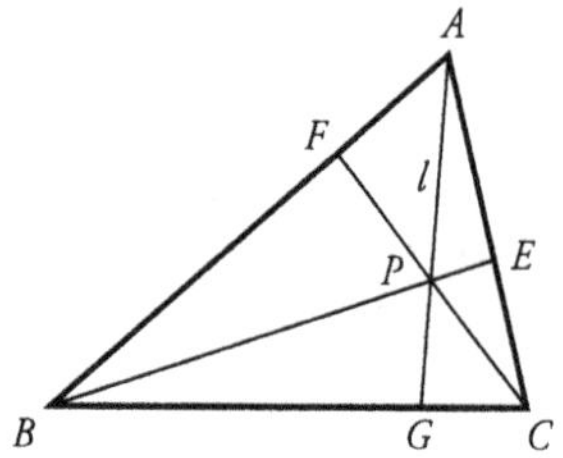

$$\frac{\overline{AF}}{\overline{FB}} \cdot \frac{\overline{BG}}{\overline{GC}} \cdot \frac{\overline{CE}}{\overline{EA}} = 1.$$

Since we also have

$$\frac{\overline{AF}}{\overline{FB}} \cdot \frac{\overline{BD}}{\overline{DC}} \cdot \frac{\overline{CE}}{\overline{EA}} = 1,$$

when l is known not to be parallel to BC, the proof may be completed as in case (i).

To completely finish case (ii), we must show that AP cannot be parallel to BC, and we will prove this by contradiction.

Let us suppose that AP is parallel to BC. It then follows that

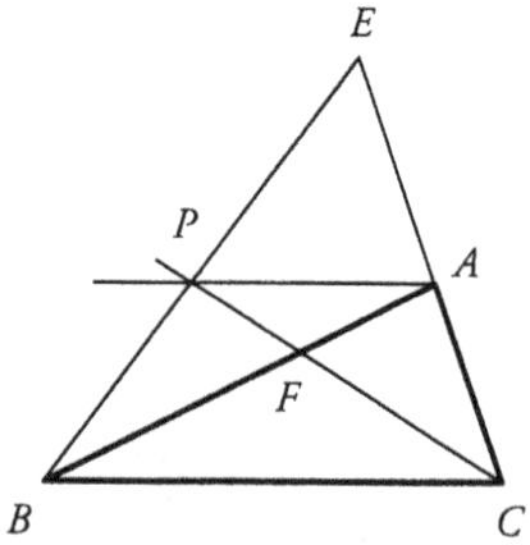

$$\frac{\overline{AF}}{\overline{FB}} = \frac{\overline{AP}}{\overline{CB}}, \quad \text{because } \triangle AFP \sim \triangle BFC,$$

$$\frac{\overline{CE}}{\overline{EA}} = \frac{\overline{CB}}{\overline{PA}}, \quad \text{because } \triangle CEB \sim \triangle AEP,$$

and, consequently, that

$$\frac{\overline{AF}}{\overline{FB}} \cdot \frac{\overline{BD}}{\overline{DC}} \cdot \frac{\overline{CE}}{\overline{EA}} = \frac{\overline{AP}}{\overline{CB}} \cdot \frac{\overline{BD}}{\overline{DC}} \cdot \frac{\overline{CB}}{\overline{PA}} = -\frac{\overline{BD}}{\overline{DC}}.$$

Since

$$\frac{\overline{AF}}{\overline{FB}} \cdot \frac{\overline{BD}}{\overline{DC}} \cdot \frac{\overline{CE}}{\overline{EA}} = 1,$$

it follows that

$$\frac{\overline{BD}}{\overline{DC}} = -1.$$

But there is no point D on the line BC for which

$$\frac{\overline{BD}}{\overline{DC}} = -1.$$

This shows that AP cannot be parallel to BC and completes the proof of case (ii).

□

Proof of Menelaus' Theorem

As with Ceva's Theorem, we will treat the "necessary" and "sufficient" parts separately. There are a variety of different proofs for the first part, and we will give additional proofs later in this chapter.

Theorem 4.5.3. *(Menelaus' Theorem: Necessity)*
If the menelaus points D, E, and F on the (extended) sides BC, CA, and AB of a triangle are collinear, then

$$\frac{\overline{AF}}{\overline{FB}} \cdot \frac{\overline{BD}}{\overline{DC}} \cdot \frac{\overline{CE}}{\overline{EA}} = -1.$$

Proof. The proof uses the fact that a transversal meets either two sides internally or no sides internally. Thus, either exactly one or all three of the ratios

$$\frac{\overline{AF}}{\overline{FB}}, \quad \frac{\overline{BD}}{\overline{DC}}, \quad \text{and} \quad \frac{\overline{CE}}{\overline{EA}}$$

are numerically negative, and consequently the sign of

$$\frac{\overline{AF}}{\overline{FB}} \cdot \frac{\overline{BD}}{\overline{DC}} \cdot \frac{\overline{CE}}{\overline{EA}}$$

is guaranteed to be negative. So, to prove the "necessary" part, we need only show that

$$\frac{AF}{FB} \cdot \frac{BD}{DC} \cdot \frac{CE}{EA} = 1.$$

Drop perpendiculars AR, BS, and CT from the vertices to the transversal. Then,

$$\frac{AF}{FB}\cdot\frac{BD}{DC}\cdot\frac{CE}{EA}=\frac{AR}{BS}\cdot\frac{BS}{CT}\cdot\frac{CT}{AR}=1\,.$$

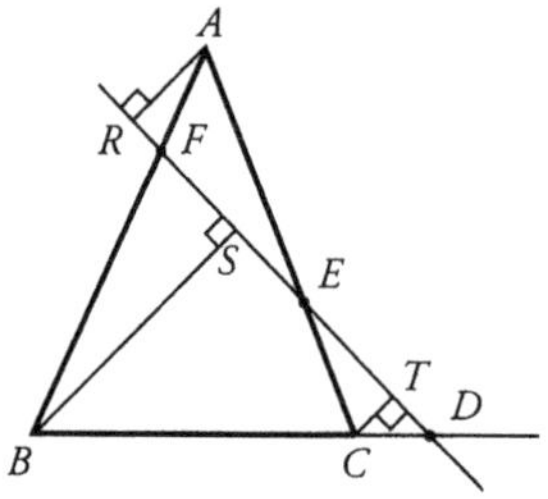

The reasons are

$$\frac{AF}{FB}=\frac{AR}{BS}\quad(\triangle AFR\sim\triangle BFS),$$
$$\frac{BD}{DC}=\frac{BS}{CT}\quad(\triangle BSD\sim\triangle CTD),$$

and

$$\frac{CE}{EA}=\frac{CT}{AR}\quad(\triangle ARE\sim\triangle CTE).$$

This completes the proof of the necessity.

□

For the sufficiency part of Menelaus' Theorem, we need to prove the following:

Theorem 4.5.4. *(Menelaus' Theorem: Sufficiency)*

Let D, E, and F be three menelaus points on sides BC, CA, and AB of a triangle. If

$$\frac{\overline{AF}}{\overline{FB}}\cdot\frac{\overline{BD}}{\overline{DC}}\cdot\frac{\overline{CE}}{\overline{EA}}=-1,$$

then D, E, and F are collinear.

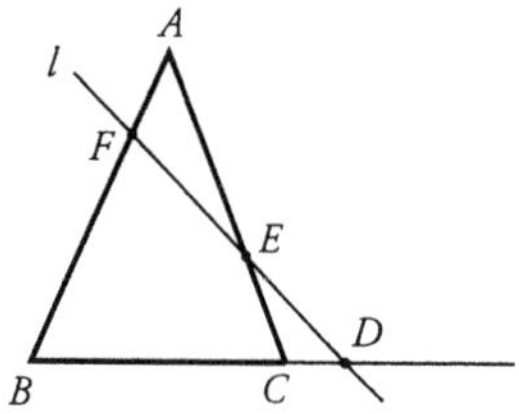

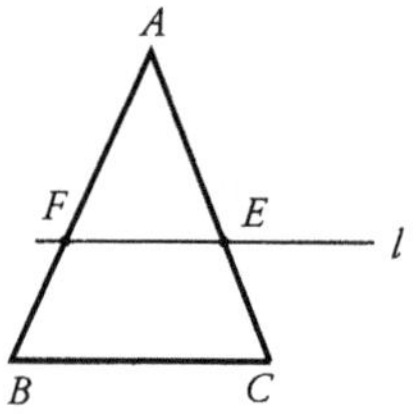

Proof. Suppose that

$$\frac{\overline{AF}}{\overline{FB}}\cdot\frac{\overline{BD}}{\overline{DC}}\cdot\frac{\overline{CE}}{\overline{EA}}=-1.$$

Let l be the line through F and E. Then there are two possibilities:

(i) l meets BC at some point D' or

(ii) l is parallel to BC.

Case (i). l meets BC at some point P.

In this case, it suffices to show that $P = D$. By the "necessary" part of Menelaus' Theorem,

$$\frac{\overline{AF}}{\overline{FB}} \cdot \frac{\overline{BP}}{\overline{PC}} \cdot \frac{\overline{CE}}{\overline{EA}} = -1.$$

This, together with the hypothesis, means that

$$\frac{\overline{AF}}{\overline{FB}} \cdot \frac{\overline{BP}}{\overline{PC}} \cdot \frac{\overline{CE}}{\overline{EA}} = \frac{\overline{AF}}{\overline{FB}} \cdot \frac{\overline{BD}}{\overline{DC}} \cdot \frac{\overline{CE}}{\overline{EA}},$$

and so

$$\frac{\overline{BP}}{\overline{PC}} = \frac{\overline{BD}}{\overline{DC}},$$

implying that $P = D$ and completing case (i).

Case (ii). l is parallel to BC.

We will show that this case cannot arise, for if l is parallel to BC, then

$$\frac{\overline{AF}}{\overline{FB}} = \frac{\overline{AE}}{\overline{EC}}.$$

Now, by the hypothesis,

$$\frac{\overline{AF}}{\overline{FB}} \cdot \frac{\overline{BD}}{\overline{DC}} \cdot \frac{\overline{CE}}{\overline{EA}} = -1\,,$$

and since

$$\frac{\overline{AF}}{\overline{FB}} = \frac{\overline{AE}}{\overline{EC}},$$

it follows that

$$\frac{\overline{BD}}{\overline{DC}} = -1.$$

But it is impossible for any point D on the line BC to satisfy

$$\frac{\overline{BD}}{\overline{DC}} = -1.$$

This finishes the proof of the "sufficient" part of the theorem and completes the proof of Menelaus' Theorem.

□

Additional Proofs

The similarity between Menelaus' Theorem and Ceva's Theorem is not just in the language. Here is how Menelaus' Theorem can be used to prove the first part of Ceva's Theorem.

Example 4.5.5. *Menelaus' Theorem implies Ceva's Theorem.*

Solution. In the figure on the right, assuming that the cevians are concurrent, our objective is to show that

$$\frac{\overline{AF}}{\overline{FB}} \cdot \frac{\overline{BD}}{\overline{DC}} \cdot \frac{\overline{CE}}{\overline{EA}} = +1 .$$

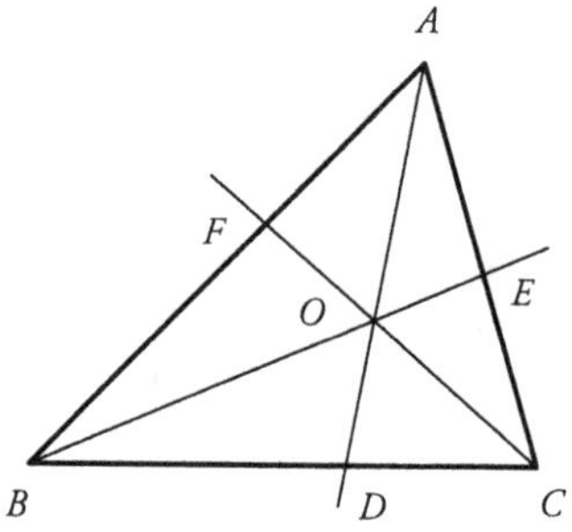

The idea is to use the cevian AD of triangle ABC to decompose the triangle ABC into two "subtriangles" ABD and ADC. We then use the facts that CF is a transversal for $\triangle ABD$ and BE is a transversal for $\triangle ADC$.

Applying Menelaus' Theorem to $\triangle ABD$ and transversal COF,

$$\frac{\overline{AF}}{\overline{FB}} \cdot \frac{\overline{BC}}{\overline{CD}} \cdot \frac{\overline{DO}}{\overline{OA}} = -1 ,$$

and also applying Menelaus' Theorem to $\triangle ADC$ and transversal BOE,

$$\frac{\overline{AO}}{\overline{OD}} \cdot \frac{\overline{DB}}{\overline{BC}} \cdot \frac{\overline{CE}}{\overline{EA}} = -1 .$$

Multiplying the two equations together and cancelling a few terms, we get

$$\frac{\overline{AF}}{\overline{FB}} \cdot \frac{\overline{DB}}{\overline{CD}} \cdot \frac{\overline{CE}}{\overline{EA}} = (-1)^2 = +1.$$

Thus,

$$\frac{\overline{AF}}{\overline{FB}} \cdot \frac{\overline{BD}}{\overline{DC}} \cdot \frac{\overline{CE}}{\overline{EA}} = +1,$$

which completes the proof.

□

The following two proofs of Menelaus' Theorem are from a 1937 letter that the physicist Albert Einstein wrote to the psychologist Max Wertheimer.

Einstein's Ugly Proof

Einstein called this his "ugly proof." It is based on similar triangles but uses a single parallel line rather than three perpendicular lines to create them.

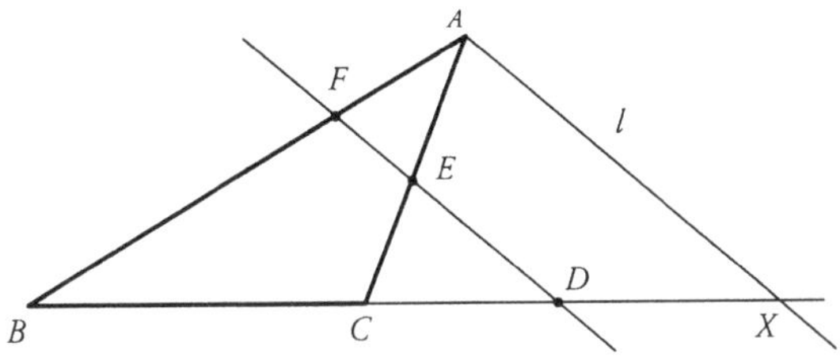

Given $\triangle ABC$ and a transversal, draw l through A parallel to the transversal hitting BC at X. Then

$$\frac{AF}{FB} \cdot \frac{BD}{DC} \cdot \frac{CE}{EA} = \frac{DX}{DB} \cdot \frac{BD}{DC} \cdot \frac{CD}{DX} = 1,$$

and the theorem follows.

□

Einstein's Area Proof

Einstein called this next proof his "area proof." It uses the following lemma.

Lemma 4.5.6. *The ratio of the areas of two triangles that have a common or supplementary angle is equal to the ratio of the corresponding products of their adjacent sides.*

Proof. As in the figure on the following page,

$$\frac{\text{area}(ABC)}{\text{area}(DEF)} = \frac{\frac{1}{2}BC \cdot BA \sin\alpha}{\frac{1}{2}EF \cdot ED \sin\alpha} = \frac{BA \cdot BC}{ED \cdot EF}.$$

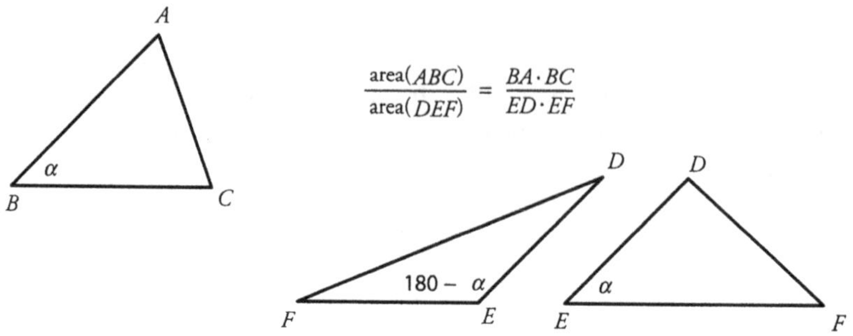

Since $\sin(180 - \alpha) = \sin\alpha$, the same result holds for supplementary angles.

□

Einstein's Area Proof.

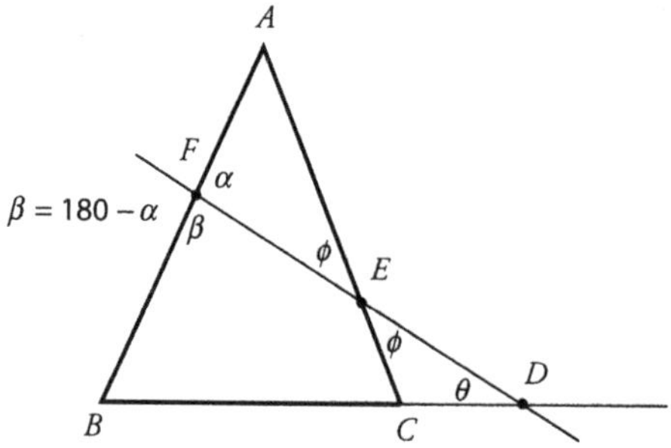

Given $\triangle ABC$ and a transversal, then by the lemma,

$$\frac{\text{area}(AFE)}{\text{area}(BFD)} = \frac{AF \cdot FE}{BF \cdot FD},$$
$$\frac{\text{area}(BFD)}{\text{area}(CDE)} = \frac{BD \cdot FD}{CD \cdot DE},$$
$$\frac{\text{area}(CDE)}{\text{area}(AFE)} = \frac{CE \cdot ED}{AE \cdot FE},$$

from which it follows that

$$\frac{AF}{FB} \cdot \frac{BD}{DC} \cdot \frac{CE}{EA} = \frac{FD \cdot \text{area}(AFE)}{FE \cdot \text{area}(BFD)} \cdot \frac{DE \cdot \text{area}(BFD)}{FD \cdot \text{area}(CDE)} \cdot \frac{FE \cdot \text{area}(CDE)}{DE \cdot \text{area}(AFE)}$$
$$= 1,$$

and the theorem follows.

□

4.6 Extended Versions of the Theorems

Example 4.6.1. *(A Cevian Parallel to a Side)*

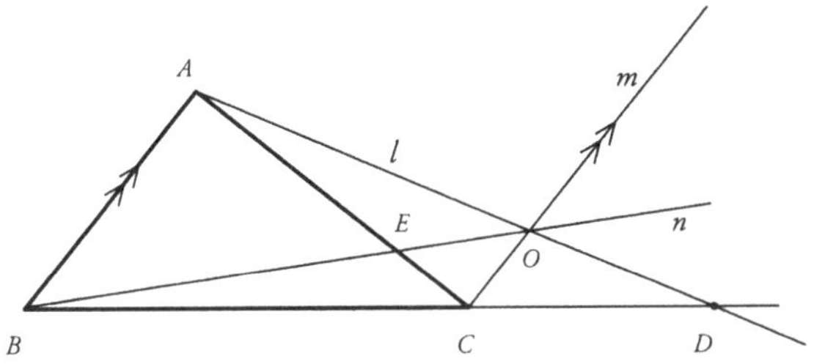

In the figure above there are three concurrent cevians, l, m, and n, so the cevian product

$$\frac{\overline{AF}}{\overline{FB}} \cdot \frac{\overline{BD}}{\overline{DC}} \cdot \frac{\overline{CE}}{\overline{EA}}$$

should be $+1$. But we have a problem: where is F? It should be the point where m meets the extended side AB, but m and AB are parallel and so that point does not exist. □

We have dealt with the problem by defining the word "cevian" so that it excludes lines like m that are parallel to a side. This approach is unsatisfactory. It seems like we are playing with semantics rather than doing geometry. After all, the "cevians" in the example are concurrent, and it is a legitimate source of concern that Ceva's Theorem cannot handle this case.

In this section, we will show that the problem can be remedied by introducing certain ideal elements called "points at infinity" or "ideal points" at which two parallel lines meet. (If you have ever looked down a long straight prairie railway track, you may have used exactly the same language and said that the rails "meet at infinity.")

When we introduce ideal points, we must be careful that we do not create more problems than we remove. The following conventions and definitions will help us avoid complications.

We append to the plane a collection of ***ideal points*** or ***points at infinity***. This collection of ideal points is called the ***ideal line*** or the ***line at infinity***. Points that are not ideal points are called ***ordinary points***; lines that are not ideal lines are called ***ordinary lines***. (The words "point" and "line" by themselves do not distinguish between the ordinary and the ideal versions.)

We postulate that:

(i) Every ordinary line of the plane contains exactly one ideal point.

(ii) Parallel lines have the same ideal point.

(iii) Nonparallel lines do not have the same ideal point.

The plane together with all of the ideal points is called the ***extended Euclidean plane*** or simply the ***extended plane***. The plane without the ideal line is called the ***Euclidean plane*** to distinguish it from the extended plane.[3]

We designate an ideal point by an arrow indicating the direction in which it lies, and an arrow pointing in exactly the opposite direction designates the same ideal point. In the figure below, parallel lines l_1, l_2, and l_3 all meet at the ideal point I. The lines m_1 and m_2 meet at a different ideal point J.

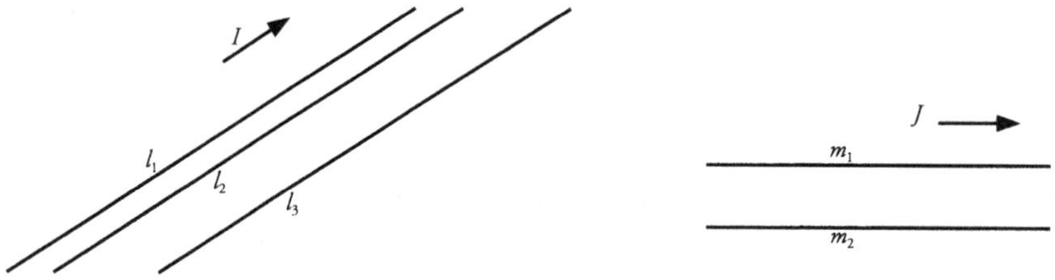

In the extended plane, every two distinct points determine a unique line, and every two distinct lines determine a unique point. Technically, in the extended plane there are no parallel lines. Nevertheless, we will continue to use the word *parallel* to describe ordinary lines in the extended plane that meet only at an ideal point.

If I is the ideal point on the line AB, we cannot assign a real number to the directed distance AI. However, Ceva's Theorem deals with directed ratios, and we adopt the convention that

$$\frac{\overline{AI}}{\overline{IB}} = -1$$

for the ideal point on the line AB.

[3]The extended plane is also called the ***projective plane*** and is the setting for what is called "projective geometry" in Part III.

This makes a certain amount of sense, for two reasons:

1. We have already observed that there is no ordinary point X on the line for which

$$\frac{\overline{AX}}{\overline{XB}} = -1.$$

2. As the ordinary point X on the ordinary line AB moves along the line away from A, the ratio

$$\frac{\overline{AX}}{\overline{XB}}$$

tends to -1.

4.6.1 Ceva's Theorem in the Extended Plane

In the extended Euclidean plane, a ***cevian*** is defined to be any line that contains exactly one vertex of an *ordinary* triangle. (An **ordinary triangle** is one whose edge segments do not contain ideal points.) Thus, the notion of a cevian has been modified to include lines through a vertex that in the Euclidean plane would be parallel to the opposite side.

The following two examples show that Ceva's Theorem is valid in the extended Euclidean plane with the modified notion of a cevian.

Example 4.6.2. *Given an ordinary triangle ABC and three concurrent cevians l, m, and n, with m parallel to BC, show that the cevian product is* 1.

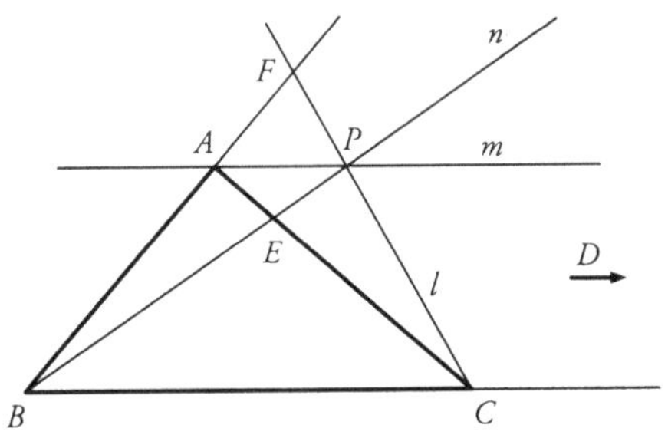

Solution. We have

$$\frac{\overline{AF}}{\overline{FB}} \cdot \frac{\overline{BD}}{\overline{DC}} \cdot \frac{\overline{CE}}{\overline{EA}} = \left(-\frac{AF}{FB}\right) \cdot (-1) \cdot \frac{CE}{EA}$$
$$= \frac{AF}{FB} \cdot \frac{CE}{EA},$$

and by similar triangles,

$$\frac{\overline{AF}}{\overline{FB}} \cdot \frac{\overline{BD}}{\overline{DC}} \cdot \frac{\overline{CE}}{\overline{EA}} = \frac{AP}{BC} \cdot \frac{BC}{AP}$$
$$= 1\,,$$

showing that the cevian product is 1.

□

Example 4.6.3. *Given $\triangle ABC$ and three concurrent cevians l, m, and n, with l parallel to BC and n parallel to AB, show that the cevian product is* 1.

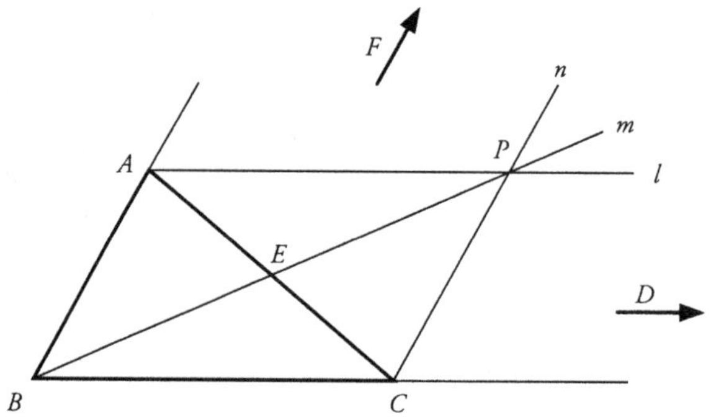

Solution.

$$\frac{\overline{AF}}{\overline{FB}} \cdot \frac{\overline{BD}}{\overline{DC}} \cdot \frac{\overline{CE}}{\overline{EA}} = (-1) \cdot (-1) \cdot \frac{CE}{EA} = +1\,,$$

since the diagonals of parallelogram $ABCP$ bisect each other.

□

The previous two examples show that in the extended plane, Ceva's Theorem can be stated more succinctly.

Theorem 4.6.4. *(Ceva's Theorem for the Extended Plane)*

A necessary and sufficient condition that the cevians AD, BE, and CF of a triangle ABC be concurrent is that

$$\frac{\overline{AF}}{\overline{FB}} \cdot \frac{\overline{BD}}{\overline{DC}} \cdot \frac{\overline{CE}}{\overline{EA}} = +1.$$

The theorem includes the possibility that the point of concurrency is an ideal point; that is, it includes the possibility that the three cevians are parallel in the Euclidean plane. It also includes the possibility that some of the vertices are ideal points, but one must be careful about the sufficiency part of the theorem.

4.6.2 Menelaus' Theorem in the Extended Plane

In the extended Euclidean plane, the notion of a transversal is modified to include those that are parallel to a side of an ordinary triangle: any ordinary line that does not contain a vertex of the triangle is called a ***transversal*** of the triangle.

Example 4.6.5. *Show that if a transversal l is parallel to side BC of an ordinary triangle ABC, then the corresponding cevian product is -1.*

Solution. Let us suppose that l meets sides BC, CA, and AB at D, E, and F, respectively. Then D is an ideal point and, since EF is parallel to BC,

$$\frac{\overline{AF}}{\overline{FB}} = \frac{\overline{AE}}{\overline{EC}},$$

so that

$$\frac{\overline{AF}}{\overline{FB}} \cdot \frac{\overline{BD}}{\overline{DC}} \cdot \frac{\overline{CE}}{\overline{EA}} = \frac{\overline{AF}}{\overline{FB}} \cdot (-1) \cdot \frac{\overline{CE}}{\overline{EA}} = -1.$$

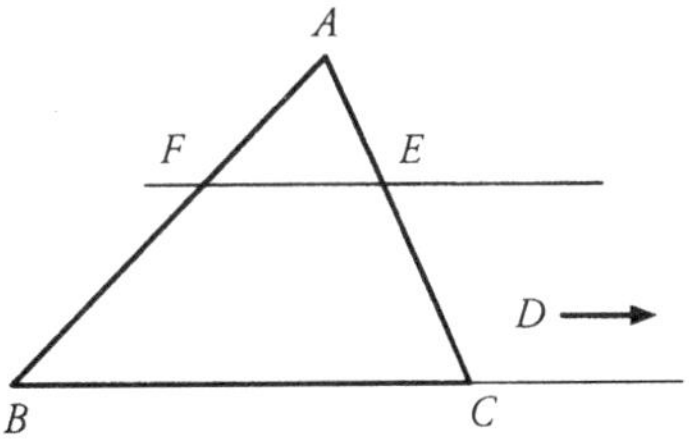

An alternate proof: drop perpendiculars from the vertices, as in the figure on the right. Then

$$\begin{aligned}\frac{\overline{AF}}{\overline{FB}} \cdot \frac{\overline{BD}}{\overline{DC}} \cdot \frac{\overline{CE}}{\overline{EA}} &= \frac{\overline{AF}}{\overline{FB}} \cdot (-1) \cdot \frac{\overline{CE}}{\overline{EA}} \\ &= \frac{\overline{AX}}{\overline{YB}} \cdot (-1) \cdot \frac{\overline{CZ}}{\overline{XA}} \\ &= -1,\end{aligned}$$

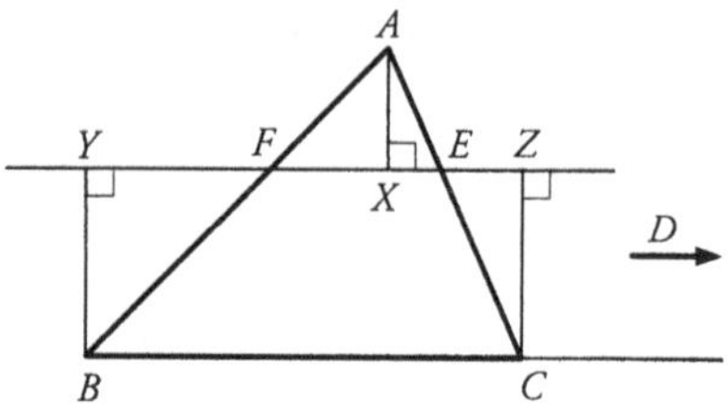

since $\overline{YB} = -\overline{CZ}$.

□

In the extended Euclidean plane, the statement of Menelaus' Theorem is the same as before.

Theorem 4.6.6. *(Menelaus' Theorem for the Extended Plane)*

Let ABC be an ordinary triangle, and let D, E, F be nonvertex points on the (possibly extended) sides BC, CA, and AB of the triangle. Then the points D, E, and F are collinear if and only if

$$\frac{\overline{AF}}{\overline{FB}} \cdot \frac{\overline{BD}}{\overline{DC}} \cdot \frac{\overline{CE}}{\overline{EA}} = -1 .$$

Although the statement is the same, the meaning is a bit different, for in the extended version neither the vertices nor the menelaus points are restricted to being ordinary points.

Remark. The increased generality of the theorems does have some cost. If we wish to apply the extended versions to Euclidean problems, we must first restate the problem in the extended plane. For example, the theorem about the concurrency of the medians of a triangle would, in the extended plane, be stated as follows:

> *The medians of an ordinary triangle are concurrent at an ordinary point.*

To prove this by the extended version of Ceva's Theorem, we would have to show that the point of concurrency is not an ideal point; that is, we still have to show that the cevians in question are not parallel.

4.7 Problems

1. Show that the lines drawn from a vertex to a point halfway around the perimeter of a triangle are concurrent.

   This point of concurrency is called the ***Nagel point*** of the triangle. In the diagram, a, b, and c denote the lengths of the sides.

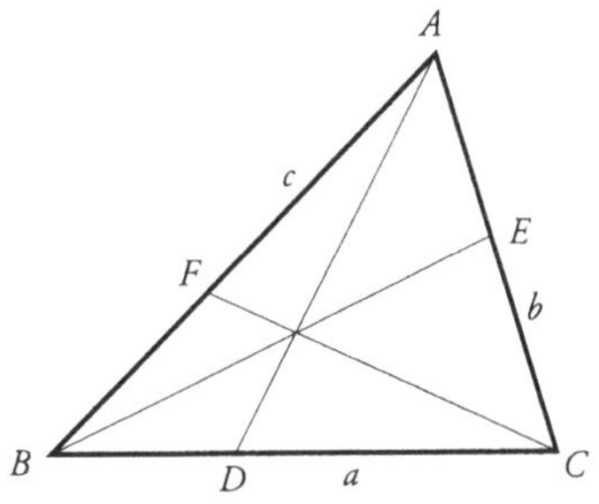

2. Given $\triangle ABC$ with its incircle, the lines drawn from the vertices to the opposite points of tangency are concurrent.

   This point of concurrency is known as the ***Gergonne point*** of the triangle.

   The reader is cautioned that the Gergonne point and the incenter are usually different points, and the associated cevians are neither altitudes nor angle bisectors.

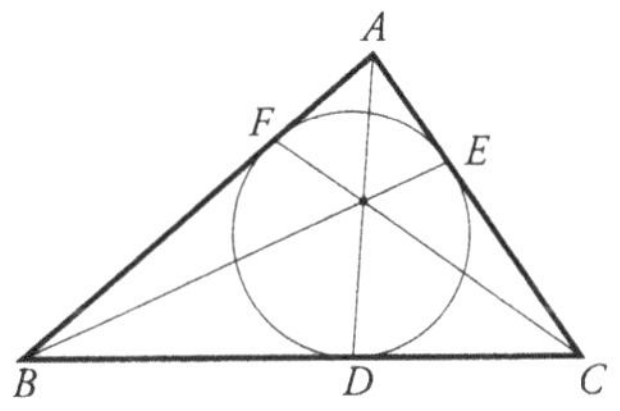

3. Prove the necessary part of Ceva's Theorem for parallel cevians; that is, prove the following:

   If the cevians AD, BE, and CF are parallel cevians for triangle ABC, then

$$\frac{\overline{AF}}{\overline{FB}} \cdot \frac{\overline{BD}}{\overline{DC}} \cdot \frac{\overline{CE}}{\overline{EA}} = 1.$$

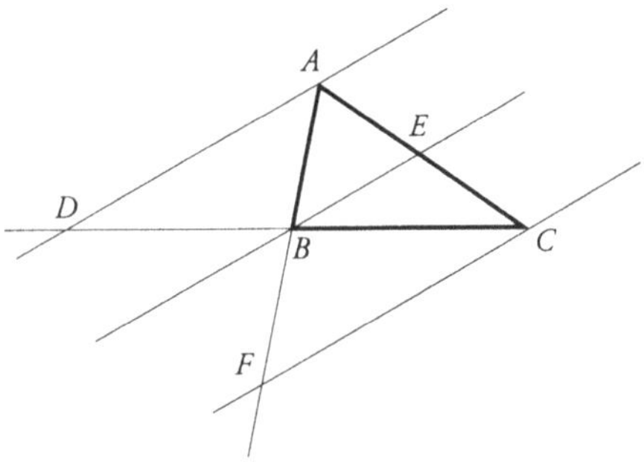

4. Find the length d of the segment BD in the following figure.

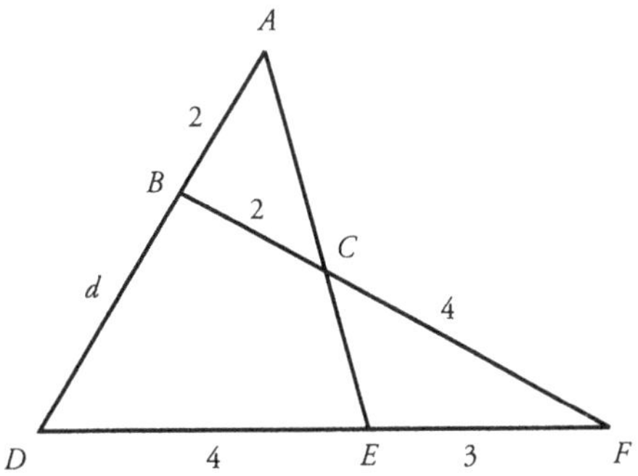

5. Prove that AD is an angle bisector in the figure below.

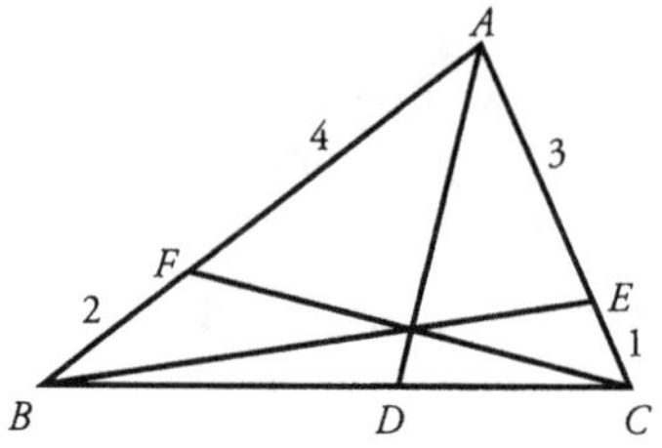

6. In the figure below, find d, the length of the segment BC.

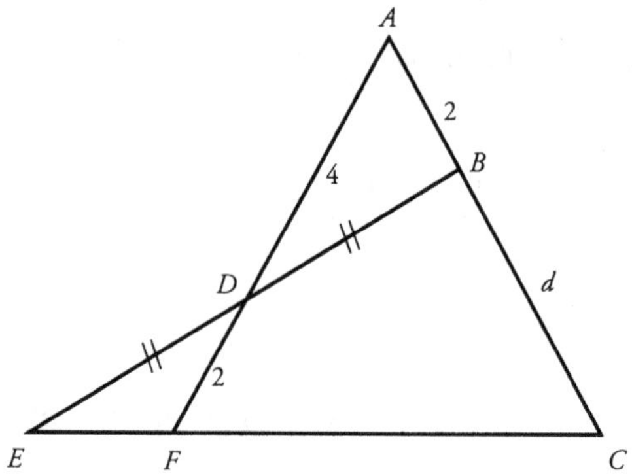

7. (a) Prove that an external angle bisector is parallel to the opposite side if and only if the triangle is isosceles.
   (b) Show that in a nonisosceles triangle, the external angle bisectors meet the opposite sides in three collinear points.

8. In the figure, show that $EP \cdot AB = FP \cdot AC$.

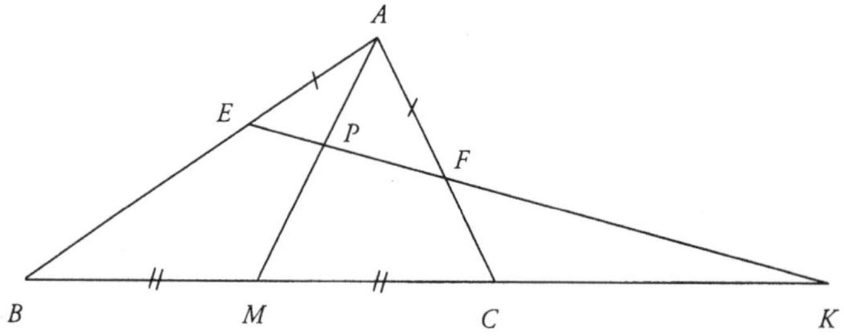

CHAPTER 5

AREA

5.1 Basic Properties

In this chapter we will use the word ***polygon*** to refer to a polygon together with its interior, even though properly we should use the term ***polygonal region***. This should not cause any confusion.

Suppose one polygon is inside another. When treated as wire frames, the polygons would be considered as being disjoint; in the present context, they overlap. In general, if two figures share interior points, they will be considered as ***overlapping***; otherwise, they will be considered as ***nonoverlapping***.

Classical Geometry: Euclidean, Transformational, Inversive and Projective, First Edition. By
I. E. Leonard, J. E. Lewis, A. C. F. Liu, G. W. Tokarsky
Copyright © 2014 John Wiley & Sons, Inc.

5.1.1 Areas of Polygons

We will associate with each simple polygon a nonnegative number called its ***area***, and we will assume that area has certain reasonable properties.

Postulates for polygonal areas:

(i) To each simple polygon is associated a nonnegative number called its ***area***.

(ii) ***Invariance Property***: Congruent polygons have equal area.

(iii) ***Additivity Property***: The area of the union of a finite number of nonoverlapping polygons is the sum of the areas of the individual polygons.

(iv) ***Rectangular Area***: The area of an $a \times b$ rectangle is ab.

Square brackets will be used to denote area. So, for example, the area of a quadrilateral $ABCD$ will be denoted $[ABCD]$.

Properties (ii) and (iii) certainly conform to our preconceived notions about area. We expect figures to have the same area if they have the same shape and size, and we also expect to be able to find the area of a large shape by summing the areas of the individual pieces making up the shape.

To develop a concrete theory of area, we must specify how it is to be measured, and our specification must satisfy (ii) and (iii). The easiest (and most useless) way would be to designate the area of every figure to be zero. To avoid this, a fundamental shape is chosen and defined to have a positive area, which is what statement (iv) accomplishes.

As an alternative to (iv), we could have used the ***unit square***[4] as the fundamental region, defined its area to be one square unit, and derived statement (iv) from it. This approach, although logical, presents some obstacles, especially when the rectangle has sides of irrational length.[5] Because of this, we have chosen to use the set of ***all rectangles*** as a family of fundamental regions on which to base the computation of areas.

[4] The **unit square** is a 1×1 square.

[5] A **rational** number is a real number that can be expressed as one integer divided by another nonzero integer. A number that cannot be expressed this way is called **irrational**.

Exercise 5.1.1. *The following figure is a regular five-pointed star. Which is larger, the area of the shaded part or the total area of the unshaded parts?*

Since a square is a special case of a rectangle, we immediately have the following formula.

Theorem 5.1.2. *(Area of a Square)*

The area of an $a \times a$ square is a^2.

Now let us consider a parallelogram, of which a rectangle is a special case. We may choose any of its sides as the ***base***. We will use the same symbol to designate both the length of the base and the base itself. The context will make it clear which meaning is intended. The distance from the base to the opposite side is called the ***altitude*** on that base.

Theorem 5.1.3. *(Area of a Parallelogram)*

The area of a parallelogram with altitude h on a base b is bh.

Proof. Let $ABCD$ be a parallelogram with $AB = b$. Drop perpendiculars AE and BF to CD from A and B, respectively. Then $ABFE$ is a rectangle with $AE = h$, so that $[ABFE] = bh$. Now, triangles ADE and BCF are congruent. Hence, $[ADE] = [BCF]$ by the Invariance Property. By the Additivity Property,

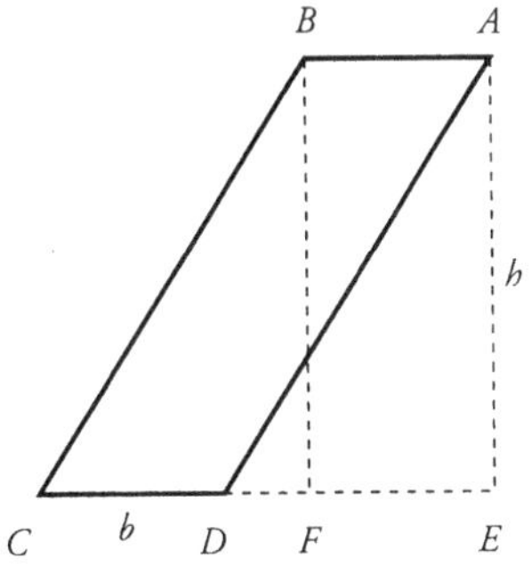

$$\begin{aligned}[ABCD] &= [ABCE] - [ADE] \\ &= [ABCE] - [BCF] \\ &= [ABFE] \\ &= bh,\end{aligned}$$

which completes the proof. □

Any side of a triangle may be designated as the ***base***. The perpendicular from the opposite vertex to the base is called the ***altitude*** on that base. As with parallelograms, the word *base* has a dual meaning, referring to a specific side of a triangle and also to the length of that side. For triangles, the word *altitude* has a similar double meaning.

Theorem 5.1.4. *(Area of a Triangle: Base-Altitude Formula)*

The area of a triangle with altitude h on a base b is $\frac{1}{2}bh$.

Proof. Let ABC be the triangle. Complete the parallelogram $ABCD$. Then the area of parallelogram $ABCD$ is bh. Now, since $\triangle ABC \equiv \triangle CDA$, by the Invariance Property, we have

$$[ABC] = [CDA],$$

and, by the Additivity Property,

$$[ABC] + [CDA] = [ABCD],$$

so that $[ABC] = \frac{1}{2}bh$.

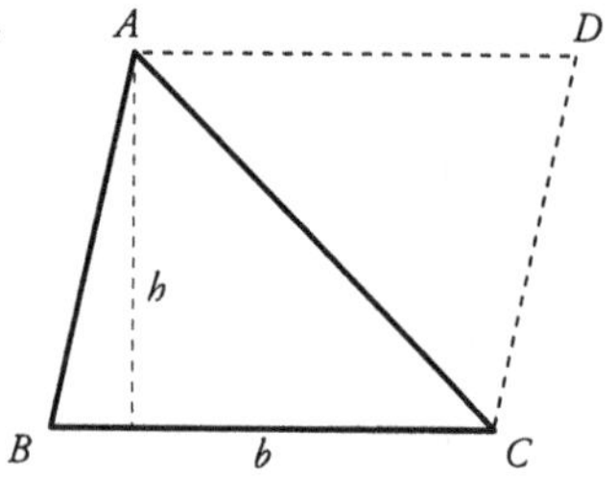

□

Theorem 5.1.5. *(Area of a Triangle: Inradius Formula)*

The area of a triangle with inradius r and semiperimeter s is rs.

(The ***semiperimeter*** is half the perimeter.)

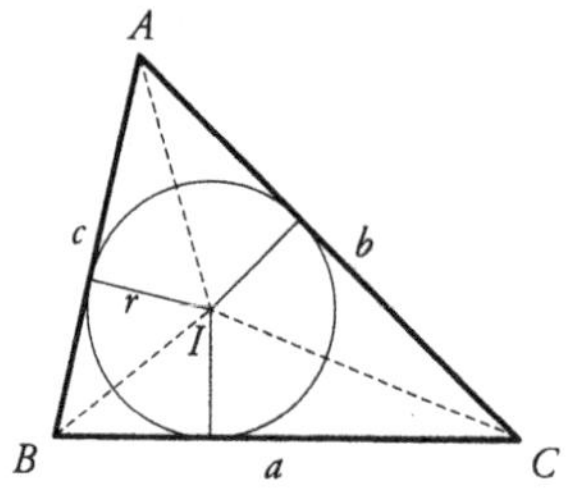

Proof. Let I be the incenter of triangle ABC. Then

$$\begin{aligned}[ABC] &= [IBC] + [ICA] + [IAB] \\ &= \frac{r}{2}(BC + CA + AB) \\ &= rs.\end{aligned}$$

□

Example 5.1.6. *Find the area of a trapezoid whose parallel bases have lengths b and t and whose altitude is h using:*

(1) the formula for the area of a parallelogram;

(2) the formula for the area of a triangle.

Solution.

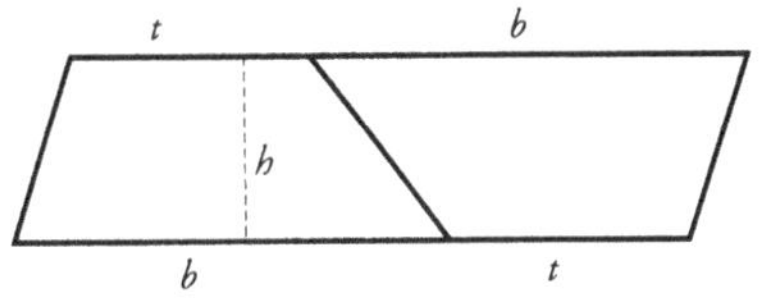

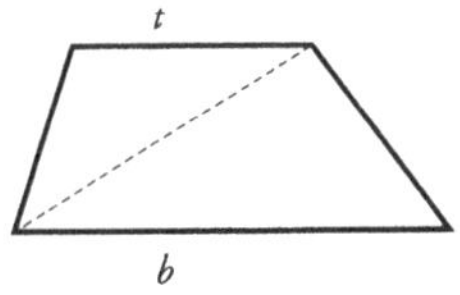

(1) Two copies of the trapezoid will form a parallelogram with base $t + b$ and altitude h. Its area is $(t + b)h$, so the area of the trapezoid is $\frac{1}{2}(t + b)h$.

(2) A diagonal divides the trapezoid into two triangles with altitudes h and respective bases t and b. The areas of the triangles are $\frac{1}{2}th$ and $\frac{1}{2}bh$, from which the desired result follows.

□

The following example shows how area can be used to give a simple proof of Pythagoras' Theorem.

Example 5.1.7. *(Pythagoras' Theorem)*

Given a right triangle with sides of length a and b and with hypotenuse of length c,

$$a^2 + b^2 = c^2.$$

Proof. Draw a square with side c and place four copies of the triangle around it, as shown below.

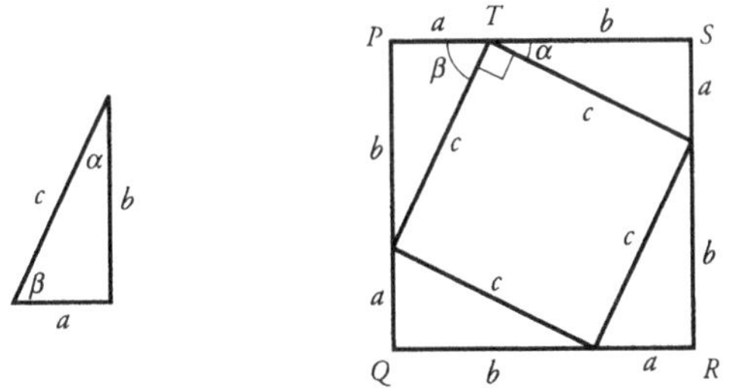

Since $\alpha + \beta = 90°$, it follows that PTS is a straight line, and so $PQRS$ is a square. Using the Additivity Property of area, we have

$$(a+b)^2 = c^2 + 4 \cdot \frac{ab}{2},$$

so that

$$a^2 + 2ab + b^2 = c^2 + 2ab,$$

and so

$$a^2 + b^2 = c^2.$$

□

5.1.2 Finding the Area of Polygons

The area of an arbitrary polygon can be found by decomposing the polygon into triangular regions. If the polygon is convex, all diagonals are internal, so we can choose an arbitrary vertex and join it to all others by diagonals, as in figure (a) below, thereby dividing the polygon into triangles. Such a process is called ***triangulation***. Since we can determine the area of each triangle, the Additivity Property yields the area of the polygon.

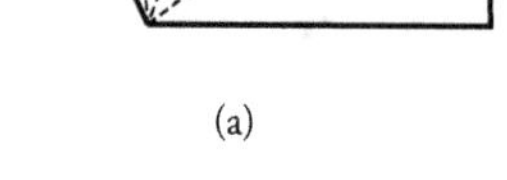

(a)

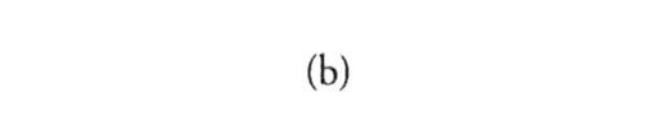

(b)

If the polygon is not convex, the following process can be used: extend the sides of all reflex angles until they meet the perimeter of the polygon. This will divide the polygon into convex polygons, each of which can be triangulated as in the preceding paragraph. Again, the Additivity Property yields the desired result.

The methods above show that the problem of finding the area of any polygon can always be reduced to the problem of finding the area of triangles.

However, from a computational point of view, triangulation methods are extremely clumsy, and we should find a better method. There are many different ways to triangulate a given polygon, and a good choice can ease the computation of the area. For example, we would likely prefer the triangulation of Figure (c) below rather than that of Figure (d), because the former decomposes the pentagon into congruent triangles.

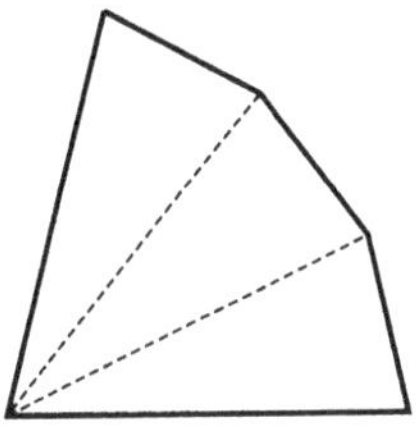

(c)

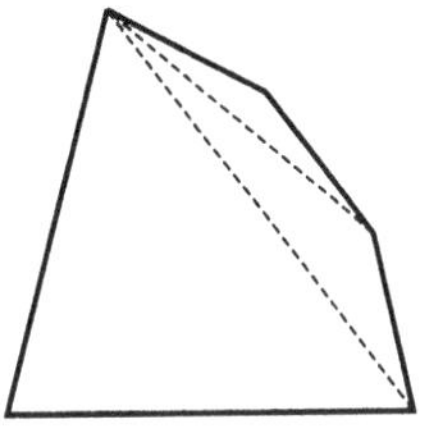

(d)

Also, we should keep in mind the possibility of using other methods that do not involve triangulation. In the figure on the right, it is surely more convenient to view the area of the 7-gon as the difference in the areas of triangle ABG and the square $CDEF$.

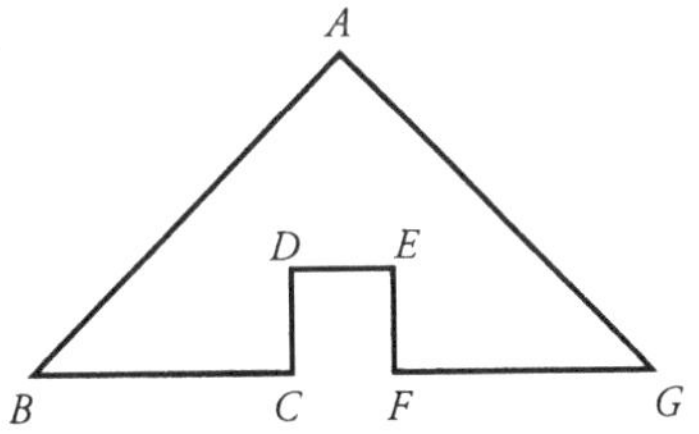

5.1.3 Areas of Other Shapes

Is it possible to extend the notion of area to *all bounded regions* of the plane in a way that satisfies both the Invariance Property and the Additivity Property?

The answer to this question is yes—although the proof of this fact is well beyond the scope of this book. The positive answer means that the properties of polygonal area postulated earlier can therefore be applied to all bounded figures in the plane.

This does not mean, however, that the computation of areas of nonpolygonal regions is straightforward. In fact, one must resort to a limiting process even to find the area of such a basic region as a circular disk. The figure on the following page shows how to approximate a circle of radius r by polygons, with successive approximations coming closer and closer to the circle.

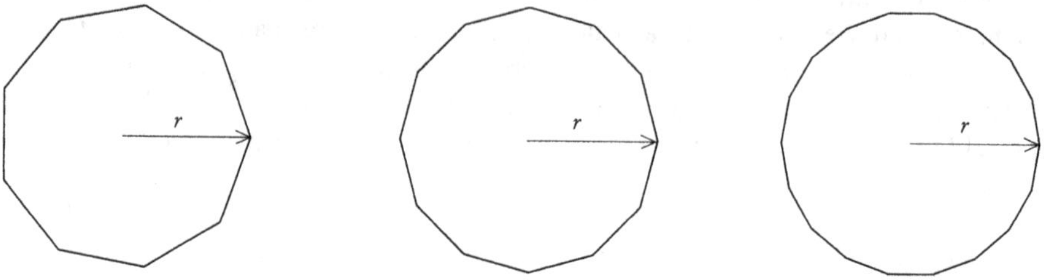

If the approximations in the figure above are cut along the dotted lines as shown below, the pie-shaped regions can be reassembled to form a parallelogram. As the approximations improve, the parallelogram comes closer and closer to becoming a rectangle whose altitude is the radius of the circle and whose base is half the circumference, which yields the following:

Theorem 5.1.8. *The area of a circle is half the product of its radius and its circumference.*

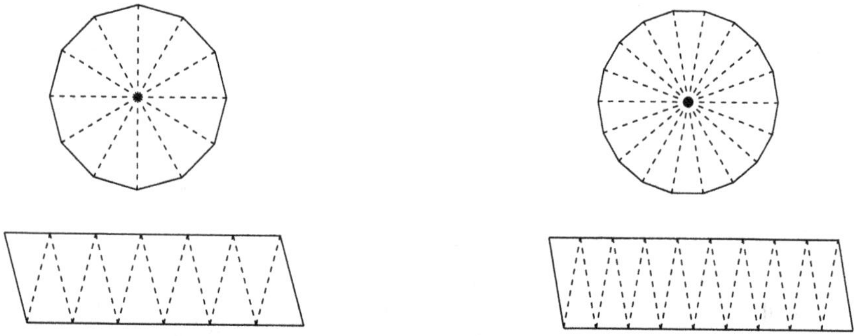

5.2 Applications of the Basic Properties

In this section we will develop several tools using the notion of area and give examples to show how these new tools can be used to solve a variety of problems. Some of the examples in this section may seem to have no connection with the notion of area . . . but they do.

Theorem 5.2.1. *Let l and m be parallel lines. Let $ABCD$ and $EFGH$ be parallelograms with AB and EF on m and CD and GH on l. Then*

$$\frac{[ABCD]}{[EFGH]} = \frac{AB}{EF}.$$

Proof. The two parallelograms have equal altitude h on the respective bases AB and EF. Hence,

$$\frac{[ABCD]}{[EFGH]} = \frac{h \cdot AB}{h \cdot EF} = \frac{AB}{EF}.$$

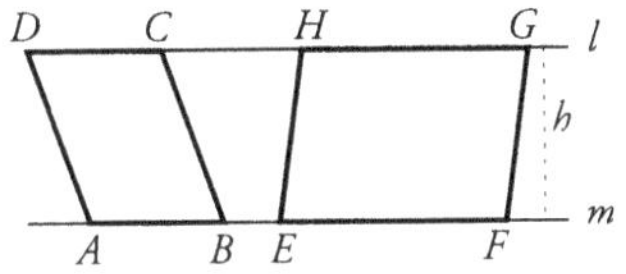

□

The same result holds for triangles.

Corollary 5.2.2. *If ABC and DEF are triangles such that A and D are on a line l and BC and EF are on a parallel line m, then*

$$\frac{[ABC]}{[DEF]} = \frac{BC}{EF}.$$

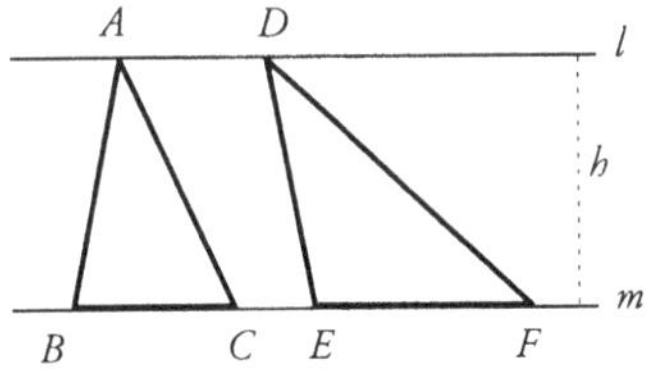

Two special cases of this result that will be used frequently are:

Corollary 5.2.3. *If AB and CD are parallel, then*

$$[ACD] = [BCD].$$

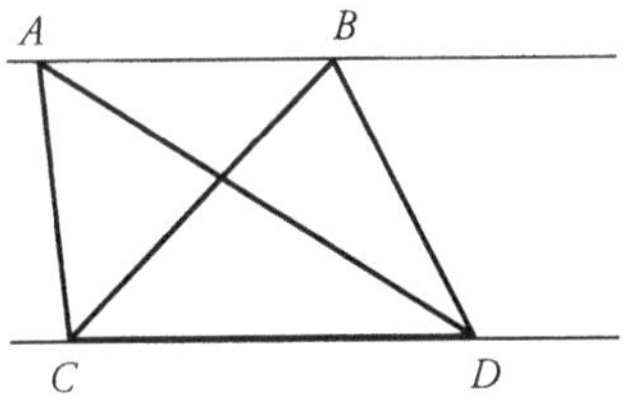

Corollary 5.2.4. *If B, C, and D are collinear points and if A is a point not collinear with them, then*

$$\frac{[ABD]}{[ADC]} = \frac{BD}{DC}.$$

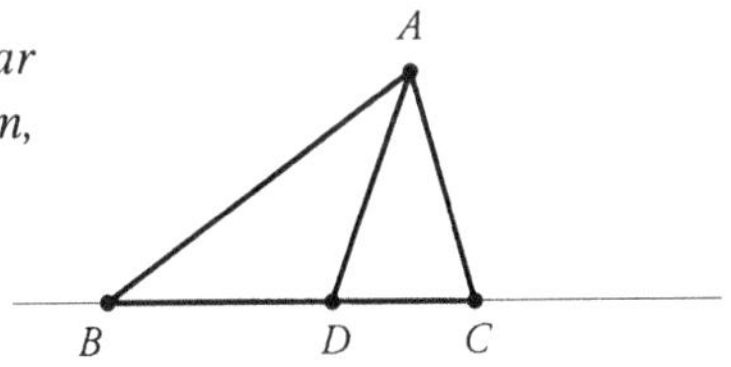

The following is another proof that the three medians of a triangle are concurrent at the centroid.

Theorem 5.2.5. *The medians of a triangle are concurrent.*

Proof. Let D be the midpoint of the side BC of triangle ABC, and let G be a point on AD such that $AG = 2GD$. Suppose that the extension of BG meets CA at E. We will show that $AE = EC$ and from this deduce that the three medians of ABC are concurrent at G.

Let $[DEG] = x$ and $[BDG] = y$. Since

$$AG = 2GD,$$

then by Corollary 5.2.4, we have

$$[AGE] = 2x,$$
$$[BAG] = 2y,$$

and

$$[CDE] = [BDE] = x + y.$$

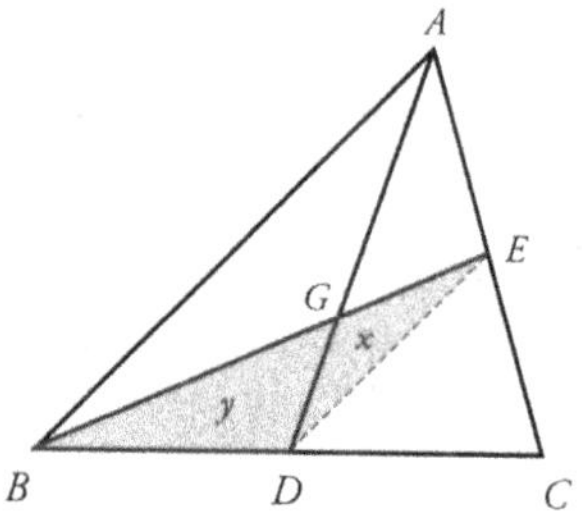

Hence,

$$[ABE] = 2x + 2y = [BEC],$$

so that $AE = EC$; that is, BE is a median.

Similarly, if we let the extension of CG meet AB at F, then $AF = FB$ and AF is a median. Hence, the three medians are concurrent at G.

□

Example 5.2.6. *Let D be a point on the side BC of triangle ABC. Construct a line through D which bisects the area of ABC.*

Solution. If D is the midpoint of BC, then clearly AD is the desired line. Suppose D is between B and the midpoint M of BC. Draw a line through M parallel to AD, cutting CA at E.

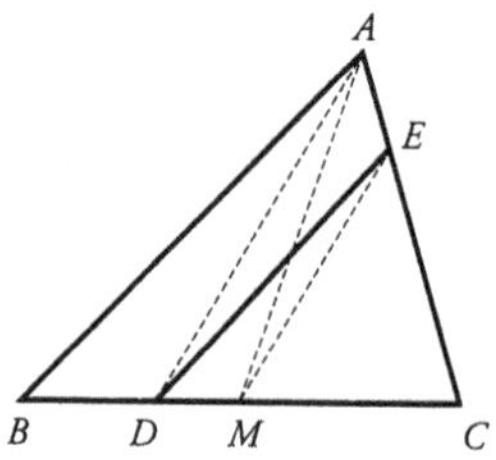

We claim that DE is the desired line. By Corollary 5.2.3, $[ADE] = [MAD]$, and so

$$\begin{aligned}[ABDE] &= [BAD] + [ADE] \\ &= [BAD] + [MAD] \\ &= [BAM] \\ &= \frac{1}{2}[ABC].\end{aligned}$$

□

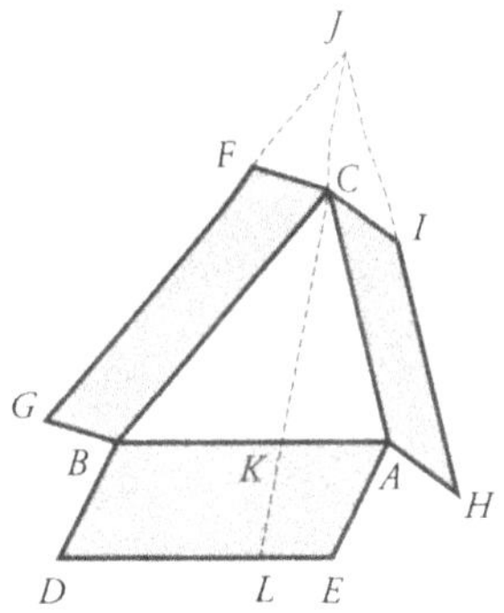

Example 5.2.7. *ABDE, BCFG, and CAHI are three parallelograms drawn outside $\triangle ABC$. The lines FG and HI meet at J. The extension of JC meets AB at K and the line DE at L. If $JC = KL$, prove that*

$$[ABDE] = [BCFG] + [CAHI].$$

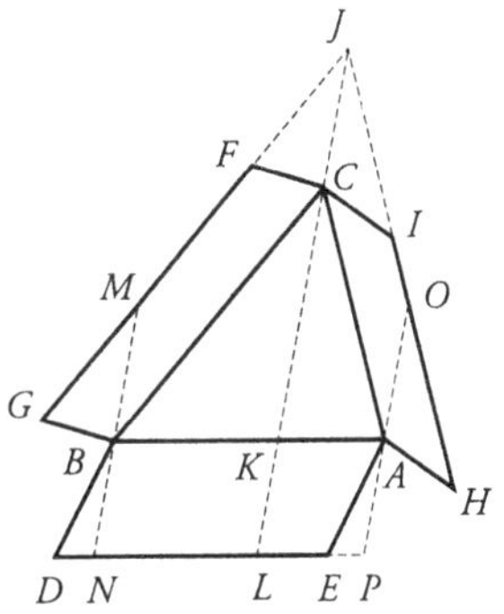

Solution. Draw a line through B parallel to JL, cutting the line FG at M and the line DE at N. Draw a line through A parallel to JL, cutting the line HI at O and the line DE at P. By Theorem 5.2.1 and the Additivity Property of area, we get

$$\begin{aligned}[ABDE] = [ABNP] &= [KBNL] + [AKLP] \\ &= [JMBC] + [JOAC] \\ &= [BCFG] + [CAHI],\end{aligned}$$

which completes the proof.

□

Example 5.2.7 is due to Pappus and has Pythagoras' Theorem as a corollary:

Corollary 5.2.8. *(Pythagoras' Theorem)*

Let ABC be a triangle. If $\angle C$ is a right angle, then

$$AB^2 = BC^2 + CA^2.$$

Proof. Suppose $\angle ACB$ is a right angle. Draw squares $ABDE$, $BCFG$, and $CAHI$ outside ABC. Let the extensions of GF and HI meet at J. Let the extension of JC meet AB at K and DE at L. Since triangles ABC and JCF are congruent, $JC = AB = AE = KL$. By Example 5.2.7,

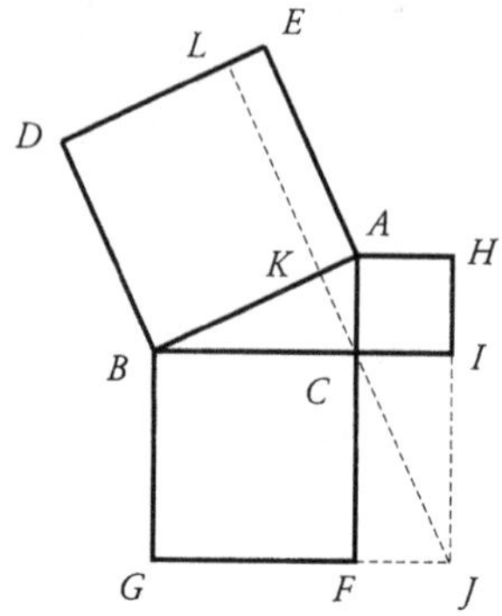

$$\begin{aligned} AB^2 &= [ABDE] \\ &= [BCFG] + [CAHI] \\ &= BC^2 + CA^2. \end{aligned}$$

□

The next example uses Corollary 5.2.3 to obtain another proof of the Angle Bisector Theorem. In the proof, we make use of the fact that an angle bisector is characterized by each point on the bisector being equidistant from the arms of the angle. Before stating the theorem, we recall that the notation $\overline{XY}$ refers to the *directed distance* from X to Y. This enables us to distinguish between interior and exterior angle bisectors.

Theorem 5.2.9. *(The Angle Bisector Theorem)*

Let D be a point on the line BC and A a point not on it. Then

(1) $AB/AC = \overline{BD}/\overline{DC}$ if and only if AD bisects angle A of triangle ABC and

(2) $AB/AC = -\overline{BD}/\overline{DC}$ if and only if AD bisects the exterior angle A of triangle ABC.

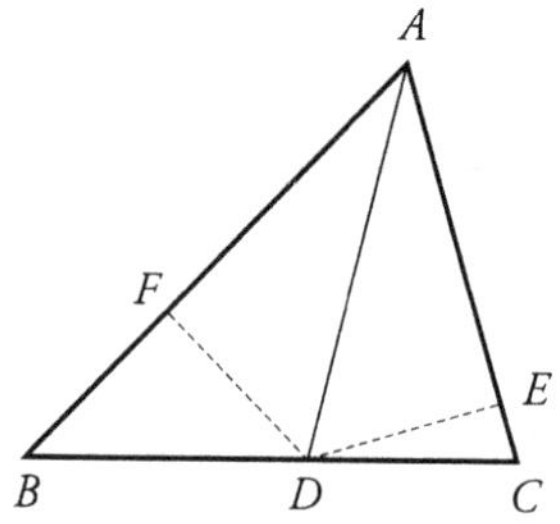

(i)

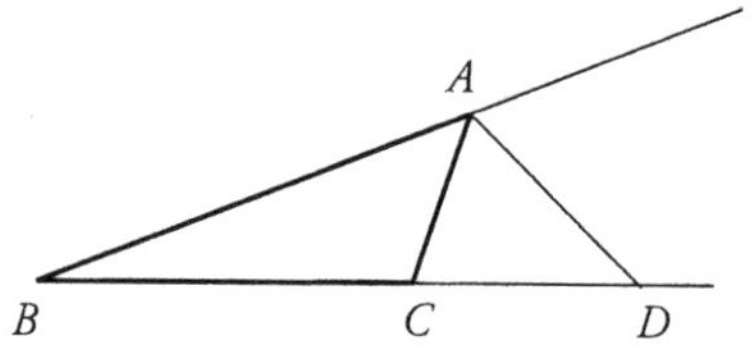

(ii)

Proof.

(1) Drop perpendiculars DE and DF from D to CA and AB, respectively. Suppose AD bisects $\angle CAB$. Since D is on the angle bisector, we have $DE = DF$. By Corollary 5.2.3,

$$\overline{BD}/\overline{DC} = [BAD]/[CAD] = \left(\tfrac{1}{2}AB \cdot DF\right)/\left(\tfrac{1}{2}AC \cdot DE\right) = AB/AC.$$

Conversely, suppose $\overline{BD}/\overline{DC} = AB/AC$. Reversing the argument, we have $DE = DF$, so D is equidistant from the arms of $\angle BAC$. It follows that AD bisects $\angle CAB$.

(2) Statement (2) can be proved in an analogous manner.

□

As the previous examples show, Corollary 5.2.3 is very useful. It can be generalized to the following result.

Theorem 5.2.10. *Let D be a point on the line BC and A a point not on it. If E is a point on the line AD, then*

$$\frac{[ABE]}{[ACE]} = \frac{BD}{CD}.$$

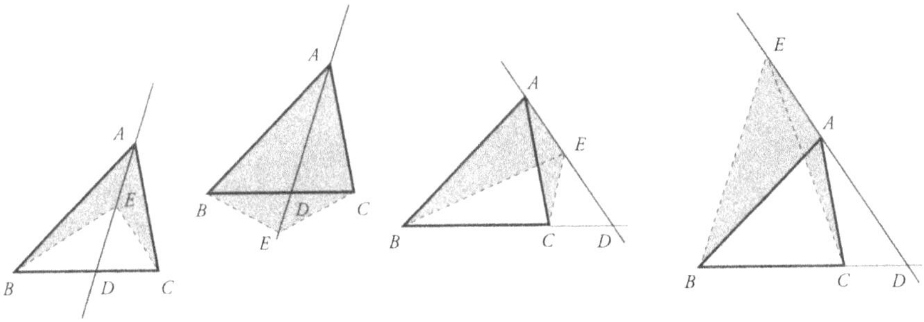

Proof. The figure above shows that there are many possibilities for the location of the points D and E. We will consider one subcase of the situation where D is between B and C, namely, the case where E is between A and D as in the figure on the right.

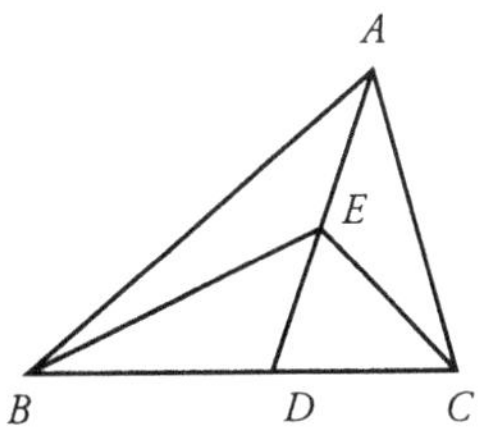

Let $BD/CD = t$. Corollary 5.2.3 implies that

$$\frac{[BAD]}{[CAD]} = t$$

and that

$$\frac{[BED]}{[CED]} = t.$$

Then

$$[ABE] = [BAD] - [BED] = t([CAD] - [CED]) = t[ACE],$$

so that

$$\frac{[ABE]}{[ACE]} = t = \frac{BD}{CD}.$$

□

We now apply Theorem 5.2.10 to solve the following, a special case of a result due to Routh.[6]

Example 5.2.11. *Let D, E, and F lie respectively on the sides BC, CA, and AB of triangle ABC such that*

$$DC = 2BD, \quad EA = 2CE, \quad \text{and} \quad FB = 2AF.$$

Suppose also that AD meets BE at G and CF at H and that BE meets CF at I.

Determine $\dfrac{[GHI]}{[ABC]}$.

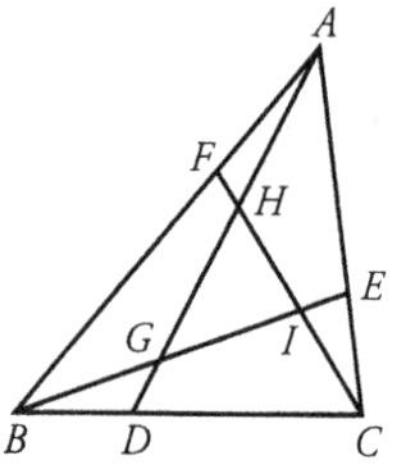

Solution.

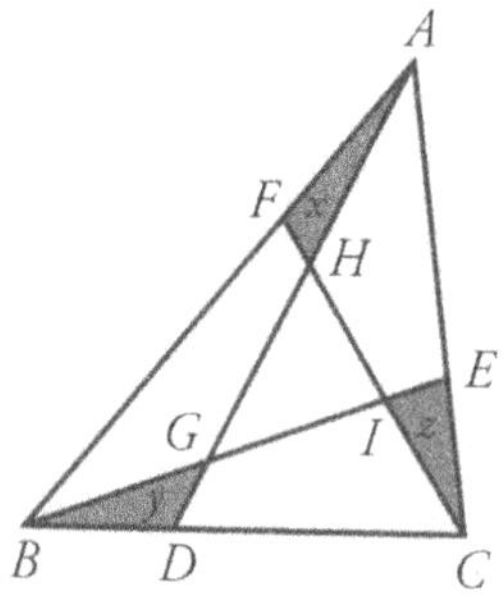

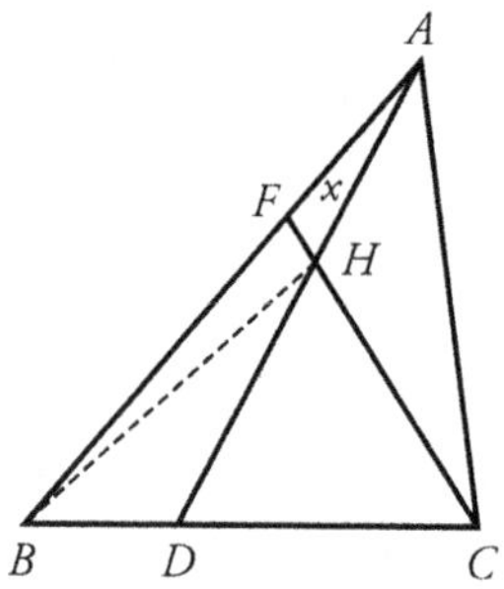

Let the areas of triangles HAF, GBD, and ICE be x, y, and z, respectively. We will first show that $[ABC] = 21x$ and that $x = y = z$.

[6]A generalization of this example was given by E. J. Routh in 1891 (without proof) who needed this ratio in his analysis of the stresses and tensions in mechanical frameworks.

By Corollary 5.2.3,

$$\frac{[HBF]}{[HAF]} = \frac{BF}{FA} = 2,$$

so that

$$[HBF] = 2x.$$

Similarly, since $DC = 2BD$, Theorem 5.2.10 gives us

$$[HAC] = 2[HAB] = 6x,$$

and so

$$[FAC] = [HAF] + [HAC] = 7x.$$

Since $BF = 2AF$, from Corollary 5.2.3, we get $[FBC] = 14x$ so that

$$[ABC] = 21x.$$

In the same way, we can prove that $[ABC] = 21y = 21z$. Hence $x = y = z$ and so

$$[BAG] = [CBI] = [ACH] = 6x.$$

It follows that

$$[GHI] = 21x - 18x = 3x,$$

so that

$$\frac{[GHI]}{[ABC]} = \frac{1}{7}.$$

□

5.3 Other Formulae for the Area of a Triangle

The **SAS** congruency condition asserts that a triangle is uniquely determined given two of its sides and the included angle. All other congruency conditions also describe geometric data that determine a unique triangle. Since these conditions do not explicitly specify the altitude of the triangle, the formula for the area of a triangle cannot be used without some preliminary work. It would be much more convenient to be able to compute the area of a triangle directly from the data that describes it. In this section we develop a list of formulae for the area of a triangle based on the various congruency conditions. As you might expect, some of the formulae involve trigonometric functions.

In this section we will continue to follow the practice of denoting the sides of a triangle by the lowercase letters that correspond to the labels for the opposite vertices.

We start by proving a useful trigonometric tool.

Theorem 5.3.1. *(Law of Sines)*

In triangle ABC,

$$\frac{a}{\sin A} = \frac{b}{\sin B} = \frac{c}{\sin C} = 2R,$$

where R is the circumradius.

Proof.

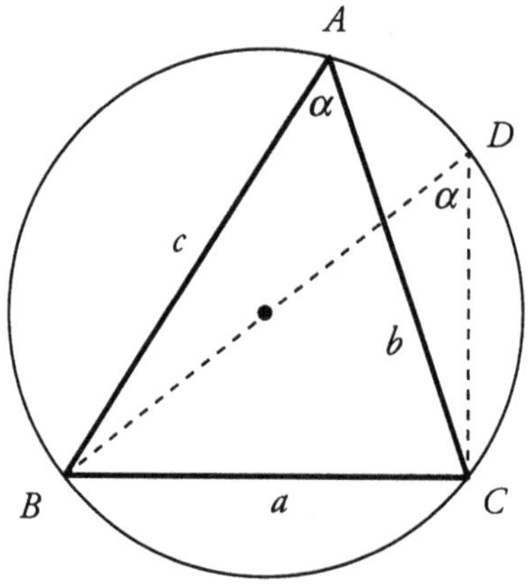

Let BD be the diameter of the circumcircle. Then $\angle BCD = 90°$ and

$$\angle A = \angle BAC = \angle BDC$$

by Thales' Theorem. Hence,

$$a = BC = BD \sin A = 2R \sin A,$$

so that

$$\frac{a}{\sin A} = 2R.$$

Similarly,

$$\frac{b}{\sin B} \quad \text{and} \quad \frac{c}{\sin C}$$

are both equal to $2R$.

Theorem 5.3.2. *(***SAS** *Case)*

The area of a triangle with sides a, b and angle C is

$$[ABC] = \frac{1}{2}ab\sin C.$$

Proof.

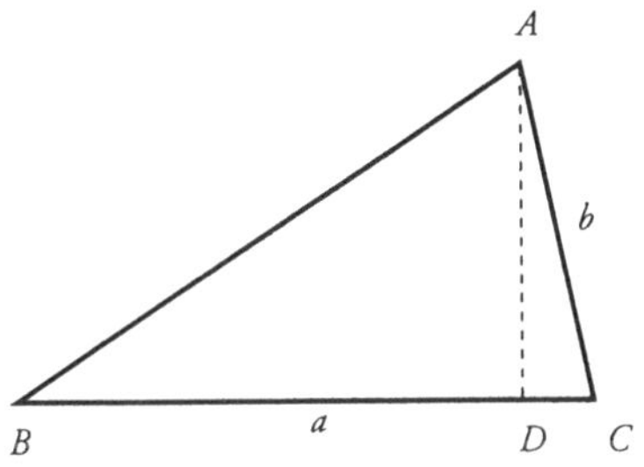

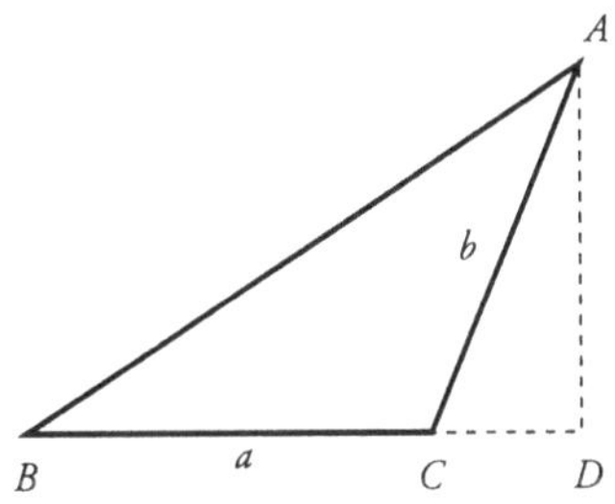

Let AD be the altitude on BC. Then $AD = b\sin C = b\sin(180° - C)$. Hence,

$$[ABC] = \frac{1}{2}a\,AD = \frac{1}{2}ab\sin C\,.$$

□

Example 5.3.3. *Prove that*

$$[ABC] = \frac{abc}{4R},$$

where R is the circumradius of $\triangle ABC$.

Solution. We have

$$[ABC] = \frac{1}{2}ab\sin C = \frac{abc}{4R}$$

by the Law of Sines.

□

For the **ASA** case, we make use of all three angles.

Theorem 5.3.4. *(ASA Case)*

Given angles B and C and included side a of $\triangle ABC$, then

$$[ABC] = \frac{a^2 \sin B \sin C}{2 \sin A}.$$

Proof. By the Law of Sines, $a \sin B = b \sin A$. Hence,

$$\begin{aligned}[ABC] &= \frac{1}{2}ab \sin C \\ &= \frac{a^2 \sin B \sin C}{2 \sin A}.\end{aligned}$$

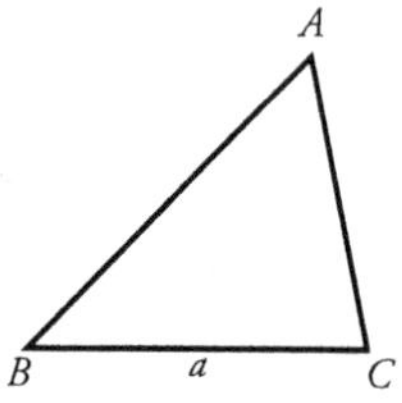

□

We remark that the Law of Sines is also useful when one needs to find the remaining dimensions of a triangle given a side and two angles.

The proof of the following uses the compound angle formula for the sine function, that is, $\sin(A + B) = \sin A \cos B + \cos A \sin B$.

Theorem 5.3.5. *(SSA Case)*

Assuming that $a > b$, the area of $\triangle ABC$ given sides a and b and angle A is

$$[ABC] = \frac{1}{2}b \sin A\big(\sqrt{a^2 - b^2 \sin^2 A} + b \cos A\big).$$

Proof. Since $a > b$, $A > B$, so that $\cos B > 0$.
By the Law of Sines, $a \sin B = b \sin A$. Hence,

$$\begin{aligned}[ABC] &= \frac{1}{2}ab \sin C = \frac{1}{2}ab \sin(A + B) \\ &= \frac{1}{2}ab \sin A \cos B + \frac{1}{2}ab \cos A \sin B \\ &= \frac{1}{2}b \sin A(a \cos B + b \cos A) \\ &= \frac{1}{2}b \sin A\big(\sqrt{a^2 - a^2 \sin^2 B} + b \cos A\big) \\ &= \frac{1}{2}b \sin A\big(\sqrt{a^2 - b^2 \sin^2 A} + b \cos A\big).\end{aligned}$$

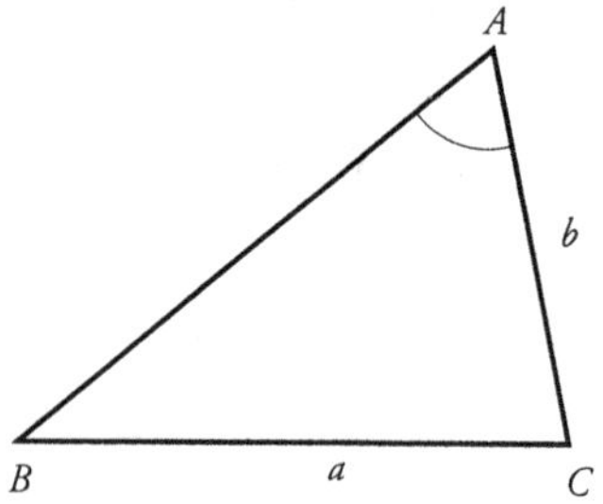

□

In the following, s denotes the semiperimeter of a triangle; that is, for a triangle with sides a, b, and c,

$$s = \frac{a+b+c}{2}.$$

Theorem 5.3.6. *(***SSS** *Case: Heron's Formula)*

For a triangle with sides a, b, and c,

$$[ABC] = \sqrt{s(s-a)(s-b)(s-c)}.$$

Proof. Let BC be the longest side. Then the foot D of the altitude AD lies between B and C. Let $CD = x$ and $AD = y$, so that $BD = a-x$. By Pythagoras' Theorem, $x^2 + y^2 = b^2$ and $(a-x)^2 + y^2 = c^2$. Subtraction yields

$$2ax - a^2 = b^2 - c^2 \quad \text{or} \quad x = (a^2 + b^2 - c^2)/2a.$$

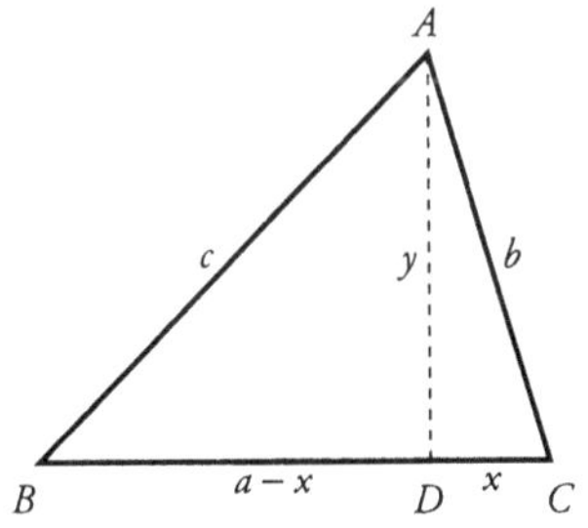

Now,

$$\begin{aligned}[ABC] &= \frac{1}{2}ay = \frac{1}{2}a\sqrt{b^2 - x^2} \\ &= \frac{1}{2}a\sqrt{b^2 - [\,(a^2+b^2-c^2)/2a\,]^2} \\ &= \frac{1}{4}\sqrt{4a^2b^2 - (a^2+b^2-c^2)^2} \\ &= \frac{1}{4}\sqrt{(2ab + a^2 + b^2 - c^2)(2ab - a^2 - b^2 + c^2)} \\ &= \frac{1}{4}\sqrt{((a+b)^2 - c^2)(c^2 - (a-b)^2)} \\ &= \frac{1}{4}\sqrt{(a+b+c)(a+b-c)(a-b+c)(-a+b+c)} \\ &= \sqrt{s(s-a)(s-b)(s-c)}.\end{aligned}$$

□

Heron's Formula was derived from Pythagoras' Theorem, but it is possible to reverse directions and derive Pythagoras' Theorem from Heron's Formula, meaning that the two theorems are equivalent. (To derive Pythagoras' Theorem, apply Heron's Formula to a right triangle. The details are left as an exercise.)

The next result, which is far more familiar than Heron's Formula, is also equivalent to Pythagoras' Theorem.

Theorem 5.3.7. *(The Law of Cosines)*

In triangle ABC,

$$c^2 = a^2 + b^2 - 2ab\cos C.$$

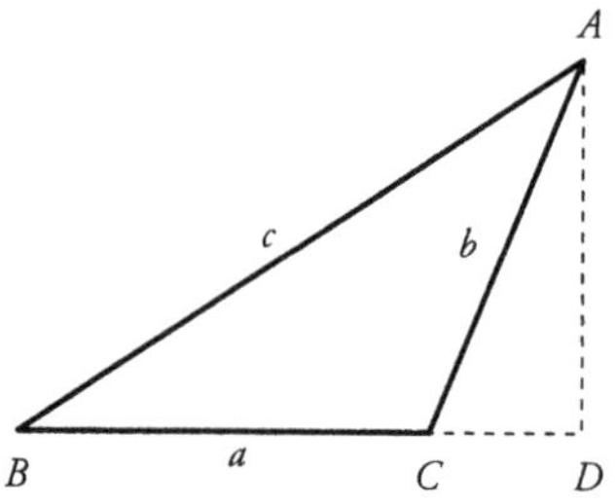

Proof. If $\angle C = 90°$, then $\cos C = 0$ and the result is just Pythagoras' Theorem. Suppose $\angle C > 90°$. Then the foot D of the altitude AD lies on the extension of BC. By Pythagoras' Theorem,

$$\begin{aligned} c^2 &= AD^2 + BD^2 = AD^2 + (a + CD)^2 \\ &= AD^2 + CD^2 + a^2 + 2a \cdot CD \\ &= b^2 + a^2 - 2ab\cos C, \end{aligned}$$

since $\cos C < 0$.

If $\angle BCA < 90°$, the argument is similar.

□

The Law of Cosines is not used to find the area of a triangle. It is indispensable, however, when one needs to find the remaining sides and angles in the **SAS** and **SSS** cases.

5.4 Solutions to the Exercises

Solution to Exercise 5.1.1

In the figure below, the shaded region has been divided into four parts, each of which is congruent to a part of the unshaded region. Therefore, the shaded region is equal in area to the total of the unshaded regions.

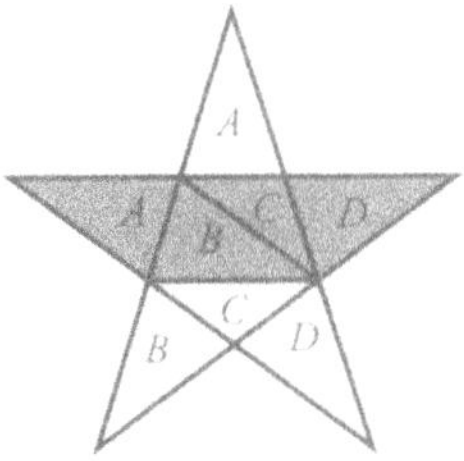

5.5 Problems

1. In the regular octagon shown below, is the area of the shaded region larger or smaller than the total area of the unshaded regions?

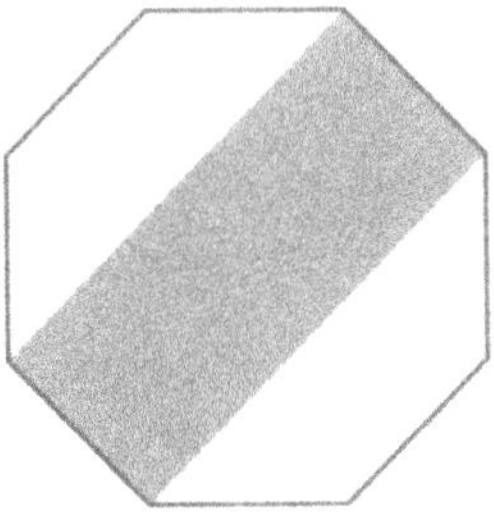

2. P is a point inside square $ABCD$. Show that

$$[APB] + [DPC] = [APD] + [BPC].$$

3. A square is divided into five nonoverlapping rectangles, with four of the rectangles completely surrounding the fifth rectangle, as shown in the diagram. The outer rectangles are the same area. Prove that the inner rectangle is a square.

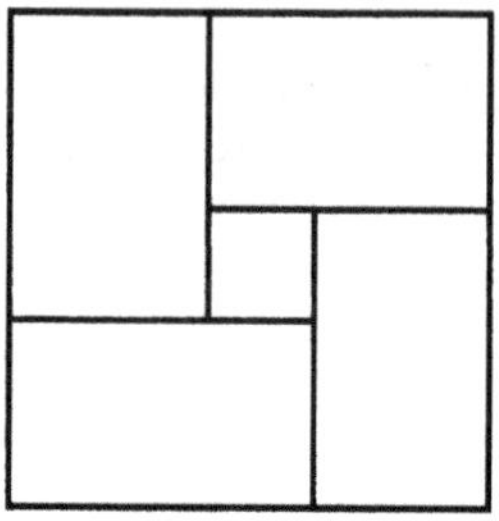

4. $ABCD$ is a quadrilateral with perpendicular diagonals inscribed in a circle with center O. Prove that $[ADCO] = [ABCO]$.

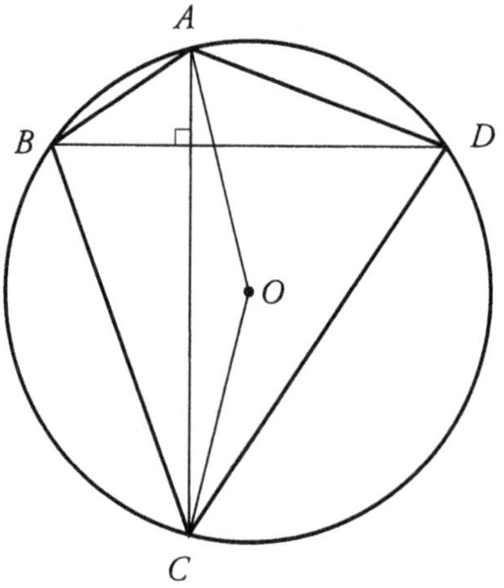

5. A paper rectangle $ABCD$ of area 1 is folded along a straight line so that C coincides with A. Prove that the area of the pentagon obtained is less than $\frac{3}{4}$.

6. $ABCD$ is a parallelogram. E is a point on BC and F a point on CD. AE cuts BF at G, AF cuts DE at H, and BF cuts DE at K. Prove that

$$[AGKH] = [BEG] + [CEKF] + [DFH].$$

7. $ABCDEF$ is a convex hexagon in which opposite sides are parallel. Prove that $[ACE] = [BDF]$.

8. G is a point inside triangle ABC such that $[GBC] = [GCA] = [GAB]$. Show that G is the centroid of ABC.

9. Seven children share a square pizza whose crust may be considered to consist only of the perimeter. Show how they make straight cuts to divide the pizza into seven pieces such that all pieces have the same amount of pizza and the same amount of crust.

10. AB and CD are four points such that $AB = CD$. Find the set of all points P such that $[PAB] = [PCD]$.

11. $ABCDEF$ is a convex hexagon with side AB parallel to CF, side CD parallel to BE, and side EF parallel to AD. Prove that $[ACE] = [BDF]$.

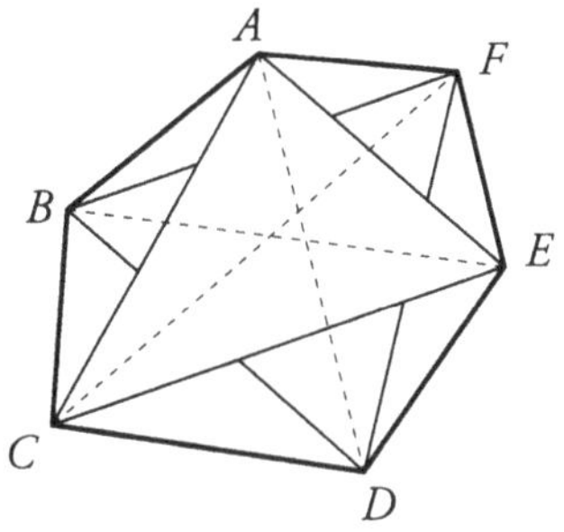

12. $ABCD$ is a convex quadrilateral, and F and H are the midpoints of BC and AD, respectively. If AC cuts FH at the midpoint K of FH, show that $[ABC] = [ADC]$.

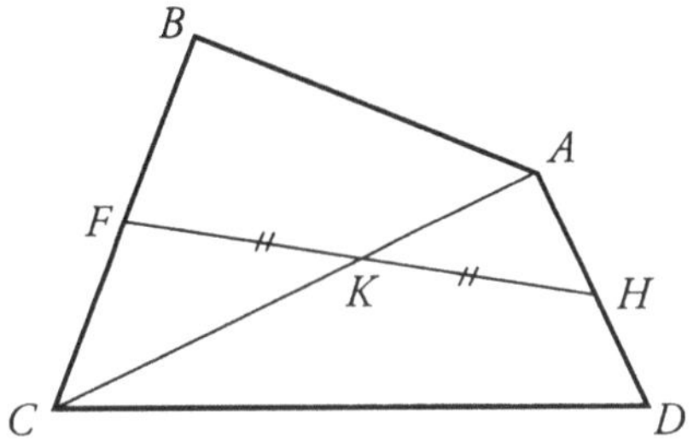

13. Given a convex quadrilateral $ABCD$, construct a point E on the extension of BC such that the area $[ABCD] = [ABE]$.

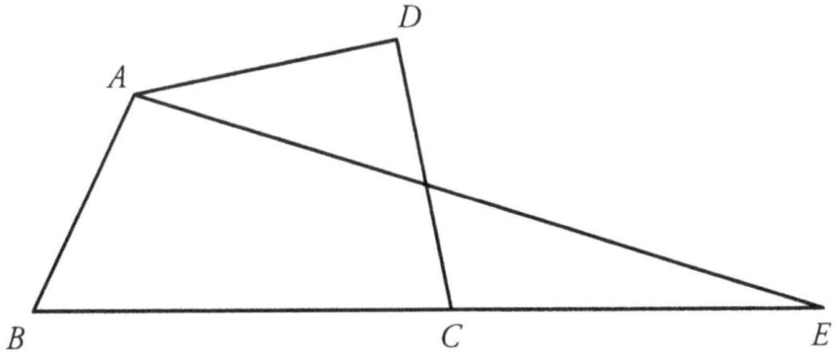

14. $ABCDE$ is a regular pentagon. The points P, Q, and R are the feet of the perpendiculars from A to BC, CD, and DE, respectively. The center O of the pentagon lies on AQ. If $OQ = 1$, compute $AP + AQ + AR$.

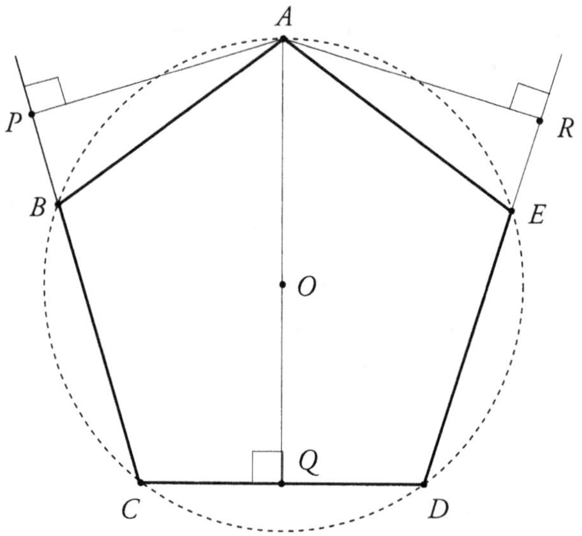

15. Given the figure below with $3BF = 2FC$, $AE = 2EF$, and $[DEF] = 1$:

(a) Find $[DFC]$.

(b) Find $\dfrac{AC}{AD}$.

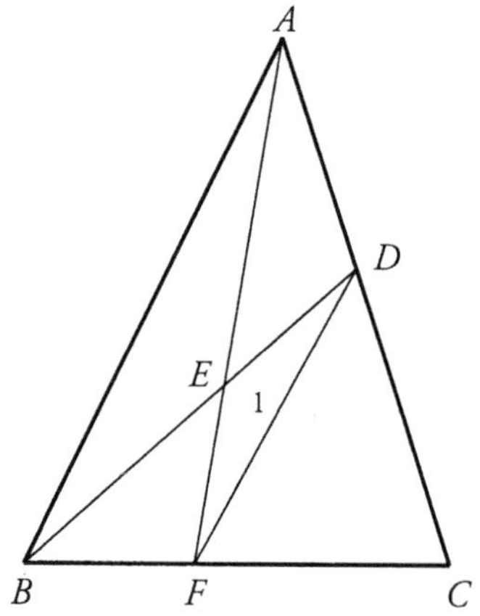

16. M is a point interior to the rectangle $ABCD$. Prove that $AM \cdot CM + BM \cdot DM > ABCD$.

17. Given a rectangle, construct a square having the same area.

18. A triangle is inside a parallelogram. Prove that the area of the triangle is at most half that of the parallelogram.

19. P is a point inside an equilateral triangle ABC. Perpendiculars PD, PE, and PF are dropped from P onto BC, CA, and AB, respectively. Prove that

$$[PAF] + [PBD] + [PCE] = [PAE] + [PCD] + [PBF].$$

20. Find the area of a triangle of sides 13, 18, and 31.

21. A triangle has sides 13, 14, and 15. Find its altitude on the base of length 14.

22. The sides of a triangle are 5, 7, and 8.

    (a) Calculate its area.

    (b) Calculate its inradius.

    (c) Calculate its circumradius.

23. $ABDE$, $BCGF$, and $CAHI$ are three squares drawn on the outside of triangle ABC, which has a right angle at C. Prove that

$$GD^2 - HE^2 = 3([BCFG] - [CAHI]).$$

24. M is a point interior to the rectangle $ABCD$. Prove that

$$AM \cdot CM + BM \cdot DM \geq [ABCD].$$

25. $ABCDE$ is a convex pentagon such that

$$AB = AC, \quad AD = AE, \quad \text{and} \quad \angle CAD = \angle ABE + \angle AEB.$$

If M is the midpoint of BE, prove that $CD = 2AM$.

CHAPTER 6

MISCELLANEOUS TOPICS

6.1 The Three Problems of Antiquity

In some of the earlier chapters we had sections on construction problems. In this chapter, we expand further and describe some useful techniques.

The object is to draw geometric figures in the plane using two simple ***tools***:

1. A ***straightedge***. This is a device for drawing a straight line through any two given points. It is of arbitrary length—that is, you can draw a line as long as you need that passes through the given points. Note that this means that given a segment AB you can extend this segment: take two points on the segment and draw the line through those two points. Note also that a straightedge is *not a ruler*. You cannot use a straightedge to measure distances. A ruler is a different tool.

Classical Geometry: Euclidean, Transformational, Inversive and Projective, First Edition. By
I. E. Leonard, J. E. Lewis, A. C. F. Liu, G. W. Tokarsky
Copyright © 2014 John Wiley & Sons, Inc.

2. A ***modern compass***. This is a device for drawing arcs and circles given any point as center and the length of any given segment as radius. The modern compass holds its radius when it is lifted from the page, as opposed to the ***classical compass*** which collapses to zero radius when removed from the page.

The constructions that can be accomplished using these two basic tools are called ***Euclidean constructions***.

There are certain famous construction problems that have been shown to be impossible in the sense that they cannot be accomplished with a straightedge and compass. These are:

1. ***Squaring the circle***. The problem is to construct a square of the same area as a given circle.

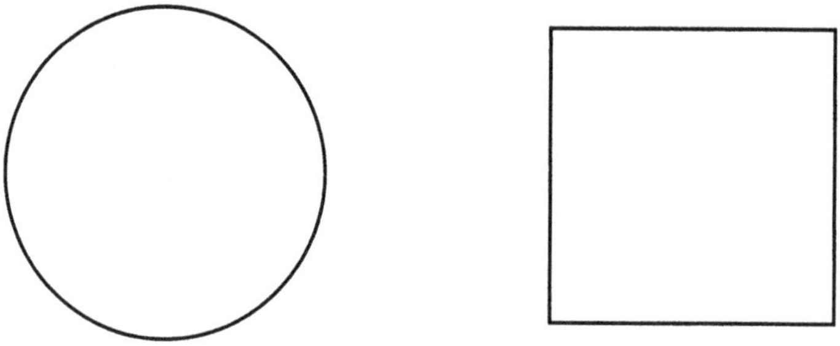

   Squaring the circle amounts to the following: given a segment of length 1, construct a segment of length $\sqrt{\pi}$.

2. ***Doubling the cube***. Given a cube, construct another cube of twice its volume.

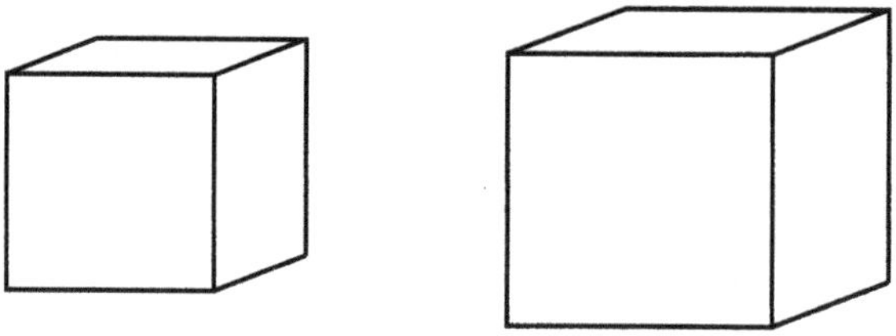

   Doubling the cube amounts to the following: given a segment of length 1, construct a segment of length $2^{1/3}$.

3. ***Trisecting any general angle***. Given an arbitrary angle, construct an angle that is one-third its size.

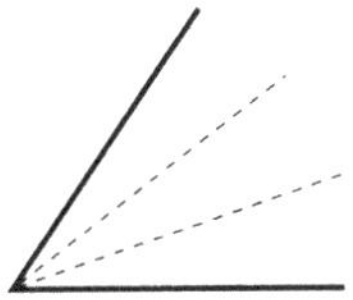

Note that some angles *can* be trisected—for example, it is possible to trisect a right angle (because we can construct an angle of $30°$). We can construct an angle of $60°$. Trisecting this angle would amount to constructing an angle of $20°$, and once we have an angle of $20°$ and a segment of length 1 unit, we could construct a segment equal in length to the cosine of $20°$. The proof that there is no general method for trisecting an angle is accomplished by proving that, given only a segment of length 1, it is impossible to construct a segment equal in length to $\cos 20°$.

6.2 Constructing Segments of Specific Lengths

Given a segment of length 1 unit, what other lengths can we construct? The construction must use only a straightedge and a compass and must be accomplished in a finite number of steps. This section describes what lengths we can construct. A complete discussion would prove that these are the only lengths that we can construct, but that is well beyond the scope of this book.

It is clear that, given a segment of length 1, we can construct a segment of length n where n is any positive integer.

Given segments of length p and q, we can construct segments of length $p+q$ and $p-q$.

What about constructing segments of length p/q or pq?

Example 6.2.1. *Given segments of lengths* 1, *p, and q, construct a segment of length* p/q.

Solution.

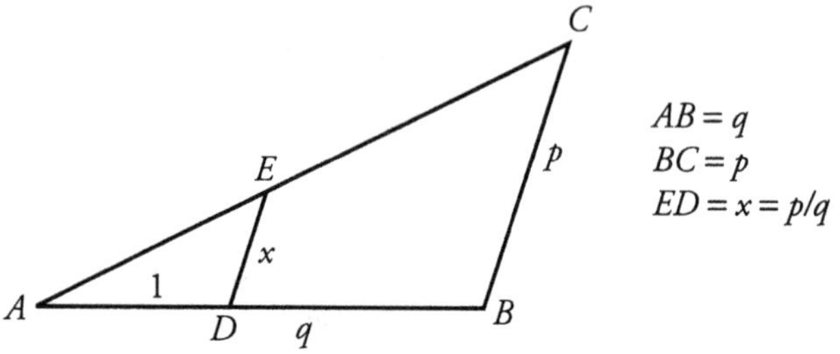

The diagram above reveals how to use similar triangles to accomplish the task. The actual construction is as follows:

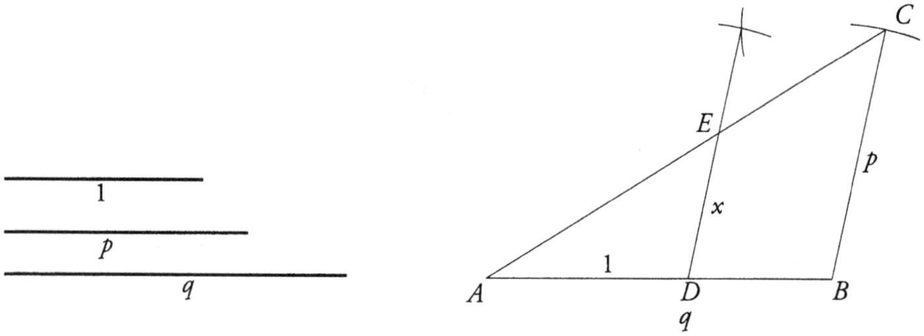

(1) Construct a segment AB equal in length to the given length q.

(2) Construct a segment BC equal in length to the given length p so that ABC is a triangle.

(3) On AB, cut off a segment AD of length 1.

(4) Through D, construct a line parallel to BC cutting AC at E; then DE has length p/q.

The proof that DE has length p/q follows from the fact that $\triangle ABC \sim \triangle ADE$.

□

Exercise 6.2.2. *Given segments of lengths* 1, p, *and* q, *construct a segment of length* pq.

Example 6.2.3. *Given segments of length* a *and* b, *show how to construct a segment of length* $\sqrt{ab}$.

Solution. The key to this is to somehow use Pythagoras' Theorem. It is actually used in conjunction with Thales' Theorem, which tells us that the angle inscribed in a semicircle is $90°$. Here is the analysis figure:

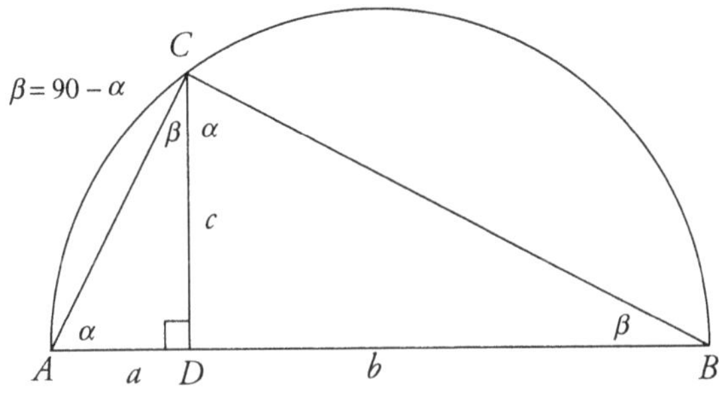

Since $\triangle ADC \sim \triangle CDB$, we have

$$\frac{DC}{AD} = \frac{DB}{CD},$$

so that

$$\frac{c}{a} = \frac{b}{c},$$

and it follows that $c = \sqrt{ab}$.

Thus the construction is as follows:

(1) Construct a segment AB of length $a + b$, with D the point on AB such that $AD = a$ and $DB = b$.

(2) Construct the semicircle with AB as diameter.

(3) Construct the perpendicular to AB at D cutting the semicircle at C; then CD has length $\sqrt{ab}$.

□

Note. Using the *power of a point* also yields the same construction:

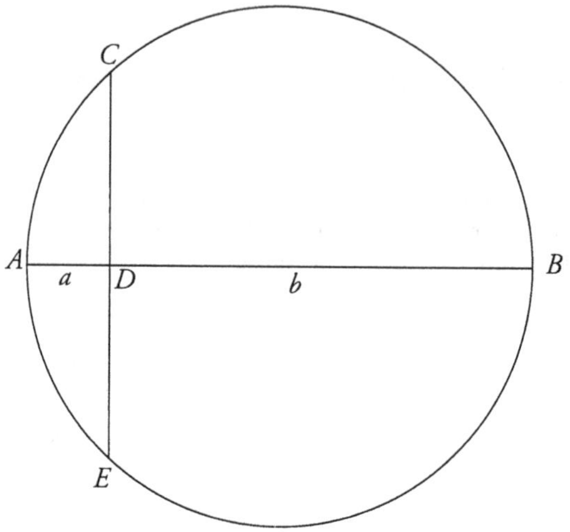

(1) Construct a segment AB of length $a + b$, with D the point on AB such that $AD = a$ and $DB = b$.

(2) Construct the circle with AB as diameter.

(3) Construct the chord CE perpendicular to AB at D; then CD has length $\sqrt{ab}$.

Since CE is perpendicular to the diameter AB, we must have $CD = DE$. By the power of the point D, we have $CD \cdot DE = AD \cdot DB$, that is, $CD = \sqrt{ab}$.

Constructible Numbers

The ancient Greeks described a number x as being ***constructible*** if, starting with a segment of length 1, you could construct a segment of length x. For example, $3/5$ is a constructible number: starting with a segment of length 1, construct segments of length 3 and 5, and then using Example 6.2.1, construct a segment of length $3/5$.

Combining Examples 6.2.1, 6.2.2, and 6.2.3 we have the following theorem:

Theorem 6.2.4. *(Constructible Numbers)*

If the nonnegative numbers a and b are constructible, then so are the following numbers:

$$a + b, \qquad a - b, \qquad \frac{a}{b} \ \ (\textit{if } b \neq 0), \qquad ab, \qquad \sqrt{a}.$$

We can build many constructible numbers by taking a succession of these operations. For example:

5 and 6 are constructible, so $\sqrt{5}$ and $\sqrt{6}$ are constructible.

3 and $\sqrt{5}$ are constructible, so $3+\sqrt{5}$ is constructible.

$3+\sqrt{5}$ is constructible, so $\sqrt{3+\sqrt{5}}$ is constructible.

4 and $\sqrt{6}$ are constructible, so $4\sqrt{6}$ is constructible.

$4\sqrt{6}$ and $\sqrt{3+\sqrt{5}}$ are constructible, so $\dfrac{4\sqrt{6}}{\sqrt{3+\sqrt{5}}}$ is constructible.

And so on.

Example 6.2.5. *Show that* $(1+\sqrt{2})^{1/4}$ *is constructible.*

Solution. The numbers 1 and $\sqrt{2}$ are constructible, so $1+\sqrt{2}$ is constructible, thus $(1+\sqrt{2})^{1/2}$ is also constructible and then so is $\left((1+\sqrt{2})^{1/2}\right)^{1/2}$. Since

$$(1+\sqrt{2})^{1/4} = \left((1+\sqrt{2})^{1/2}\right)^{1/2},$$

we are finished.

□

Starting with the number 1, and by taking a finite succession of additions, subtractions, ratios, products, and square roots, with repetitions allowed, we can obtain ***all*** of the constructible numbers. For example, the number

$$\sqrt{\frac{2}{3}+\frac{\sqrt{1+\sqrt{2}}}{\sqrt{5}}}$$

can be obtained in the way described, so it is constructible.

These are the only types of numbers that are constructible. This can be rather tricky, because there are many ways of expressing the same number. For example, at first glance

$$\left(\frac{2^6}{3^6}\right)^{1/3}$$

may not appear to be constructible, but it is constructible because

$$\left(\frac{2^6}{3^6}\right)^{1/3} = \frac{4}{9}.$$

Some numbers that are known ***not*** to be constructible are

$$\pi, \quad e, \quad \sqrt[3]{2}, \quad \cos 20^\circ.$$

This explains why the Three Problems of Antiquity cannot be solved.

Remark. To reiterate, a number a is ***constructible*** if and only if, given a segment of unit length, it is possible to construct a segment of length $|a|$ using only a straightedge and compass. It can be shown using Galois theory that the only numbers that are constructible are the following:

- Integers
- Rational numbers
- Square roots of rational numbers
- Sums, differences, ratios, and products of the above
- Sums, differences, ratios, and products of the above
- Sums, differences, ratios, and products of the above

6.3 Construction of Regular Polygons

Recall that a simple polygon is called a ***regular polygon*** if all of its sides are congruent and all of its vertex angles are congruent. This section deals with the problem of how to construct some of the regular polygons.

There is little difficulty in constructing the regular n-gon for $n = 3$ (an equilateral triangle), for $n = 4$ (a square), and for $n = 6$ (a regular hexagon).

In an earlier problem, we saw that all of the regular polygons have their vertices on a circle, so construction of the regular n-gon amounts to finding n equally spaced points on a circle. The general problem then becomes:

> Given a circle of radius 1, find n equally spaced points on that circle.

The situation for the pentagon is shown on the right.

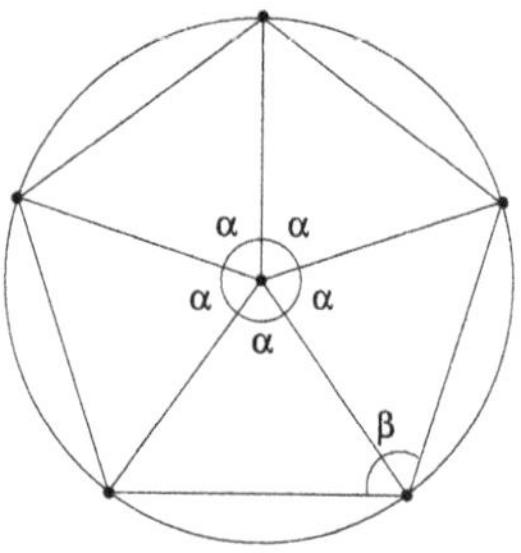

The ***central angle*** α is given by

$$\alpha = \frac{360}{n},$$

while the ***vertex angle*** β is given by

$$\beta = 180 - \alpha = 180\left(\frac{n-2}{n}\right),$$

since $\alpha + 2 \cdot (\beta/2) = 180$.

Construction Tips:

(i) We can construct a regular n-gon if and only if we can construct its central angle

$$\alpha = 360/n.$$

(ii) We can construct a regular n-gon if and only if we can construct the vertex angle

$$\beta = 180\left(\frac{n-2}{n}\right).$$

(iii) If we can construct a regular n-gon, then we can construct a regular $2n$-gon by bisecting the central angles. In general, we can construct a regular $2^k \cdot n$-gon by continually bisecting the central angles.

(iv) If we can construct a regular $2n$-gon, then we can construct a regular n-gon by joining alternate vertices.

It should be noted that nonsimple polygons can also have all of their sides congruent and all of their vertex angles congruent. The vertices for these polygons also lie on a circle. Such polygons are known as ***regular star polygons***.

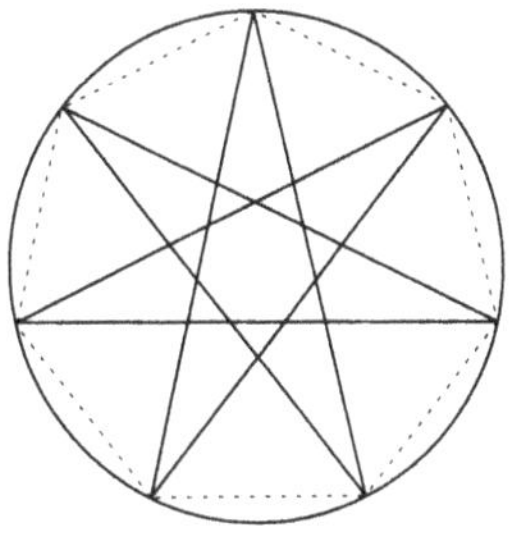

The regular $\begin{Bmatrix} n \\ d \end{Bmatrix}$ star polygon can be obtained from the regular n-gon by joining every dth point.

The regular $\begin{Bmatrix} 7 \\ 3 \end{Bmatrix}$ star polygon is shown on the right.

6.3.1 Construction of the Regular Pentagon

To construct the regular pentagon, we actually first construct the regular decagon and then join alternate vertices. The analysis figure for the regular decagon is as follows:

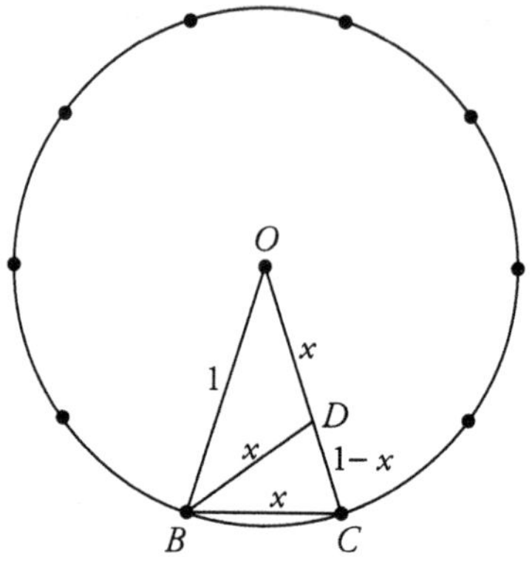

The circle is of radius 1 and center O, and there are 10 vertices, each at distance x from its immediate neighbours. If B and C are two consecutive vertices, then $\angle BOC = 360/10$; that is, in the isosceles triangle OBC, the angles are 36°, 72°, and 72°. Thus, if BD bisects the base angle OBC, we have $\angle OBD = 36$, so triangles OBD and CBD are isoceles. Thus, $OD = BD = BC = x$.

Since $\triangle OBC \sim \triangle BCD$ we have

$$\frac{BC}{OB} = \frac{CD}{BC},$$

so that

$$\frac{x}{1} = \frac{1-x}{x};$$

that is,

$$x^2 + x - 1 = 0.$$

Solving for x, we get the roots

$$x = \frac{\sqrt{5}-1}{2} \quad \text{and} \quad x = \frac{-\sqrt{5}-1}{2}.$$

The positive number

$$\frac{\sqrt{5}-1}{2}$$

is constructible, so given the radius 1 of the circle, we can construct the segment of length x and therefore we can construct the regular decagon and the regular pentagon.

Exercise 6.3.1. *Given a circle with radius $OP = 1$, construct the segment of length*

$$\frac{\sqrt{5}-1}{2}$$

and complete the construction of the regular pentagon.

Remark. The number

$$\frac{\sqrt{5}+1}{2}$$

(note the plus sign) is called the ***golden ratio*** or ***golden section*** or ***golden mean*** and is usually denoted by the Greek letter ϕ.

Note that

$$\frac{\sqrt{5}+1}{2} \cdot \frac{\sqrt{5}-1}{2} = 1,$$

so that

$$\frac{\sqrt{5}-1}{2}$$

is the reciprocal of ϕ. Note also that

$$\frac{\sqrt{5}+1}{2} \quad \text{and} \quad \frac{\sqrt{5}-1}{2}$$

differ by 1.

6.3.2 Construction of Other Regular Polygons

We can construct a regular n-gon for $n = 3, 4, 5$, and it is not very difficult to construct a $2n$-gon if we can construct an n-gon. For $n \leq 10$, this leaves $n = 7$ and $n = 9$. Unfortunately, we cannot construct these regular n-gons with straightedge and compass alone. In fact, there are very few n-gons that are constructible, and this section describes all of them. First, we need the following definitions.

A ***Fermat number*** is a number of the form $2^{2^n} + 1$, where $n \geq 0$.

A ***prime number*** is a positive integer $p > 1$ such that p has exactly two positive divisors, namely 1 and p.

A ***Fermat prime*** is a Fermat number that is also a prime number.

Here are the first few Fermat numbers:

n	F_n	$2^{2^n}+1$	Prime?
0	F_0	3	yes
1	F_1	5	yes
2	F_2	17	yes
3	F_3	257	yes
4	F_4	65,537	yes
5	F_5	4,294,967,297	no

Remark. As can be seen from the table, the first five Fermat numbers

$$F_0 = 3, \quad F_1 = 5, \quad F_2 = 17, \quad F_3 = 257, \quad F_4 = 65,537$$

are all primes, and this led Fermat to conjecture that every Fermat number F_n is a prime.

In 1732, almost 100 years later, Euler showed that this conjecture was false, and he gave the following counterexample:

$$F_5 = 4,294,967,297 = 641 \cdot 6,700,417.$$

Even today we do not know if there are an infinite number of Fermat primes. In fact, the only known Fermat primes are the ones in the table above.

In 1796, Gauss found what is probably the most important aspect of the Fermat numbers, the connection between the Fermat primes and the straightedge and compass construction of regular polygons. His result is as follows:

Theorem 6.3.2. *(Gauss' Theorem)*

A regular n-gon is constructible if and only if

$$n = 2^{k+2} \qquad or \qquad n = 2^k \cdot p_1 \cdot p_2 \cdots p_s,$$

where $k \geq 0$ and $p_1, p_2, \ldots, p_s$ are distinct Fermat primes.

The early Greeks knew how to construct regular polygons with 2^k, $3 \cdot 2^k$, $5 \cdot 2^k$, and $15 \cdot 2^k = 2^k \cdot 3 \cdot 5$ sides. They also knew how to construct regular polygons with 3, 4, 5, 6, 8, 10, 12, 15, and 16 sides, but not one with 17 sides. Gauss, however, did this at age 19 and so reportedly decided to devote the rest of his life to mathematics. He also requested that a 17-sided regular polygon be engraved on his tombstone (it wasn't).

Corollary 6.3.3. *The regular 7-gon and the regular 9-gon are not constructible.*

Proof. When $n = 7$, n is a prime, but not a Fermat prime. When $n = 9$, n is the product of Fermat primes since $n = 3 \cdot 3$, but it is not the product of *distinct* Fermat primes.

□

Example 6.3.4. *Is an angle of* 3° *constructible?*

Solution. This is the central angle formed by the edges of a 120-gon, since

$$n = \frac{360}{3} = 120.$$

The question amounts to asking whether we can construct a 120-gon.

Since

$$120 = 2^3 \cdot 3 \cdot 5,$$

and since 3 and 5 are distinct Fermat primes, the construction is possible.

□

6.4 Miquel's Theorem

Now we return to the ideas of concurrency and collinearity. First we note that two circles intersect in zero, one, or two points, or they coincide, as in the figure, and therefore:

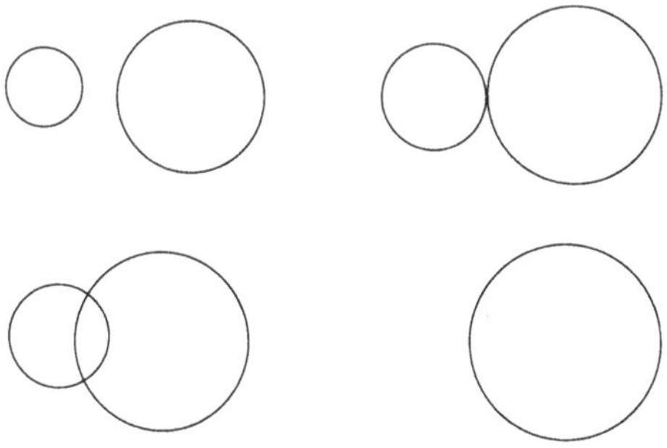

(i) If two circles have three distinct points of intersection, then they must coincide.

(ii) The circumcircle of any triangle is unique.

Theorem 6.4.1. *(Miquel's Theorem)*

Given $\triangle ABC$ and three menelaus points X, Y, and Z, one on each side (possibly extended) of the triangle, then the circles formed using a vertex and its two adjacent menelaus points are concurrent at a point M.

The point of concurrency is called the ***Miquel point***.

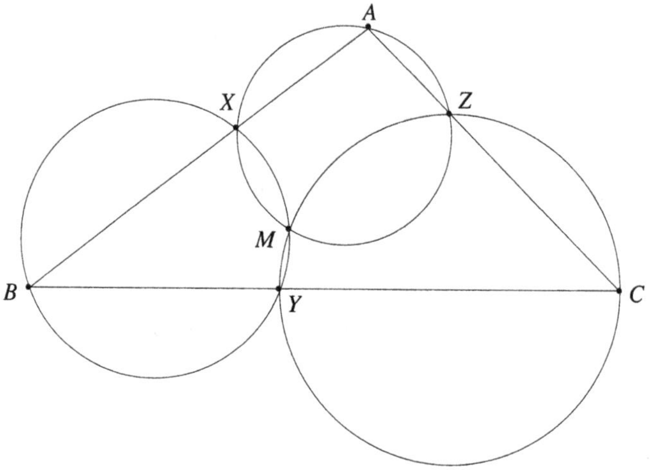

Proof. Let two of the circles have a second point of intersection M. We want to show that the third circle also goes through the point M.

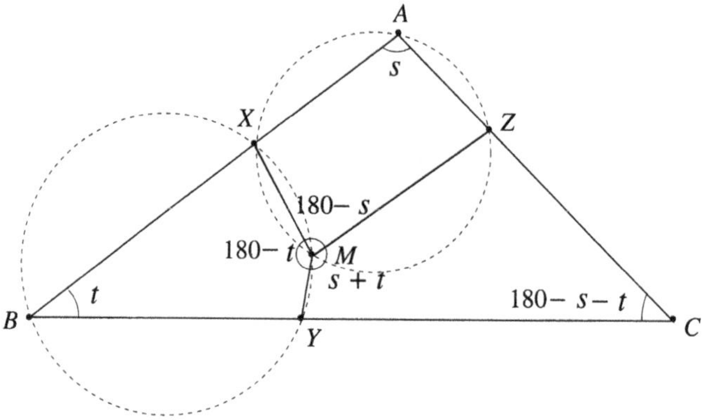

We know that the quadrilateral $AXMZ$ is cyclic and that $BYMX$ is also cyclic, and this implies that the angles are as shown in the figure.

Since

$$\angle YMZ = s + t \qquad \text{and} \qquad \angle YCZ = 180 - s - t,$$

then

$$\angle YMZ + \angle YCZ = s + t + 180 - s - t = 180,$$

and the quadrilateral $YCZM$ is also cyclic. Therefore, the circumcircle of $\triangle YZC$ also passes through M, and the three circles are concurrent at the Miquel point M.

□

Corollary 6.4.2. *Given $\triangle ABC$ and three menelaus points X, Y, and Z, one on each side (possibly extended) of the triangle, if X, Y, and Z are collinear, then the circumcircle of $\triangle ABC$ passes through the Miquel point M.*

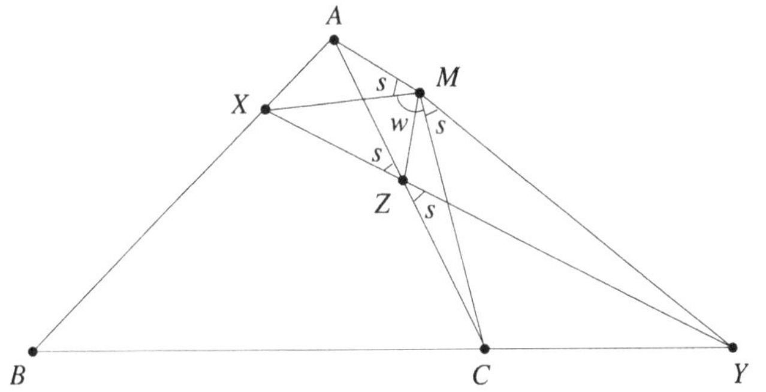

Proof. We recall that the Miquel point M lies on each of the circumcircles of $\triangle AXZ$, $\triangle BXY$, and $\triangle CYZ$, and therefore:

(1) Since $AXZM$ is cyclic, from Thales' Theorem we have

$$\angle AMX = \angle AZX = s.$$

(2) Since $BXMY$ is cylic, then

$$\angle XBY + \angle XMY = 180.$$

(3) Since $CZMY$ is cyclic, from Thales' Theorem we have

$$\angle CMY = \angle CZY = s.$$

Therefore, if we let $\angle XMC = w$, then

$$\angle B + \angle XMY = 180;$$

that is,

$$\angle B + w + s = 180,$$

so that

$$\angle B + \angle AMC = 180,$$

and thus $ABCM$ is cyclic.

□

The converse is also true.

Corollary 6.4.3. *Given $\triangle ABC$ and three menelaus points X, Y, and Z, one on each side (possibly extended) of the triangle, if the circumcircle of $\triangle ABC$ also goes through the Miquel point M, then the three menelaus points X, Y, and Z must be collinear.*

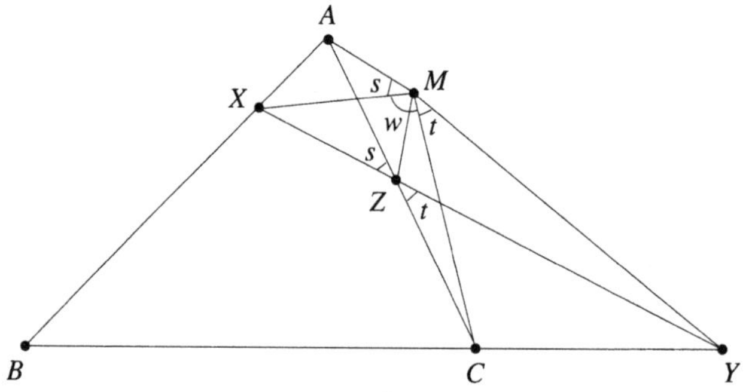

Proof. In the figure we let $\angle XMC = w$. Now:

(1) Since M is on the circumcircle of $\triangle ABC$, then $ABCM$ is cyclic.

(2) Since M is on the circumcircle of $\triangle AXY$, then $AXZM$ is cyclic.

(3) Since M is on the circumcircle of $\triangle CZY$, then $CZMY$ is cyclic.

(4) Since M is on the circumcircle of $\triangle BXY$, then $BXMY$ is cyclic.

From (2), using Thales' Theorem, we have

$$\angle AZX = \angle AMX = s,$$

and from (3), using Thales' Theorem, we have

$$\angle CZY = \angle CMY = t,$$

and to show that X, Y, and Z are collinear, it is enough to show that $s = t$.

However, from (1), since $ABCM$ is cyclic, the opposite angles are supplementary, so that

$$\angle B + w + s = 180,$$

while from (4), since $BXMY$ is cyclic, the opposite angles are supplementary, so that

$$\angle B + w + t = 180.$$

Therefore $s = t$, and we are done.

□

Example 6.4.4. *Given $\triangle ABC$ and three menelaus points X, Y, and Z, one on each side, where M is the Miquel point, as in the figure, show that if P, Q, and R are the centers of the circumcircles of $\triangle AXZ$, $\triangle BXY$, and $\triangle CYZ$, respectively, then $\triangle PQR$ is similar to $\triangle ABC$.*

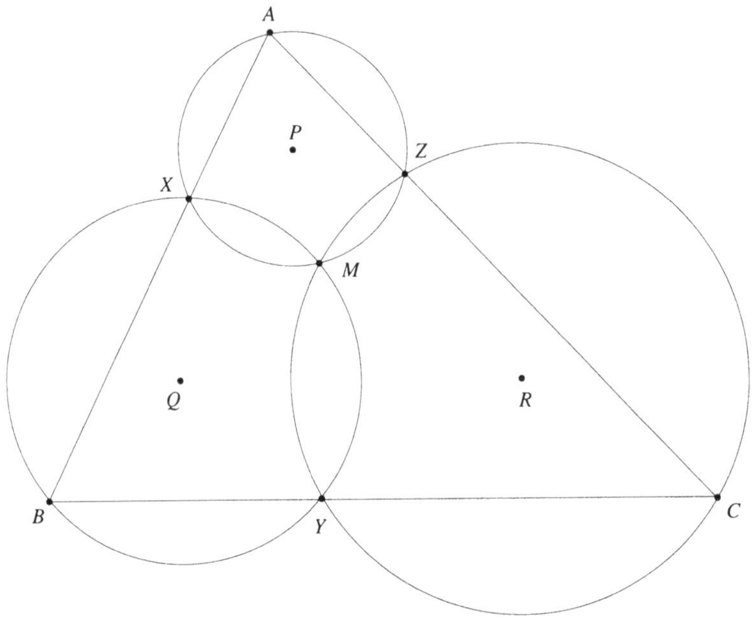

Solution. First, observe that quadrilateral $XPMQ$ is a kite. Let ω be the circumcircle of triangle BXY and draw the common chords $\overline{XM}$, $\overline{YM}$, and $\overline{ZM}$. Let the side $\overline{PQ}$ meet ω at S and the side $\overline{QR}$ meet ω at T, as in the figure on the following page.

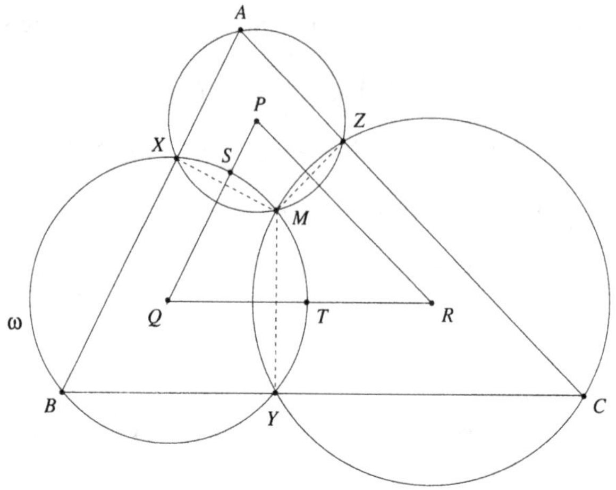

Since the line joining the centers of two circles is the perpendicular bisector of their common chord, $\overline{PQ}$ is the perpendicular bisector of $\overline{XM}$ and therefore $\overline{PQ}$ also bisects $\angle XQM$, so that

$$\angle XQS = \angle MQS.$$

Similarly, $\overline{QR}$ bisects $\angle MQY$, so that

$$\angle MQT = \angle TQY.$$

Since

$$\angle SQT = \angle SQM + \angle MQT = \frac{1}{2}\angle XQY,$$

and from Thales' Theorem we have

$$\angle XBY = \frac{1}{2}\angle XQY,$$

then

$$\angle B = \angle XBY = \frac{1}{2}\angle XQY = \angle SQT = \angle Q.$$

Similarly,

$$\angle A = \angle P \qquad \text{and} \qquad \angle C = \angle R,$$

and therefore $\triangle PQR \sim \triangle ABC$.

□

Example 6.4.5. *Prove Johnson's Theorem (1916).*

Given three circles concurrent at O, all with the same radius r, as in the figure below, then the circumcircle of the other three intersection points A, B, and C has radius r also.

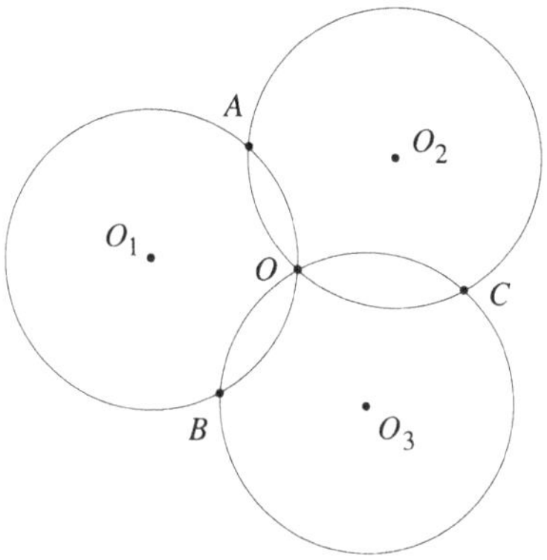

Solution. Let O_1, O_2, and O_3 be the centers of the three circles. Then the quadrilaterals

$$AO_1OO_2, \qquad O_1OO_3B, \qquad CO_2OO_3$$

are rhombii, since all the sides have length r, as in the figure below.

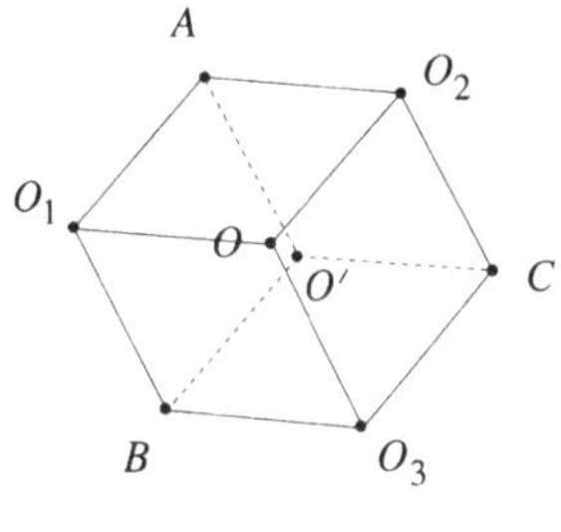

Let AO_1BO' be the completed parallelogram of triangle AO_1B, which is in fact a rhombus. Since

$$\overline{BO'} \parallel \overline{O_1A} \parallel \overline{OO_2} \parallel \overline{O_3C} \qquad \text{and} \qquad BO' = O_1A = OO_2 = O_3C,$$

then $BO'CO_3$ is also a rhombus, and therefore the circumcircle of triangle ABC is $\mathcal{C}(O', r)$.

Note. In general, the center O' of the circumcircle of ABC is different from O.

□

6.5 Morley's Theorem

The following result was discovered by Frank Morley in about 1900. He mentioned it to friends in Cambridge and published it about 20 years later in Japan.

Morley's Theorem states that the points of intersection of the adjacent trisectors of the angles of any triangle are the vertices of an equilateral triangle, as in the figure below.

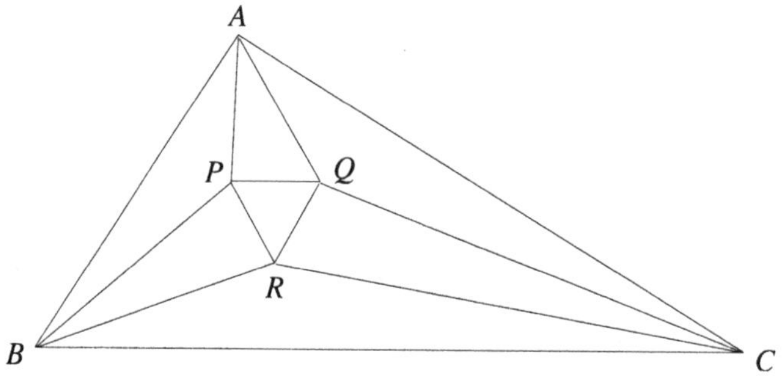

Before proving this theorem we need a lemma, which is yet another characterization of the incenter of a triangle.

Lemma 6.5.1. *(Another Characterization of the Incenter)*

The incenter of a triangle $\triangle ABC$ is the unique point I interior to the triangle which satisfies the following two properties:

(1) it lies on an angle bisector of one of the angles (say at A) and

(2) it subtends an angle $90 + \frac{1}{2}\angle A$ with the side BC.

Proof. The incenter has property (1) by definition, since it is the intersection of the internal angle bisectors of the triangle.

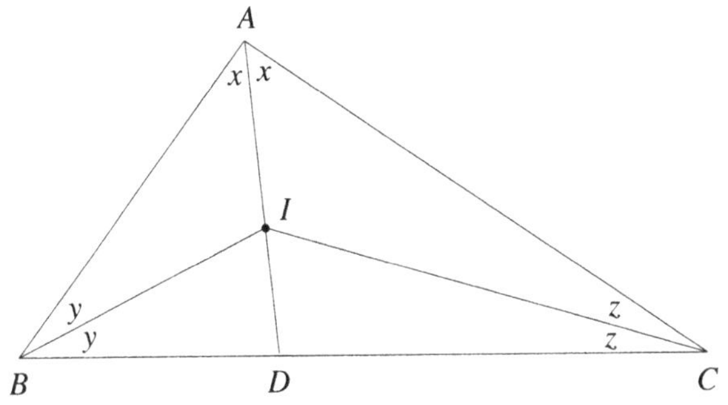

Also, from the External Angle Theorem, we have

$$\begin{aligned}\angle BIC &= \angle BID + \angle CID \\ &= x + y + x + z \\ &= x + \frac{1}{2}(2x + 2y + 2z) \\ &= x + \frac{1}{2}180 \\ &= 90 + x \\ &= 90 + \frac{1}{2}\angle A,\end{aligned}$$

and property (2) holds also.

To prove uniqueness, suppose that the point $I' \neq I$ lies on the angle bisector of $\angle A$, and suppose that I' also subtends an angle $90 + \frac{1}{2}\angle A$ with the side BC.

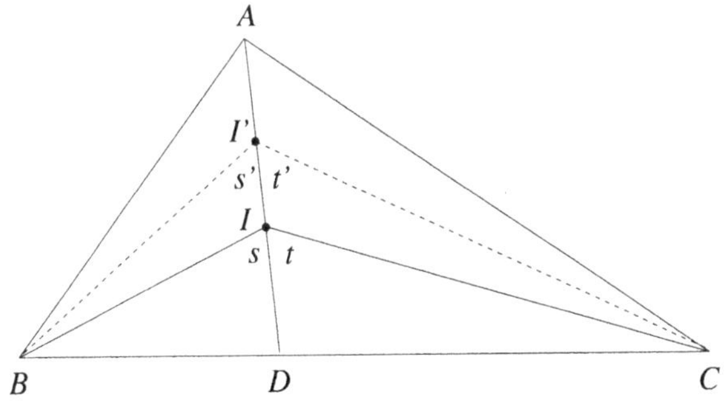

Note that in the figure on the previous page, where I' is between A and I, the External Angle Inequality implies that

$$s > s' \qquad \text{and} \qquad t > t',$$

so that

$$90 + \frac{1}{2}\angle A = \angle BIC = s + t > s' + t' = \angle BI'C = 90 + \frac{1}{2}\angle A,$$

which is a contradiction. Similarly, if I is between A and I', we again get a contradiction. Therefore, the point I satisfying (1) and (2) is unique.

□

Theorem 6.5.2. *(Morley's Theorem)*

The points of intersection of adjacent trisectors of the angle of any triangle form an equilateral triangle.

Proof. Let $\triangle ABC$ be a fixed triangle, so that the angles at the vertices A, B, and C are fixed angles. It is enough to prove Morley's Theorem for a triangle $\triangle A'B'C'$ that is similar to the given triangle, since scaling the sides by a proportionality factor k does not change the angles, and so an equilateral triangle remains equilateral.

In the proof we start with an equilateral triangle $\triangle PQR$ and then construct a triangle $\triangle A'B'C'$ that is similar to $\triangle ABC$ and has the property that the adjacent trisectors form the given equilateral triangle $\triangle PQR$.

Construction Phase

Step 1. Construct an equilateral triangle $\triangle PQR$.

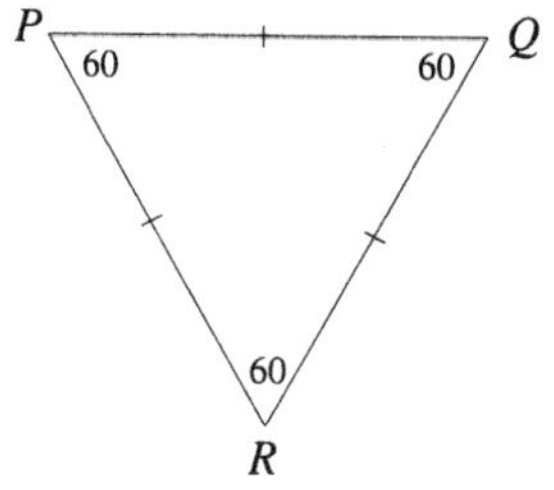

Step 2. Construct three isosceles triangles on the sides of $\triangle PQR$, with angles as shown,

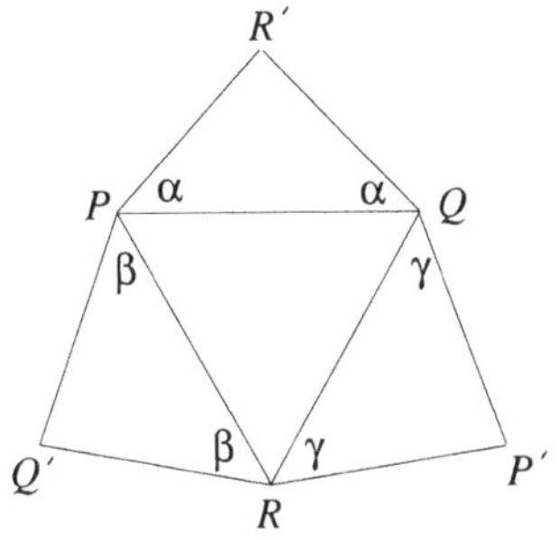

where

$$\alpha = 60 - \frac{1}{3}\angle A,$$
$$\beta = 60 - \frac{1}{3}\angle B,$$
$$\gamma = 60 - \frac{1}{3}\angle C.$$

Observations:

(1) $0 < \alpha, \beta, \gamma < 60$.

Since, for example, $\angle A > 0$ implies that $60 > 60 - \frac{1}{3}\angle A = \alpha$, while $\angle A < 180$ implies that $\alpha = 60 - \frac{1}{3}\angle A > 60 - \frac{1}{3}180 = 60 - 60 = 0$.

(2) $\alpha + \beta + \gamma = 120$.

Since $\alpha+\beta+\gamma = 180 - \frac{1}{3}(\angle A + \angle B + \angle C) = 180 - \frac{180}{3} = 180 - 60 = 120$.

(3) The sum of any two of the angles α, β, γ is greater than 60.

Since, for example,

$$\alpha + \beta = 120 - \tfrac{1}{3}(\angle A + \angle B) > 120 - \tfrac{180}{3} = 120 - 60 = 60.$$

Step 3. Extend the sides of the isosceles triangles until they meet as shown to produce a larger triangle $\triangle A'B'C'$.

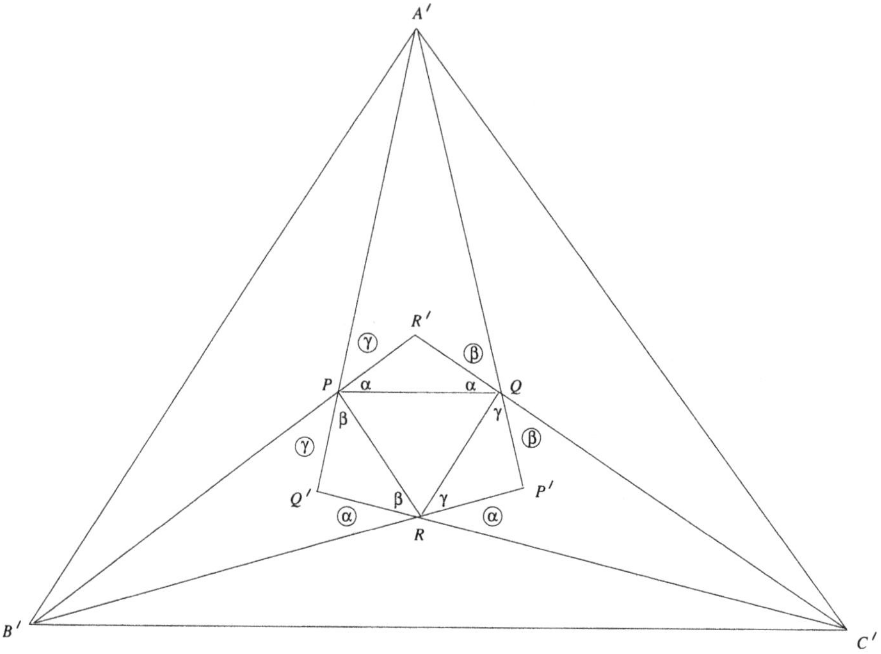

We claim that $\triangle A'B'C'$ is similar to $\triangle ABC$ and that the lines

$$A'P, \quad A'Q, \quad B'P, \quad B'R, \quad C'Q, \quad C'R$$

are the angle trisectors of the angles at A', B', and C'.

Argument Phase

Step 1. $\alpha + \beta + \gamma + 60 = 180$, so that the angles are as shown on the figure.

For example, the angle $\angle B'PR'$ is a straight angle and $\angle QPR = 60$ so that $\alpha + 60 + \beta + \angle B'PQ' = 180$, which implies that $\gamma = \angle B'PQ'$ and that the vertically opposite angle $\angle A'PR' = \gamma$ also.

We should also show that the extensions of the sides of the isosceles triangles actually do intersect at the points A', B', and C'.

For example, if we isolate part of the figure, we can show that QR' and $Q'R$ intersect at a point C', as shown in the figure on the following page.

If we consider the sum $\angle RQR' + \angle Q'RQ$, then

$$\angle RQR' + \angle Q'RQ = \alpha + 60 + \beta + 60 = \alpha + \beta + 120 > 60 + 120 = 180,$$

and therefore the parallel postulate says that RQ' and $R'Q$ intersect on the side of the transversal QR where the sum of the interior angles is less than 180; that is, RQ' and $R'Q$ intersect on the side opposite P at some point C', as shown.

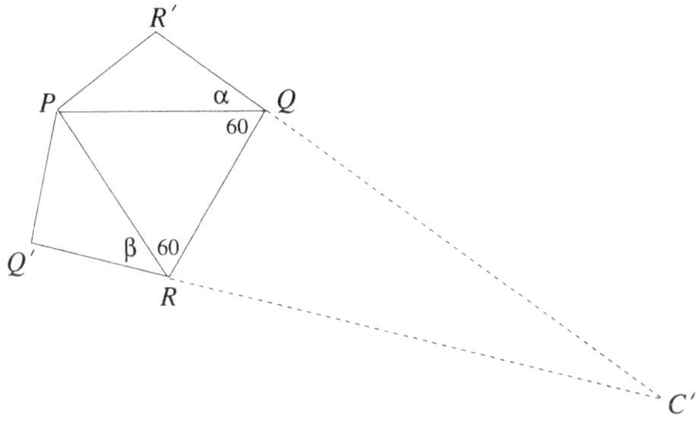

Similarly, PQ' and $P'Q$ intersect at some point A', while PR' and $P'R$ intersect at some point B', as shown.

Step 2. $\alpha + \beta + \gamma = 120$, and therefore

$$\angle PA'Q = 60 - \alpha = \frac{1}{3}\angle A,$$
$$\angle PB'R = 60 - \beta = \frac{1}{3}\angle B,$$
$$\angle QC'R = 60 - \gamma = \frac{1}{3}\angle C.$$

For example, in $\triangle PB'R$, the sum of the interior angles is

$$\angle PB'R + \gamma + \beta + \beta + \alpha = 180,$$

so that

$$\angle PB'R = 180 - \beta - (\alpha + \beta + \gamma) = 180 - \beta - 120 = 60 - \beta,$$

and

$$\angle PB'R = 60 - \beta = \frac{1}{3}\angle B.$$

Similarly,

$$\angle PA'Q = 60 - \alpha = \frac{1}{3}\angle A \quad \text{and} \quad \angle QC'R = 60 - \gamma = \frac{1}{3}\angle C.$$

Step 3. R is the incenter of $\triangle B'R'C'$. Similarly, P is the incenter of $\triangle A'B'P'$, while Q is the incenter of $\triangle A'C'Q'$.

We will show that R is the incenter of $\triangle B'R'C'$. The other two results follow in the same way.

(a) Note that R lies on the angle bisector of $\angle B'R'C'$ since $\triangle PR'R$ is congruent to $\triangle QR'R$ by the **SSS** congruency theorem.

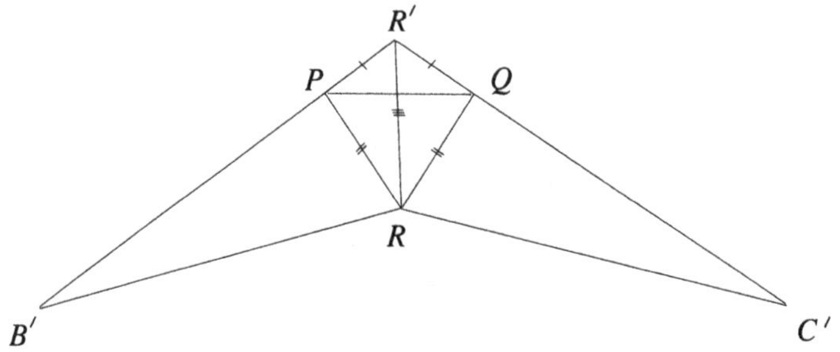

(b) Note that

$$\angle B'RC' = 180-\alpha = 90+(90-\alpha) = 90+\frac{1}{2}(180-2\alpha) = 90+\frac{1}{2}\angle B'R'C'.$$

By the characterization theorem for the incenter proven in the lemma, (1) and (2) imply that R is the incenter of $\triangle B'R'C'$.

Therefore,

$$\angle PB'R = \angle RB'C' = \frac{1}{3}\angle B$$

so that PB' and RB' are angle trisectors of $\angle B'$. Similarly, PA' and QA' are angle trisectors of $\angle A'$, and QC' and RC' are angle trisectors of $\angle C'$.

Therefore, $\angle A' = \angle A$, $\angle B' = \angle B$, and $\angle C' = \angle C$, and by the **AAA** similarity theorem, $\triangle A'B'C'$ is similar to $\triangle ABC$, and the corresponding segments in $\triangle ABC$ are the angle trisectors. Therefore, in $\triangle ABC$ the points of intersection of adjacent trisectors of the angles form an equilateral triangle.

□

6.6 The Nine-Point Circle

In any triangle $\triangle ABC$, the following nine points all lie on a circle, called the ***9-point circle***, and they occur naturally in three groups.

(a) The three feet of the altitudes: F_1, F_2, F_3.

(b) The three midpoints of the sides: M_1, M_2, M_3.

(c) The three midpoints of the segments from the vertices to the orthocenter: O_1, O_2, O_3.

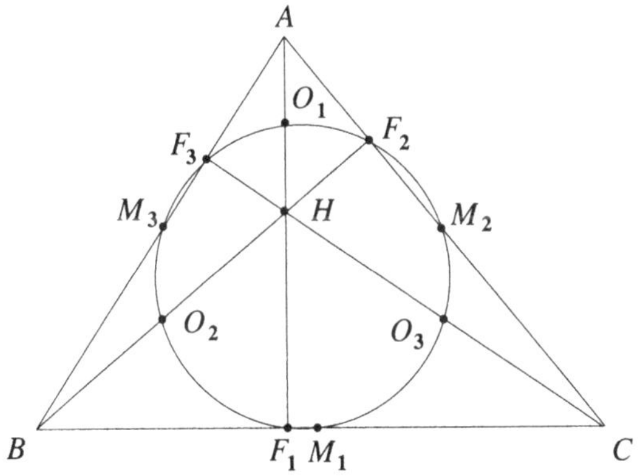

Note. In general, the orthocenter H is *not* the center of the 9-point circle.

Recall the following facts:

Fact 1. In a rectangle,

(a) the diagonals have the same length and

(b) the diagonals bisect each other.

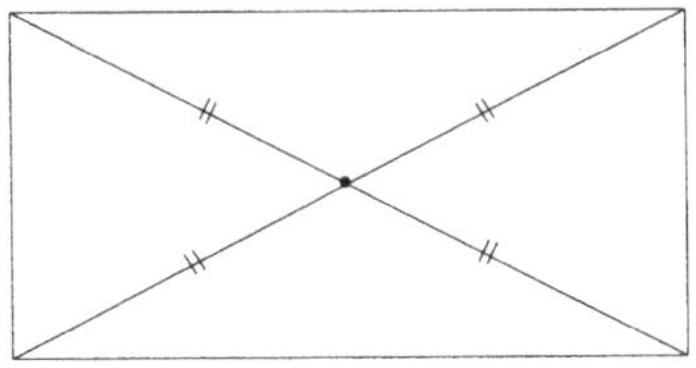

Conclusion. The center of the circumcircle of a rectangle is the intersection of the diagonals and is also the midpoint of either diagonal.

Fact 2. The Thales' Locus of points subtending an angle of 90° with a segment $\overline{AB}$ is exactly the circle with $\overline{AB}$ as a diameter.

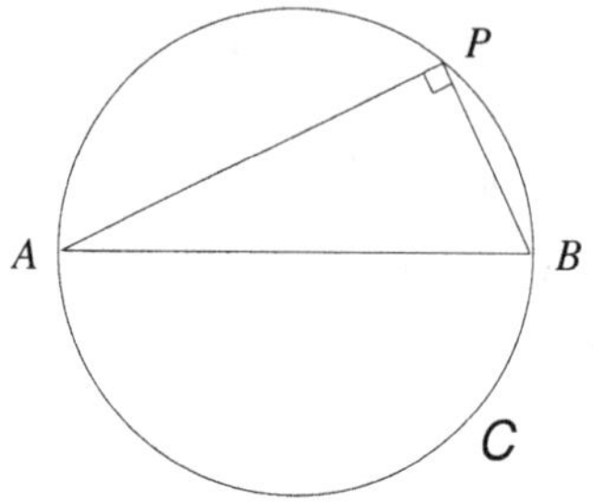

Conclusion. If a point P subtends an angle of 90° with any diameter of a circle $\mathcal{C}$, then P must be on the circle.

The 9-point circle theorem was first proved by Feuerbach (1800 – 1834).

Theorem 6.6.1. *(Feuerbach's Theorem)*

In any triangle ABC, the following nine points all lie on a circle, called the **9-point circle***:*

(i) The three feet of the altitudes: F_1, F_2, F_3.

(ii) The three midpoints of the sides: M_1, M_2, M_3.

(iii) The three midpoints of the segments from the vertices to the orthocenter: O_1, O_2, O_3.

The center of the 9-point circle is denoted by N.

Proof. In the figure on the following page, the following constructions have been performed.

Step 1.

Join M_3 to M_2, so that M_2M_3 is parallel to BC, and $M_2M_3 = \frac{1}{2}BC$ by the Midline Theorem.

Join O_2 to O_3, so that O_2O_3 is parallel to BC, and $O_2O_3 = \frac{1}{2}BC$ by the Midline Theorem.

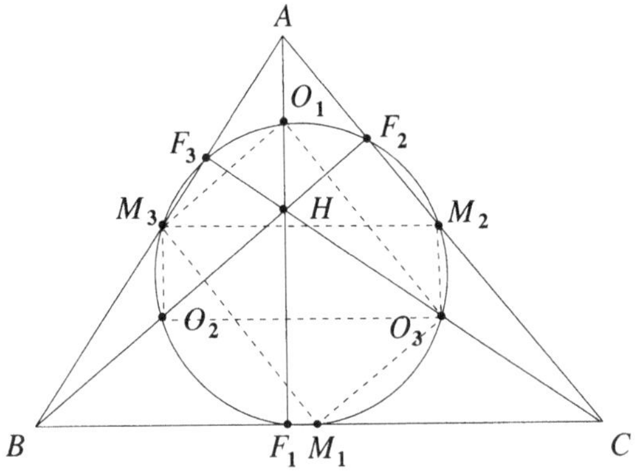

Step 2.

Join M_3 to O_2, so that M_3O_2 is parallel to AH, and $M_3O_2 = \frac{1}{2}AH$ by the Midline Theorem.

Join M_2 to O_3, so that M_2O_3 is parallel to AH, and $M_2O_3 = \frac{1}{2}AH$ by the Midline Theorem.

Since AH is an altitude and is perpendicular to the side BC, then $M_3M_2O_3O_2$ is a rectangle. Similarly, $M_3M_1O_3O_1$ is a rectangle.

Now note the following:

(1) M_3O_3 is a common diagonal of both rectangles so that the circumcircles of both rectangles coincide. This means that M_1, M_2, M_3 and O_1, O_2, O_3 all lie on the same circle $\mathcal{C}$.

(2) M_3O_3, M_2O_2, and M_1O_1 are all diameters of $\mathcal{C}$.

$$(3) \begin{cases} F_1 \quad \text{is on} \quad \mathcal{C} \quad \text{since} \quad \angle M_1F_1O_1 = 90^\circ, \\ F_2 \quad \text{is on} \quad \mathcal{C} \quad \text{since} \quad \angle M_2F_2O_2 = 90^\circ, \\ F_3 \quad \text{is on} \quad \mathcal{C} \quad \text{since} \quad \angle M_3F_3O_3 = 90^\circ. \end{cases}$$

□

Note. In the proof, we joined

$$O_2 M_3, \quad O_3 M_2, \quad O_2 O_3, \quad M_2 M_3$$

and

$$O_1 O_3, \quad O_1 M_3, \quad M_3 M_1, \quad M_1 O_3$$

to get rectangles with common diagonals.

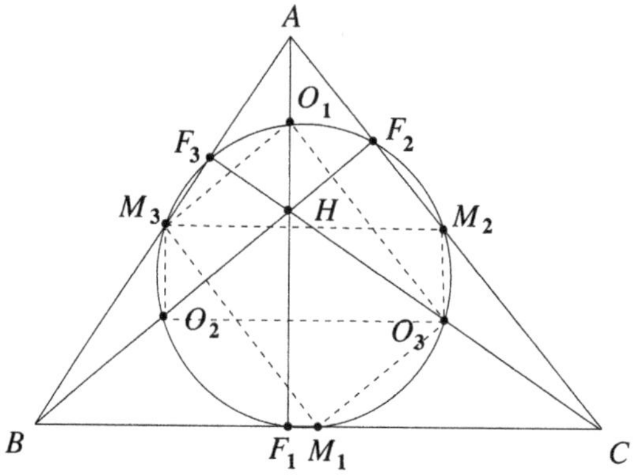

The segments O_1M_1, O_2M_2, and O_3M_3 are diameters of the 9-point circle and are also the diagonals of these rectangles. Thus, the point N, the center of the 9-point circle, is the midpoint of each of these segments.

6.6.1 Special Cases

Isosceles Triangle

For an isosceles triangle which is not an equilateral triangle, only eight of the nine points on the 9-point circle are distinct.

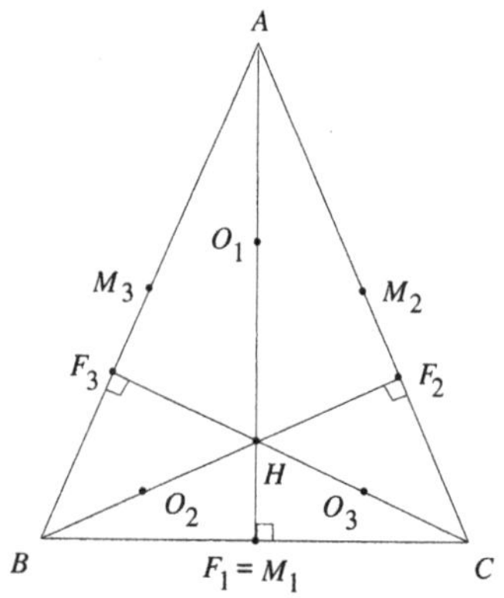

Equilateral Triangle

For an equilateral triangle, only six of the nine points on the 9-point circle are distinct. The center N of the 9-point circle is H, and the circle is just the *incircle*.

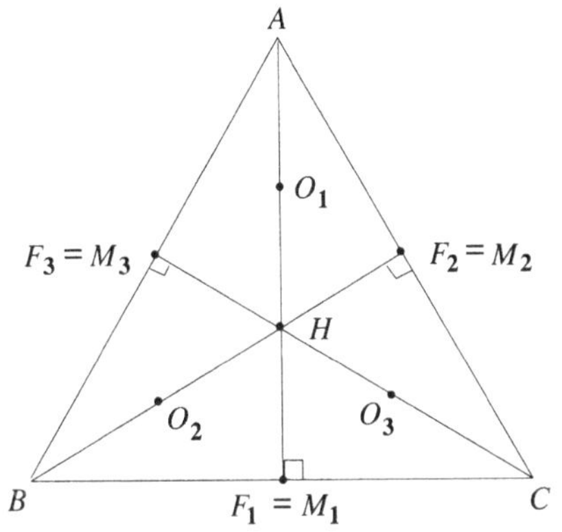

Right-Angled Isosceles Triangle

For a right-angled isosceles triangle, as below, where $AB = BC$ and $\angle B$ is a right angle, only four of the nine points on the 9-point circle are distinct.

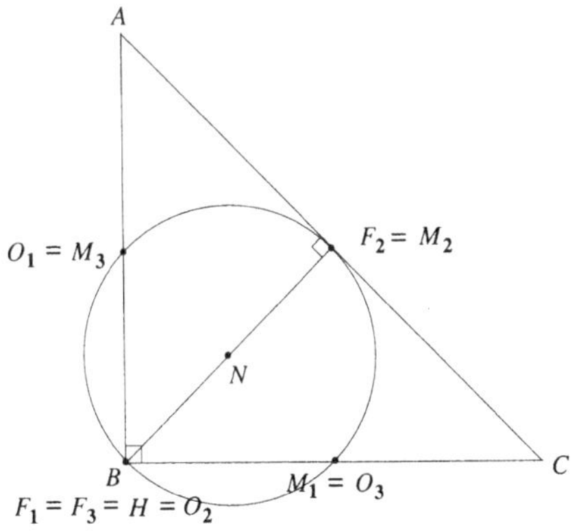

The Euler Line

Theorem 6.6.2. *(Euler Line)*

Given a nonequilateral triangle ABC, the circumcenter S, the centroid G, and the orthocenter H are collinear and form the **Euler line**.

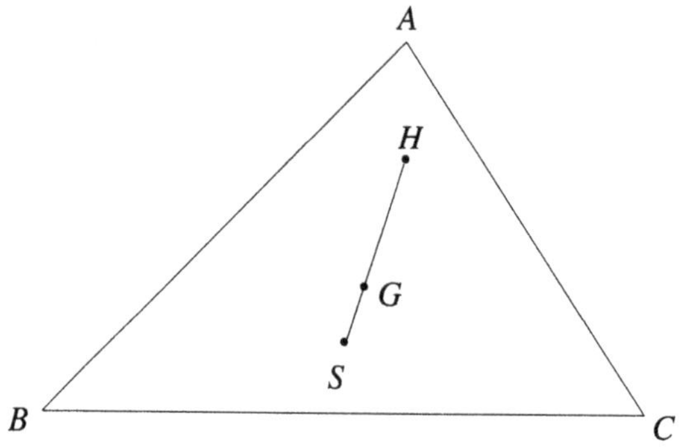

In fact, G is a trisection point of HS; that is, G is between H and S and $GH = 2GS$.

Note. If $\triangle ABC$ is equilateral, then $S = G = H$, and conversely, if $S = G = H$, then $\triangle ABC$ is equilateral.

Proof. Since $\triangle ABC$ is nonequilateral, then $G \neq S$. Extend SG to SO, with $S - G - O$ and $GO = 2GS$. If we can show that $O = H$, then we are done.

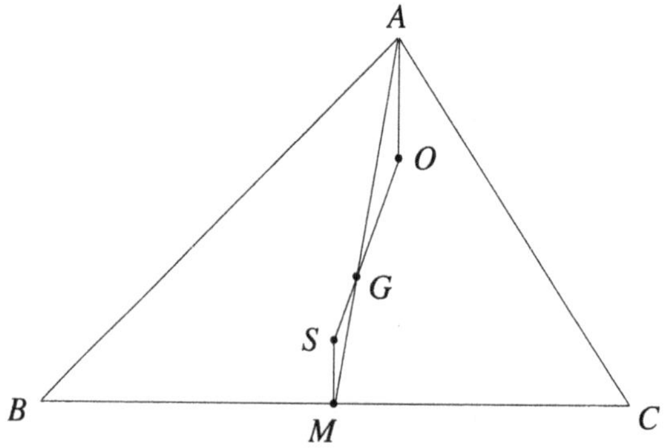

Let M be the midpoint of BC. Since S is the circumcenter of $\triangle ABC$, then SM is perpendicular to BC.

Now join A and M. Then AM is a median and so passes through G. Next, join A to O, and by construction $GO = 2GS$.

Also, since G is the centroid of $\triangle ABC$ and AM is a median, $GA = 2GM$, and since vertically opposite angles are equal, $\angle AGO = \angle MGS$.

By the **sAs** similarity theorem, $\triangle AGO \sim \triangle MGS$, with proportionality constant $k = 2$, so that $SG = 2GO$ and $AO = 2SM$.

Now $\angle SMG = \angle OAG$, and the alternate interior angles formed by the transversal AM of SM and AO are equal so that SM is parallel to AO, and if AO is extended it hits BC at $90°$. Thus, the altitude from A goes through O. Similarly, the altitudes from B and C also pass through O, and $O = H$.

We note for future reference that $AH = 2SM$.

□

Theorem 6.6.3. *The center N of the 9-point circle of $\triangle ABC$ lies midway between the circumcenter S and the orthocenter H.*

Note. This gives us four special points on the Euler line

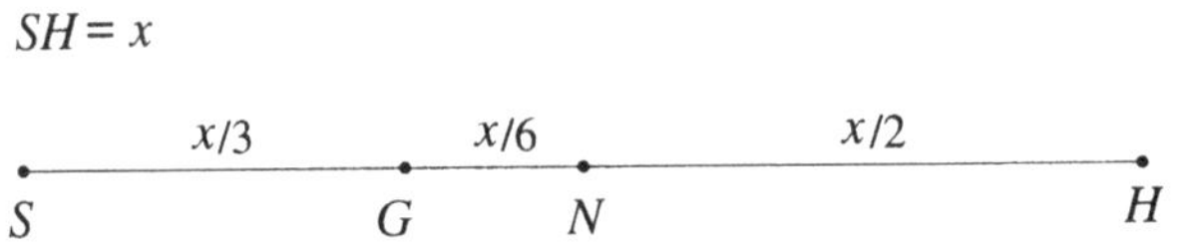

and we can compute all of these distances.

Proof. Note that if the triangle is equilateral, then $S = N = G = H$ and there is nothing to prove.

Suppose that $\triangle ABC$ is nonequilateral, so that $S \neq H$. Introduce the midpoint M_1 of BC so that SM_1 is perpendicular to BC.

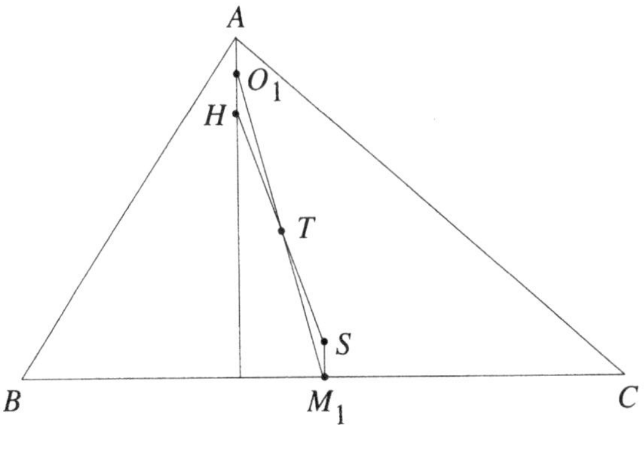

Join AH and let O_1 be the midpoint of AH, and then join O_1M_1 intersecting SH at T. Since H is the orthocenter and AH (extended) is perpendicular to BC, then AH is parallel to SM_1.

Since SH and O_1M_1 are criss-crossing transversals of the parallel lines AH and SM_1, then

$$\triangle STM_1 \sim \triangle HTO_1.$$

However,

$$AH = 2SM_1 \qquad \text{and} \qquad O_1H = \tfrac{1}{2}AH = SM_1,$$

so the proportionality constant is 1, and

$$\triangle STM_1 \equiv \triangle HTO_1.$$

Therefore, $ST = TH$, and T is the midpoint of SH.

Also, $TO_1 = TM_1$ and T is also the midpoint of O_1M_1, but O_1M_1 is a diameter of the 9-point circle, and therefore $T = N$, the center of the 9-point circle.

□

Theorem 6.6.4. *The radius of the* 9*-point circle is one half the radius of the circumcircle.*

Proof. In the figure below, NO_1 is the radius of the 9-point circle, AS is the radius of the circumcircle, and

$$NO_1 = \tfrac{1}{2}AS$$

by the Midline Theorem.

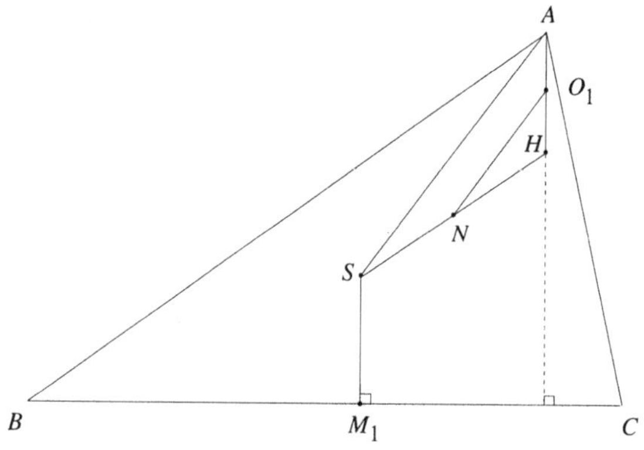

□

Theorem 6.6.5. *The* 9*-point circle of* $\triangle ABC$ *bisects every segment connecting the orthocenter* H *to a point* P *on the circumcircle.*

Proof. In the figure below, let M be the midpoint of PH. We will show that the 9-point circle passes through M.

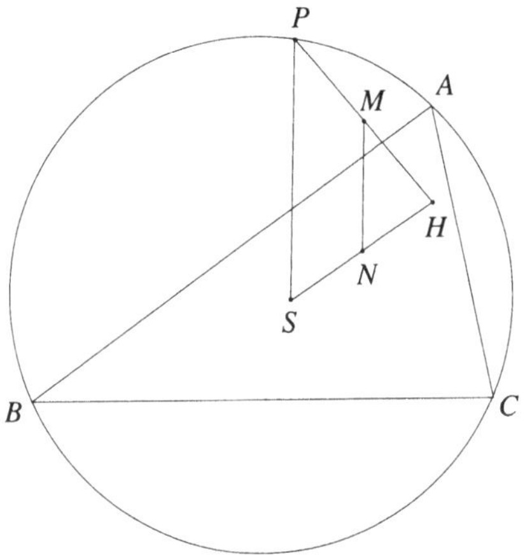

By the Midline Theorem, $MN = \frac{1}{2}PS = \frac{1}{2}R$, where R is the radius of the circumcircle, but MN is the radius of the 9-point circle by the previous theorem, so M is on the 9-point circle.

□

6.7 The Steiner-Lehmus Theorem

We note the following two theorems.

Theorem 6.7.1. *If the altitudes of a triangle are congruent, then the triangle is isosceles.*

Proof. In the figure on the following page, if $BD = EC$, then $\triangle BDC \equiv \triangle CEB$ by the **HSR** congruence theorem.

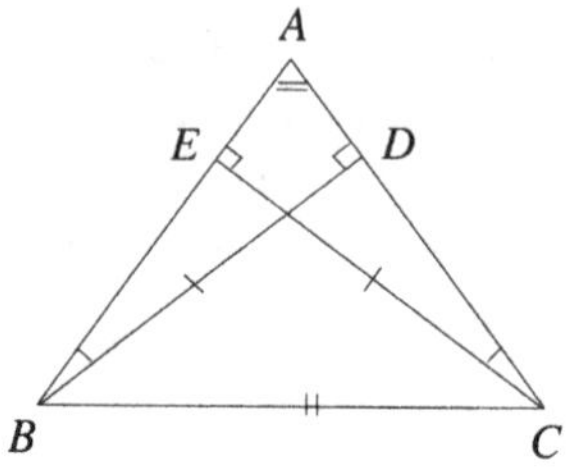

Therefore, $\triangle ADB \equiv \triangle AEC$ by the **ASA** congruence theorem, so that $AB = AC$.

Therefore, $\angle B = \angle C$ and $\triangle ABC$ is isosceles.

□

Theorem 6.7.2. *If two medians of a triangle are congruent, then the triangle is isosceles.*

Proof. In the figure below, G is the centroid of $\triangle ABC$ and medians BD and CE are equal.

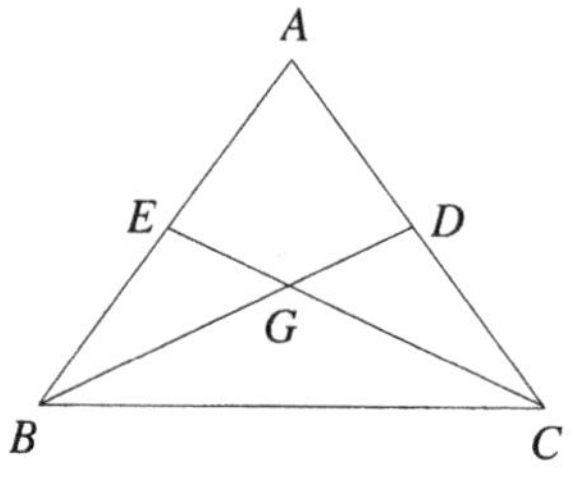

Since G is the centroid, we have

$$BG = \tfrac{2}{3}BD = \tfrac{2}{3}CE = CG$$

and

$$DG = \tfrac{1}{3}BD = \tfrac{1}{3}CE = EG,$$

and since $\angle EGB = \angle DGC$, then by the **SAS** congruency theorem we have

$$\triangle EGB \equiv \triangle DGC.$$

Therefore, $BE = DC$ and $2BE = 2DC$; that is, $AB = AC$, so $\triangle ABC$ is isosceles.

□

Thus, we see that if two altitudes of a triangle are congruent, or if two medians of a triangle are congruent, then the triangle must be isosceles.

In 1880, Lehmus conjectured that the result was also true for the angle bisectors, and in about 1884 Steiner proved that Lehmus' conjecture was true. Today almost 60 proofs have been given. We give below one of the simplest proofs.

Theorem 6.7.3. *(Steiner-Lehmus Theorem)*

If two internal angle bisectors of a triangle are congruent, then the triangle is isosceles.

Proof. In the figure below, let BD and EC be the internal angle bisectors at B and C, respectively, and suppose that $BD = CE$ but that $\triangle ABC$ is not isosceles, so that $x < y$.

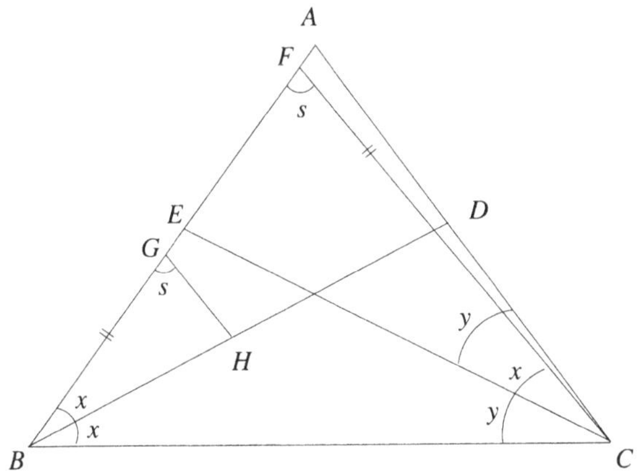

Transfer x to C, as shown. Then by the Angle-Side Inequality applied to $\triangle BFC$, since $2x < x + y$, we have $FC < FB$.

Now transfer CF to BG as shown, and draw GH, making an angle $s = \angle EFC$ at G. Then by the **ASA** congruency theorem, we have

$$\triangle BGH \equiv \triangle CFE,$$

so that $BH = CE$. However, $BH < BD$, which implies that $CE = BH < BD$, a contradiction.

Similarly, $x > y$ is impossible. Hence $x = y$, and $\angle B = \angle C$, so that $\triangle ABC$ is isosceles.

□

Example 6.7.4. *Let ABC be a triangle, and let L and M be points on $\overline{AB}$ and $\overline{AC}$, respectively, such that $AL = AM$. Let P be the intersection of $\overline{BM}$ and $\overline{CL}$. Prove that $PB = PC$ if and only if $AB = AC$.*

Solution. Suppose that $AB = AC$, so that $\triangle ABC$ is isosceles. Then

$$BL = AB - AL = AC - AM = CM,$$

and by the **SAS** congruency theorem, $\triangle LBC \equiv \triangle MCB$.

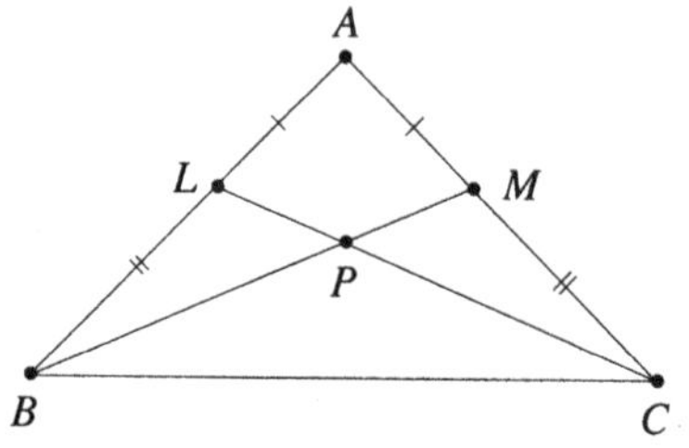

Therefore, $\angle PCB = \angle PBC$ and $\triangle BPC$ is isosceles, and thus, $BP = PC$.

Conversely, suppose that $AL = AM$ but $AB > AC$. Extend the side AC to D so that $AB = AD$, as in the figure below. We will show that this implies $PB > PC$.

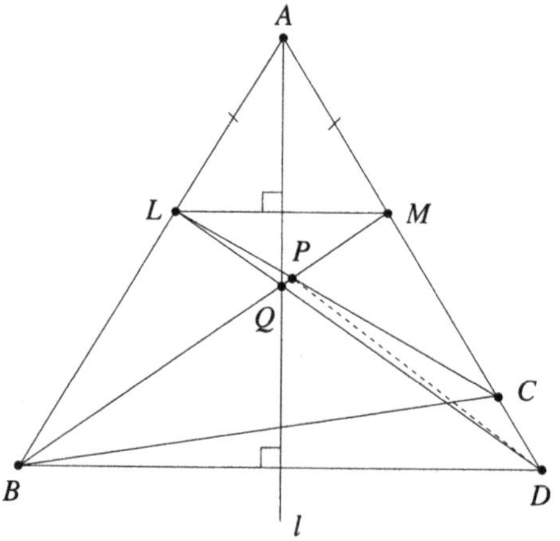

Let l be the common perpendicular bisector of LM and BD. By symmetry, BM and DL intersect at a point Q on l.

Since C is between M and D, the point P is on the same side of l as D so that $PB > PD$.

Now,

$$\angle DCP > \angle ALC > \angle ALM = \angle ADB > \angle CDP.$$

Hence, $PD > PC$ and the conclusion follows.

We have used the fact that an exterior angle of a triangle is greater than either of the interior angles and the fact that the larger angle is opposite the longer side. □

6.8 The Circle of Apollonius

In this section we will prove the Apollonian circle theorem, but first we give a characterization of points P on a line ℓ determined by distinct points A and B in terms of the ratio

$$\gamma = \gamma(P) = \frac{PA}{PB}.$$

We give the line an orientation such that the positive direction corresponds to going from A towards B, and we use the notation $A - P - B$ to mean that P is between A and B.

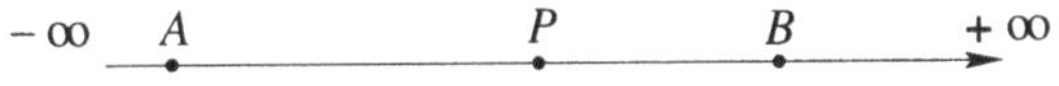

We have the following result.

Lemma 6.8.1. *If A, B, and P are distinct points on the line ℓ, and*

$$\gamma = \gamma(P) = \frac{PA}{PB},$$

then:

(1) *For $A - P - B$, we have $0 < \gamma(P) < \infty$ and $\gamma(P) \uparrow +\infty$ as P goes from A to B.*

(2) *For $A - B - P$, we have $1 < \gamma(P) < \infty$ and $\gamma(P) \downarrow 1$ as P goes from B to $+\infty$.*

(3) *For $P - A - B$, we have $0 < \gamma(P) < 1$ and $\gamma \uparrow 1$ as P goes from A to $-\infty$.*

Proof.

(1) For $A - P - B$, we have $PB = AB - PA$, so that

$$\gamma(P) = \frac{PA}{PB} = \frac{PA}{AB - PA} = \frac{\dfrac{PA}{AB}}{1 - \dfrac{PA}{AB}}.$$

As P approaches B, PA/AB approaches 1 through positive values so that the denominator goes to 0 through positive values. Therefore, $\gamma(P) \uparrow +\infty$ as P goes from A to P.

(2) For $A - B - P$, we have $PA = AB + PB$, so that

$$\gamma(P) = \frac{PA}{PB} = \frac{PB + AB}{PB} = 1 + \frac{AB}{PB} > 1.$$

As P approaches $+\infty$, AB/PB approaches 0 through positive values so that $\gamma(P) \downarrow 1$ as P goes from B to $+\infty$.

(3) For $P - A - B$, we have $PB = PA + AB$, so that

$$\gamma(P) = \frac{PA}{PB} = \frac{PB - AB}{PB} = 1 - \frac{AB}{PB} < 1.$$

As P approaches $-\infty$, AB/PB approaches 0 through positive values so that $\gamma(P) \uparrow 1$ as P goes from A to $-\infty$.

□

Theorem 6.8.2. *(Circle of Apollonius)*

Given two fixed points A and B, with $A \neq B$, together with a fixed positive constant $\gamma \neq 1$, then the locus of points P that satisfy

$$\frac{PA}{PB} = \gamma$$

is a circle with center on the line joining A and B, called the **circle of Apollonius**.

Note. We have avoided $\gamma = 1$, since

$$\{P \in \mathbb{R}^2 \,|\, \gamma = PA/PB = 1\}$$

is the perpendicular bisector of the segment AB.

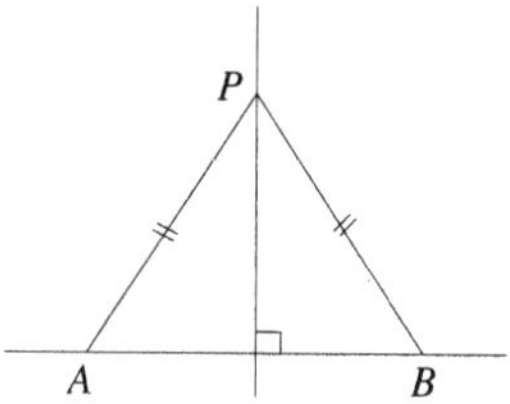

Proof. Let C be an internal point of the segment AB such that

$$\frac{CA}{CB} = \gamma.$$

The previous lemma implies that there is only one such point C.

Let D be an external point to the segment AB such that

$$\frac{DA}{DB} = \gamma.$$

Again, the lemma implies that there is only one such point D.

Let $\mathcal{C}$ be the circle with CD as diameter. Then we claim that $\mathcal{C}$ is the circle of Apollonius, that is,

$$\mathcal{C} = \{P \in \mathbb{R}^2 \mid \gamma = PA/PB\}.$$

(1) Suppose first that $P \in \mathbb{R}^2$ and $PA/PB = \gamma$. Then

$$\frac{PA}{PB} = \gamma = \frac{CA}{CB},$$

and C is internal to the segment AB.

Thus, in $\triangle APB$, PC is the internal bisector of $\angle APB$ by the converse of the internal bisector theorem.

Also,

$$\frac{PA}{PB} = \gamma = \frac{DA}{DB},$$

and D is external to the segment AB.

Thus, in $\triangle APB$, PD is the external bisector of the external angle at P.

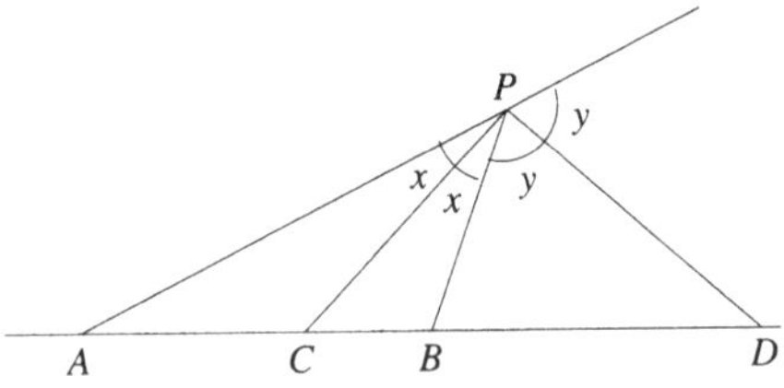

Now, $2x + 2y = 180$, so that $x + y = 90$, and $\angle CPD = 90$. Therefore, P is on the circle $\mathcal{C}$ with diameter CD.

(2) Suppose now that p is on $\mathcal{C}$, and assume for a contradiction that

$$\frac{PA}{PB} < \gamma = \frac{CA}{CB}.$$

From the lemma, the internal bisector of $\angle P$ divides the side opposite P in the ratio

$$\frac{PA}{PB} = \frac{C_1A}{C_1B} < \gamma,$$

at a point C_1 to the *left* of C.

Again, from the lemma, the extenal bisector of $\angle P$ divides the side opposite P in the ratio

$$\frac{PA}{PB} = \frac{D_1A}{D_1B} < \gamma$$

at a point D_1 to the *right* of D.

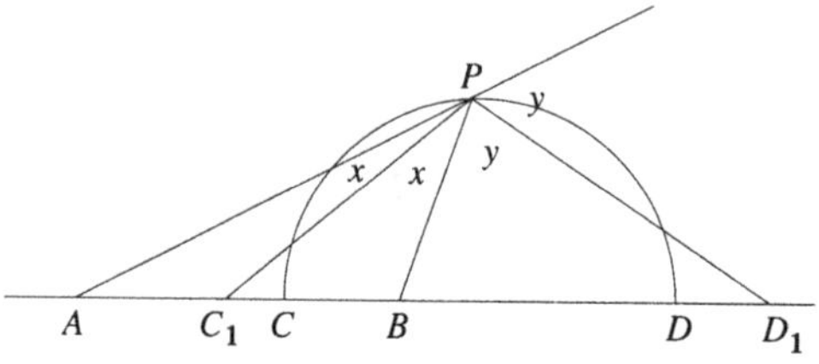

However, $\angle C_1PD_1 = 90$ since the internal and external angle bisectors are perpendicular, and $\angle CPD = 90$ since P is on the circle $\mathcal{C}$ with diameter CD. This is a contradiction since $\angle CPD < \angle C_1PD_1$. Thus, it is not true that $PA/PB < \gamma$. Similarly, if we assume that $PA/PB > \gamma$, we get a contradiction. Therefore, we must have $PA/PB = \gamma$.

□

6.9 Solutions to the Exercises

Analysis Figure for Exercise 6.2.2

In the following figure, triangles ABC and ADE are similar, so

$$\frac{w}{x} = \frac{z}{y}.$$

If we choose $y = 1$ we get $w = xz$. Let one of x or z be p, the other be q, and then $w = pq$.

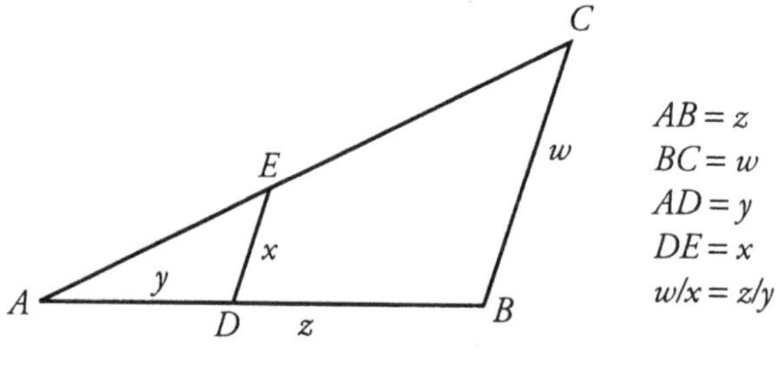

Solution for Exercise 6.3.1

Here is a fairly efficient construction.

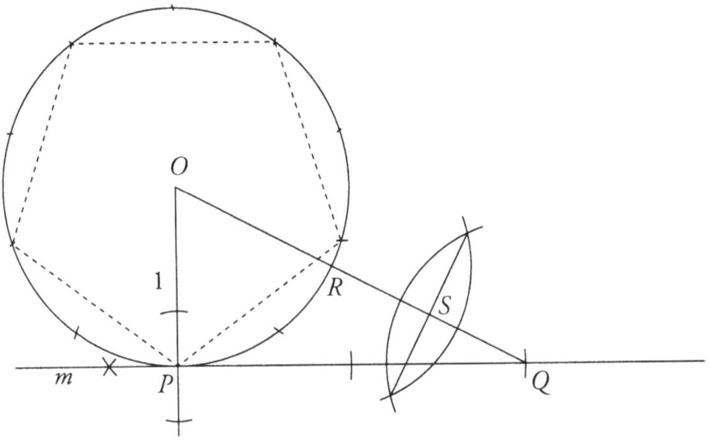

(1) Construct a line m perpendicular to OP at P.

(2) Strike off a point Q such that $PQ = 2$. By Pythagoras' Theorem, $OQ = \sqrt{5}$.

(3) Let R be the point where the circle intersects OQ so that $RQ = \sqrt{5} - 1$.

(4) Bisect RQ at S. Then $RS = x = \frac{1}{2}(\sqrt{5} - 1)$.

(5) Beginning at P, strike off the 10 successive points around the circle at distances x from each other. Connect every second point to construct the regular pentagon.

6.10 Problems

1. Construct a triangle given the three midpoints of its sides.

2. Construct a triangle given the length of one side, the size of an adjacent angle, and the length of the median from that angle.

3. Construct a triangle given the length of one side, the distance from an adjacent vertex to the incenter, and the radius of the incircle.

4. Construct $\triangle ABC$ given the length of side BC and the lengths of the altitudes from B and C.

5. Construct a triangle given the measure of one angle, the length of the internal bisector of that angle, and the radius of the incircle.

6. Given segments of length 1 and a, construct a segment of length $1/a$.

7. Given segments of length 1, a, and b, explain how to solve the following geometrically:

$$x^2 = a + b^2.$$

Explain how to construct the segment of length x.

8. Which of the following are constructible numbers?

   (a) 1

   (b) 3.1416

   (c) π

   (d) $\sqrt{203}$

   (e) $\sqrt{3+\sqrt{2}}$

   (f) $4^{1/3}$

9. Can we construct an angle of $2°$?

10. Given a unit line segment $\overline{AB}$ as shown:

   (a) Construct a segment of length $\dfrac{1+\sqrt{5}}{2}$.

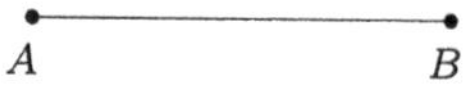

   (b) Construct a segment of length $\sqrt{\dfrac{2}{3}}$.

11. Prove Miquel's Theorem for the case where two of the circles are tangent. That is, given $\triangle ABC$ with menelaus points X, Y, and Z, as shown, where the circumcircles of $\triangle AXZ$ and $\triangle BXY$ are tangent at X, show that the quadrilateral $XYCZ$ is cyclic.

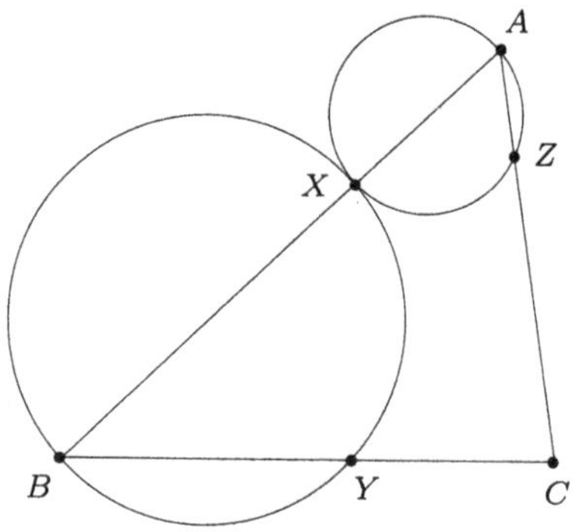

Hint. Tangent circles at X have the same tangent line to both circles at X.

12. Construct a triangle given the foot F of an altitude, the orthocenter H, and the center N of the 9-point circle.

13. Given a diameter $\overline{AB}$ of a circle and a point P as shown, construct a perpendicular from P to $\overleftrightarrow{AB}$, *with a straightedge alone.*

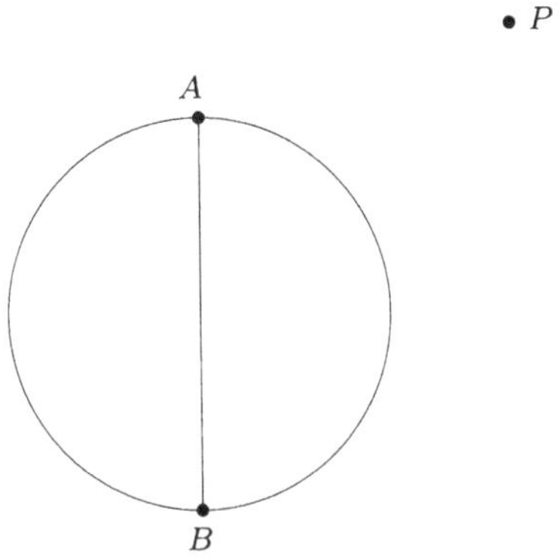

14. (a) List all the regular n-gons with $n \leq 100$ sides that are constructible with a straightedge and compass.
    (b) Use Gauss' Theorem to prove that an angle of $20°$ is not constructible.
    (c) Use Gauss' Theorem to decide whether or not an angle of $6°$ is constructible.

15. Construct $\triangle ABC$ given a line segment $\overline{XY}$ and two adjacent angles, as in the figure, where the length of the perimeter is XY, $\alpha = \angle B$, and $\beta = \angle C$.

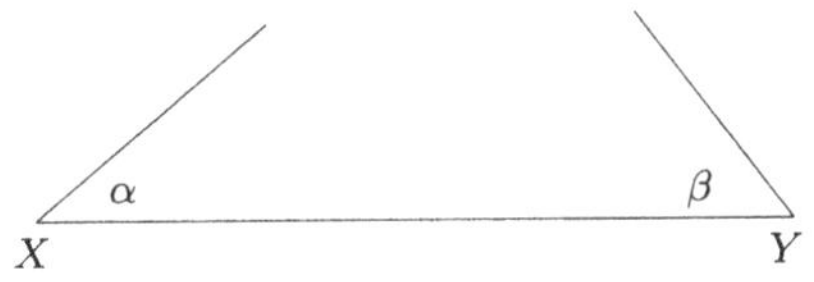

16. Construct triangle ABC given the location of its circumcenter S, the location of its orthocenter H, and the foot of an altitude F.

17. Construct a triangle given the foot F of an altitude, the circumcenter S, and the center N of the 9-point circle.

18. Construct triangle ABC given the location of vertex A, the location of its circumcenter S, and the center N of the 9-point circle.

19. Show that the incenter of a triangle is the Nagel point of its medial triangle.

PART II

TRANSFORMATIONAL GEOMETRY

CHAPTER 7

THE EUCLIDEAN TRANSFORMATIONS OR ISOMETRIES

7.1 Rotations, Reflections, and Translations

Two children want to share a piece of cake of an unusual shape, as shown in the figure on the following page, where $BCDE$ is a square, the curve AE is a circular arc with center C, and the points A, B, and E are collinear. Being children, they are not enthusiastic about receiving pieces that differ in shape. In other words, they want to cut the cake into two congruent pieces. How can this be done?

Classical Geometry: Euclidean, Transformational, Inversive and Projective, First Edition. By
I. E. Leonard, J. E. Lewis, A. C. F. Liu, G. W. Tokarsky
Copyright © 2014 John Wiley & Sons, Inc.

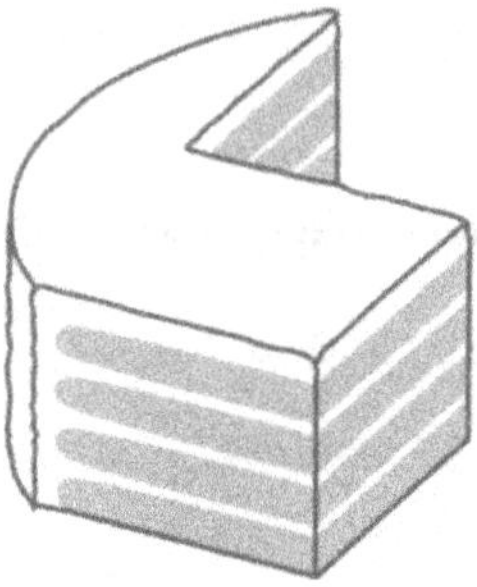

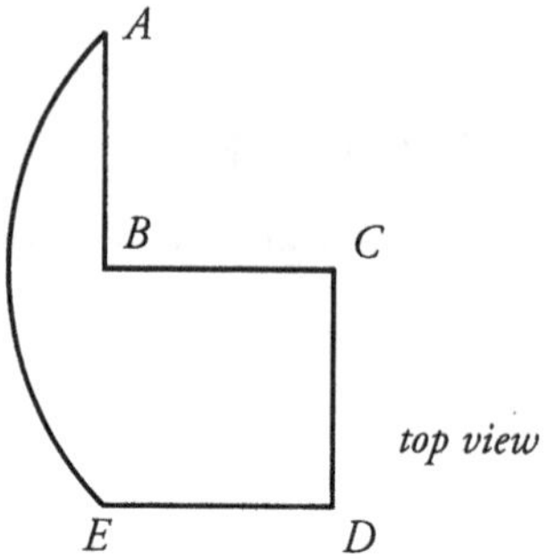

If the cake included the square space above $BCDE$, so that it was of the shape shown in the figure on the right, the task would be very simple. The cake could be cut down the middle along the line l.

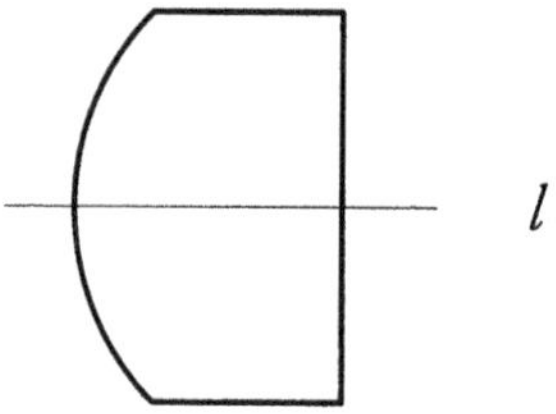

The cake in this figure has ***reflectional symmetry*** about the line l: if you hold a mirror vertical to the page with one edge of the mirror along l, the reflection of one half of the figure will coincide with the other half (this type of symmetry will be defined more precisely later).

When we wish to divide cakes and pies equally, we tend to look for reflectional symmetry and often overlook other possibilities. This might be why many people try to divide the cake by cutting it into two pieces and then dividing each of the pieces along an axis of reflectional symmetry, as in the figure on the right.

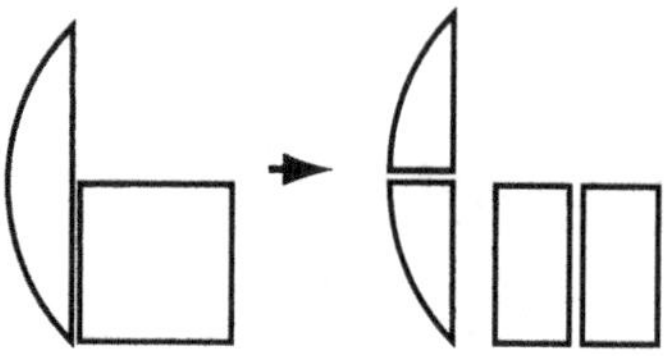

Although it can be cut into two pieces, each of which has reflectional symmetry, the cake itself does not have reflectional symmetry. But this does not mean that there is no solution to the original problem.

The cake was probably one quarter of a round cake, as shown in the figure below. A spin around the center C through a $45°$ angle shows how to cut the cake into two congruent pieces.

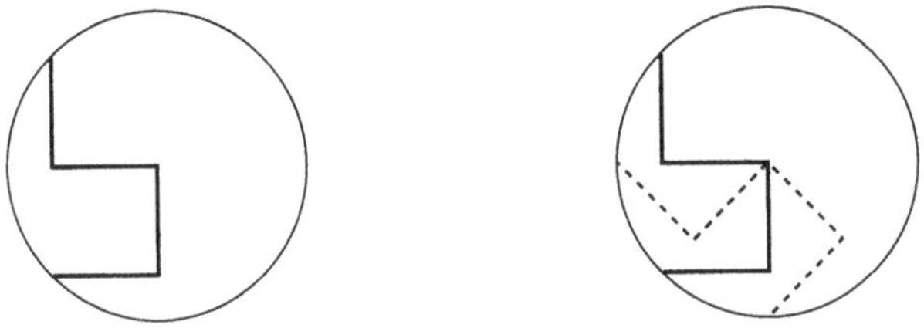

Directed Angles

The cake problem was solved by using a counterclockwise rotation through an angle of $45°$. A rotation in the clockwise direction would also have led to a solution. The usual way to distinquish between clockwise and counterclockwise rotations is to use ***directed*** or ***signed*** angles.

Angles that are measured in a counterclockwise direction are considered positive, while those measured in a clockwise direction are negative, as shown in the figure below. For a directed angle, the symbol $\angle ABC$ is interpreted as the angle *from* the ray $\overrightarrow{BA}$ *to* the ray $\overrightarrow{BC}$.

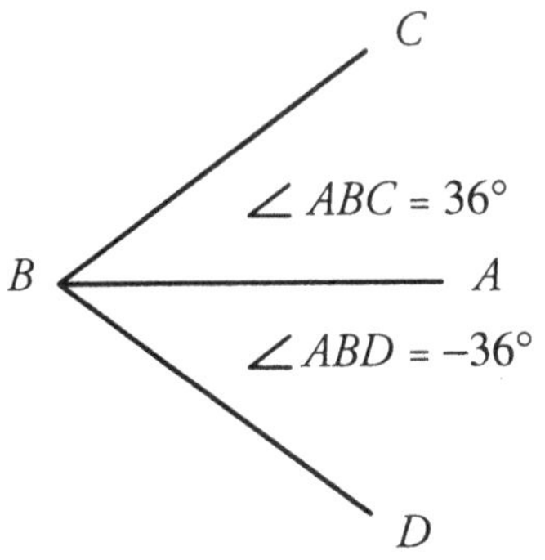

Rotations

Let O be a point and θ be a directed angle. The ***rotation*** about O through the angle θ, denoted by $\mathbf{R}_{O,\theta}$, maps each point P of the plane, where $P \neq O$, into another point P', where

$$|OP'| = |OP| \quad \text{and} \quad \angle POP' = \theta.$$

The point O, which is called the ***center of rotation***, is mapped onto itself. Since it does not move, it is called a ***fixed point*** or an ***invariant point*** under $\mathbf{R}_{O,\theta}$.

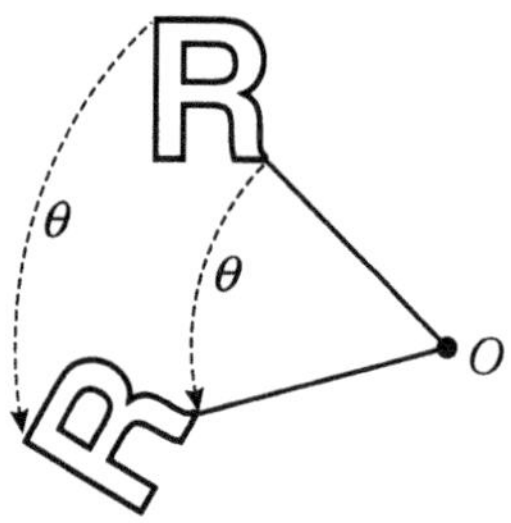

Reflections

Let l be a line in the plane. The ***reflection*** about l, denoted by $\mathbf{R}_l$, maps a point P not on l to the point P' such that l is the perpendicular bisector of PP'.

Under the reflection $\mathbf{R}_l$, every point on l is mapped to itself, so every point on l is a fixed point.

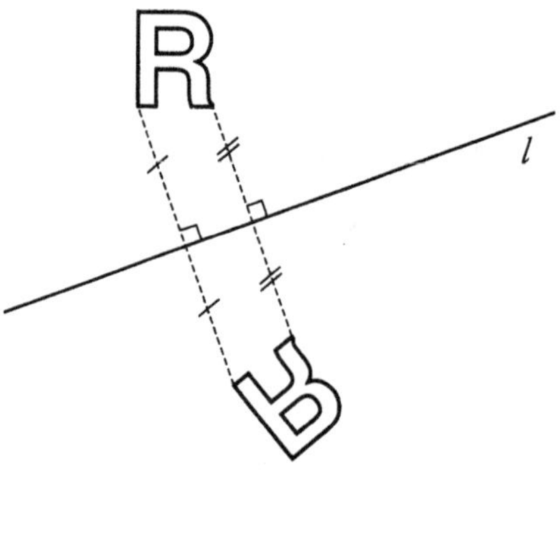

Translations

Let $\overline{AB}$ be a directed line segment. A ***translation*** by $\overline{AB}$, denoted by $\mathbf{T}_{AB}$, maps a point P to the point P' such that the directed segment $\overline{PP'}$ is congruent to $\overline{AB}$, parallel to $\overline{AB}$, and in the same direction as $\overline{AB}$.

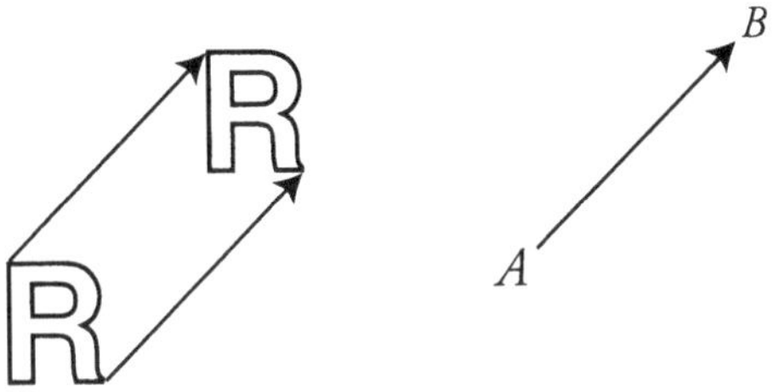

7.2 Mappings and Transformations

Mappings

We use the word ***mapping*** or ***function*** to describe an association between two sets $\mathcal{X}$ and $\mathcal{Y}$ which has the property that each point of $\mathcal{X}$ is associated with one and only one point of $\mathcal{Y}$.

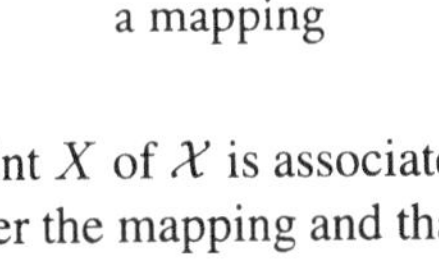

a mapping

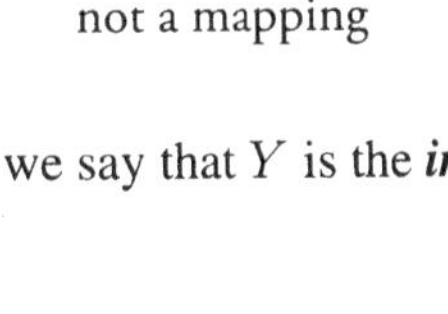

not a mapping

If a point X of $\mathcal{X}$ is associated with the point Y of $\mathcal{Y}$, we say that Y is the ***image*** of X under the mapping and that X is a ***preimage*** of Y.

There are two things that should be mentioned in connection with images and preimages. The first is that the definition of a mapping forbids that a point X of $\mathcal{X}$ have more than one image in $\mathcal{Y}$. However, it is quite permissible that a point Y of $\mathcal{Y}$ have more than one preimage in $\mathcal{X}$—many different points of $\mathcal{X}$ can have the same image in $\mathcal{Y}$. In other words, we can say that a mapping may be ***many-to-one***. Such terminology, although harmless, is a bit redundant because the definition allows such behavior. A mapping of a set $\mathcal{X}$ to a set $\mathcal{Y}$ is called ***one-to-one*** or an ***injection*** if no point of $\mathcal{Y}$ has more than one preimage in $\mathcal{X}$ or, equivalently, if distinct points of $\mathcal{X}$ have distinct images in $\mathcal{Y}$.

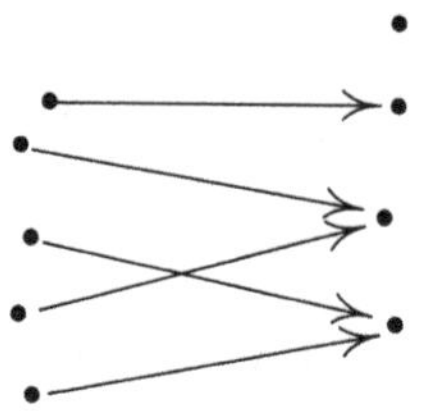

a *many-to-one* mapping

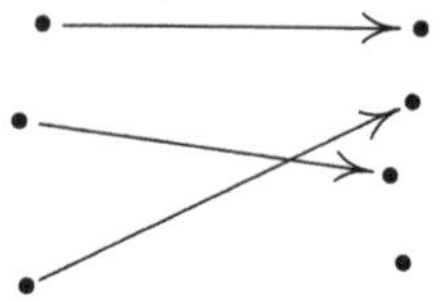

a *one-to-one* mapping

The second thing to note is that although the definition of a mapping from $\mathcal{X}$ to $\mathcal{Y}$ says that *every* point of $\mathcal{X}$ must have an image in $\mathcal{Y}$, it does *not* say that every point of $\mathcal{Y}$ must have a preimage in $\mathcal{X}$. When every point of $\mathcal{Y}$ has a preimage in $\mathcal{X}$, we say that the mapping is ***onto*** $\mathcal{Y}$ or that it is a ***surjection***.

When $\mathcal{X}$ and $\mathcal{Y}$ are the same set, it sometimes occurs that a point is its own image. Such a point is called a ***fixed point*** or an ***invariant point*** of the mapping. If all of the points in $\mathcal{X}$ are fixed points, the mapping is called the ***identity mapping***, or more simply the ***identity***, and it is denoted by **I**.

Transformations

A mapping that is both one-to-one and onto is called a ***bijection***, and if $\mathcal{X}$ and $\mathcal{Y}$ are the same set, then the bijection is called a ***transformation***. In other words, when we use the word *transformation* we mean a mapping with the following properties:

The mapping is from one set into the same set.

The mapping is one-to-one.

The mapping is onto.

It is easy to see that rotations, reflections, and translations have all three properties, and therefore that all three are transformations.

The Inverse of a Transformation

The ***inverse*** of a mapping T from $\mathcal{X}$ to $\mathcal{Y}$ is another mapping S from $\mathcal{Y}$ to $\mathcal{X}$ such that for every point x in $\mathcal{X}$ the point $T(x)$ is mapped back onto x by S. In other words, if $T(x) = y$, then $S(y) = x$. A mapping that is not one-to-one cannot have an inverse. However, in geometry, a *transformation* must be one-to-one and onto, and so every transformation automatically has an inverse, and the inverse mapping is itself a transformation.

The three fundamental mappings—rotations, reflections, and translations—are easily seen to be transformations whose inverses are the same type of transformation:

Theorem 7.2.1. *(Inverses of Transformations)*

(1) The inverse of the rotation $\mathbf{R}_{O,\alpha}$ *is the rotation* $\mathbf{R}_{O,-\alpha}$.

(2) The inverse of the reflection $\mathbf{R}_l$ *is the same reflection.*

(3) The inverse of the translation $\mathbf{T}_{AB}$ *is the translation* $\mathbf{T}_{BA}$.

7.2.1 Isometries

A transformation that preserves distances is called an ***isometry***.

Theorem 7.2.2. *Rotations, reflections, and translations are isometries.*

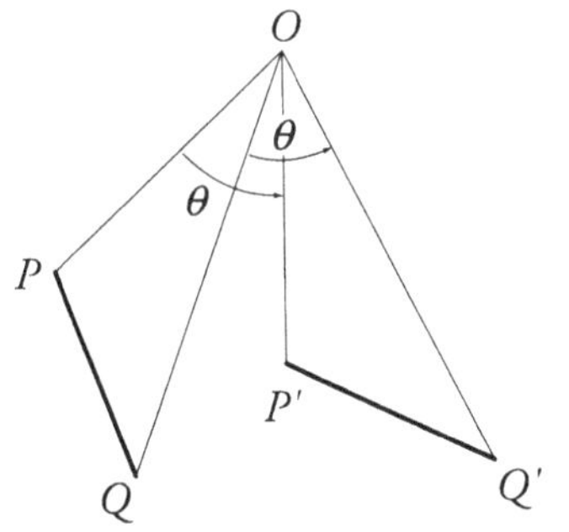

Proof. We will show that a rotation $\mathbf{R}_{O,\theta}$ is actually an isometry (the proofs that reflections and translations are also isometries are similar).

Consider the figure above, which shows a typical case. Since

$$\angle POP' = \theta = \angle QOQ',$$

we must have

$$\angle POQ = \theta - \angle QOP' = \angle P'OQ'.$$

Since

$$OP = OP' \qquad \text{and} \qquad OQ = OQ',$$

then by the **SAS** congruency theorem we have

$$\triangle OPQ \cong \triangle OP'Q',$$

and therefore $PQ = P'Q'$.

□

Composition of Isometries

What happens if an isometry T is applied to the plane and then followed by another isometry S? When a transformation T is followed by another one S, the combined result is called the ***composition*** of the two transformations and is written $S \circ T$. Notice that the first transformation is on the right, while the second is on the left.[7]

Suppose that we start with points P and Q at distance d from each other. When T is applied, these points are mapped into P' and Q', and

$$\text{dist}(P', Q') = \text{dist}(P, Q) = d.$$

When S is applied to P' and Q', these points are mapped to P'' and Q'', respectively, and

$$\text{dist}(P'', Q'') = \text{dist}(P', Q') = d.$$

The combined effect is that P and Q are mapped to P'' and Q'', and the distance is preserved; that is, $S \circ T$ is itself an isometry.

Since we can create new isometries by composing known isometries, it seems like there is an unlimited supply of different types of isometries. For example, we could create a new isometry by first doing a rotation, then a reflection about some line, then a reflection about another line, then a translation. We will see later that we cannot really get too much that is new, and, in fact, there are only four different types of isometries in the plane. In addition to rotations, reflections, and translations, the only other type is a *glide reflection*.

[7]This is the conventional notation in geometry. It is a common convention in algebra texts to write the first transformation on the left.

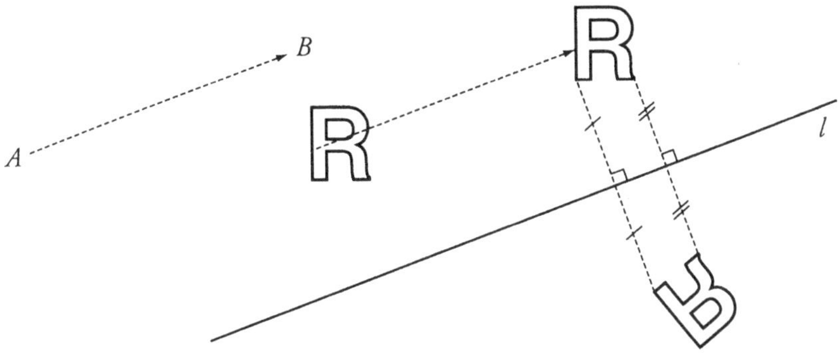

A ***glide reflection*** $\mathbf{G}_{l,AB}$ is simply a translation $\mathbf{T}_{AB}$ followed by a reflection $\mathbf{R}_l$ about a line l that is parallel to $\overline{AB}$. We will prove that this is the only other additional isometry later.

It is obvious that all isometries have inverses and that the inverses themselves must also be isometries. Not quite so obvious is the fact that an isometry also preserves straight lines.

Theorem 7.2.3. *(Isometries Preserve Straight Lines)*

(1) Let P, Q, and R be three points, and let P', Q', and R' be their images under an isometry. The points P, Q, and R are collinear, with Q between P and R, if and only if the points P', Q', and R' are collinear, with Q' between P' and R'.

(2) Let l be a straight line, and let l' be the image of l under an isometry. Then l' is a straight line.

Proof. Here we write $|AB|$ for dist(A, B).

(1) We will show that if Q is between P and R, then Q' must be between P' and R' (the proof of the converse may be obtained by interchanging P, Q, and R with P', Q', and R').

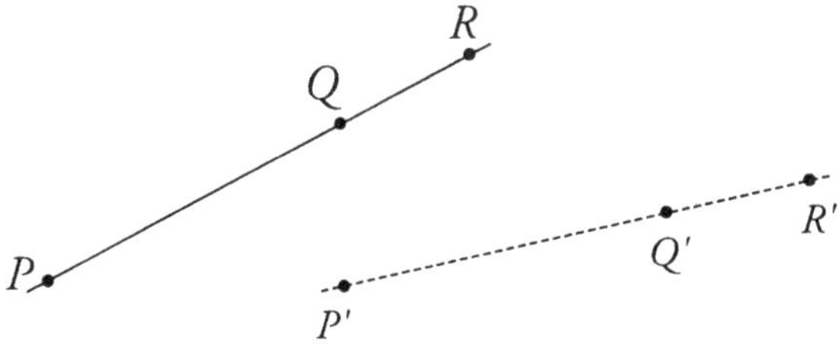

If Q is between P and R, then

$$|PQ| + |QR| = |PR|.$$

Since an isometry preserves distances, we must have

$$|P'Q'| = |PQ|, \qquad |Q'R'| = |QR|, \qquad \text{and} \qquad |P'R'| = |PR|.$$

Hence,

$$|P'Q'| + |Q'R'| = |P'R'|,$$

and the Triangle Inequality shows that P', Q', and R' are collinear with Q' between P' and R'.

(2) Let P and Q be two points on l, and let P' and Q' be their images under an isometry. Let m be the line passing through P' and Q'. We will show that m is the image of l under the isometry. We must check two things:

(a) that every point R on l has its image R' on m and

(b) that every point S' on m has its preimage S on l.

It follows from statement (1) above that if R is a point on l other than P or Q, then P', Q', and R' must be collinear, so R' is a point on m. Conversely, if S' is on m, then P', Q', and S' are collinear, and it follows again from statement (1) that P, Q, and S are on l.

□

The next theorem tells us that an isometry preserves the shape and size of all of the geometric figures.

Theorem 7.2.4. *Under an isometry,*

(1) the image of a triangle is a congruent triangle;

(2) the image of an angle is a congruent angle;

(3) the image of a polygon is a congruent polygon;

(4) the image of a circle is a congruent circle.

Proof. We will prove statement (1) and leave the rest as exercises. Let P, Q, and R be the vertices of a triangle. It follows from Theorem 7.2.3 that their images P', Q', and R' are the vertices of a triangle and that the edges

$$P'Q', \qquad Q'R', \qquad \text{and} \qquad P'R'$$

are the images of the edges

$$PQ, \qquad QR, \qquad \text{and} \qquad PR.$$

Since the isometry preserves distances, congruency now follows from the **SSS** congruence property.

□

Recall that the notion of congruency is defined in different ways for different figures. For example, two triangles are congruent if the three angles and the three corresponding sides of one triangle are the same size as the three angles and the three corresponding sides of the other, while two circles are said to be congruent if they have the same radius. Theorem 7.2.4 shows that the notion of isometry encompasses and generalizes the notion of congruency.

7.3 Using Rotations, Reflections, and Translations

Example 7.3.1. *Cut each of the following figures into two congruent pieces using a single cut.*

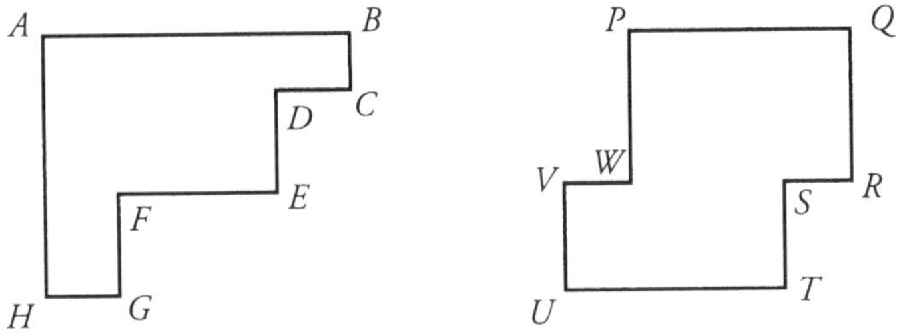

Solution. After the cut we must end up with two pieces that are congruent to each other. This means that one of the pieces must be obtainable from the other by an isometry—either a rotation, a reflection, a translation, or some combination. One way to attack the problem is to use a "trace and fit" method: trace the figure on tracing paper and place the traced figure on the original one in various positions until the two overlapping figures create the outline of the two congruent shapes (much as was done at the beginning of the chapter to cut the cake). When this is done, you can see that the solution for the polygon $ABCDEFGH$ is obtained by applying the rotation $\mathbf{R}_{O,90^\circ}$, where O is the midpoint of the edge EF. The solution for $PQRSTUVW$ may be obtained via the translation $\mathbf{T}_{UM}$, where M is the midpoint of TU.

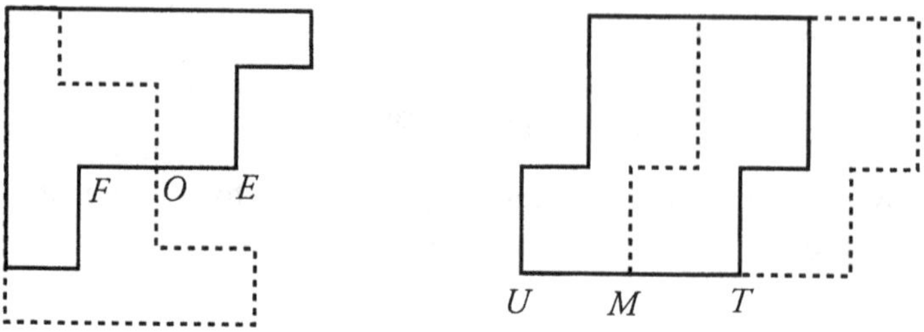

□

It is interesting to note that a slight change in the problem can result in a different solution. For example, consider the problem of cutting the polygon in the figure below into two congruent halves. It resembles the polygon $ABCDEFGH$ from the previous problem, but it is solved by a glide reflection, not a rotation.

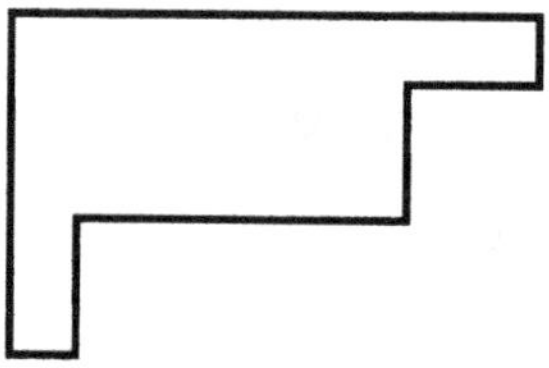

Puzzles of this type can get quite complicated, and the trace and fit method does not always lead to a solution. Here is a more complicated example.

Example 7.3.2. *Cut the polygon in the figure below into two congruent pieces using a single cut. The shaded region is a hole, and cutting through the hole still counts as a single cut.*

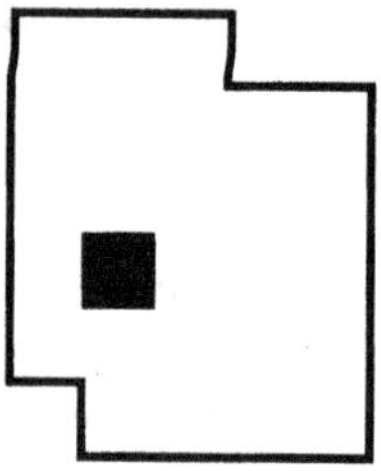

Solution. One way to approach this problems is to divide the shape into congruent squares suggested by the shape of the hole as in diagram (1) in the figure on the following page, then try to gradually build up the congruent pieces by coloring the squares. One possible sequence of events is illustrated by diagrams (2) through (7) in the figure.

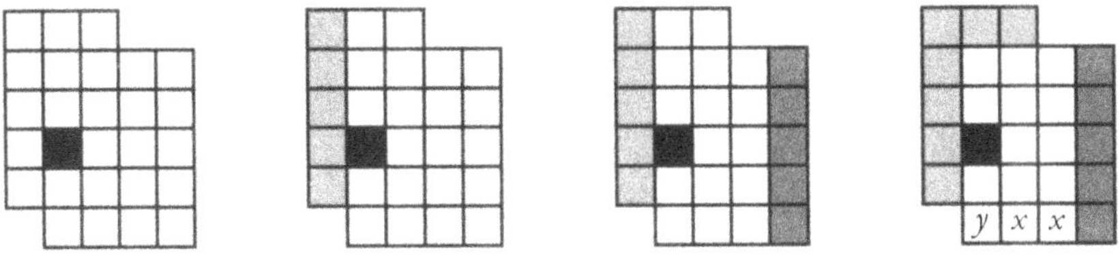

(1) (2) (3) (4)

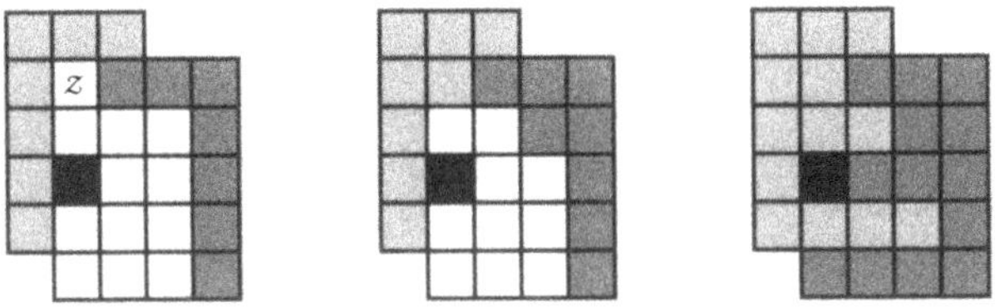

(5) (6) (7)

Here is an explanation of each step.

(2) We might imagine that squares along the left side all belong to the same piece, so we color them light gray.

(3) There must be a matching set of squares in the other piece, probably the five squares along the right side. Color them dark gray.

(4) A little exploration will convince us that the squares along the top cannot be dark gray, since we would not be able to find matching light gray squares, so they must be light gray.

(5) We might think that the light gray squares in (4) should be matched by dark gray ones at x and x. However, this leaves y as a light gray square and then we cannot find a match for the light gray square at y. Thus, the dark gray squares must be added as in (5).

(6) This forces the square at z to be light gray, which in turn forces the corresponding dark gray one.

(7) Continuing in this manner, we eventually reach a solution.

□

Example 7.3.3. *Given a square $ABCD$, a line l, and a point P, find all points X and Y with X in an edge of $ABCD$, Y in l, and with P being the midpoint of the segment XY.*

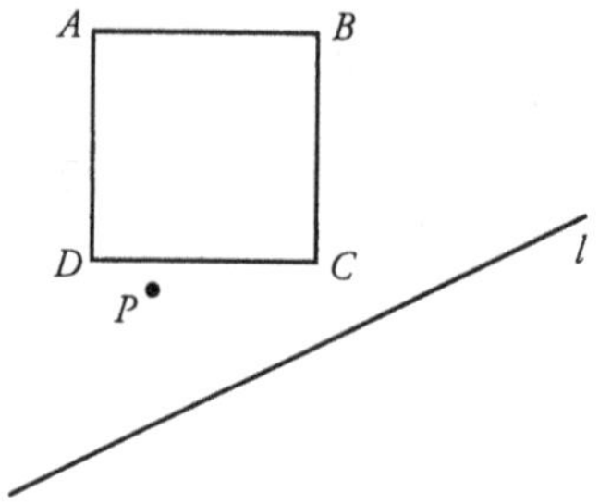

Solution. We will take a trial-and-error approach. Let the point Y be located on l and find the corresponding point Y' so that YY' has P as its midpoint. If Y' happens to land on an edge of the square, we have found a solution. It is more likely that Y' is not on an edge of the square, so we will try several positions for Y and see what happens to Y', as in the figure below.

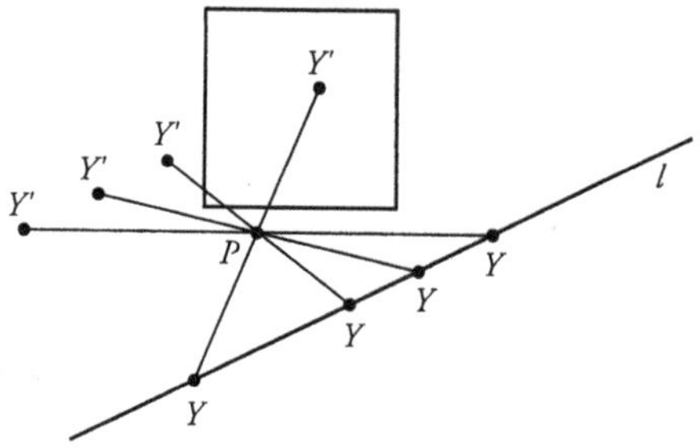

Note that each Y' is obtained from the corresponding Y by a rotation of 180° around P. Thus, we can get the solution by applying $\mathbf{R}_{P,180°}$ to the line l. The points X, if there are any, are the points where the image l' meets the square $ABCD$, as in the figure below.

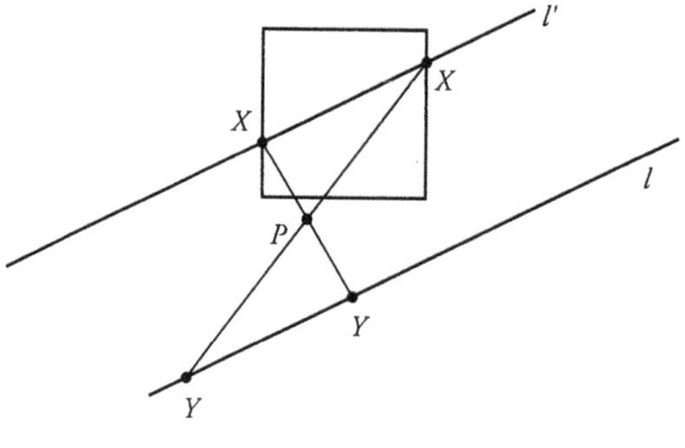

□

Example 7.3.4. *Given two circles E and F separated by a line l, find all squares $ABCD$ with vertex A in E, the opposite vertex C in F, and the remaining vertices on l.*

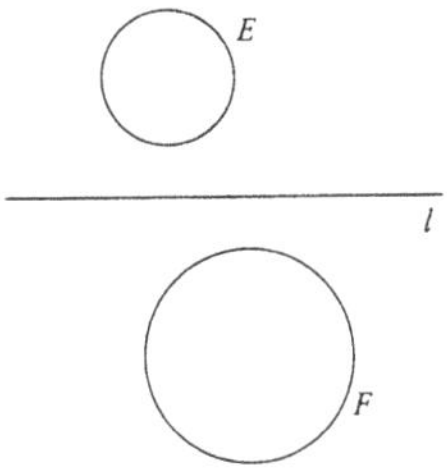

Solution. Using a trial-and-error approach again, let a point C be taken on the circle F, and find the corresponding point C' so that C and C' are opposite vertices of a square whose other two vertices are on l. Observe that l is the right bisector of the diagonal CC' of the square. As C takes up different positions in F, C' must therefore be a point on the circle F' that is obtained by reflecting F through the line l, as in the figure below.

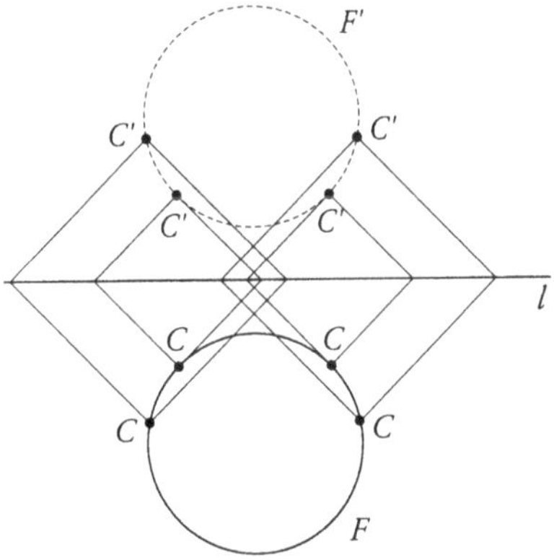

Thus, we obtain the solution by applying $\mathbf{R}_l$ to the circle F, and the points where the image F' intersects E give us the point or points A of the desired square. Having found A, we can now construct the square to complete the solution, as in the following figure.

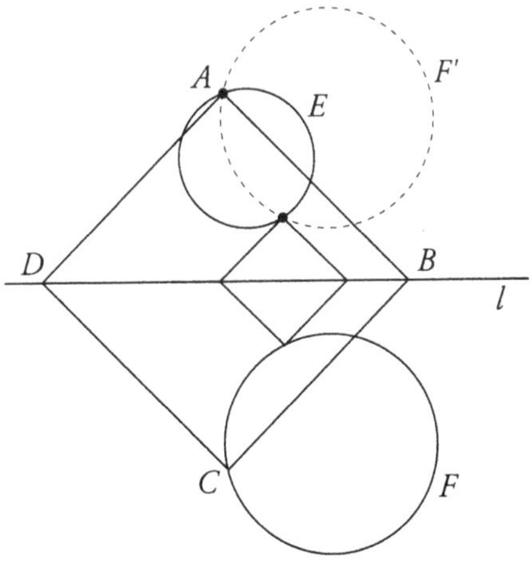

□

Example 7.3.5. *A highway is bounded by two parallel lines l and m. A horse and rider at point R wish to return to camp at a point C on the other side of the highway, and the rider wishes to cross the highway in a perpendicular direction. What is the shortest route that will fulfill both wishes?*

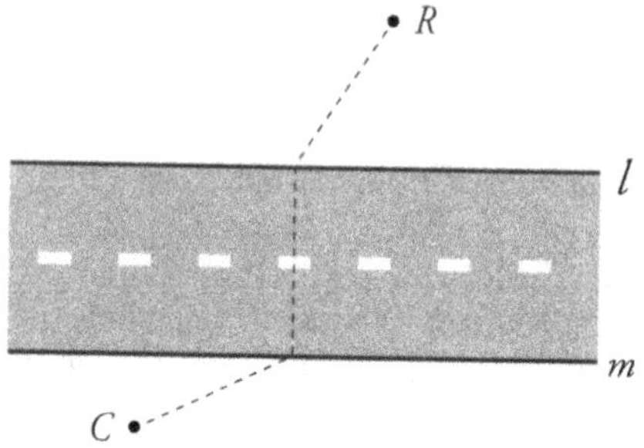

Solution. Let $\overline{AB}$ be the directed segment whose magnitude is equal to the distance between l and m and whose direction is perpendicular to them, going from l toward m. Let R' be the image of R under $\mathbf{T}_{AB}$, and let CR' meet m at Q. Let P be the foot of the perpendicular from Q to l, and note that Q is the image of P under $\mathbf{T}_{AB}$, as in the figure below.

The horse and rider should go from R to P, cross the highway to Q, and proceed to C. This route is the shortest. To see why, let S be any other point on l, and let T be the foot of the perpendicular from S to m.

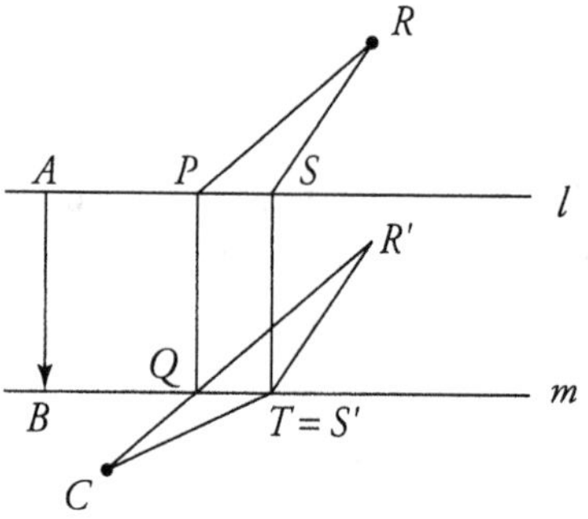

Then T is the image of S under $\mathbf{T}_{AB}$, and

$$\begin{aligned} RS + ST + TC &= R'T + PQ + SC \\ &> PQ + R'C \qquad \text{(by the Triangle Inequality)} \\ &= PQ + R'Q + QC \\ &= PQ + RP + QC, \end{aligned}$$

since $R'Q = RP$.

□

The previous problem is what is known as an ***extremal problem***; that is, it is a problem where the objective is to find either the largest or smallest value of some quantity.

The following is another extremal problem whose solution is obtained by a reflection rather then a translation.

Example 7.3.6. *A horse and rider at a point R wish to return to camp at a point C, but the horse wishes to take a drink from a straight river l before doing so. If C and R are on the same side of l, what is the shortest route that will fulfill both wishes?*

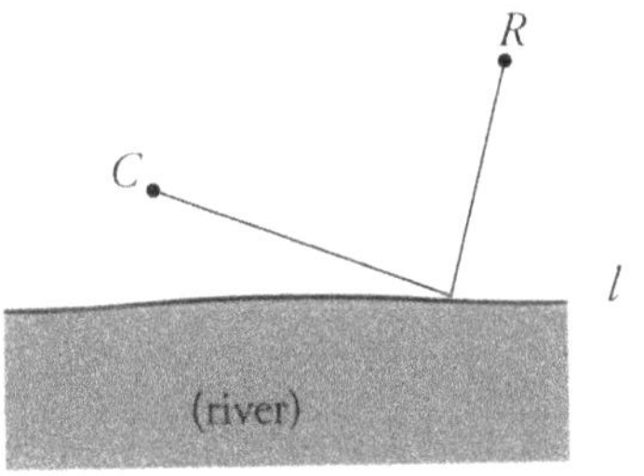

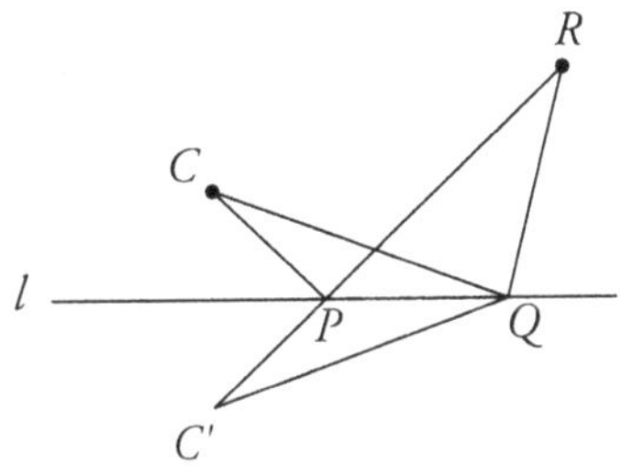

Solution. Let C' be the image of C under $\mathbf{R}_l$. Draw the line $C'R$, and let P be the points where it cuts l. We claim that if the horse and rider go from R to P and then from P to C, this route is the shortest. If Q is any other point on l, then by the Triangle Inequality we have

$$RQ + QC = RQ + QC' > RC' = RP + PC' = RP + PC,$$

which proves our claim.

□

Example 7.3.7. *(The Fermat Point)*

Let ABC be a triangle with $AB = 4$, $AC = 3$, and $\angle BAC = 30^\circ$. Determine the minimum value of

$$PA + PB + PC$$

where P is any point inside the triangle ABC, and find the location of the point P that yields this minimum value.

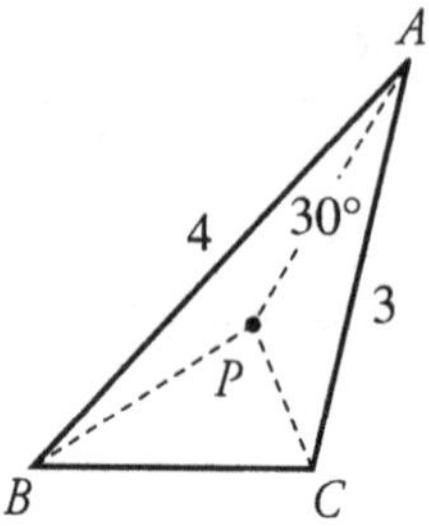

Solution. Consider the effect of $\mathbf{R}_{A,60°}$ on $\triangle ABC$, as in the figure below. This maps $\triangle ABC$ onto $\triangle A'B'C'$ and takes the point P to the point P'. Since $\angle PAP' = 60°$, triangle APP' must be equilateral. Hence

$$\begin{aligned} PB + PA + PC &= BP + PP' + P'C' \\ &\geq BP' + P'C' \\ &\geq BC' \end{aligned}$$

by the Triangle Inequality.

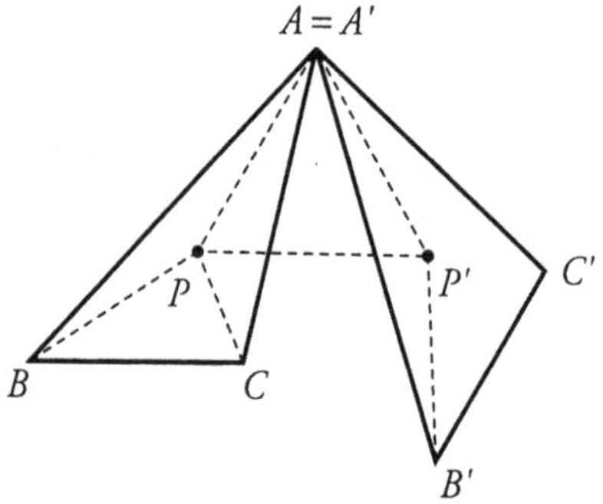

Now, $\angle BAC' = \angle BAC + \angle CAC' = 90°$, so $BC' = 5$ by Pythagoras' Theorem. Therefore, the value of $PA + PB + PC$ is at least 5, and the value will be exactly 5 if B, P, P', and C' are collinear. Thus, to attain the minimum value, P and P' would have to lie on BC'.

If this happens, then

$$\angle APB = 180° - \angle P'PA = 120° \quad \text{and} \quad \angle CPA = \angle C'P'A' = 120°.$$

Now, this defines the point P uniquely, and since all three angles of triangle ABC are less than $120°$, P is indeed inside $\triangle ABC$.

□

Example 7.3.8. *A and B are the centers of two circles on the same side of a line l. Construct a line m, parallel to l, such that the circles cut off equal segments from m.*

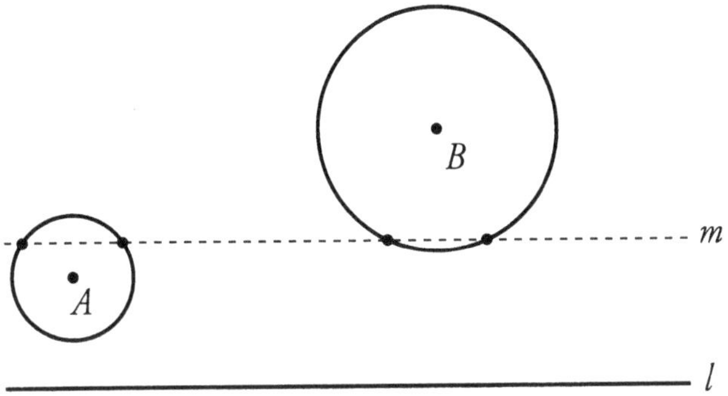

Solution. Let α and β denote circles centered at A and B, respectively. Let P and Q be the feet of the perpendiculars from A and B upon l, respectively. The solution here is obtained by using the translation $\mathbf{T}_{PQ}$. Under this translation, the circle α is mapped to the circle α' centered at A', as in the figure below.

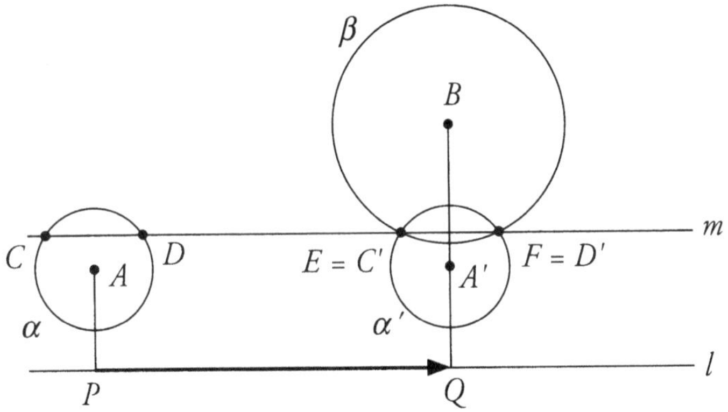

Since the image of P is Q, we have $AA' = PQ$, and so $A'APQ$ is a rectangle. Thus, A' must be on BQ. Let E and F be the points at which the translated circle intersects β, and let m be the line EF, cutting α at C and D. Then $E = C'$ and $F = D'$, so that

$$EF = C'D' = CD.$$

□

In the next example, we use a reflection to give a solution to a difficult construction problem.

Example 7.3.9. *Let $ABCD$ be a quadrilateral with right angles at C and D and with $BC < CD$. Construct the point E on CD such that*

$$\angle AED = 2\angle BEC.$$

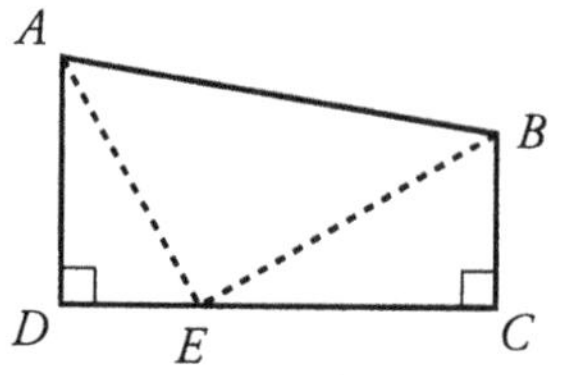

Solution. Let A' be the image of A under $\mathbf{R}_{CD}$, as in the figure below.

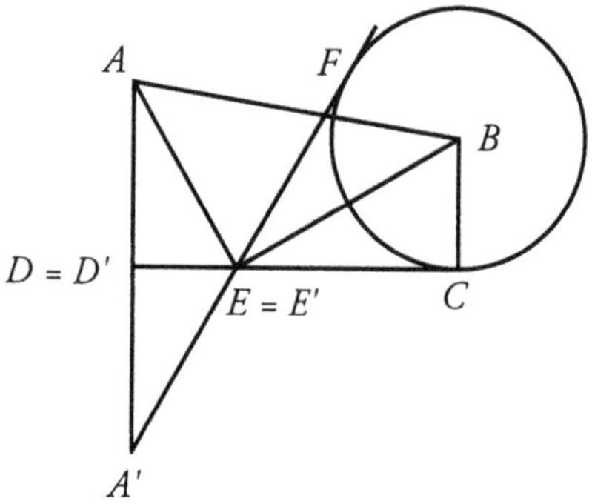

Draw the circle with center B and radius BC so that DC is tangent to the circle at C. There are two tangents from A' to the circle. Draw the tangent $A'F$ that meets CD at the point E between C and D.

Note that $D' = D$ and $E' = E$, since D and E are fixed points of the reflection $\mathbf{R}_{CD}$. Therefore, the triangles ADE and $A'D'E'$ are congruent, and hence

$$\angle AED = \angle A'E'D' = \angle FEC = 2\angle BEC.$$

□

7.4 Problems

1. A swimming pool is posting notices to inform visitors about their Monday closure. All the notices are identical except for the one by the diving board. It reads: "NOW NO SWIMS ON MON." What could be the reason for this?

2. The capital letters have been classified into the following subsets:

   (a) A, M, T, U, V, W, Y

   (b) B, C, D, E, K

   (c) F, G, J, L, P, Q, R

   (d) H, I, O, X

   (e) N, S, Z

   On what basis could this classification be made?

3. The diagram below shows five symbols in a sequence. What could the sixth symbol be?

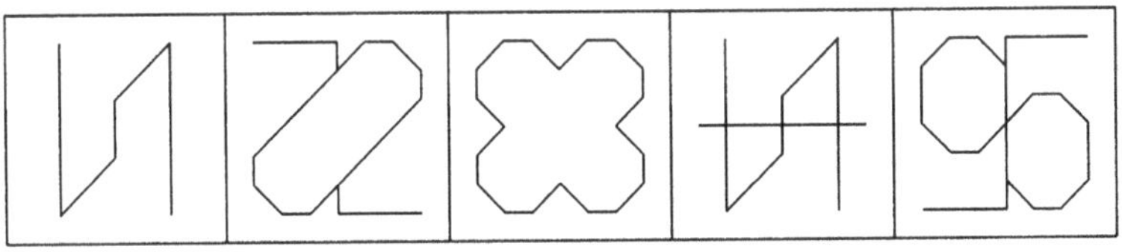

4. The diagram below shows five symbols in a sequence. What could the sixth symbol be?

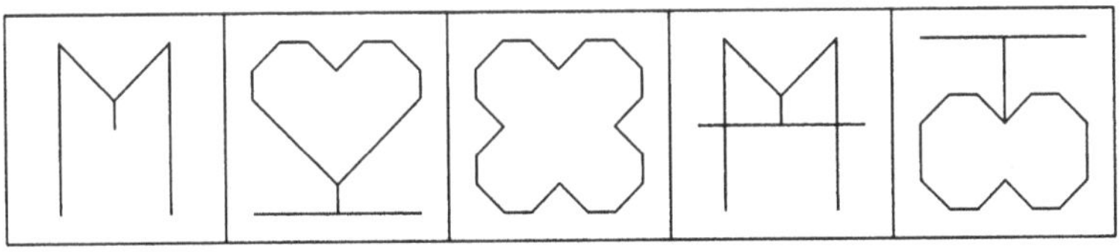

5. Dissect each of the figures below into two congruent pieces.

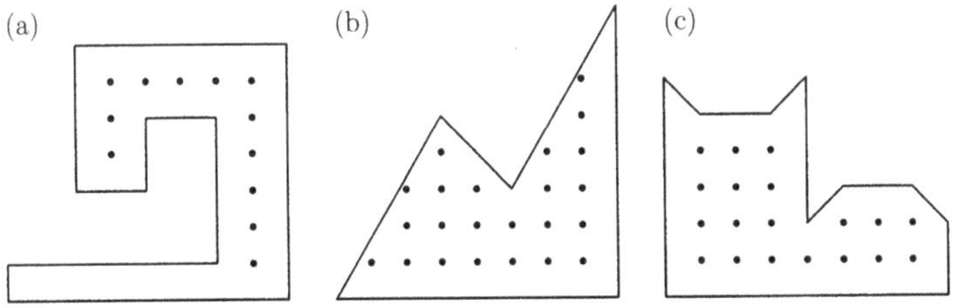

6. Dissect each of the figures in the diagram below into two congruent pieces.

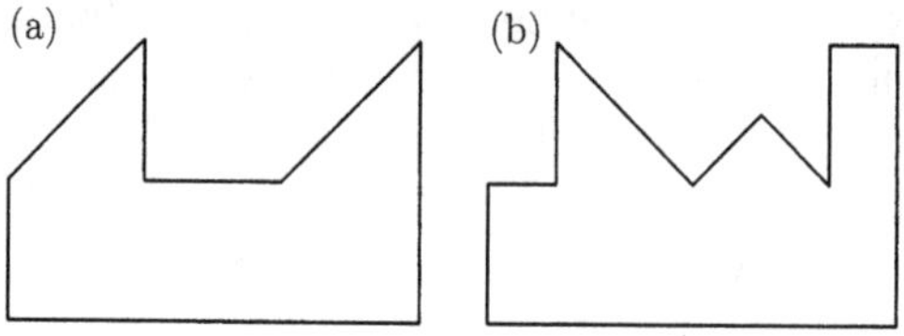

7. Dissect each of the figures below into two congruent pieces.

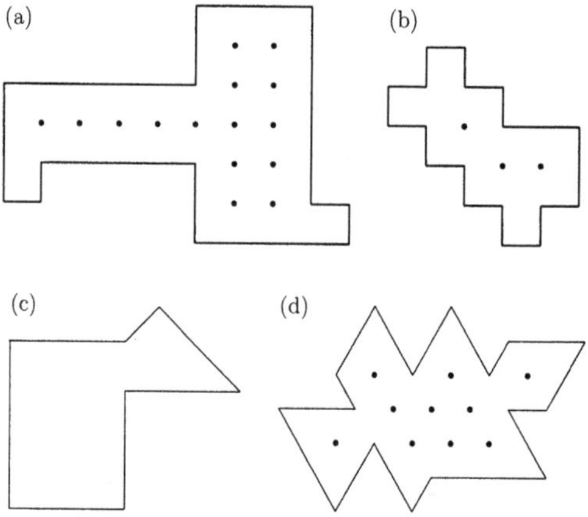

8. Dissect each of the figures in the diagram below into two congruent pieces.

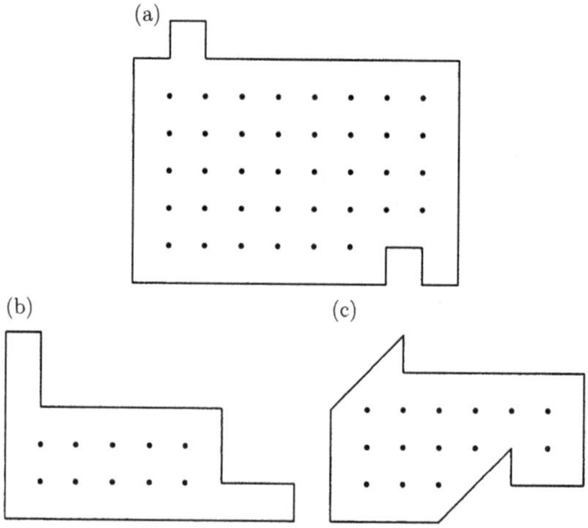

9. Dissect each of the figures in the diagram below into two congruent pieces.

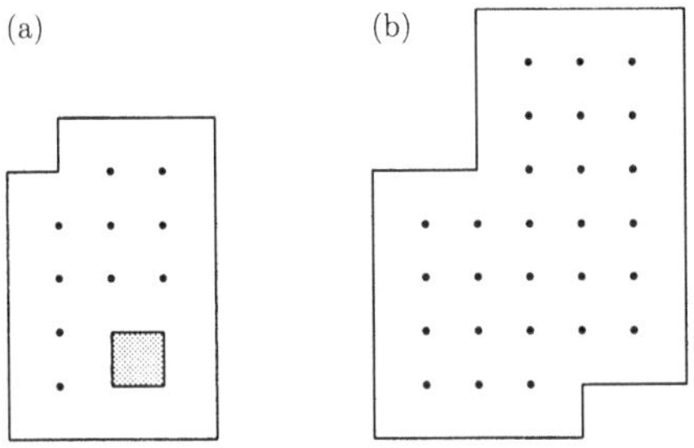

10. A square is divided into four smaller squares, and each small square is divided into four right isosceles triangles, as shown in the diagram on the left. One of the triangles is slid over, resulting in the figure shown in the diagram in the middle. This figure is to be dissected into four congruent pieces, and one solution is shown in the diagram on the right. Find an alternative solution.

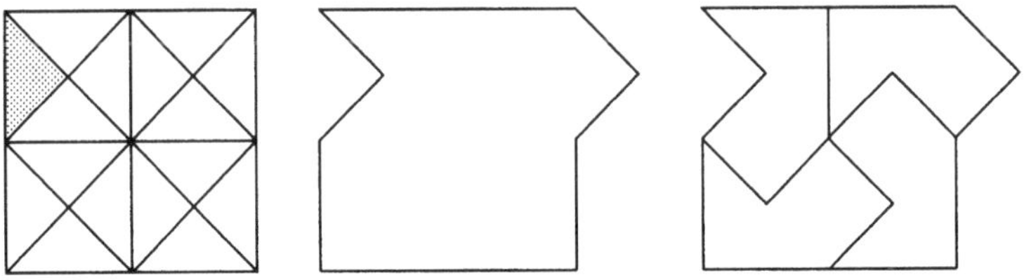

11. Dissect a circle into a number of congruent pieces such that at least one piece does not include the center of the circle, either in its interior or on its boundary. Find two solutions.

12. Find an integer root of each of the following equations:

(a) $$\frac{9(8-x)}{(9-8)x} + \frac{8-x}{9-8} + \frac{11-x}{x-1} = x.$$

(b) $$x = \frac{1-x}{x-11} + \frac{8-6}{x-8} + \frac{x(8-6)}{(x-8)6}.$$

13. Describe how to inscribe a parallelogram with center P in the quadrilateral $ABCD$ shown below.

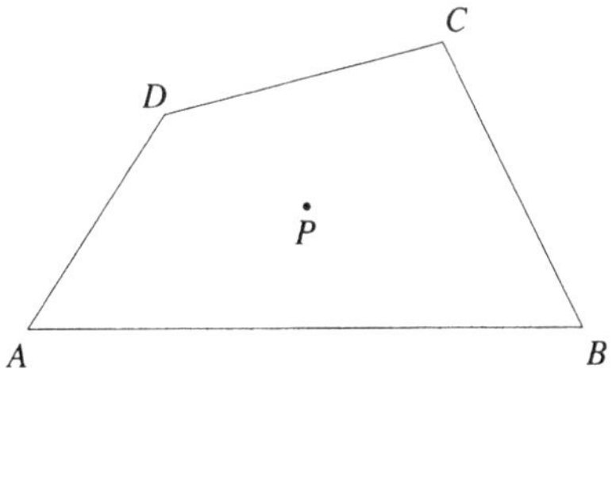

14. Given three parallel lines ℓ, m, and n construct an equilateral triangle with A on ℓ, B on m, and C on n.

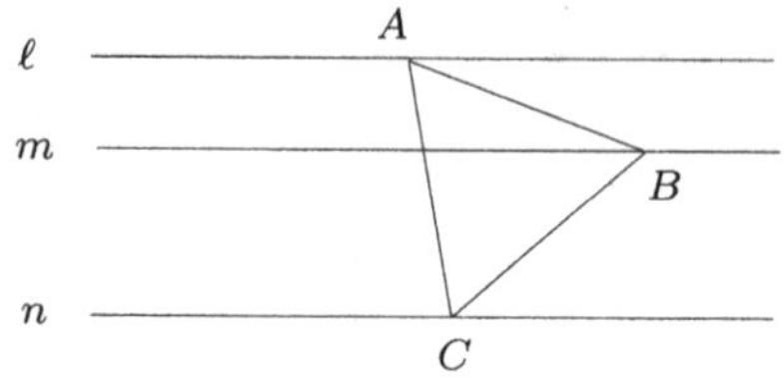

15. Given three concentric circles construct an equilateral triangle with one vertex on each circle.

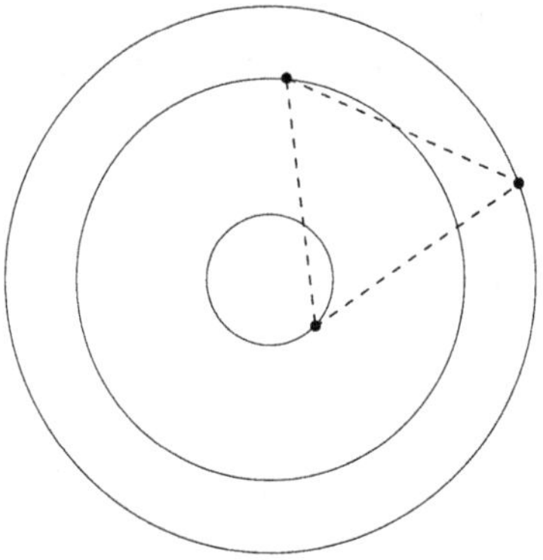

16. Describe how to find all points X on circle $\mathcal{C}_1$ and Y on circle $\mathcal{C}_2$ so that the segment $\overline{XY}$ is parallel and congruent to $\overline{PQ}$.

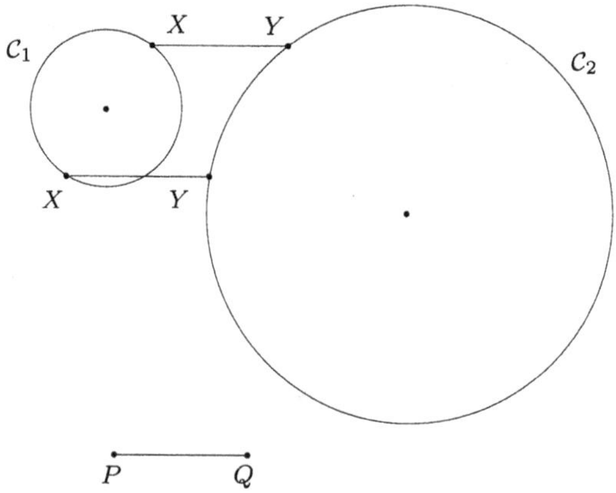

17. Describe how to find all points X on circle $\mathcal{D}_1$ and Y on circle $\mathcal{D}_2$ so that X, P, and Y are collinear and $XP = YP$.

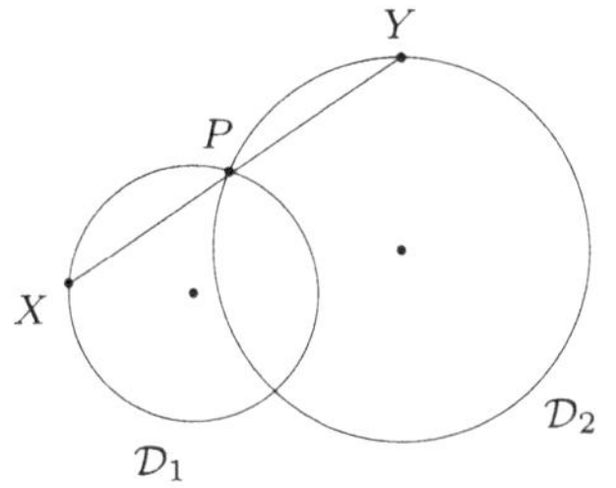

18. In the figure below, the points A, B, and C are collinear, AB is a diameter of $\mathcal{C}_0$, AC is a diameter of $\mathcal{C}_1$, and CB is a diameter of $\mathcal{C}_2$.

    (a) Construct a perpendicular to $\overline{AB}$ from C hitting $\mathcal{C}_0$ at D. Join $\overline{AD}$ and $\overline{DB}$ hitting $\mathcal{C}_1$ and $\mathcal{C}_2$ at E and F, respectively.

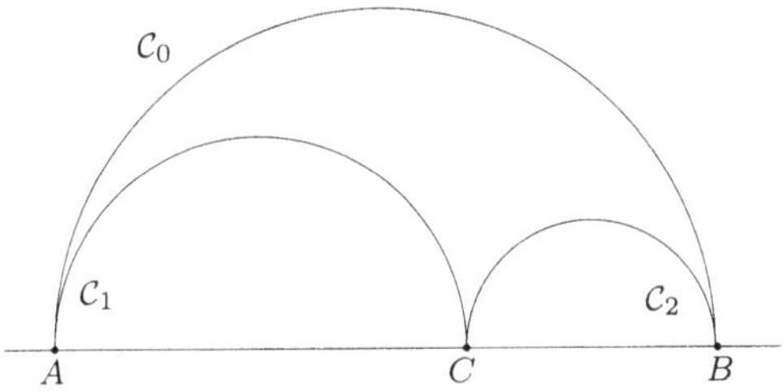

    (b) Explain why the quadrilateral $DECF$ is a rectangle and why the line through E and F is a common tangent to the circles $\mathcal{C}_1$ and $\mathcal{C}_2$.

19. Two facing mirrors OX and OY form an angle at O. A light ray $ABCD$ reflects off each mirror once, as shown in the figure on the right. If the ray's final direction is opposite to its initial direction, what is the measure of the angle between the mirrors?

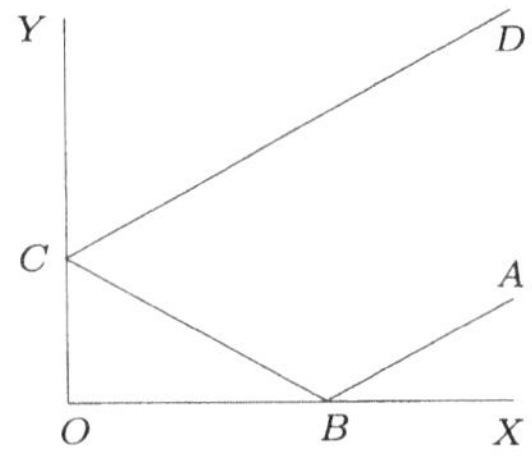

20. Two facing mirrors OX and OZ form an angle at O. A light ray $ABCDEF$ reflects off each mirror twice, as shown in the figure on the right. If the ray's final direction is opposite to its initial direction, what is the measure of the angle between the mirrors?

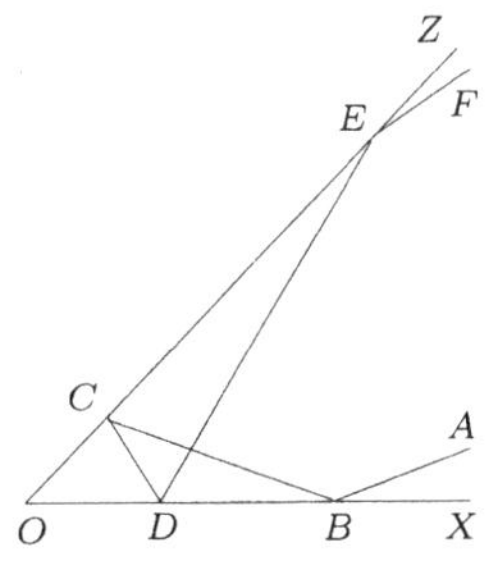

21. Two facing mirrors OX and OU form an angle at O. A light ray $ABCDEFGH$ reflects off each mirror thrice, as shown in the diagram below (G and H not shown). If the ray's final direction is opposite to its initial direction, what is the measure of the angle between the mirrors?

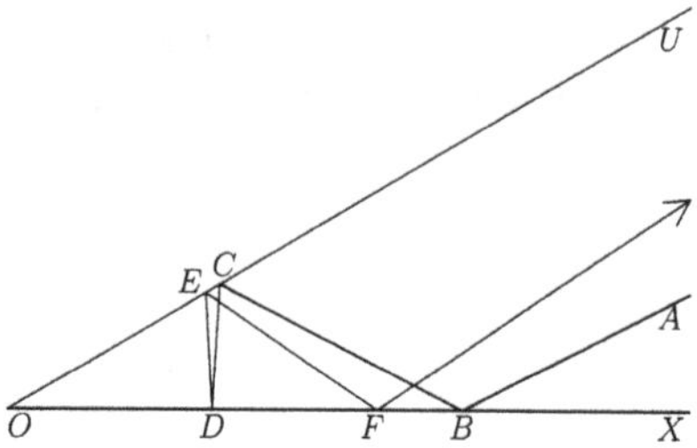

22. Prove that the perimeter of a quadrilateral with one vertex on each side of a rectangle has a perimeter no shorter than the sum of the diagonals of the rectangle.

23. A woodsman's hut is in the interior of a peninsula which has the form of an acute angle. The woodsman must leave his hut, walk to one shore of the peninsula, then to the other shore, then return home. How should he choose the shortest such path?

24. A woodsman's hut is in the interior of a peninsula which has the form of an obtuse angle. The woodsman must leave his hut, walk to one shore of the peninsula, then to the other shore, then return home. How should he choose the shortest such path?

25. The figure below represents one hole on a mini golf course. The ball B is 5 units from the west wall and 16 units from the south wall. The hole H is 4 units from the east wall and 8 units from the north wall. What is the length of the shortest path for the ball to go into the hole in one stroke, bouncing off a wall only once?

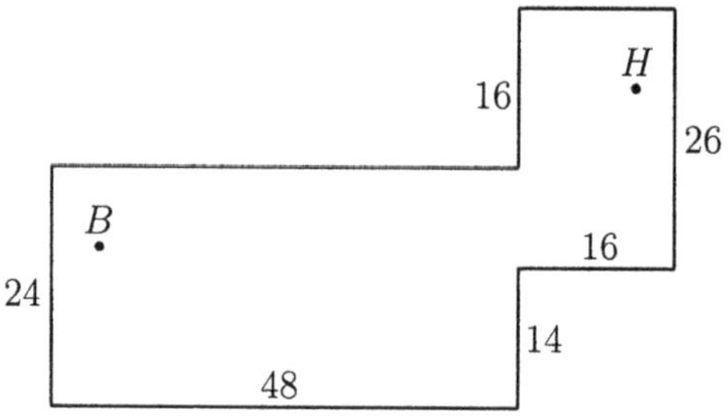

26. The diagram below represents a one hole on a mini golf course. The ball B is 5 units from the west wall and 16 units from the south wall. The hole H is 4 units from the east wall and 8 units from the north wall. What is the length of the shortest path for the ball to go into the hole in one stroke, bouncing off a wall only once?

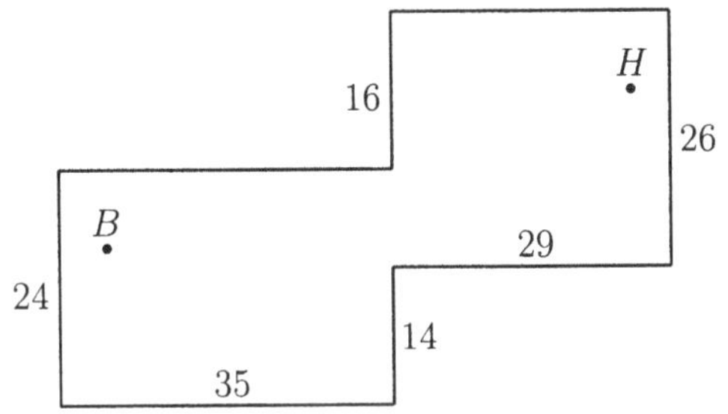

27. Two circles intersect at two points. Through one of these points P, construct a straight line intersecting the circles again at A and B such that $PA = PB$.

28. A point P lies outside a circle ω and on the same side of a straight line ℓ as ω. Construct a straight line through P intersecting ℓ at A and ω at B such that $PA = PB$. Find all solutions.

29. Two disjoint circles are on the same side of a straight line ℓ. Construct a tangent to each circle so that they intersect on ℓ and make equal angles with ℓ. Find all solutions.

30. Initially, only three points on the plane are painted: one red, one yellow and one blue. In each step, we choose two points of different colours. A point is painted in the third colour so that an equilateral triangle with vertices painted red, yellow, and blue in clockwise order is formed with the two chosen points. Note that a painted point may be painted again, and it retains all of its colours. Prove that after any number of moves, all points of the same colour lie on a straight line.

CHAPTER 8

THE ALGEBRA OF ISOMETRIES

8.1 Basic Algebraic Properties

Consider a large iron grate (G) in the shape of a right triangle, as shown in the figure below. The grate has to move from its current position to cover the hole (H). It must fit exactly within the dotted lines.

Classical Geometry: Euclidean, Transformational, Inversive and Projective, First Edition. By
I. E. Leonard, J. E. Lewis, A. C. F. Liu, G. W. Tokarsky
Copyright © 2014 John Wiley & Sons, Inc.

The grate is far too heavy to be lifted by hand, so a machine has been rigged that can lift the grate and flip it around any desired axis, thereby placing it on the other side of the turning axis. The action of the machine is shown in the figure below. If necessary, the axis can pass through the grate.

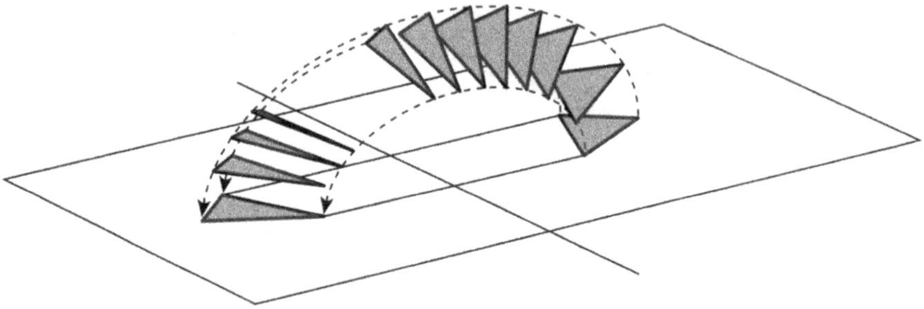

Find a sequence of flips that will move the grate to cover the hole. What is the minimum number of flips it will take to cover the hole?

The machine is a "reflecting" device—after the machine does its work, the new position of the grate is the reflection through the axis of the original position of the grate. The solution to the problem is that the grate can be moved in a step-by-step manner to cover the hole in three flips, and the minimum number of flips necessary is three. A sequence of flips is depicted in the figure below.

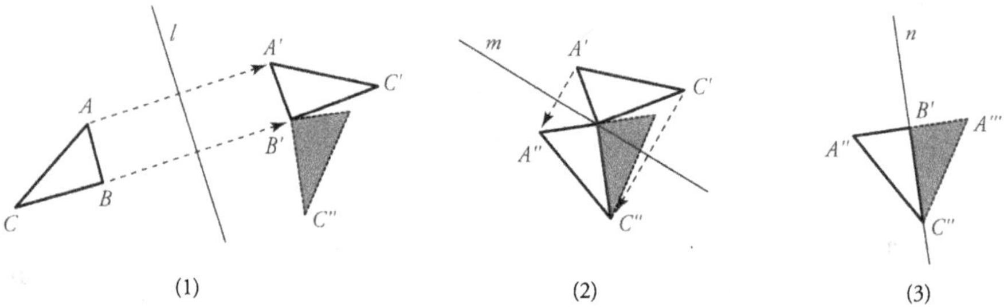

The first step uses a reflection through the line l, which is the right bisector of the segment joining the right angle vertex B of the grate with the corresponding vertex of the hole. With this reflection, the grate ABC is moved into position $A'B'C'$.

We now have one vertex in the correct position. Vertices A' and C' are not in the correct position, and the next step is to do another flip to get one of them into the correct place.

In (2), vertex C'' of the hole corresponds to vertex C' of the moved grate, and by relecting $A'B'C'$ about the right bisector of $C'C''$ we can move C' to the correct position.

Note that the right bisector of $C'C''$ passes through B' because $B'C' = B'C''$. After the second step, the grate is in position $A''B'C''$. The third and final step is to reflect $A''B'C''$ about the line $B'C''$.

The Composition of Transformations

Earlier we defined the composition of two isometries. This terminology is used for any collection of transformations. If T and S are two transformations, the ***composition*** or ***product*** of T and S is denoted by either $S \circ T$ or ST. In terms of the individual points in the plane, ST acts as follows. First, the point X is mapped by T to $T(X)$, then $T(X)$ is mapped by S to $S(T(X))$. In other words, for each point X in the plane:

$$ST(X) = S(T(X)).$$

To move the grate to its desired position, we applied the product $\mathbf{R}_n\mathbf{R}_m\mathbf{R}_l$ of three reflections. Again, we emphasize that we follow the "right-to-left" rule. When evaluating $\mathbf{R}_n\mathbf{R}_m\mathbf{R}_l(X)$, first find $\mathbf{R}_l(X)$, then $\mathbf{R}_m(\mathbf{R}_l(X))$, and finally $\mathbf{R}_n(\mathbf{R}_m(\mathbf{R}_l(X)))$.

Equal Transformations

In the iron grate problem, we used the product of three reflections to shift the grate to its desired position. If we had a better machine—one that could perform rotations as well as reflections—we could accomplish the same thing by performing just two operations, as shown in the figure below.

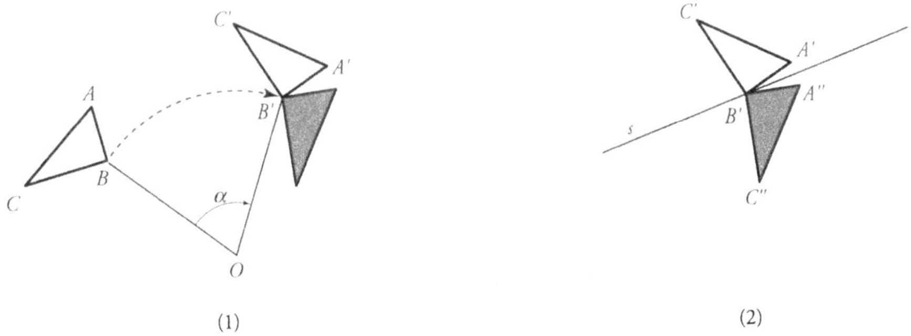

In this case, the grate is moved by applying the product $\mathbf{R}_s\mathbf{R}_{O,\alpha}$. In the first solution, we used $\mathbf{R}_n\mathbf{R}_m\mathbf{R}_l$. The individual transformations that make up the two products are quite different, but the net effect is the same. So we can write:

$$\mathbf{R}_s\mathbf{R}_{O,\alpha} = \mathbf{R}_n\mathbf{R}_m\mathbf{R}_l.$$

Two transformations are said to be ***equal*** if they have the same effect on every point in the plane. In other words, saying that $T = S$ means that $T(X) = S(X)$ for *every* point X in the plane.

As the example shows, $T = \mathbf{R}_s\mathbf{R}_{O,\alpha}$ and $S = \mathbf{R}_n\mathbf{R}_m\mathbf{R}_l$ does not mean that T and S necessarily have the same description. The situation is similar to equality of functions in trigonometry: if

$$f(x) = 1 - 2\sin^2 x \qquad \text{and} \qquad g(x) = \cos 2x,$$

then the functions f and g are equal (since we have pointwise equality), although their descriptions are quite different.

Other examples of equal transformations that are defined differently are the *halfturn* and a *reflection in a point.*

The ***halfturn*** about a point O, denoted by $\mathbf{H}_O$, is the transformation $\mathbf{R}_{O,180^\circ}$. The ***reflection in the point*** O is the transformation that takes each point P to the point P' so that O is the midpoint of PP'. The point O is called the ***center of the halfturn*** or the ***center of reflection***. In the plane, since a reflection through a point is identical to a halfturn, there is no further need to talk about reflections through a point, and there is really no need for a special symbol to denote it.

Closure

In Chapter 7, we mentioned that the composition of two isometries results in a transformation that is also an isometry. In algebra, we would describe the situation by saying that the set of isometries of the plane is *closed* under the operation of composition.

More generally, if $\mathcal{S}$ is a set of elements and if $\circ$ is a binary operation on $\mathcal{S}$, we say that $\mathcal{S}$ is ***closed*** under the operation $\circ$ if for every pair of elements a and b in $\mathcal{S}$ the product $a \circ b$ is also in $\mathcal{S}$. For example, the set of positive integers is closed under addition.

Associativity

If R, S, and T are three transformations, the product TSR means first apply R, then S, and then T. For a point X, the notation $TSR(X)$ means $T\left(S\left(R(X)\right)\right)$.

We can overrule this by using parentheses since the operations inside parentheses are carried out first. The notation $(TS)\,R$ is interpreted as follows: first, determine what (TS) is. It will be some transformation, call it H, and H is usually different than either T or S. The notation $(TS)\,R$ means HR; that is, first apply the transformation R, then apply the transformation H.

For example, let us suppose that S and T are reflections about two different parallel lines and R is a rotation. Then, as we will see later, TS is some translation H. So we interpret $(TS)\,R$ as meaning first do the rotation R, then follow it by the translation H.

In a similar way, the notation $T\,(SR)$ means that we should first determine what SR is, namely, some different isometry L, and then take the product of L and T: $T\,(SR) = TL$. Note, however, that the parentheses in this case could be omitted, because in the absence of parentheses, the expression is evaluated from right to left.

The following theorem shows that $(TS)\,R$ and $T\,(SR)$ are equal.

Theorem 8.1.1. *(Associative Law)*

The **associative law** *holds for the product of transformations; that is, given three transformations* T, S, *and* R *in the plane,*

$$T\,(SR) = (TS)\,R.$$

Proof. Let X be a point in the plane. We will evaluate $T\,(SR)\,(X)$ and $(TS)\,R(X)$.

The notation $T\,(SR)\,(X)$ means "first evaluate $SR(X)$, then evaluate $T\,(SR(X))$." By definition, $SR(X) = S\,(R(X))$, so

$$T\,(SR)\,(X) = T\,(S\,(R(X)))\,.$$

The notation $(TS)\,R(X)$ tells us to evaluate $TS\,(R(X))$. Now, $TS(Z) = T\,(S(Z))$ for all Z in the plane. In particular, when $Z = R(X)$, we get

$$TS(Z) = T\,(S(Z)) = T\,(S\,(R(X)))\,.$$

Thus,

$$(TS)\,R(X) = T\,(S\,(R(X)))\,.$$

Since $(TS)\,R$ and $T\,(SR)$ have the same effect on every point X in the plane, we conclude that they are equal transformations.

□

8.2 Groups of Isometries

In algebra, a set of elements $\mathcal{G}$ together with a binary operation $\cdot$ is called a ***group*** and is denoted by $(\mathcal{G}, \cdot)$ if it possesses the following properties:

1. The set $\mathcal{G}$ is closed under the binary operation.
2. The associative law holds: $(a \cdot b) \cdot c = a \cdot (b \cdot c)$ for all a, b, and c in $\mathcal{G}$.
3. $\mathcal{G}$ has an ***identity element***: there is some element e in $\mathcal{G}$ such that $e \cdot a = a \cdot e = a$ for every a in $\mathcal{G}$.
4. For every a in $\mathcal{G}$, there is an ***inverse element*** a': there is an element a' in $\mathcal{G}$ such that $a \cdot a' = a' \cdot a = e$.

If it is also true that the commutative law holds (that is, if $a \cdot b = b \cdot a$ for all a and b in $\mathcal{G}$), then $\mathcal{G}$ is called an ***Abelian group***.

The notation for the binary operation $\cdot$ is usually omitted, so that we write ab instead of $a \cdot b$.

We learned in Chapter 7 that every isometry has an inverse. This, along with the results of the previous section, show that the family of isometries in the plane, together with the composition operation, satisfy all four of the conditions listed above. We can summarize this with the following theorem:

Theorem 8.2.1. *The set of all isometries of the plane, together with the operation of composition, forms a group.*

Given an isometry $\mathbf{T}$, we denote its inverse by $\mathbf{T}^{-1}$. Theorem 7.2.1 stated that

$$(\mathbf{R}_{P,\theta})^{-1} = \mathbf{R}_{P,-\theta},$$

$$(\mathbf{R}_l)^{-1} = \mathbf{R}_l,$$

$$(\mathbf{T}_{AB})^{-1} = \mathbf{T}_{BA}.$$

Example 8.2.2. *What are the inverses of* $\mathbf{G}_{l,AB}$ *and* $\mathbf{H}_O$*?*

Solution. We have

$$(\mathbf{G}_{l,AB})^{-1} = \mathbf{G}_{l,BA} \qquad \text{and} \qquad (\mathbf{H}_O)^{-1} = \mathbf{H}_O.$$

□

In any group, an element a other than the identity e such that $a^2 = a \cdot a = e$ is called an ***involution***. In other words, an involution is an element other than the identity that is its own inverse.

Any transformation that is an involution in a group of transformations is said to be an ***involutory*** or ***involutoric*** transformation; that is, T is involutory if and only if $TT = T \circ T = I$. Among the transformations

$$\mathbf{R}_{P,\theta}, \quad \mathbf{R}_l, \quad \mathbf{T}_{AB}, \quad \mathbf{G}_{l,AB}, \quad \text{and} \quad \mathbf{H}_P,$$

the involutions are $\mathbf{R}_l$, $\mathbf{H}_P$, and $\mathbf{R}_{P,\theta}$ when θ is a multiple of 180°.

8.2.1 Direct and Opposite Isometries

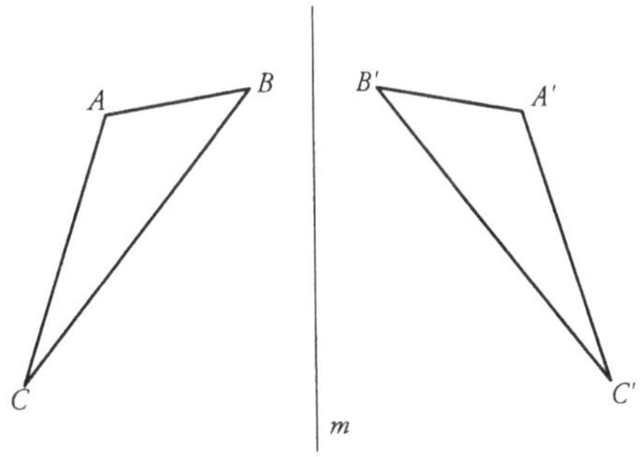

In the figure above, triangle ABC has been carried into $A'B'C'$ by a reflection in the line m. The orientation of the triangle has changed—the path ABC proceeds in a clockwise (or negative) direction around the triangle, while the image path $A'B'C'$ goes in a counterclockwise (or positive) direction. This reversal of orientation will obviously happen to every triangle to which $\mathbf{R}_m$ is applied, and $\mathbf{R}_m$ is therefore called an *opposite* isometry. In general, any isometry T that reverses the orientation of every triangle is called an ***opposite isometry***. On the other hand, an isometry T that preserves the orientation of every triangle is said to be a ***direct isometry***.

There is a strong analogy between the products of direct and opposite isometries and the products of positive and negative numbers: direct isometries are like positive numbers and opposite isometries are like negative numbers. Some people prefer to draw an analogy to the addition of even and odd numbers, with direct isometries corresponding to the even numbers, so that the product of a direct and opposite isometry is like the sum of an even and an odd number. Some texts use the terms *even* and *odd* isometries instead of *direct* and *opposite* isometries.

Theorem 8.2.3. *The product of two direct isometries is a direct isometry. The product of two opposite isometries is a direct isometry. The product of a direct isometry and an opposite isometry is an opposite isometry.*

Theorem 8.2.4. *The direct isometries of the plane form a group $\mathcal{D}$.*

Proof. It is clear that $\mathcal{D}$ is closed (under multiplication). Since the associative law holds for all transformations, it also holds for $\mathcal{D}$. The identity map is a direct isometry; that is, I is in $\mathcal{D}$. Every isometry has an inverse, and an isometry and its inverse are either both direct or both indirect. Therefore, whenever T is in $\mathcal{D}$, so is T^{-1}. This completes the proof.

□

It is relatively easy to invent a transformation of the plane that is neither direct nor opposite. For example, let B and C be two different points in the plane, and let T be the transformation that interchanges B and C but leaves every other point where it is. The transformation T is almost the identity map, and there are many triangles PQR that remain unchanged under T. However, if A is a point that forms a triangle with B and C, the transformation maps A, B, and C to A, C, and B, respectively, thereby reversing the orientation. In other words, T preserves the orientation of some triangles while reversing the orientation of others. Thus, T is neither direct nor opposite.

The situation where T is an isometry is quite different as a consequence of two fundamental theorems about isometries in the plane, namely, that a plane isometry is completely determined by its action on three noncollinear points and that every plane isometry is the product of at most three reflections.

We begin the proofs of these facts with a theorem that characterizes the identity transformation.

Theorem 8.2.5. *If T is an isometry that fixes each of three noncollinear points, then T is the identity.*

Proof. Let A, B, and C be three collinear points that are fixed by T, so that

$$T(A) = A, \qquad T(B) = B, \qquad \text{and} \qquad T(C) = C.$$

We want to show that $T(X) = X$ for every point X in the plane. Suppose that this is not the case. Then there is a point P such that

$$P' = T(P) \neq P.$$

Since T is an isometry with $T(A) = A$ and $T(P) = P'$, we have

$$\text{dist}(A, P) = \text{dist}\left(T(A), T(P)\right) = \text{dist}(A, P').$$

This means that A is on the right bisector of PP'. Similarly, B and C must also be on the right bisector of PP', contradicting the fact that A, B, and C are noncollinear. This contradiction shows that our supposition that T is not the identity must be false and completes the proof.

□.

When we say that a transformation T *fixes* a point X we mean that $T(X) = X$, and the point X is called a ***fixed point*** of T. In Martin's text [50], he uses the phrase "T *fixes* a set $\mathcal{S}$" to mean that under T the image of $\mathcal{S}$ is $\mathcal{S}$. He uses the phrase "T *fixes* $\mathcal{S}$ *pointwise*" to means that $T(X) = X$ for each point X of $\mathcal{S}$. Another way of saying the same thing is to say that the set $\mathcal{S}$ is *invariant* under the transformation T.

As an example of the difference in the meaning of this language, consider the mapping of an equilateral triangle ABC onto itself by a rotation of $120°$ around the centroid. The triangle ABC is fixed by the transformation, but it is *not fixed pointwise*.

Question 8.2.6. *Suppose that an isometry fixes each of two different points. Can we assume that the isometry is the identity?*

One of the virtues of group theory is that once you have proven a theorem about groups, it is valid for every group and does not have to be proven separately for each different group. Here is a very simple yet useful result:

Theorem 8.2.7. *Let a and b be elements of a group $(\mathcal{G}, \cdot)$ such that $a \cdot b = e$, where e is the identity. Then $a = b^{-1}$ and $b = a^{-1}$.*

Proof. Since $\mathcal{G}$ is a group, the element a has an inverse a^{-1}. Multiplying each side of the equation

$$a \cdot b = e$$

by a^{-1} on the left, we have

$$a^{-1} \cdot (a \cdot b) = a^{-1} \cdot e,$$

so that

$$\left(a^{-1} \cdot a\right) \cdot b = a^{-1},$$

and so

$$e \cdot b = a^{-1},$$

and thus, $b = a^{-1}$.

The proof that $a = b^{-1}$ can be obtained by multiplying the equation $a \cdot b = e$ on the right by b^{-1}.

□

The following theorem can be proven in much the same way as Theorem 8.2.5, but we can give a more satisfying proof by using the fact that the isometries of the plane form a group. This proof illustrates how nicely algebra fits with geometry.

Theorem 8.2.8. *An isometry of the plane is completely determined by its action on three noncollinear points.*

Proof. Let A, B, and C be three noncollinear points in the plane, and let S and T be two isometries such that

$$S(A) = T(A), \qquad S(B) = T(B), \qquad \text{and} \qquad S(C) = T(C).$$

We want to show that $S = T$. Now, since the isometries of the plane form a group, we know that T has an inverse T^{-1} so that

$$T^{-1}S(A) = A, \qquad T^{-1}S(B) = B, \qquad \text{and} \qquad T^{-1}S(C) = C.$$

Thus, the isometry $T^{-1}S$ fixes each of the points A, B, and C, and by Theorem 8.2.5, this means that $T^{-1}S$ must be the identity. By the previous theorem, S must be equal to the inverse of T^{-1}; that is,

$$S = \left(T^{-1}\right)^{-1} = T.$$

□

Theorem 8.2.9. *Every isometry of the plane that is not the identity can be decomposed into the product of at most three reflections.*

Proof. Let T be a given isometry, and let A, B, and C be three noncollinear points.

From the iron grate example, we know how to map A, B, and C to $T(A)$, $T(B)$, and $T(C)$, respectively, by a sequence of at most three reflections.

Suppose that it actually took three reflections, say,

$$\mathbf{R}_l, \qquad \mathbf{R}_m, \qquad \text{and} \qquad \mathbf{R}_n,$$

in that order. Then

$$\mathbf{R}_n\mathbf{R}_m\mathbf{R}_l(A) = A' = T(A),$$

$$\mathbf{R}_n\mathbf{R}_m\mathbf{R}_l(B) = B' = T(B),$$

$$\mathbf{R}_n\mathbf{R}_m\mathbf{R}_l(C) = C' = T(C).$$

Thus, $\mathbf{R}_n\mathbf{R}_m\mathbf{R}_l$ and T both have exactly the same effect on A, B, and C, so $\mathbf{R}_n\mathbf{R}_m\mathbf{R}_l = T$, by Theorem 8.2.8.

□

Theorem 8.2.9, together with the fact that a reflection in a line is an opposite isometry, now allows us to show that every isometry in the plane is either a direct isometry or an opposite isometry.

Theorem 8.2.10. *Every isometry in the plane is either a direct isometry or an opposite isometry.*

Proof. If the isometry T is the identity or the product of two reflections, it is a direct isometry. If T is a reflection or the product of three reflections, it is an opposite isometry.

□

8.3 The Product of Reflections

In this section, we use reflections to show that there are only four different types of plane isometries: reflections (in a line), rotations, translations, and glide reflections. We begin by examining the product of two reflections.

Example 8.3.1. *Let l and m be two distinct parallel lines. Show that $\mathbf{R}_m\mathbf{R}_l = \mathbf{T}_{XY}$, where $\overline{XY}$ is a directed segment perpendicular to l and m and twice the distance from l to m.*

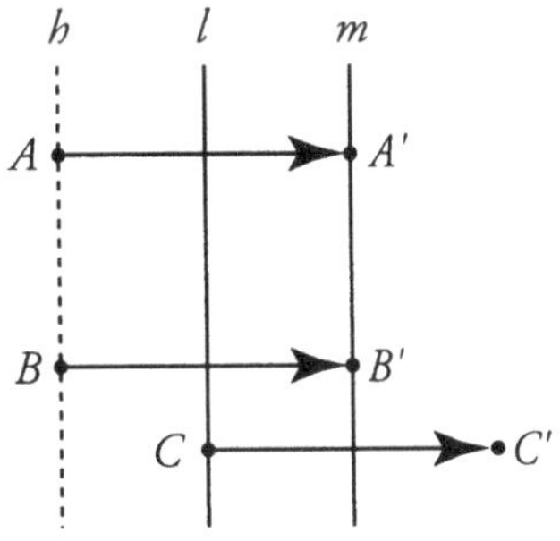

Solution. Let h be a line parallel to l and m such that l is midway between h and m, as in the figure above.

Let A and B be points on h and let C be a point on l. $\mathbf{R}_l$ maps A and B to points A' and B' on m and leaves C where it is.

Note that the two directed segments $\overline{AA'}$ and $\overline{BB'}$ are parallel and twice the distance from l to m.

The reflection $\mathbf{R}_m$ leaves A' and B' where they are and maps C to C', where $\overline{CC'}$ is parallel to and equal in length and direction to $\overline{AA'}$.

Thus, the effect of $\mathbf{R}_m\mathbf{R}_l$ is to map A, B, and C to A', B', and C', respectively.

However, the translation $\mathbf{T}_{AA'}$ also does the same thing, and by Theorem 8.2.8,

$$\mathbf{R}_m\mathbf{R}_l = \mathbf{T}_{AA'},$$

which completes the proof. □

Example 8.3.2. *Let l and m be two different lines intersecting at a point P. Show that $\mathbf{R}_m\mathbf{R}_l = \mathbf{R}_{P,\theta}$, where θ is twice the directed angle α from l to m.*

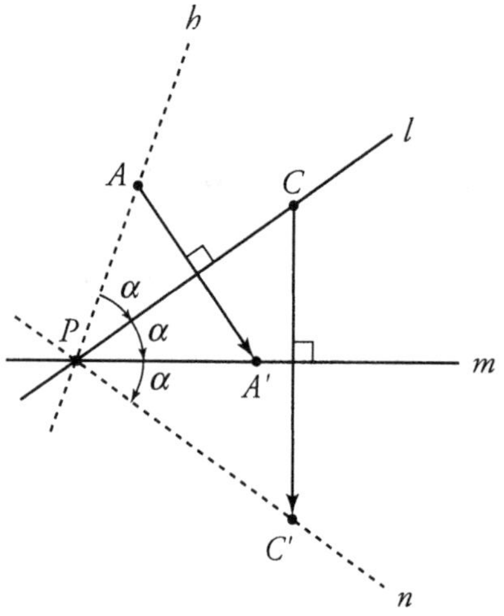

Solution. Let h be the line through P such that the directed angle from h to l is α, and let n be the line through P such that the directed angle from m to n is α, as in the figure above.

Let A be a point on h and let C be a point on l. Note that A, C, and P cannot be collinear. If they were collinear, then $h = l$, and it would follow that l, h, and m are all the same line, which is a contradiction.

Consider the effect of $\mathbf{R}_m\mathbf{R}_l$ on the points A, P, and C. It is clear that $\mathbf{R}_{P,\theta}$ has exactly the same effect, so by Theorem 8.2.8, $\mathbf{R}_m\mathbf{R}_l = \mathbf{R}_{P,\theta}$.

□

If l and m are the same line, then $\mathbf{R}_m\mathbf{R}_l$ is the identity. This fact, together with the previous two examples, proves the following theorem:

Theorem 8.3.3. *The only direct isometries of the plane are the identity, the translations, and the rotations.*

We next turn our attention to the product of three reflections.

Theorem 8.3.4. *Let l, m, and n be three lines with a common point P. Then $\mathbf{R}_n\mathbf{R}_m\mathbf{R}_l$ is a reflection.*

Proof. If l and m coincide, then

$$\mathbf{R}_n\mathbf{R}_m\mathbf{R}_l = \mathbf{R}_n\left(\mathbf{R}_m\mathbf{R}_l\right) = \mathbf{R}_n\,\mathbf{I} = \mathbf{R}_n.$$

If n and m coincide, then

$$\mathbf{R}_n\mathbf{R}_m\mathbf{R}_l = \left(\mathbf{R}_n\mathbf{R}_m\right)\mathbf{R}_l = \mathbf{I}\,\mathbf{R}_l = \mathbf{R}_l.$$

If neither of these two cases occurs, we proceed as follows: let α be the directed angle from l to m. Now, there is a unique line h through P such that the angle from h to n is α.

Thus,

$$\mathbf{R}_m\mathbf{R}_l = \mathbf{R}_n\mathbf{R}_h = \mathbf{R}_{P,2\alpha},$$

and, therefore,

$$\mathbf{R}_n\mathbf{R}_m\mathbf{R}_l = \mathbf{R}_n\mathbf{R}_n\mathbf{R}_h = \mathbf{I}\,\mathbf{R}_h = \mathbf{R}_h.$$

□

To handle the combination $\mathbf{R}_n\mathbf{R}_m\mathbf{R}_l$ when the lines l, m, and n are all different and have no point in common, we break it down into separate cases. First, we consider the cases where l and m are both perpendicular to n, then the cases where m and n are both perpendicular to l.

We note first that if all three lines are parallel, then $\mathbf{R}_n\mathbf{R}_m\mathbf{R}_l$ is a reflection. We leave the proof of this as an exercise.

Theorem 8.3.5. *Let l, m, and n be three different lines.*

(1) *If l and m are perpendicular to n, then $\mathbf{R}_n\mathbf{R}_m\mathbf{R}_l$ is the glide reflection $\mathbf{G}_{n,AB}$, where AB is a line segment parallel to n and twice the directed distance from l to m.*

(2) *If m and n are perpendicular to l, then $\mathbf{R}_n\mathbf{R}_m\mathbf{R}_l$ is the glide reflection $G_{l,CD}$, where CD is a line segment parallel to l and twice the directed distance from m to n.*

Proof. For (1), we have $\mathbf{R}_m\mathbf{R}_l = \mathbf{T}_{AB}$ by Example 8.3.1, and so $\mathbf{R}_n\mathbf{R}_m\mathbf{R}_l$ is the glide reflection $\mathbf{R}_n\mathbf{T}_{AB} = \mathbf{G}_{n,AB}$, as claimed.

For (2), Let $A = l \cap m$, let $B = l \cap n$, and let C be any point on m other than A. We leave it as an exercise to show that

$$\mathbf{R}_n\mathbf{R}_m\mathbf{R}_l(X) = \mathbf{R}_l\mathbf{R}_n\mathbf{R}_m(X)$$

for each of the points $X = A, B, C$ so that

$$\mathbf{R}_n\mathbf{R}_m\mathbf{R}_l = \mathbf{R}_l\mathbf{R}_n\mathbf{R}_m$$

by the fundamental Theorem 8.2.8, and $\mathbf{R}_l\mathbf{R}_n\mathbf{R}_m$ is the glide reflection $G_{l,CD}$ by (1).

□

The cases not covered by the previous theorem are depicted in the figures below, which show the possible arrangements of the reflecting lines.

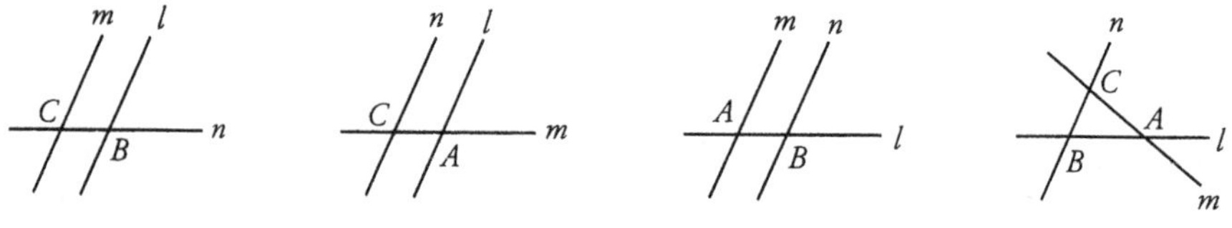

(1) (2) (3) (4)

Although arrangements (1), (2), and (3) appear to be identical, they are not the same because the order of the reflections matters. It may appear that these cases are complicated, but the associative law comes to the rescue in a rather magnificent way.

Theorem 8.3.6. *Let l, m, and n be three nonconcurrent lines. Then $\mathbf{R}_n\mathbf{R}_m\mathbf{R}_l$ is a glide reflection.*

Proof. The proof of each of the four cases is very similar. Here is how it is done for case (4).

The combination $\mathbf{R}_m\mathbf{R}_l$ is a rotation about A through the angle 2α, where α is the directed angle from l to m, as in (1) below. Let m' be a line through A perpendicular to n, intersecting n at C'. Let l' be the line through A such that the angle from l' to m' is α, as in (2) below. Then

$$\mathbf{R}_n\mathbf{R}_m\mathbf{R}_l = \mathbf{R}_n\left(\mathbf{R}_m\mathbf{R}_l\right) = \mathbf{R}_n\left(\mathbf{R}_{m'}\mathbf{R}_{l'}\right).$$

Now apply the associative law:

$$\mathbf{R}_n\left(\mathbf{R}_{m'}\mathbf{R}_{l'}\right) = \left(\mathbf{R}_n\mathbf{R}_{m'}\right)\mathbf{R}_{l'}.$$

The transformation $\mathbf{R}_n\mathbf{R}_{m'}$ is the rotation $\mathbf{H}_{C'}$. Let n'' be the line through C' perpendicular to l', and let m'' be the line through C' perpendicular to n'', as in (3) below. Then

$$\mathbf{R}_{n''}\mathbf{R}_{m''} = \mathbf{H}_{C'}.$$

Therefore,

$$\left(\mathbf{R}_n\mathbf{R}_{m'}\right)\mathbf{R}_{l'} = \left(\mathbf{R}_{n''}\mathbf{R}_{m''}\right)\mathbf{R}_{l'} = \mathbf{R}_{n''}\left(\mathbf{R}_{m''}\mathbf{R}_{l'}\right),$$

and since m'' and l' are parallel and both perpendicular to n'', we have a glide reflection.

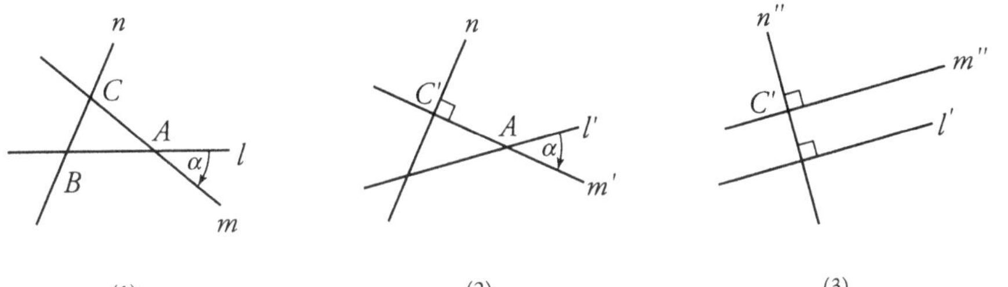

(1) (2) (3)

□

8.4 Problems

1. Let S and T be two involutive transformations of the plane.

   (a) Prove that ST is involutive if and only if $ST = TS$.

   (b) Assume that S, T, and I are distinct transformations, where I is the identity, such that

   $$ST = TS = X.$$

   Let $\Gamma = \{I, S, T, X\}$. Prove that Γ is a commutative subgroup of $\mathcal{G}$, the group of all transformations on the plane, by constructing the multiplication table.

2. Let P, Q, and R be three points in the plane, and let P', Q', and R', respectively, be their images under an isometry T. Show that the points P, Q, and R are collinear, with Q between P and R, if and only if the points P', Q', and R' are collinear, with Q' between P' and R'.

   Hint: When does equality hold in the Triangle Inequality?

3. Let T be an isometry of the plane. Show that if P and Q are fixed points of T, then every point X on the line through P and Q is a fixed point of T.

4. Let T be an isometry of the plane. Show that if T has three fixed points that are not collinear, then $T = I$, the identity.

5. Let S and T be isometries and let A, B, and C be three noncollinear points for which

   $$S(A) = T(A), \qquad S(B) = T(B), \qquad \text{and} \qquad S(C) = T(C).$$

   Show that $S = T$.

6. Let H_A be a halfturn about a point A so that

   $$H_A(P) = P',$$

   where A, P, and P' are collinear and $d(A, P) = d(A, P')$.

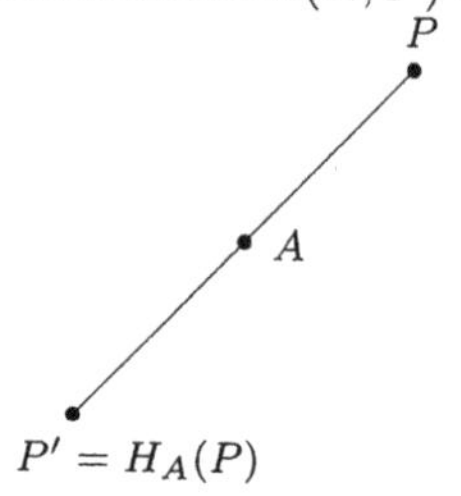

   (a) Show that H_A is an isometry.

   (b) Show that H_A is an involution; that is, $H_A = H_A^{-1}$.

   (c) Show that if ℓ is a line in the plane, then $H_A(\ell)$ is a line parallel to ℓ.

7. Let D, E, and F be the midpoints of the sides BC, AC, and AB, respectively, of $\triangle ABC$ and let T be the transformation of the plane that is the product of the three halfturns
$$T = H_E\, H_D\, H_F.$$
Show that the vertex A of $\triangle ABC$ is a fixed point of T; that is, that $T(A) = A$.

Hint. Draw the picture.

8. Let T be an isometry of the plane that admits an invariant line ℓ (that is, $T(\ell) = \ell$) and a fixed point P. Prove that there is a point $Q \in \ell$ such that $T(Q) = Q$ and a line ℓ' through P such that $T(\ell') = \ell'$.

9. Show that if a circle is invariant under the isometry T, then its center is a fixed point of T.

10. Let $T \neq I$ be an involutive isometry. Show that T has at least one fixed point.

11. Let T be an isometry that is an involution and has **exactly** one fixed point O in the plane. Show that T is the halfturn H_O about the point O.

12. If the isometry T is an involution, show that for any point P in the plane the midpoint of the line segment joining P and $T(P)$ is a fixed point of T.

13. Let T be an isometry of the plane and let ℓ be the perpendicular bisector of the segment $\overline{AB}$. Prove that $T(\ell)$ is the perpendicular bisector of the segment $\overline{T(A)\,T(B)}$.

14. Let R_ℓ be a reflection in the line ℓ so that
$$R_\ell(P) = P',$$
where either ℓ is the perpendicular bisector of the segment PP' for each P or P and P' coincide on the line ℓ for each P.

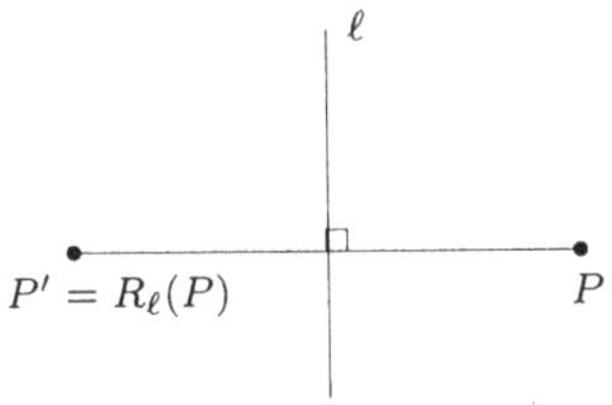

(a) Show that R_ℓ is an isometry.

(b) Show that R_ℓ is an involution; that is, that $R_\ell = R_\ell^{-1}$.

(c) Show that if m is a line distinct from ℓ, then $R_\ell(m)$ is distinct from ℓ.

15. Show that if m and n are perpendicular lines that intersect at a point P in the plane, then

$$R_n\, R_m = H_P.$$

16. Given a point O and a directed segment $\overline{AB}$.

    (a) Find the point Q such that

    $$T_{AB} H_O T_{AB}^{-1} = H_Q.$$

    (b) What is the product $H_O T_{AB}$?

17. Let A and C be distinct points in the plane. Show that B is the midpoint of the segment $\overline{AC}$ if and only if

$$H_C\, H_B = H_B\, H_A.$$

18. In the triangle $\triangle ABC$, show that G is the centroid if and only if

$$H_G H_C H_G H_B H_G H_A = I,$$

where I is the identity.

19. Using halfturns, prove that the diagonals of a parallelogram bisect each other.

    Hint. Show that if N is the midpoint of the diagonal AC of parallelogram $ABCD$ so that $H_A\, H_N = H_N\, H_C$, then N is the midpoint of BD also, that is, $H_D\, H_N = H_N\, H_B$.

20. Using halfturns, prove that if $ABCD$ and $EBFD$ are parallelograms, $EAFC$ is also a parallelogram.

21. Find all triangles such that three given noncollinear points are the midpoints of the sides of the triangle.

    Hint. Given P, Q, and R, $H_R H_Q H_P$ fixes a vertex of a unique triangle $\triangle P'Q'R'$, as in the figure below.

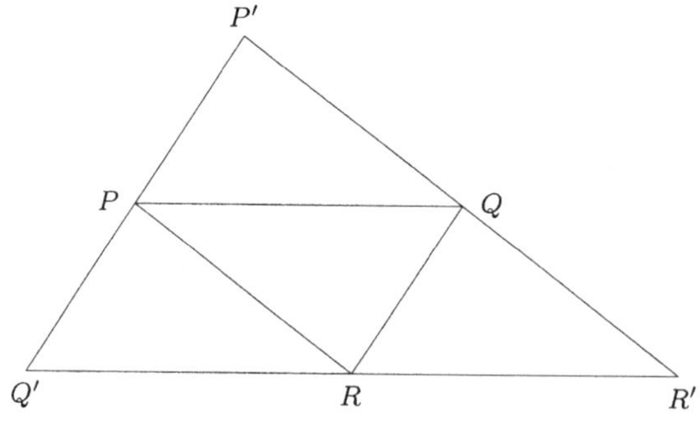

22. Given $\angle ABC$, construct a point P on AB and a point Q on BC such that $PQ = AB$, and the line PQ intersects the line BC at an angle of 60°.

    Hint. Take a point D such that $AB = BD$ and BD intersects BC at an angle of 60°.

23. Prove that if $R_n R_m$ fixes the point P and $m \neq n$, then the point P is on both lines m and n.

24. Show that if m and n are distinct lines in the plane, then

$$R_n R_m = R_m R_n$$

    if and only if m and n are perpendicular.

25. Let m be a line with equation $2x + y = 1$. Find the equations of the transformation R_m.

26. Suppose that the lines ℓ and m have equations $x + y = 0$ and $x - y = 1$, respectively. Find the equations for the transformation $R_\ell R_m$.

27. Given triangles ABC and DEF, where $\triangle ABC \equiv \triangle DEF$ and where

$$A(0,0),\ B(5,0),\ C(0,10),\ D(4,2),\ E(1,-2),\ F(12,-4),$$

    find the equations of the lines such that the product of reflections in the lines maps $\triangle ABC$ to $\triangle DEF$.

28. Let A_0 be a given point and $\ell_1, \ell_2, \ldots, \ell_n$ be given lines. For $1 \leq k \leq n$, let A_k be obtained from A_{k-1} by a reflection across ℓ_k, and let A_{n+k} be obtained from A_{n+k-1} by a reflection across ℓ_k.

    (a) Prove that A_{2n} will coincide with A_0 if n is odd.

    (b) Can the same conclusion be drawn if n is even?

29. Let $A_0 = B_0$ be a given point and $\ell_1, \ell_2, \ldots, \ell_n$ be given lines. For $1 \leq k \leq n$, let A_k be obtained from A_{k-1} by a reflection across ℓ_k, and let B_k be obtained from B_{k-1} by a reflection across ℓ_{n-k+1}. For what values of n will A_n coincide with B_n?

CHAPTER 9

THE PRODUCT OF DIRECT ISOMETRIES

Given two direct isometries, say the rotation $\mathbf{R}_{P,\phi}$ and the translation $\mathbf{T}_{AB}$, we know that their product $\mathbf{R}_{P,\phi}\mathbf{T}_{AB}$ is a direct isometry, and we may even suspect that it is a rotation $\mathbf{R}_{Q,\theta}$. The question is, what is Q and what is θ? This chapter will describe some of the ways that we can determine the values of the parameters that describe the result, in this case, Q and θ.

9.1 Angles

We first need to clear up some possible ambiguities about directed angles.

In Example 8.3.2, we showed that the product of two reflections in nonparallel lines is a rotation whose center is the intersection point of the two lines and whose angle of rotation is twice the angle from the first line to the second line; that is,

$$\mathbf{R}_m\mathbf{R}_l = \mathbf{R}_{Q,\alpha},$$

Classical Geometry: Euclidean, Transformational, Inversive and Projective, First Edition. By I. E. Leonard, J. E. Lewis, A. C. F. Liu, G. W. Tokarsky
Copyright © 2014 John Wiley & Sons, Inc.

where $Q = l \cap m$ and where α is twice the directed angle from l to m. If l and m are rays, as in (a) below, there is no ambiguity in the meaning of the phrase "the directed angle from l to m." We can think of it as being 36° or −324°, since we identify the angles of θ and $\theta + 360n$ for all integers n.[8]

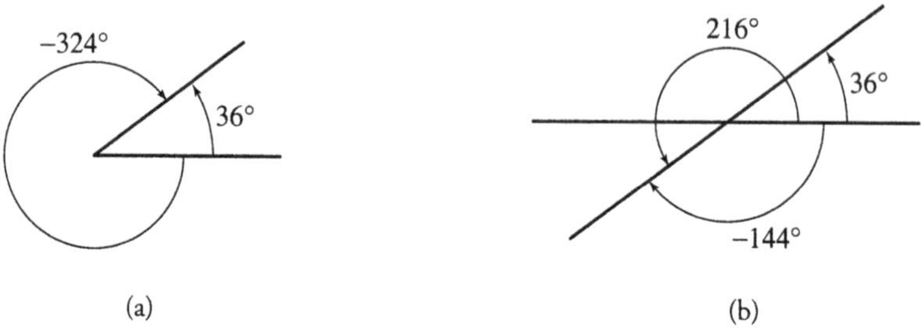

When we talk about the directed angle from one line to another, however, there are other possible interpretations. In (b) above, in addition to 36° and −324°, the directed angle from l to m can be legitimately interpreted as 216° or −144°. Does this affect the validity of the example? In fact, the answer is no, it does not matter which of the four angles we use. This is illustrated in the following table:

θ	2θ	modulo 360
36	72	72
−324	−648	72 $[= -648 + 360(2)]$
216	432	72 $[= 432 + 360(-1)]$
−144	−288	72 $[= -288 + 360(1)]$

In each case, 2θ is always 72 plus or minus some integral multiple of 360, so the entries in the 2θ column all represent exactly the same angle.

[8]This is the same as saying that α and β differ by a multiple of 360. In number theory, this is written mathematically as follows:

$$\alpha \equiv \beta \pmod{360}.$$

The expression is called a *congruence* and is expressed verbally by saying "α is congruent to β modulo 360." Although the two notions coincide in this specific case, generally there does not have to be any connection between geometric congruence and number theoretical congruence.

9.2 Fixed Points

Under the identity, every point of the plane is fixed. Under a reflection, each point in the line of reflection is fixed. Under a rotation, the center of rotation is the only fixed point. Translations and glide reflections have no fixed points.

Conversely, given an isometry T, we know from Theorem 8.2.5 that if T fixes each of three noncollinear points, then T must be the identity. The following theorem tells us what we can conclude if we know that either one or two points are fixed.

Theorem 9.2.1. *(Fixed Points)*

(1) An isometry that fixes a point P is either the identity, a rotation centered at P, or a reflection in a line that passes through P.

(2) An isometry that fixes each of two given points is either the identity or a reflection in the line determined by those points.

Proof. In each case, these are the only isometries that have (at least) the specified number of fixed points.

□

It is possible to distinquish between translations and glide reflections in terms of fixed sets.

Theorem 9.2.2. *An isometry that fixes exactly one line but does not fix any points is a glide reflection.*

Proof. By the previous theorem, the isometry is either a translation or a glide reflection, since it does not fix any points. A translation $\mathbf{T}_{AB}$ fixes all lines parallel to AB, and since this is not the case, we must conclude that the isometry is a glide reflection.

□

It should be noted that halfturns and translations each fix infinitely many lines. A halfturn fixes each line that contains the center of the turn, and a translation fixes each line parallel to the direction of translation. A halfturn also fixes the center of rotation, as the point common to all of the lines that it fixes.

9.3 The Product of Two Translations

The effect of a translation $\mathbf{T}_{UV}$ is to map each point P to a point P' such that $\overline{PP'}$ is congruent to $\overline{UV}$. This means that

$$\mathbf{T}_{UV} = \mathbf{T}_{PP'}.$$

Thus, if we know that an isometry T is a translation, we can write

$$T = \mathbf{T}_{PP'},$$

where $P' = T(P)$. This handy fact is used in the next theorem.

Theorem 9.3.1. *The product of two translations is a translation or the identity.*

Proof. Let $\mathbf{T}_{AB}$ and $\mathbf{T}_{CD}$ be the two translations. The product $\mathbf{T}_{CD}\mathbf{T}_{AB}$ must be a direct isometry, so it is either the identity, a translation, or a rotation. We will show that the only way the product can be a rotation is if $\mathbf{T}_{CD}$ and $\mathbf{T}_{AB}$ are inverses of each other, and so the rotation is actually the identity.

Suppose that $\mathbf{T}_{CD}\mathbf{T}_{AB} = \mathbf{R}_{P,\theta}$. Then

$$\mathbf{T}_{CD}\mathbf{T}_{AB}(P) = P.$$

Let $P' = \mathbf{T}_{AB}(P)$. Then $\mathbf{T}_{AB} = \mathbf{T}_{PP'}$. Similarly, $\mathbf{T}_{CD} = \mathbf{T}_{P'P}$, but then

$$\mathbf{T}_{CD}\mathbf{T}_{AB} = \mathbf{T}_{P'P}\mathbf{T}_{PP'} = \mathbf{I}.$$

Thus, the rotation would actually be the identity.

□

Note that if $\overline{AB}$ is not congruent to $\overline{CD}$, then $\mathbf{T}_{CD}\mathbf{T}_{AB}$ must be a translation T. To pin down this translation geometrically, consider the effect of $\mathbf{T}_{CD}\mathbf{T}_{AB}$ on a point P. $\mathbf{T}_{AB}$ maps P to P' and $\mathbf{T}_{CD}$ maps P' to P''. Therefore, $T = \mathbf{T}_{PP''}$. We can construct a directed line segment that is congruent to PP'' by completing the parallelogram $EFGH$, where $\overline{EF}$ and $\overline{EH}$ are congruent to $\overline{AB}$ and $\overline{CD}$, respectively. Thus, $\mathbf{T}_{CD}\mathbf{T}_{AB}$ maps E to G, and the diagonal $\overline{EG}$ is congruent to PP''.

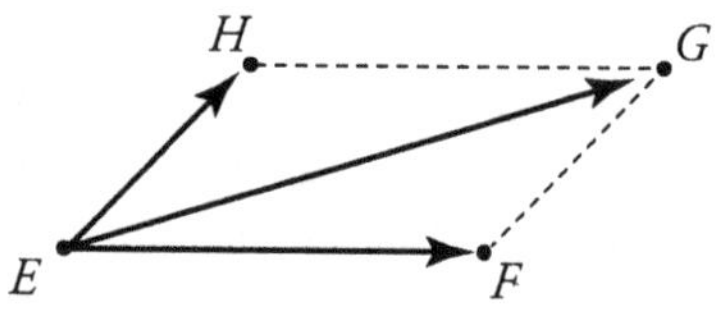

9.4 The Product of a Translation and a Rotation

We will determine the product of a rotation and a translation in two stages. First, we will show that the product is a rotation, and then we will describe how to find the center of rotation and the angle of rotation.

The product of a translation and rotation is a direct isometry, so it is either the identity, a translation, or a rotation. Supposing that $\overline{AB}$ is not of length zero and that the angle of rotation is θ, where θ is not a multiple of $360°$, a little bit of algebra shows that the product $\mathbf{T}_{AB}\mathbf{R}_{O,\theta}$ cannot be either the identity or a translation.

- It cannot be the identity. The reason is that

$$\mathbf{T}_{AB}\mathbf{R}_{O,\theta}(O) = \mathbf{T}_{BA}(O) = O',$$

  and since $\text{dist}(O, O') = \text{dist}(A, B) \neq 0$, it follows that $O \neq O'$. Since O is not fixed, the product cannot be the identity.

- It cannot be a translation. Let $S = \mathbf{T}_{AB}\mathbf{R}_{O,\theta}$. Then multiplying both sides of this equation by $\mathbf{T}_{BA}$ (the inverse of $\mathbf{T}_{AB}$), we get

$$\mathbf{T}_{BA}S = \mathbf{T}_{BA}\left(\mathbf{T}_{AB}\mathbf{R}_{O,\theta}\right) = \left(\mathbf{T}_{BA}\mathbf{T}_{AB}\right)\mathbf{R}_{O,\theta} = \mathbf{R}_{O,\theta}.$$

  If S were a translation, then Theorem 9.3.1 tells us that $\mathbf{T}_{BA}S$ (and hence $\mathbf{R}_{O,\theta}$) is the identity or a translation, which contradicts our assumptions about $\overline{AB}$ and θ.

The only possibility left is that the product is a rotation.

In a similar way, we can show that $\mathbf{R}_{O,\theta}\mathbf{T}_{AB}$ cannot be a translation unless $\mathbf{R}_{O,\theta}$ is the identity or a translation, which is again a contradiction. Thus, we have shown:

Theorem 9.4.1. *The product of a nontrivial translation and a rotation is a rotation, unless the angle of rotation is a multiple of* $360°$.

Although we know that the result has to be a rotation $\mathbf{R}_{Q,\phi}$, the theorem does not tell us how to find Q or ϕ. The next example does this.

Example 9.4.2. *Given $\overline{AB}$, the point O, and the angle θ, where $\theta \neq n \cdot 180°$, find Q and ϕ such that*

$$\mathbf{T}_{AB}\mathbf{R}_{O,\theta} = \mathbf{R}_{Q,\phi}.$$

Solution. The key here is to look for the fixed point Q. In other words, we are looking for the point Q such that

$$\mathbf{T}_{AB}\mathbf{R}_{O,\theta}(Q) = Q.$$

The point Q is transported to Q' by $\mathbf{R}_{O,\theta}$ and then back to Q by $\mathbf{T}_{AB}$. The solution can be derived from the figure on the right. The points Q and Q' are on a circle centered at O, with $\angle QOQ' = \theta$. The segment QQ' is a chord of a circle, and so its right bisector passes through O.

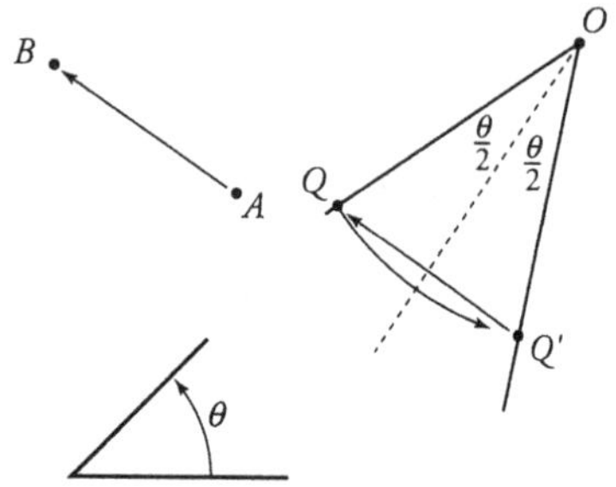

The center Q can be constructed as follows (see the figure on the right).

Through O, draw a line parallel to $\overline{AB}$, and let P be the point such that $\overline{OP} \cong \overline{AB}$. Construct the right bisector l of $\overline{OP}$, and construct the line m through O so that the directed angle from $\overline{OP}$ to m is $90° - \theta/2$. The point where m intersects l is Q.

This constructs the center Q of the rotation $\mathbf{R}_{Q,\phi}$. The angle of rotation ϕ is equal to the directed angle θ, and this can be confirmed by noting that $\angle OQP = \theta$ and that O is mapped to P by $\mathbf{T}_{AB}\mathbf{R}_{O,\theta}$.

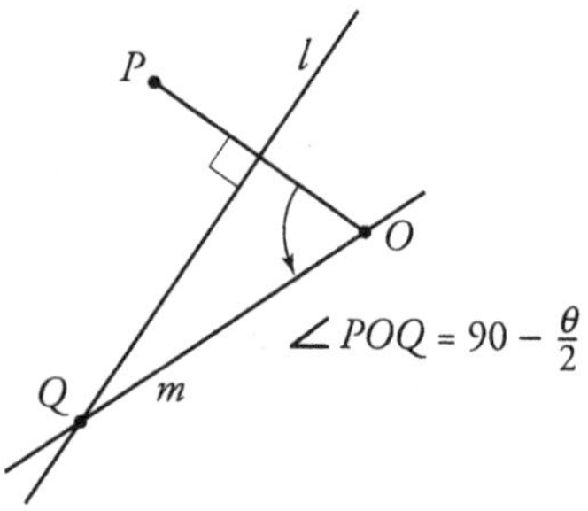

□

The product $\mathbf{R}_{O,\theta}\mathbf{T}_{AB}$ can be analyzed in a similar way. Here, the center would be a point X such that $\mathbf{T}_{AB}$ maps X to Z and $\mathbf{R}_{O,\theta}$ maps Z back to X. In the first figure above, this would be the point Q.

Care must also be taken when the angle of rotation is negative.

Example 9.4.3. *The figure below represents the centers of the rotations resulting from the four products*

$$\mathbf{T}_{AB}\mathbf{R}_{Q,30^\circ}, \quad \mathbf{T}_{AB}\mathbf{R}_{Q,-30^\circ}, \quad \mathbf{R}_{Q,30^\circ}\mathbf{T}_{AB}, \quad \mathbf{R}_{Q,-30^\circ}\mathbf{T}_{AB}.$$

The directed segment $\overline{AB}$ is as shown. The points W, X, Y, and Z are the centers of rotation, although not necessarily in that order. Determine the correct center for each rotation.

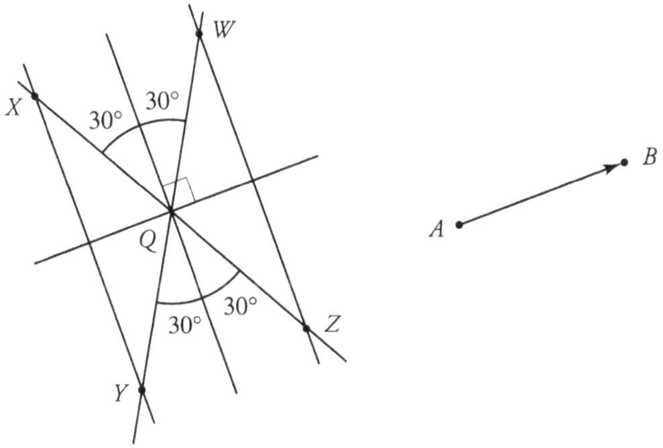

Solution. The centers are as follows:

W is the center of $\mathbf{T}_{AB}\mathbf{R}_{Q,30^\circ}$,

X is the center of $\mathbf{R}_{Q,30^\circ}\mathbf{T}_{AB}$,

Y is the center of $\mathbf{R}_{Q,-30^\circ}\mathbf{T}_{AB}$,

Z is the center of $\mathbf{T}_{AB}\mathbf{R}_{Q,-30^\circ}$.

□

Example 9.4.4. *Given the point Q and rotation $\mathbf{R}_{O,\theta}$, where θ is not a multiple of 360°, as in the figure on the left on the following page, find a translation $\mathbf{T}_{CD}$ such that*

$$\mathbf{T}_{CD}\mathbf{R}_{O,\theta} = \mathbf{R}_{Q,\theta}.$$

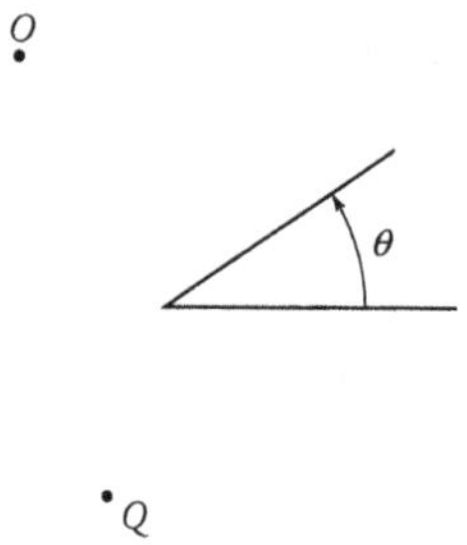

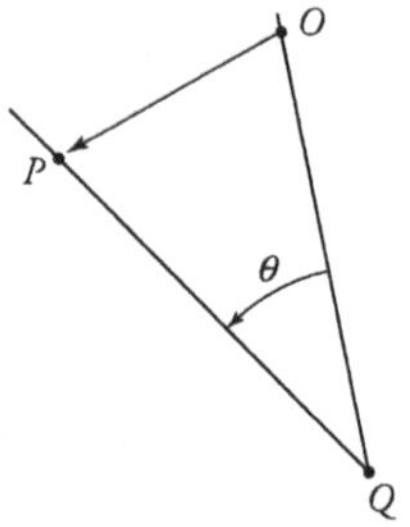

Solution. Again, we exploit the idea that if for some point X we can find its image X' under $\mathbf{T}_{CD}$, then $\mathbf{T}_{CD} = \mathbf{T}_{XX'}$.

We first solve for $\mathbf{T}_{CD}$. Multiply the equation

$$\mathbf{T}_{CD}\mathbf{R}_{O,\theta} = \mathbf{R}_{Q,\theta}$$

on the right by the inverse of $\mathbf{R}_{O,\theta}$, as follows:

$$(\mathbf{T}_{CD}\mathbf{R}_{O,\theta})\,\mathbf{R}_{O,-\theta} = \mathbf{R}_{Q,\theta}\mathbf{R}_{O,-\theta}.$$

Next use the associative law and the fact that $\mathbf{R}_{O,\theta}\mathbf{R}_{O,-\theta} = \mathbf{I}$ to obtain $\mathbf{T}_{CD}$:

$$\mathbf{T}_{CD} = \mathbf{R}_{Q,\theta}\mathbf{R}_{O,-\theta}.$$

Now apply $\mathbf{T}_{CD}$ to the point O (we chose O because it is fixed under $\mathbf{R}_{O,-\theta}$). Then

$$\mathbf{T}_{CD}(O) = \mathbf{R}_{Q,\theta}\mathbf{R}_{O,-\theta}(O) = \mathbf{R}_{Q,\theta}(O).$$

Letting P be the point $\mathbf{R}_{Q,\theta}(O)$, we therefore have $\mathbf{T}_{CD} = \mathbf{T}_{OP}$, as in the figure on the right at the top of the page.

□

9.5 The Product of Two Rotations

The product of two rotations with the same center is a third rotation, also with the same center, through an angle that is the sum of the two angles of rotation; that is,

$$\mathbf{R}_{A,\beta}\mathbf{R}_{A,\alpha} = \mathbf{R}_{A,\alpha+\beta}.$$

To analyze $\mathbf{R}_{B,\beta}\mathbf{R}_{A,\alpha}$ when A and B are different, we must first show the following:

Theorem 9.5.1. *When A and B are different, $\mathbf{R}_{B,\beta}\mathbf{R}_{A,\alpha}$ is a rotation through an angle $\alpha + \beta$ about some point P.*

Proof. We rely on the fact that we can use a translation to "move" the center of a rotation $\mathbf{R}_{Q,\theta}$ to any point we like, as in Example 9.4.4.

Let $\mathbf{T}_{CD}$ be the translation such that

$$\mathbf{T}_{CD}\mathbf{R}_{B,\beta} = \mathbf{R}_{A,\beta}.$$

Let S be the product $\mathbf{R}_{B,\beta}\mathbf{R}_{A,\alpha}$; that is,

$$S = \mathbf{R}_{B,\beta}\mathbf{R}_{A,\alpha}.$$

Multiply this equation on the left by $\mathbf{T}_{CD}$ to get

$$\mathbf{T}_{CD}S = \mathbf{T}_{CD}\left(\mathbf{R}_{B,\beta}\mathbf{R}_{A,\alpha}\right),$$

which implies that

$$\begin{aligned}\mathbf{T}_{CD}S &= \left(\mathbf{T}_{CD}\mathbf{R}_{B,\beta}\right)\mathbf{R}_{A,\alpha} \\ &= \mathbf{R}_{A,\beta}\mathbf{R}_{A,\alpha} \\ &= \mathbf{R}_{A,\alpha+\beta}.\end{aligned}$$

Now multiply this equation on the left by $\mathbf{T}_{DC}$, the inverse of $\mathbf{T}_{CD}$, to get

$$S = \mathbf{T}_{DC}\mathbf{R}_{A,\alpha+\beta}.$$

From the previous section, we know that $\mathbf{T}_{DC}\mathbf{R}_{A,\alpha+\beta}$ is a rotation $\mathbf{R}_{P,\alpha+\beta}$ for some point P (unless $\alpha + \beta$ is a multiple of $360°$, in which case $\mathbf{T}_{DC}\mathbf{R}_{A,\alpha+\beta} = \mathbf{T}_{DC}$).

□

To find the center P when $\alpha + \beta$ is not a multiple of $360°$, we can trace our progress through the preceding equations. For $\overline{CD}$, we can take $C = A$ and $D = \mathbf{R}_{A,\beta}(B)$, as in the discussion following Example 9.4.4. Applying the inverse transformation $\mathbf{T}_{DC}$ to $\mathbf{R}_{A,\alpha+\beta}$ then allows us to geometrically construct the center P. A better option is to look for the fixed point P of the product $\mathbf{R}_{B,\beta}\mathbf{R}_{A,\alpha}$, as in the following example.

Example 9.5.2. *Given $\mathbf{R}_{B,\beta}$ and $\mathbf{R}_{A,\alpha}$, with α and β as shown in the figure below, construct the center Q of the rotation $\mathbf{R}_{B,\beta}\mathbf{R}_{A,\alpha}$.*

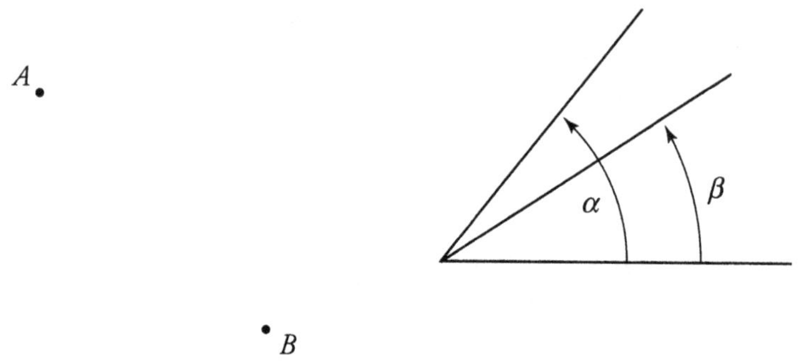

Solution. The figure below shows how to solve this problem. Construct the line AB and an angle $\angle XAX'$ of size α such that AB is the bisector of $\angle XAX'$. Construct the angle $\angle YAY'$ of size β so that AB is the bisector of $\angle YAY'$. One of the points of intersection of these two angles will be the desired point Q.

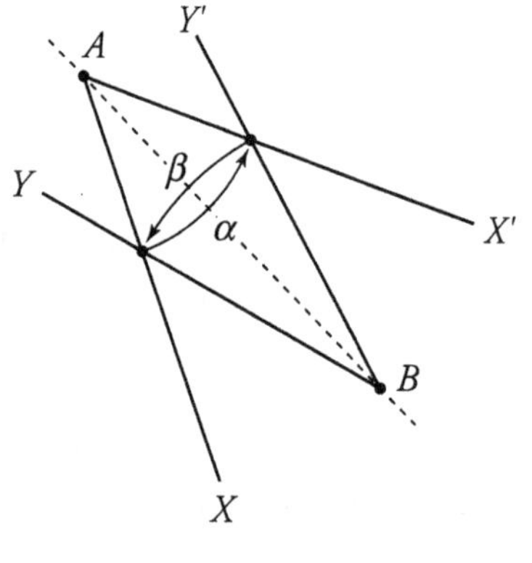

□

Remark. In general, when trying to find the center for the product $\mathbf{R}_{B,\beta}\mathbf{R}_{A,\alpha}$, the angles should be drawn so that the directed angle $\angle XAB = \alpha/2$ and the directed angle $\angle ABY = \beta/2$. The point of intersection of the line XA and the line YB will be the center Q of the product.

For example, this is what the construction looks like when $\alpha = 30°$ and $\beta = -60°$. The point Q will be mapped by $\mathbf{R}_{A,\alpha}$ to Q', and Q' will then be mapped by $\mathbf{R}_{B,\beta}$ back to Q.

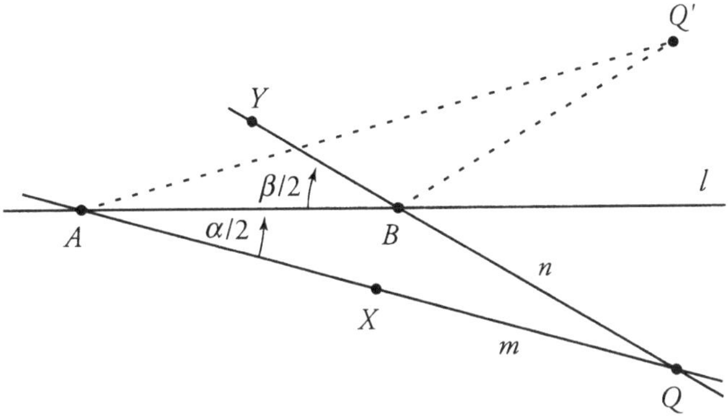

9.6 Problems

1. If ℓ, m, and n are the perpendicular bisectors of the sides AB, BC, and CA, respectively, of $\triangle ABC$, then

$$T = R_n\, R_m\, R_\ell$$

is a reflection in which line?

2. If $R_n\, R_m\, R_\ell$ is a reflection, show that the lines ℓ, m, and n are concurrent or have a common perpendicular.

3. Find Cartesian equations for lines m and n such that

$$R_m R_n(x, y) = (x + 2, y - 4).$$

4. Show that

$$H_P\, R_\ell\, H_P\, R_\ell\, H_P\, R_\ell\, H_P$$

is a reflection in a line parallel to ℓ.

5. Let C be a point on the line ℓ, and show that

$$R_\ell\, R_{C,\theta}\, R_\ell = R_{C,-\theta}.$$

6. Given nonparallel lines AB and CD, show that there is a rotation T such that

$$T(AB) = CD.$$

7. Show that if S, T, $T\,S$, and $T^{-1}\,S$ are rotations, then the centers of S, $T\,S$, and $T^{-1}\,S$ are collinear.

8. In a given acute triangle, inscribe a triangle PQR having minimum perimeter. This is called *Fagnano's problem*.

9. Prove *Thomsen's Relation*: for any lines a, b, and c, we have

$$R_c\,R_a\,R_b\,R_c\,R_a\,R_b\,R_a\,R_b\,R_c\,R_a\,R_b\,R_c \times R_b\,R_a\,R_c\,R_b\,R_a\,R_b\,R_a\,R_c\,R_b\,R_a = I,$$

where I is the identity transformation.

10. Show that Thomsen's Relation is still true if each reflection R_x is replaced by a halfturn H_X; that is, show that if A, B, and C are three distinct points in the plane, then

$$H_C\,H_A\,H_B\,H_C\,H_A\,H_B\,H_A\,H_B\,H_C\,H_A\,H_B\,H_C \times H_B\,H_A\,H_C\,H_B\,H_A\,H_B\,H_A\,H_C\,H_B\,H_A = I,$$

where I is the identity transformation.

11. If $x' = ax + by + c$ and $y' = bx - ay + d$ with $a^2 + b^2 = 1$ are the equations for an isometry T, show that T is a reflection if and only if

$$ac + bd + c = 0 \qquad \text{and} \qquad ad - bc - d = 0.$$

12. Find the Cartesian equation of the line m if the equations for a reflection in the line are

$$x' = \frac{3}{5}x + \frac{4}{5}y \qquad \text{and} \qquad y' = \frac{4}{5}x - \frac{3}{5}y.$$

13. If the equations for the rotation $R_{P,\theta}$ are

$$2x' = -\sqrt{3}x - y + 2 \qquad \text{and} \qquad 2y' = x - \sqrt{3}y - 1,$$

find the center of rotation P and and the angle of rotation θ.

14. If a and b are lines in the plane, show that the following are equivalent:

(a) $a = b$ or a and b are perpendicular,

(b) $R_a\,R_b = R_b\,R_a$,

(c) $R_b(a) = a$,

(d) $(R_b\,R_a)^2 = I$,

(e) $R_b\,R_a$ is either the identity or a halfturn.

15. If the isometry H_P is a halfturn, show that given any two perpendicular lines m and n that intersect at the point P, we have $H_P = R_m\, R_n$.

16. Let m be a line with equation $2x + y = 1$. Find the equations of the transformation R_m.

17. Given a line b and a point A, show that the following conditions are equivalent:

    (a) $A \in b$,
    (b) $H_A\, R_b = R_b\, H_A$,
    (c) $R_b(A) = A$,
    (d) $H_A(b) = b$,
    (e) $R_b\, H_A$ (or $H_A\, R_b$) is an involution,
    (f) $R_b\, H_A$ is a reflection in the line through A perpendicular to b.

18. If $A \neq C$, show that the following conditions are equivalent:

    (a) B is the midpoint of AC,
    (b) $H_C\, H_B = H_B\, H_A$,
    (c) $H_B(A) = C$,
    (d) $H_B\, H_A\, H_B = H_C$.

19. Show that nonidentity rotations of the plane with different centers do not commute.

20. Let A and B be distinct points in the plane and let S be an isometry. Show that

$$S\, T_{AB}\, S^{-1} = T_{CD},$$

    where $C = S(A)$ and $D = S(B)$.

21. Given a point A and two lines ℓ and m, construct a square $ABCD$ such that B lies on ℓ and D lies on m.

22. Given four distinct points, find a square such that each of the lines containing a side of the square passes through one of the four given points.

    Hint. Given A, B, C, and D, we want to find the lines a, b, c, and d. Take P such that $R_{P,90}(B) = C$. Let $R_{P,90}(D) = E$. Then take a to be AE.

23. Consider a triangle $\triangle ABC$ (oriented counterclockwise) with positive angles α, β, γ at A, B, C. Show that

$$R_{A,2\alpha}\, R_{B,2\beta}\, R_{C,2\gamma} = I.$$

    Are there other similar formulas?

24. Construct over each side of $\triangle ABC$ an equilateral triangle. The centroids G_1, G_2, and G_3 of these triangles form a new triangle, the so-called ***Napoleon triangle***. Show that $\triangle G_1G_2G_3$ is equilateral. This theorem is attributed to Napoleon Bonaparte.

25. Let perpendiculars erected at arbitrary points on the sides of triangle $\triangle ABC$ meet in pairs at points P, Q, and R. Show that the triangle PQR is similar to the given triangle.

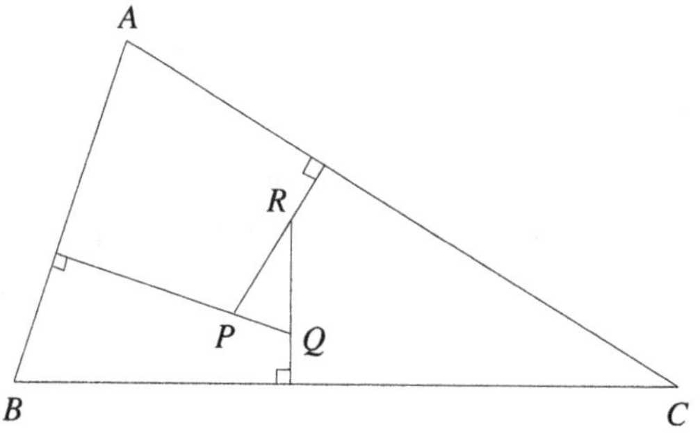

26. Let A and B be two points lying on one side of a line l. Explain how to construct one of the two circles that are tangent to ℓ and pass through A and B.

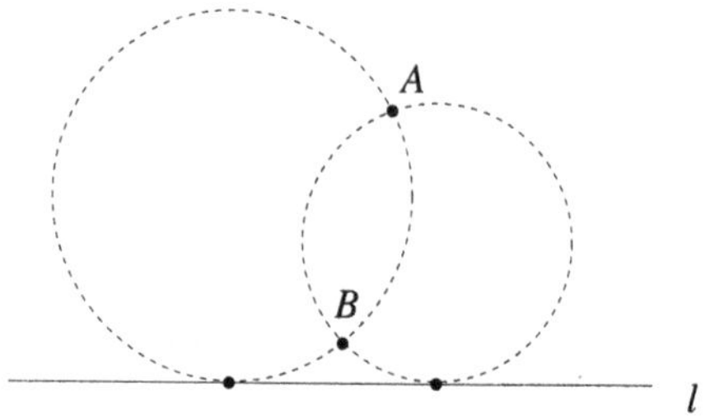

27. Given points A, B, and P in the plane, construct the reflection of P in the line AB using a Euclidean compass alone.

28. Let $ABCD$ be a square with the vertices in clockwise order. For each of the following translations, find a counterclockwise rotation that brings the image of the translation back to the same physical space as $ABCD$:

(a) distance AB, direction AB,

(b) distance $\frac{1}{2}AB$, direction AB,

(c) distance AC, direction AC,

(d) distance $\frac{1}{2}AC$, direction AC.

29. Let $A_0B_0C_0$ be an equilateral triangle with the vertices in clockwise order. We first rotate it 60° counterclockwise about A_0 to obtain $A_1B_1C_1$, then about B_1 to obtain $A_2B_2C_2$, and finally about C_2 to obtain $A_3B_3C_3$. We continue to rotate about A_3, B_4, C_5, and so on, until $A_nB_nC_n$ occupies the same physical space as $A_0B_0C_0$. What is the minimum positive value of n?

30. Let $A_0B_0C_0$ be a triangle with the vertices in counterclockwise order, where $\angle A = 40^\circ$, $\angle B = 60^\circ$, and $\angle C = 80^\circ$. We first rotate it 40° counterclockwise about A_0 to obtain $A_1B_1C_1$, then 60° counterclockwise about B_1 to obtain $A_2B_2C_2$, and finally 80° counterclockwise about C_2 to obtain $A_3B_3C_3$. We continue to rotate about A_3, B_4, C_5, and so on, until $A_nB_nC_n$ occupies the same physical space as $A_0B_0C_0$. What is the minimum positive value of n?

31. A ***halfturn*** about a point O is a 180° rotation about the point O. Prove that the composition of:

    (a) two halfturns is a translation or the identity;

    (b) a translation and a halfturn is a halfturn.

32. Prove that the composition of:

    (a) an even number of halfturns is a translation or the identity;

    (b) an odd number of halfturns is a halfturn.

33. Let A_0, B_0, O_1, O_2, $\ldots$, O_n be given points. For $1 \le k \le n$, let A_kB_k be obtained from $A_{k-1}B_{k-1}$ by a halfturn about O_k.

    (a) Prove that $A_0A_n = B_0B_n$ if n is even.

    (b) What conclusion may be drawn if n is odd?

34. Let A_0, O_1, O_2, $\ldots$, O_n be given points. For $1 \le k \le n$, let A_k be obtained from A_{k-1} by a halfturn about O_k, and let A_{n+k} be obtained from A_{n+k-1} by a halfturn about O_k.

    (a) Prove that A_{2n} will coincide with A_0 if n is odd.

    (b) Can the same conclusion be drawn if n is even?

35. Let $A_0 = B_0$, O_1, O_2, $\ldots$, O_n be given points. For $1 \le k \le n$, let A_k be obtained from A_{k-1} by a halfturn about O_k, and let B_k be obtained from B_{k-1} by a halfturn about O_{n-k+1}. For what values of n will A_n coincide with B_n?

CHAPTER 10

SYMMETRY AND GROUPS

10.1 More About Groups

Recall that a set $\mathcal{G}$ together with a binary operation $\cdot$ is called a ***group*** if the following conditions are satisfied:

1. $\mathcal{G}$ is closed under the binary operation; that is, if x and y are elements of $\mathcal{G}$, then so is $x \cdot y$.
2. The associative law holds. If x, y, and z are elements of $\mathcal{G}$, then
$$(x \cdot y) \cdot z = x \cdot (y \cdot z).$$
3. There is an identity element e in $\mathcal{G}$. For every x in $\mathcal{G}$, $e \cdot x = x \cdot e = x$.
4. $\mathcal{G}$ is closed with respect to inversion. For every member x in $\mathcal{G}$, there is another member x' also in $\mathcal{G}$ such that $x \cdot x' = x' \cdot x = e$.

Classical Geometry: Euclidean, Transformational, Inversive and Projective, First Edition. By I. E. Leonard, J. E. Lewis, A. C. F. Liu, G. W. Tokarsky
Copyright © 2014 John Wiley & Sons, Inc.

When dealing with groups in general, the binary operation is typically called ***group multiplication*** or simply ***multiplication***, and the symbol for the operation is omitted.

There are a few simple but useful facts about groups that can occasionally save us some work.

Theorem 10.1.1. *A group has only one identity element.*

Proof. Suppose that e and f are identity elements in a group $(\mathcal{G}, \cdot)$. Since $xe = x$ for every x in $\mathcal{G}$, we have $fe = f$. Since $fx = x$ for every x in $\mathcal{G}$, we have $fe = e$. Therefore, $f = fe = e$.

□

Theorem 10.1.2. *Each element of a group has only one inverse.*

Proof. Let x be an element of $\mathcal{G}$ and suppose that x' and x'' are both inverses for x. Then,

$$(x'x)x'' = ex'' = x''$$

and

$$x'(xx'') = x'e = x'.$$

However, by the associative law,

$$x'' = (x'x)x'' = x'(xx'') = x'.$$

□

Since there is exactly one inverse for each x in $\mathcal{G}$, there is no ambiguity in denoting it by x^{-1}.

An important consequence of the previous theorem is that the ***cancellation laws*** hold:

If $ax = ay$, then $x = y$; similarly, if $xa = ya$, then $x = y$.

If $ya = ax$, can we conclude that $y = x$? We certainly could if it were true that $ax = xa$ or that $ay = ya$. If a group $\mathcal{G}$ has the property that $xy = yx$ for *every* pair of elements x and y in $\mathcal{G}$, we say that the ***commutative law*** holds, and the group is called a ***commutative group*** or an ***abelian group***. The group of all isometries of the plane is *not* a commutative group, although it is true that $TS = ST$ for *some* pairs of isometries.

Returning to the question of whether $ya = ax$ implies that $x = y$, the answer is that this may not be the case. If $ya = ax$, then multiplying on the right by a^{-1} gives us

$$y = axa^{-1}.$$

This is a fairly important operation in group theory and is called ***conjugation***. More specifically, we say that y is the ***conjugate*** of x by a if $y = axa^{-1}$.

Example 10.1.3. *Let $a = \mathbf{T}_{AB}$ and let $x = \mathbf{R}_{Q,\theta}$. Find the conjugate of x by a.*

Solution. We know that a^{-1} is $\mathbf{T}_{BA}$, and so axa^{-1} is the product of three direct isometries and must therefore be a direct isometry. To figure out what it is, we know from previous exercises that we need only look at how it affects two convenient points.

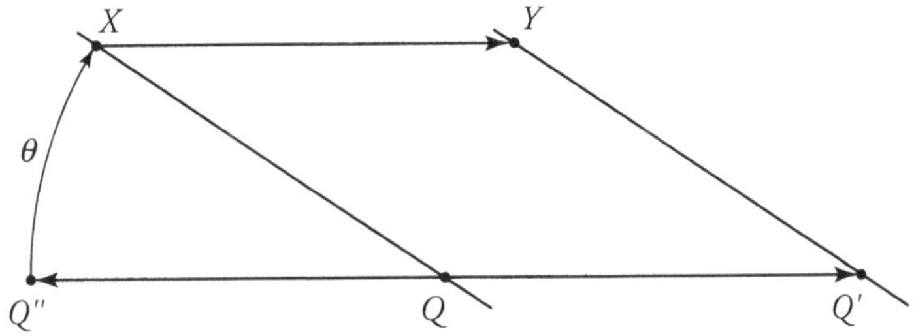

Let Q' be the point that is mapped to Q by $\mathbf{T}_{BA}$. In other words, let $Q' = \mathbf{T}_{AB}(Q)$. Then

$$axa^{-1}(Q') = \mathbf{T}_{AB}\mathbf{R}_{Q,\theta}\mathbf{T}_{BA}(Q') = Q',$$

so axa^{-1} is either the identity or a rotation about Q'.

Now apply axa^{-1} to the point Q. Let $Q'' = \mathbf{T}_{BA}(Q)$. Then Q is the midpoint of $Q'Q''$. Let $X = \mathbf{R}_{Q,\theta}(Q'')$ and let $Y = \mathbf{T}_{AB}(X)$. The segments $\overline{XY}$ and $\overline{QQ'}$ are equal in length and in the same direction, so $Q'QXY$ is a parallelogram, and so $\angle Q''QX$ is congruent to $\angle QQ'Y$. It therefore follows that $axa^{-1} = \mathbf{R}_{Q',\theta}$.

□

The image of a straight line under any isometry A is another straight line. The following theorem says that any conjugate of a reflection is again a reflection, but possibly in a different line.

Theorem 10.1.4. *Let A be any isometry in the plane and let m be any line. Then*

$$A\mathbf{R}_m A^{-1} = \mathbf{R}_{A(m)}.$$

Proof. If A is a direct isometry, so is A^{-1}, and if A is an opposite isometry, then so is A^{-1}. It follows that $A\mathbf{R}_m A^{-1}$ is an opposite isometry.

Let X be a point on $A(m)$; that is, let $X = A(Y)$ for some point Y on m. Then

$$A\mathbf{R}_m A^{-1}(X) = A\mathbf{R}_m A^{-1}A(Y) = A\mathbf{R}_m(Y) = A(Y) = X.$$

This says that $A\mathbf{R}_m A^{-1}$ fixes every point on the line $A(m)$, and since $A\mathbf{R}_m A^{-1}$ is an opposite isometry, it must be the reflection $\mathbf{R}_{A(m)}$.

□

The image of a directed line segment under any isometry A is another directed line segment of the same length. The following theorem says that any conjugate of a translation is again a translation through the same distance, although possibly in a different direction.

Theorem 10.1.5. *Let A be any isometry in the plane, and let $\overline{CD}$ be a directed line segment. Then*

$$A\mathbf{T}_{CD}A^{-1} = \mathbf{T}_{EF},$$

where the directed segment $\overline{EF} = A\left(\overline{CD}\right)$.

Proof. We first show that $A\mathbf{T}_{CD}A^{-1}$ must be a translation. The isometry $A\mathbf{T}_{CD}A^{-1}$ must be a direct isometry, and since the only direct isometries without fixed points are translations, it suffices to show that $A\mathbf{T}_{CD}A^{-1}$ has no fixed point.

Suppose to the contrary that X is a fixed point of the isometry, that is, that

$$X = A\mathbf{T}_{CD}A^{-1}(X).$$

Now, $X = A(Y)$ for some point Y, so that

$$Y = A^{-1}(X) = A^{-1}\left(A\mathbf{T}_{CD}A^{-1}\right)(X) = \left(AA^{-1}\right)\mathbf{T}_{CD}A^{-1}(X) = \mathbf{T}_{CD}(Y).$$

This says that $Y = \mathbf{T}_{CD}(Y)$; that is, Y is a fixed point of $\mathbf{T}_{CD}$. However, this contradicts the fact that a translation has no fixed points. Thus, we must conclude that $A\mathbf{T}_{CD}A^{-1}$ has no fixed points, and therefore it must be a translation.

To pin down the translation $\mathbf{T}_{EF}$, let us consider the effect of $A\mathbf{T}_{CD}A^{-1}$ upon E:

$$A\mathbf{T}_{CD}A^{-1}(E) = A\mathbf{T}_{CD}A^{-1}A(C) = A\mathbf{T}_{CD}(C) = A(D) = F,$$

so we have a translation that maps E to F. In other words,

$$A\mathbf{T}_{CD}A^{-1} = \mathbf{T}_{EF}.$$

□

Theorem 10.1.6. *Let Q be a point on the line m. Then $\mathbf{R}_l \mathbf{R}_{Q,\theta} \mathbf{R}_l = \mathbf{R}_{Q,-\theta}$.*

We leave the proof as an exercise.

10.1.1 Cyclic and Dihedral Groups

If a is an element of a group $\mathcal{G}$ and m is a positive integer, then a^m, a^{-m}, and a_0 are defined as follows:

$$a^m = \underbrace{a \cdot a \cdots a \cdot a}_{m \text{ factors}},$$

$$a^{-m} = \left(a^{-1}\right)^m,$$

$$a^0 = e \quad (e \text{ being the identity}).$$

With these definitions, it is not difficult to verify that the following two laws of exponents hold for all integers m and n:

$$a^m a^n = a^{m+n} \qquad \text{and} \qquad (a^m)^n = a^{mn}.$$

In general, we can expect that $(ab)^m \neq a^m b^m$, unless the group is commutative.

The ***order*** of an element a of a group is the smallest positive integer n such that $a^n = e$. If there is no positive integer n such that $a^n = e$, then the order of a is infinite.

For example, in the group of all isometries of the plane, the order of $\mathbf{I}$ is 1, the order of $\mathbf{R}_{Q,90^\circ}$ is 4, the order of $\mathbf{R}_{Q,180^\circ}$ is 2, the order of $\mathbf{R}_{Q,270^\circ}$ is 4, and the order of $\mathbf{T}_{AB}$ is infinite.

The ***order*** of a group $\mathcal{G}$ is the number of elements in that group.

A group $\mathcal{G}$ is said to be ***generated by a subset*** S ***of*** $\mathcal{G}$ if every element of $\mathcal{G}$ can be expressed as a product of elements of S and inverses of such elements (by this we mean a *finite* product, not something that is the limit of some infinite process). In this case, we write

$$\mathcal{G} = \langle S \rangle.$$

A group $\mathcal{G}$ is called a ***cyclic group*** if it is generated by an element $a \in \mathcal{G}$; that is, $\mathcal{G} = \langle\, a \,\rangle$. Here, the order of $\mathcal{G}$ could be finite or infinite.

- For example, the group $\mathbb{Z}$ of integers, with addition as the binary operation, is generated by the set $\{\, 1 \,\}$; that is, $(\mathbb{Z}, +) = \langle\, 1 \,\rangle$.

  Here, the group multiplication is addition of integers, and the inverse of 1 is -1. To see why $\mathbb{Z}$ is generated by $\{\, 1 \,\}$, note that, for an integer m,

$$m = \begin{cases} \underbrace{1 + 1 + \cdots + 1}_{m \text{ times}} & \text{if } m > 0, \\ \underbrace{(-1) + (-1) + \cdots + (-1)}_{m \text{ times}} & \text{if } m < 0, \\ 1 + (-1) & \text{if } m = 0. \end{cases}$$

  Thus, $\mathbb{Z} = \langle\, 1 \,\rangle$, and the order of the group is infinite.

- On the other hand, if $a \in \mathcal{G}$ and $a^n = e$ for some integer $n \geq 1$, then the group $\mathcal{G} = \langle\, a \,\rangle$ has n elements, namely

$$a,\ a^2,\ a^3,\ \ldots\ ,\ a^n = e,$$

  and the order of the group is n. This group is denoted by $\mathcal{C}_n$ and is called the ***cyclic group of order*** n.

A group $\mathcal{G}$ is called a ***dihedral group*** if it is generated by two elements a and b for which

1. $a^n = e$,
2. $b^2 = e$, and
3. $bab^{-1} = a^{-1}$.

The elements of this group are

$$\begin{array}{cccc} a, & a^2, & \ldots & a^n = e, \\ ba, & ba^2, & \ldots & ba^n = b. \end{array}$$

Consequently, the group is called the ***dihedral group of order* $2n$** and is denoted by $\mathcal{D}_{2n}$.

Unlike the situation with cyclic groups, it is not immediately obvious that the $2n$ elements listed on the previous page are the only elements of $\mathcal{D}_{2n}$.

- For example, how do we know that a product like

$$a^8b^3\left(a^{-1}\right)^{10}a^4b^5$$

is actually one of the 10 elements that are said to comprise $\mathcal{D}_{10}$ or that the inverses of such products belong to $\mathcal{D}_{10}$?

Here, $n = 5$, and the 10 elements are

$$\begin{array}{ccccc} a, & a^2, & a^3 & a^4, & a^5 = e, \\ ba, & ba^2, & ba^3 & ba^4, & ba^5 = b. \end{array}$$

Since $b^2 = e$, we can replace the factors b^k with b if k is odd or with e (that is, omit the factor) if k is even. The product $a^8b^3\left(a^{-1}\right)^{10}a^4b^5$ reduces to

$$a^8b\left(a^{-1}\right)^{10}a^4b = a^8ba^{-10}a^4b = a^8ba^{-6}b.$$

Then, recalling that $a^5 = e$, replace a^8 with $a^5a^3 = a^3$ and replace a^{-6} with $a^{-6}\left(a^{10}\right) = a^4$ to get

$$a^3ba^4b.$$

Now replace a^4 by $a\left(b^{-1}b\right)a\left(b^{-1}b\right)a\left(b^{-1}b\right)a$ to get

$$a^3ba\left(b^{-1}b\right)a\left(b^{-1}b\right)a\left(b^{-1}b\right)ab = a^3\left(bab^{-1}\right)\left(bab^{-1}\right)\left(bab^{-1}\right)bab^{-1}.$$

Since $bab^{-1} = a^{-1}$, this becomes

$$a^3a^{-4} = a^{-1} = a^4,$$

which is one of the 10 elements, as claimed.

It should be clear that by applying the same process to any finite product, we will end up with one of a^s or ba^s.

Example 10.1.7. *What is the inverse of ba^s in $\mathcal{D}_{2n}$?*

Solution. It is its own inverse. Here is the proof:

$$(ba^s)^{-1} = (a^s)^{-1}\, b^{-1} = \left(bab^{-1}\right)^s b.$$

Expanding $\left(bab^{-1}\right)^s$, several b and b^{-1} cancel each other out, and we get ba^sb^{-1}, so that

$$(ba^s)^{-1} = \left(bab^{-1}\right)^s b = ba^sb^{-1}b = ba^s,$$

as claimed.

□

Example 10.1.8. *Assuming that $0 < k < n$, what is a^kb in $\mathcal{D}_{2n}$?*

Solution. We have

$$a^kb = \left(ba^{-1}b^{-1}\right)^k b = b\left(a^{-1}\right)^k = ba^{n-k}.$$

□

Here is the prototypical concrete example of a dihedral group.

Example 10.1.9. *Show that the group of symmetries of a square is $\mathcal{D}_8$.*

Solution. Label the vertices of the square A, B, C, and D counterclockwise, and let Q be the center point of the square. Any isometry T that carries the square onto itself must map the vertex A to $T(A)$, where $T(A)$ is one of the four vertices. The vertex B is mapped to $T(B)$, which must be one of the two vertices that are adjacent to $T(A)$. Since T must also map Q to Q, the action of T on A and B completely determines the isometry.

Two particular symmetries of the square are $\mathbf{R}_{Q,90^\circ}$ and $\mathbf{R}_m$, where m is a diagonal of the square. Letting $a = \mathbf{R}_{Q,90^\circ}$ and $b = \mathbf{R}_m$, we see that

$$a^4 = e,$$

$$b^2 = e,$$

$$bab^{-1} = a^{-1}.$$

The last equality follows by Theorem 10.1.6.

Consequently, the isometries $\mathbf{R}_{Q,90^\circ}$ and $\mathbf{R}_m$ generate the dihedral group $\mathcal{D}_8$, and since there are exactly eight different symmetries of the square, we are finished.

□

The preceding example has an obvious generalization whose proof is virtually identical to the proof for the square:

Theorem 10.1.10. *The group of symmetries of the regular n-gon is the dihedral group $\mathcal{D}_{2n}$.*

Remark. This result is often used as the definition of $\mathcal{D}_{2n}$.

10.2 Leonardo's Theorem

Leonardo da Vinci apparently worked out all possible symmetries for the floor plans of many chapels. Because of this, the following theorem is known as Leonardo's Theorem.

Theorem 10.2.1. *(Leonardo's Theorem)*

Every finite group of isometries of the plane is either a cyclic group or a dihedral group.

The proof of this theorem is long but not difficult and amounts to checking what can happen. In order to prove it, we need some facts about the product

$$\mathbf{R}_{P,\phi}\mathbf{R}_{Q,\theta}$$

that were developed earlier. We marshall them here for convenience.

(1) When $P = Q$: $\mathbf{R}_{P,\phi}\mathbf{R}_{P,\theta} = \mathbf{R}_{P,\phi+\theta}$.

(2) When $P \neq Q$ and $\phi + \theta \equiv 0 \pmod{360}$: $\mathbf{R}_{P,\phi}\mathbf{R}_{Q,\theta} = \mathbf{T}_{AB}$, where $A = Q$ and $B = \mathbf{R}_{P,\phi}(Q)$.

(3) When $P \neq Q$ and $\phi + \theta \not\equiv 0 \pmod{360}$: $\mathbf{R}_{P,\phi}\mathbf{R}_{Q,\theta} = \mathbf{R}_{S,\phi+\theta}$ for some point S.

We also need the following facts about the interaction between rotations and reflections:

(4) When P is on m: $\mathbf{R}_m\mathbf{R}_{P,\phi} = \mathbf{R}_l$, where l is the line through P such that the angle from l to m is $\phi/2$.

(5) When P is not on m: $\mathbf{R}_m\mathbf{R}_{P,\phi}$ is a glide reflection.

Note that points (2) and (5) imply the following:

Lemma 10.2.2. *Suppose that $\mathcal{G}$ is a finite group of isometries.*

(1) Any two rotations that are in $\mathcal{G}$ must have the same center.

(2) For any rotation and any reflection that are in $\mathcal{G}$, the center of rotation must be on the line of reflection.

Proof. The proof uses the fact that if a group contains a translation, then it must be infinite. To see why this is true, note that if $\mathbf{T}_{AB}$ is in the group $\mathcal{G}$, then so are $(\mathbf{T}_{AB})^2$, $(\mathbf{T}_{AB})^3$, and so on. However, each of $(\mathbf{T}_{AB})^2$, $(\mathbf{T}_{AB})^3$, ..., is a translation, and all of them are different.

(1) We want to show that if $\mathbf{R}_{P,\phi}$ and $\mathbf{R}_{Q,\theta}$ are two different members of $\mathcal{G}$, then $P = Q$. To establish this, we proceed by contradiction and assume that $P \neq Q$. We consider two cases.

(a) $\phi + \theta \equiv 0 \pmod{360}$:
If $P \neq Q$, then the product $\mathbf{R}_{P,\phi}\mathbf{R}_{Q,\theta}$ would be a translation, contradicting the fact that $\mathcal{G}$ is finite.

(b) $\phi + \theta \not\equiv 0 \pmod{360}$:
The fact that $\mathcal{G}$ is a group means that the product

$$(\mathbf{R}_{P,\phi})^{-1}(\mathbf{R}_{Q,\theta})^{-1}\mathbf{R}_{P,\phi}\mathbf{R}_{Q,\theta}$$

is a member of $\mathcal{G}$. However, we would then have

$$\mathbf{R}_{P,\phi}\mathbf{R}_{Q,\theta} = \mathbf{R}_{X,\phi+\theta}$$

for some X and

$$(\mathbf{R}_{P,\phi})^{-1}(\mathbf{R}_{Q,\theta})^{-1} = \mathbf{R}_{P,-\phi}\mathbf{R}_{Q,-\theta} = \mathbf{R}_{Y,-(\phi+\theta)}$$

for some point Y. Consequently,

$$(\mathbf{R}_{P,\phi})^{-1}(\mathbf{R}_{Q,\theta})^{-1}\mathbf{R}_{P,\phi}\mathbf{R}_{Q,\theta} = \mathbf{R}_{Y,-(\phi+\theta)}\mathbf{R}_{X,\phi+\theta}.$$

This is either the identity, if $X = Y$, or a translation, if $X \neq Y$. However, it cannot be the case that $X = Y$, for this would mean that $\mathbf{R}_{Y,-(\phi+\theta)}$ is the inverse of $\mathbf{R}_{X,\phi+\theta}$, or in other words, that

$$\mathbf{R}_{P,-\phi}\mathbf{R}_{Q,-\theta} = (\mathbf{R}_{P,\phi}\mathbf{R}_{Q,\theta})^{-1}.$$

However,

$$(\mathbf{R}_{P,\phi}\mathbf{R}_{Q,\theta})^{-1} = (\mathbf{R}_{Q,\theta})^{-1}(\mathbf{R}_{P,\phi})^{-1} = \mathbf{R}_{Q,-\theta}\mathbf{R}_{P,-\phi},$$

and we would have

$$\mathbf{R}_{P,-\phi}\mathbf{R}_{Q,-\theta} = \mathbf{R}_{Q,-\theta}\mathbf{R}_{P,-\phi},$$

which contradicts the fact that rotations with different centers do not commute. Since the centers X and Y are different, it follows that the product

$$(\mathbf{R}_{P,\phi})^{-1}(\mathbf{R}_{Q,\theta})^{-1}\mathbf{R}_{P,\phi}\mathbf{R}_{Q,\theta}$$

is a translation, again contradicting the fact that $\mathcal{G}$ is finite.

(2) Suppose that $\mathbf{R}_{P,\phi}$ and $\mathbf{R}_{Q,\theta}$ both belong to $\mathcal{G}$, and suppose for a contradiction that P is not on m. Then $(\mathbf{R}_m\mathbf{R}_{P,\phi})^2$ is a translation, meaning that $\mathcal{G}$ would have to be infinite.

This completes the proof.

□

We next prove a theorem that is part of Leonardo's Theorem but which is useful in its own right.

Theorem 10.2.3. *If $\mathcal{G}$ is a finite group of isometries that consists of exactly n rotations (counting the identity as a rotation through $360°$), then $\mathcal{G}$ is the cyclic group $\mathcal{C}_n$.*

Proof. From Lemma 10.2.2, we know that $\mathcal{G}$ consists of rotations through various angles with all rotations centered at a common point Q. We may also assume that all rotations belonging to $\mathcal{G}$ are through a positive angle no greater than $360°$. Since there are only a finite number of rotations, one of them has the smallest positive angle of rotation, say θ. We claim that every other rotation, including $\mathbf{R}_{Q,360°}$, must be a multiple of θ.

We again use a proof by contradiction. Supposing that this were not the case, there would be a rotation $\mathbf{R}_{Q,\phi}$ in $\mathcal{G}$ where $\phi > 0$ and where ϕ is not a multiple of θ. Let α be the remainder when ϕ is divided by θ; that is,

$$\phi = m\theta + \alpha,$$

where m is a positive integer and $0 < \alpha < \theta$. Since $\mathcal{G}$ is a group, then

$$(\mathbf{R}_{Q,\theta})^{-m} = \mathbf{R}_{Q,-m\theta}$$

is in $\mathcal{G}$, and therefore so is the product

$$\mathbf{R}_{Q,\phi}\mathbf{R}_{Q,-m\theta} = \mathbf{R}_{Q,\phi-m\theta} = \mathbf{R}_{Q,\alpha}.$$

However, this is impossible, since $\mathbf{R}_{Q,\theta}$ is the rotation with the smallest positive angle of rotation.

This shows that all of the rotations in $\mathcal{G}$ are multiples of $\mathbf{R}_{Q,\theta}$ and, conversely, since $\mathcal{G}$ is a group, all multiples of $\mathbf{R}_{Q,\theta}$ must be in $\mathcal{G}$. Letting a denote $\mathbf{R}_{Q,\theta}$, this means that $\mathcal{G}$ consists of

$$a, \quad a^2, \quad a^3, \quad \ldots, \quad a^n$$

for some integer n, where $a^n = \mathbf{R}_{Q,360^\circ} = \mathbf{I}$. This completes the proof of Theorem 10.2.3.

□

Theorem 10.2.4. *Suppose that $\mathcal{G}$ is a finite group that contains a rotation (other than the identity) and a reflection. Then there is a rotation $\mathbf{R}_{Q,\alpha}$ in the group that generates all of the rotations in the group. That is, the set of all rotations in $\mathcal{G}$ is*

$$\mathbf{R}_{Q,\alpha}, \quad \mathbf{R}_{Q,2\alpha}, \quad \ldots, \quad \mathbf{R}_{Q,n\alpha},$$

where $n\alpha = 360^\circ$.

Furthermore, if R_m is any reflection in $\mathcal{G}$, then every reflection in $\mathcal{G}$ must be one of the following:

$$\mathbf{R}_m\mathbf{R}_{Q,\alpha}, \quad \mathbf{R}_m\mathbf{R}_{Q,2\alpha}, \quad \ldots, \quad \mathbf{R}_m\mathbf{R}_{Q,n\alpha}.$$

In particular, this means that $\mathcal{G}$ is the dihedral group $\mathcal{D}_{2n}$.

Proof. Let $\mathcal{S}$ be the set of all rotations that are in $\mathcal{G}$, including the identity. Then by the previous theorems, these rotations form a group, and all of the rotations have the same center Q. Hence, by Theorem 10.2.3, $\mathcal{S}$ is a cyclic group and so the members of $\mathcal{S}$ are

$$\mathbf{R}_{Q,\alpha}, \quad \mathbf{R}_{Q,2\alpha}, \quad \ldots, \quad \mathbf{R}_{Q,n\alpha},$$

where $n\alpha = 360^\circ$.

To complete the remainder of the proof, we have to show that if $\mathbf{R}_l$ is any reflection that is in $\mathcal{G}$, then there is some integer k such that

$$\mathbf{R}_l = \mathbf{R}_m\mathbf{R}_{Q,k\alpha}.$$

Now, consider the product $\mathbf{R}_m\mathbf{R}_l$. Since both reflections belong to $\mathcal{G}$, it follows that the lines l and m both contain Q, and so $\mathbf{R}_m\mathbf{R}_l$ is a rotation with center Q.

Since $\mathbf{R}_m\mathbf{R}_l$ is a member of $\mathcal{G}$, and since $\mathcal{S}$ contains all of the rotations in $\mathcal{G}$, it follows that

$$\mathbf{R}_m\mathbf{R}_l = \mathbf{R}_{Q,k\alpha}$$

for some integer k. Therefore, we have

$$\mathbf{R}_m\mathbf{R}_{Q,k\alpha} = \mathbf{R}_m\left(\mathbf{R}_m\mathbf{R}_l\right) = \left(\mathbf{R}_m\mathbf{R}_m\right)\mathbf{R}_l m = \mathbf{R}_l.$$

□

Leonardo's Theorem now follows from Theorem 10.2.3 and Theorem 10.2.4.

One of the consequences of Leonardo's Theorem is:

Theorem 10.2.5. *The group of symmetries of a polygon in the plane is either a cyclic group or a dihedral group.*

Proof. Given a vertex A of a polygon, together with an adjacent vertex B, any symmetry of the polygon must map A onto one of the vertices of the polygon, in which case there are at most two possible adjacent vertices that can be the image of B. In other words, there are only a finite number of symmetries. Leonardo's Theorem now tells us that the group of symmetries is either a cyclic group or a dihedral group.

□

10.3 Problems

1. Prove that a finite group of isometries cannot contain two halfturns about distinct points.
2. Prove that the set of all halfturns and all translations forms a group.
3. Prove that if a triangle is invariant under a reflection, then the triangle must be isosceles.
4. Which of the following sets of transformations form a group, and which do not form a group?
   (a) All translations.
   (b) All reflections.
   (c) All glide reflections.
   (d) All rotations.
   (e) All direct isometries.
   (f) All opposite isometries.

5. If $H_{O_1} H_{O_2} = H_{O_2} H_{O_1} = T$, prove that $T = I$, the identity transformation.

6. Find a plane figure P such that its group of symmetries equal

   (a) the cyclic group C_2 of order 2,

   (b) the cyclic group C_1 of order 1.

7. Find a plane figure P such that its group of symmetries equal

   (a) the dihedral group D_2 of order 4,

   (b) the dihedral group D_1 of order 2.

8. Find the group of symmetries of each of the following figures.

(a)

(b)

9. Let $\mathcal{G}$ be a group of isometries whose subgroup of translations is generated by T_{AB}, where $AB \neq 0$. Prove that if $R_\ell \in \mathcal{G}$, then either $\overline{AB}$ is parallel to ℓ or $\overline{AB}$ is perpendicular to ℓ.

10. Find the group of isometries of an ellipse.

11. If a and b are elements of a group $\mathcal{G}$ and

$$(ab)^2 = a^2b^2,$$

show that $ab = ba$.

12. Let a and b be elements of a group $\mathcal{G}$ such that b has order 2 and $ab = ba^{-1}$.

   (a) Show that $a^n b = ba^{-n}$ for all integers n.
   Hint: Evaluate the product $(bab)(bab)$ in two different ways to show that $ba^2b = a^{-2}$, and then extend this method.

   (b) Show that the set $S = \{a^n, ba^n \mid n \in \mathbb{Z}\}$ is closed under multiplication and in fact forms a group.

   (c) Show that $S = \langle a, b \rangle$, the dihedral group with generators a and b.

13. A *cuboid* is a rectangular parallelepiped; that is, a parallelepiped where each plane face is orthogonal to four other faces and parallel to the fifth, as in the figure.

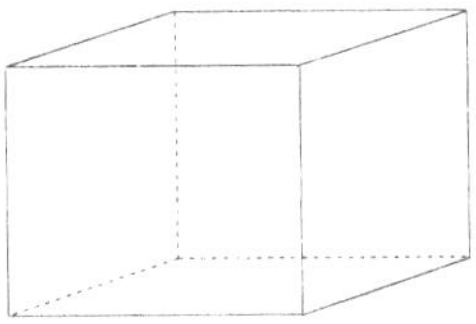

Find the group of symmetries of a cuboid with three unequal sides.

14. What is the symmetry group of a rhombus that is not a square? Find all the symmetries of the rhombus and construct the Cayley table or multiplication table for the group of symmetries.

    Hint: The diagonals of a parallelogram bisect each other, and a parallelogram is a rhombus if and only if its diagonals are perpendicular.

15. Let $\mathcal{T}$ be a (nonequilateral) isosceles triangle.

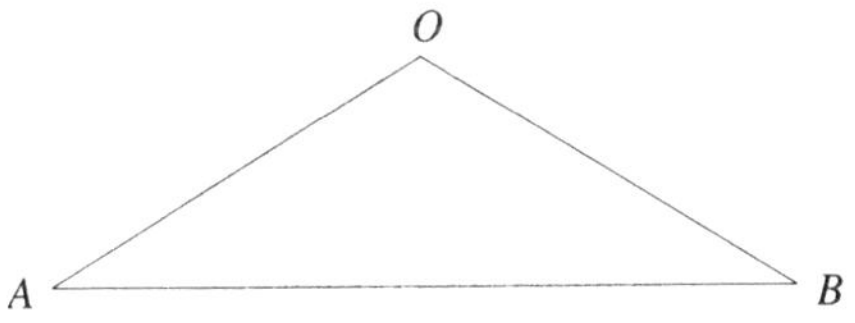

Find the group of symmetries of $\mathcal{T}$ and construct the Cayley table for the group.

- In the plane, the discrete groups fixing a line are the groups of symmetries of ribbons or friezes. To study the frieze groups, we first find an isometry fixing a line and then compose it with an isometry that has a fixed point. The group $\mathcal{F}$ generated by this isometry is called a ***frieze group***.

16. For each of the seven patterns given in the figure below, assuming each extends to infinity both to the left and to the right, name the types of isometries in the symmetry group of each pattern.

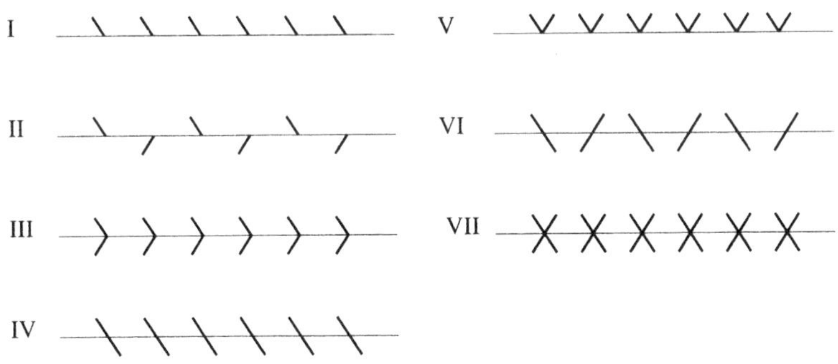

17. Prove that if $H_{O_0} \in \mathcal{F}$, a frieze group with translation subgroup

$$\mathcal{T} = \{T_{nAB}\} = \langle T_{AB} \rangle ,$$

then $H_{O_{n/2}} \in \mathcal{F}$ for every integer n, where $\overline{O_0 O_{n/2}} = \frac{n}{2}\overline{AB}$.

18. Find the group of symmetries of the following repeated pattern on an infinite horizontal strip, as shown below.

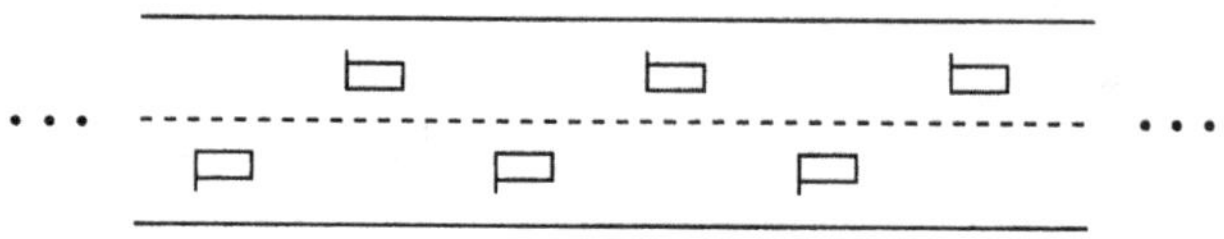

19. Find a different set of two generators for the frieze group

$$\mathcal{F} = \left\langle H_{O_{1/4}}, G_{\ell, \frac{1}{2}AB} \right\rangle ,$$

where $\ell \parallel \overline{AB}$, and O_0 and $O_{\frac{1}{4}}$ are two fixed points on ℓ with $\overline{OO_{\frac{1}{4}}} = \frac{1}{4}\overline{AB}$, as in the figure below.

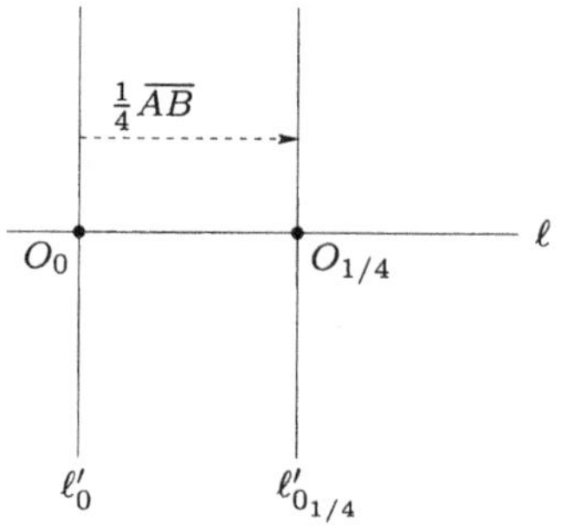

20. With repeated use of only the symbol

construct a repeated pattern on a horizontal strip whose frieze group is

$$\mathcal{F} = \left\langle H_{O_{1/4}}, G_{\ell, \frac{1}{2}AB} \right\rangle ,$$

where $\ell \parallel \overline{AB}$, and O_0 and $O_{\frac{1}{4}}$ are points on ℓ such that $\overline{OO_{\frac{1}{4}}} = \frac{1}{4}\overline{AB}$, as in the figure in Problem 10.19.

21. Find the frieze group of an infinite horizontal strip consisting of repeated I's if the I's lie above the midline of the strip as shown below.

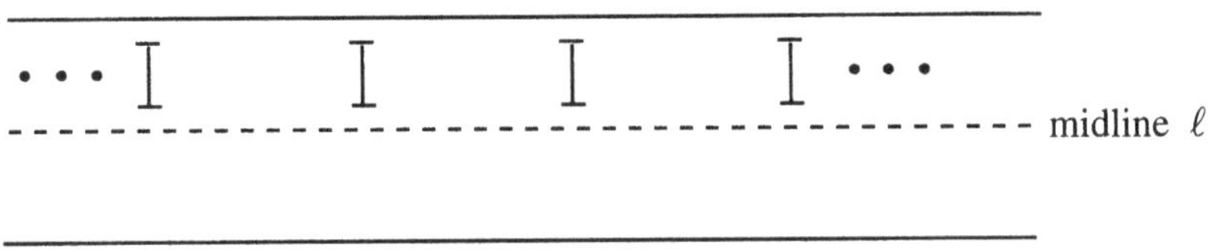

22. Consider a particular vertex of an equilateral triangle. Under a symmetry, it can land on any of the three vertices. The remaining vertices must follow in order, either counterclockwise or clockwise. Thus, there are only $3 \times 2 = 6$ symmetries for the equilateral triangle. They are: the identity **I**, a $120°$ counterclockwise rotation J, a $120°$ clockwise rotation K, and three reflections X, Y, and Z, as shown in the figure on the following page.

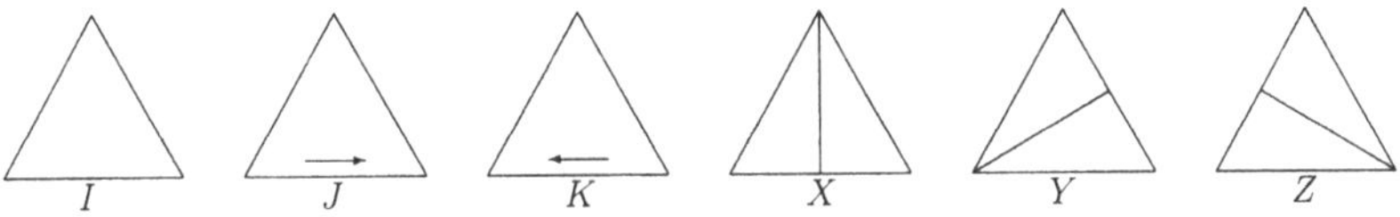

These six symmetries form a group under the operation of composition, the dihedral group of the equilateral triangle. Construct the operation table of this group.

23. Consider a particular vertex of a square. Under a symmetry of the square, this vertex can land on any one of the four vertices. The remaining vertices must follow in order, either counterclockwise or clockwise. Thus, there are only $4 \times 2 = 8$ symmetries for the square. They are: the identity I, a $180°$ rotation or halfturn R, a $90°$ counterclockwise rotation A, a $90°$ clockwise rotation C, and four reflections H, V, D, and U, as shown in the figure below.

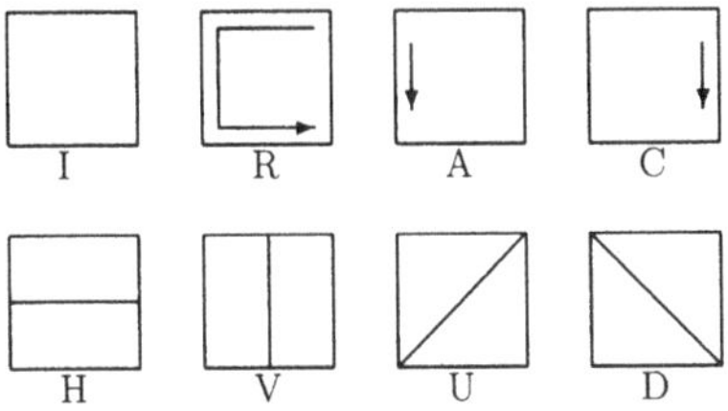

These eight symmetries of the square form a group under the operation of composition, the dihedral group of the square. Construct the operation table of this group.

24. Consider the set of symmetries of a non-square rectangle. It has only four elements: I, R, H, and V, analogous to the corresponding symmetries for the square. These four symmetries form a group with respect to composition. Construct the multiplication table of this group.

25. The complex numbers 1, -1, i, and $-i$ form a group under multiplication. Construct the operation table of this group.

26. The ***Quaternion group*** consists of the eight elements

$$1, \quad -1, \quad i, \quad -i, \quad j, \quad -j, \quad k, \quad -k,$$

with the operation of multiplication defined by

$$\begin{aligned} &i^2 = j^2 = k^2 = -1, \\ &ij = k, \quad jk = i, \quad ki = j, \\ &ji = -k, \quad kj = -i, \quad ik = -j. \end{aligned}$$

Construct the operation table.

27. Let

$$\begin{aligned} a(x) &= \frac{1}{1-x}, \\ b(x) &= \frac{x-1}{x}, \\ c(x) &= 1-x, \\ d(x) &= \frac{x}{x-1}, \\ e(x) &= x, \\ f(x) &= \frac{1}{x}. \end{aligned}$$

These functions form a group with respect to the operation of composition. Construct the operation table.

CHAPTER 11

HOMOTHETIES

11.1 The Pantograph

Isometries provide a dynamic way of dealing with congruency. In this chapter, we study a transformation which serves the same purpose for the notion of similarity.

Without employing photography, photocopying, or computer graphics, it is not a simple matter to produce an enlarged or reduced copy of a figure. There is, however, a physical instrument called a ***pantograph*** that allows us to accomplish this.

Classical Geometry: Euclidean, Transformational, Inversive and Projective, First Edition. By
I. E. Leonard, J. E. Lewis, A. C. F. Liu, G. W. Tokarsky
Copyright © 2014 John Wiley & Sons, Inc.

A pantograph is formed from four thin flat rods of equal length that are joined together by four hinge pins P, Q, A, and B so that $APBQ$ is a parallelogram and $OA = AP$. The instrument lies flat on the drawing board and is fixed to the board at the ***pivot point*** O. Pencils are attached to the instrument at points P and P'.

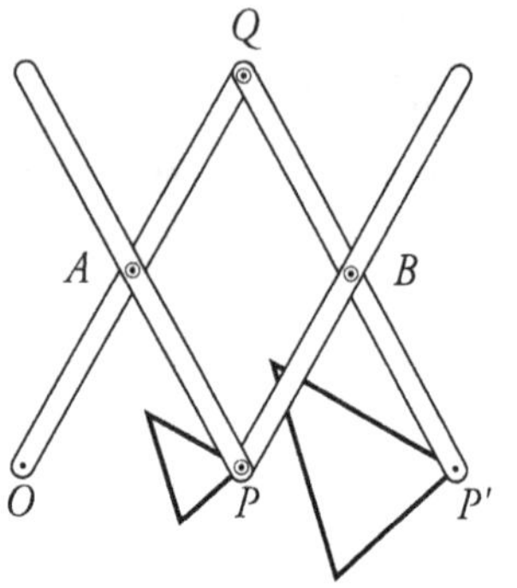

If an enlargement is desired, the pencil at P is used to trace the original feature. As this is being done, the pencil at P' draws a copy magnified by a factor equal to OQ/OA. If a reduction is desired, the pencil at P' is used to trace the figure so that the pencil at P draws the reduced copy.

To see why this works, note that OAP and OQP' are similar triangles by **sAs**. It follows that O, P, and P' are collinear. Moreover,

$$\frac{OP'}{OP} = \frac{OQ}{OA},$$

and thus the figure that is traced by P may be considered a "contraction" towards O of the figure that is traced by P'.

It should be noted that the magnification factor OQ/OA for the pantograph illustrated in the figure above is fixed. In an actual pantograph, the positions of the hinge pins at A and B may be adjusted so that OQ/OA can be set as desired. The pins A and B must be adjusted in such a manner that $APBQ$ remains a parallelogram and so that $OA = AP$.

11.2 Some Basic Properties

A similarity in which a figure contracts towards a point or expands away from a point in this manner is called a *homothety*.

More formally, let O be a point and let k be a nonzero real number. The ***homothety*** centered at O with ratio k, denoted by $\mathbf{H}_{O,k}$, maps O to O and each point $P \neq O$ onto another point P' on the line OP such that

$$\overline{OP'} = k\,\overline{OP}.$$

Since $k \neq 0$, it follows that $\mathbf{H}_{O,k}$ is a transformation, and its inverse is easily seen to be $\mathbf{H}_{O,1/k}$.

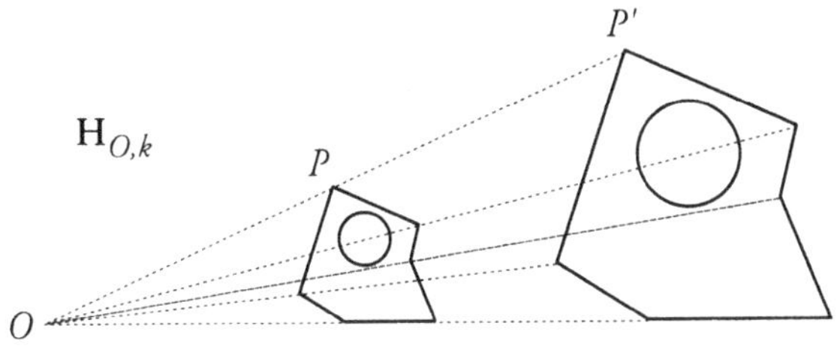

Remark. In this chapter, the word *parallel* includes the case where two lines coincide. In other words, two lines are considered to be parallel if there is a translation that maps one onto the other, including a translation through a distance of magnitude zero. Also, the notation $\overline{AB}$ is used to denote the *directed segment* from A to B.

A ***similarity*** is the composition of an isometry and a homothety. Thus, any similarity that is not an isometry is

(a) either a rotation followed by a homothety with the same center (the isometry is direct)

(b) or a reflection followed by a homothety whose center lies on the line of the reflection (the isometry is opposite).

Theorem 11.2.1. *Let A and B be two points whose images under the homothety $\mathbf{H}_{O,k}$ are A' and B', respectively. Then $A'B'$ is parallel to AB and $\overline{A'B'} = k\,\overline{AB}$.*

Proof.
Case 1. O, A, and B are collinear.
In this case, $A'B'$ and AB are both contained in the line OA, so they are parallel. Moreover,

$$\overline{A'B'} = \overline{OB'} - \overline{OA'} = k\,\overline{OB} - k\,\overline{OA} = k\,\overline{AB}.$$

Case 2. O, A, and B are noncollinear (as in the figure).
If the points O, A, and B do not lie on a line, then angle O is common to both $\triangle OA'B'$ and $\triangle OAB$. Also,

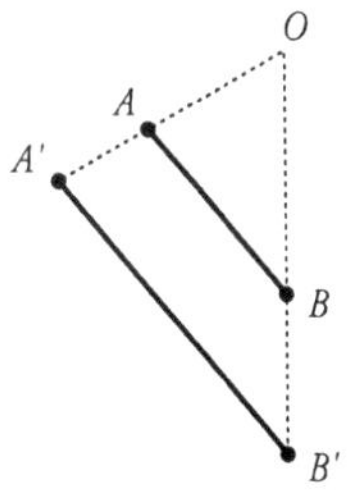

$$\overline{OA'} = k\,\overline{OA} \qquad \text{and} \qquad \overline{OB'} = k\,\overline{OB},$$

so the triangles are similar by **sAs**. It follows that

$$\overline{A'B'} = k\,\overline{AB}$$

and that the segments $A'B'$ and AB are parallel.

□

Corollary 11.2.2. *Let A', B', and C' be the images of A, B, and C, respectively, under a homothety.*

(1) *If B is between A and C, then B' is between A' and C'.*

(2) *If A, B, and C are the vertices of a triangle, then A', B', and C' are the vertices of a similar triangle.*

Proof. Suppose the homothety is $\mathbf{H}_{O,k}$.

(1) By the Triangle Inequality, $AC = AB + BC$. Using Theorem 11.2.1, we get

$$A'C' = |k|\,AC = |k|\,(AB + BC) = |k|\,AB + |k|\,BC = A'B' + B'C',$$

and so by the Triangle Inequality, B' must be between A' and C'.

(2) By Theorem 11.2.1,

$$\frac{A'B'}{AB} = \frac{A'C'}{AC} = \frac{B'C'}{BC} = |k|,$$

and the triangles are similar by **sss**.

□

A homothety preserves any geometric relationship that can be completely characterized by ratios of distances. For example, midpoints, centroids, and angle bisectors can all be characterized by distance ratios. This means that the midpoint of BC is mapped to the midpoint of $B'C'$, the centroid of $\triangle ABC$ is mapped to the centroid of $\triangle A'B'C'$, and the bisector of $\angle ABC$ is mapped to the bisector of $\angle A'B'C'$. We will use these facts freely throughout this chapter.

All of these facts can be proved using Theorem 11.2.1 and Corollary 11.2.2. For example, here is a proof that the midpoint M of BC is mapped to the midpoint M' of $B'C'$ by the homothety $\mathbf{H}_{O,k}$.

Corollary 11.2.2 tells us that B', C', and M' are collinear. Also, by Theorem 11.2.1,

$$\overline{B'M'} = k\,\overline{BM} \qquad \text{and} \qquad \overline{B'C'} = k\,\overline{BC},$$

so that

$$\frac{\overline{B'M'}}{\overline{B'C'}} = \frac{\overline{BM}}{\overline{BC}} = \frac{1}{2}.$$

Thus, the image M' of M is indeed the midpoint of $B'C'$.

□

11.2.1 Circles

Theorem 11.2.3. *The image of a circle $\mathcal{C}$ with center C and radius r under the homothety $\mathbf{H}_{O,k}$ is a circle with center $D = \mathbf{H}_{O,k}(C)$ and radius $|k|\,r$.*

Proof. The homothety maps each point X of $\mathcal{C}$ to a point Y such that

$$DY = |k|\,CX = |k|\,r,$$

so the points Y form a circle centered at D with radius $|k|\,r$.

□

If $\mathcal{C}$ and $\mathcal{D}$ are two circles, the center of any homothety that transforms one circle into the other is called a ***center of similitude***.

Example 11.2.4. *Given two circles with two different centers and two different radii, explain how to construct all centers of similitude for the circles.*

Solution. The process is illustrated in the figure below.

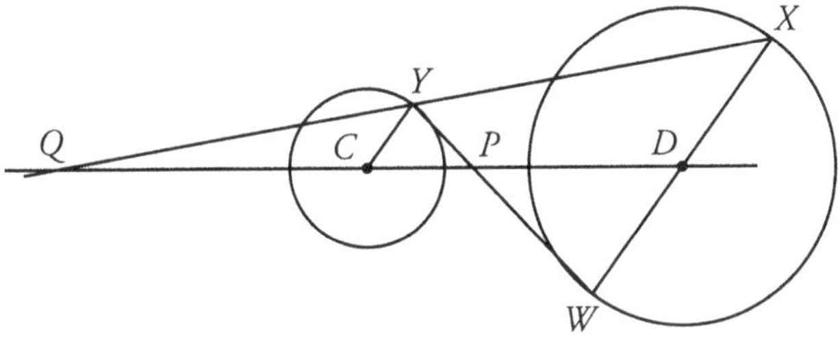

Construct any diameter of one of the circles, say WX. Then construct a radius CY of the other circle that is parallel to this diameter. The points P and Q, where the lines WY and XY intersect the line CD through the centers of the circles, will be the centers of similitude. There are no more centers of similitude for these two circles, for if O is a center of similitude, then O must be on the line CD, and the homothety $\mathbf{H}_{O,k}$ that carries C to D must transform the radius CY into a parallel radius, either DW or DX. This leaves only two possible locations for O.

□

Theorem 11.2.5. *Let P be a point on the line joining the centers A_1 and A_2 of two circles with radii r_1 and r_2, respectively. If*

$$\frac{PA_1}{PA_2} = \frac{r_1}{r_2},$$

then P is a center of similitude of the circles.

Proof. Note that there are at most only two points P on A_1A_2 such that

$$\frac{PA_1}{PA_2} = \frac{r_1}{r_2}.$$

For one of them the ratio

$$\frac{\overline{PA_1}}{\overline{PA_2}}$$

is positive, for the other it is negative. Letting

$$k = \frac{\overline{PA_1}}{\overline{PA_2}},$$

we see that the homothety $\mathbf{H}_{P,k}$ maps the circle centered at A_1 to the circle centered at A_2.

□

Example 11.2.6. *Show that any common tangent to two circles of unequal radii passes through a center of similitude.*

Solution. The tangent t cannot be parallel to the the line m joining the centers A and B. Suppose that t meets m at P. Let A' and B' be the points of tangency, as in the figure on the following page.

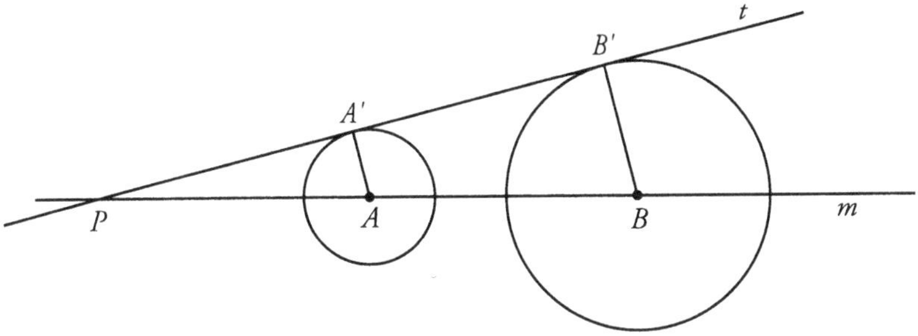

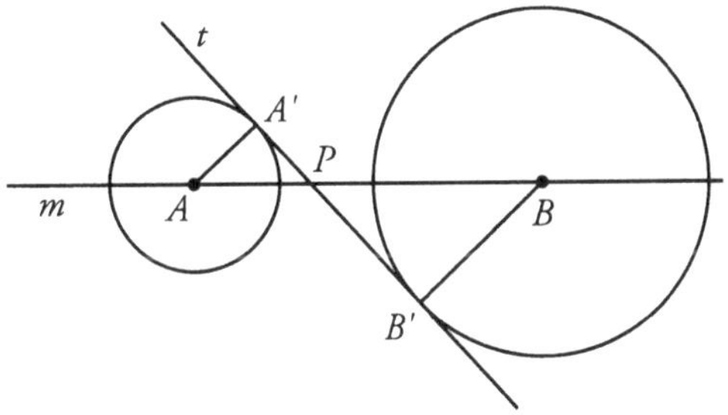

Then $\triangle PA'A \sim \triangle PB'B$ by **AAA** similarity. Consequently,

$$\frac{PA}{PB} = \frac{AA'}{BB'},$$

and it follows from Theorem 11.2.5 that P is a center of similitude.

□

11.3 Construction Problems

If one figure is the image of another under a homothety, the figures are said to be ***homothetic*** to each other. Homothetic figures are similar, but similar figures need not be homothetic, since similar figures need not be oriented the same way.

Some construction problems can be solved by first constructing a homothetic image of the figure and then enlarging or shrinking the image to get the desired solution.

Example 11.3.1. *Given an acute triangle ABC, construct a square $PQRS$ with P on AB, S on AC, and edge QR on BC.*

Solution. Let P' be any point AB. Drop the perpendicular $P'Q'$ to BC and complete the square $P'Q'R'S'$, as in figure (a) below.

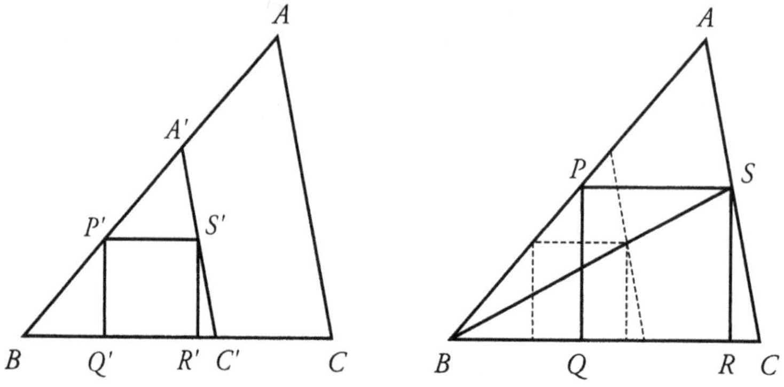

(a) (b)

If we draw the line $A'C'$ through S' parallel to AC, we note that the square $P'Q'R'S'$, together with the triangle $A'B'C'$, is homothetic to the desired solution, with B being the center of the homothety.

Thus, we need a homothet of $P'Q'R'S'$ so that the image of S' is the point S on AC. To accomplish this, draw the line BS'. Then S is the point where BS' meets AC, as in figure (b) above. Now drop the perpendicular SR to BC and complete the desired square.

□

Here is another example that uses the same idea.

Example 11.3.2. *Construct points X and Y on the sides AB and BC, respectively, of a given triangle ABC such that*

$$AX = XY = YC.$$

Solution. The desired result is shown in figure (a) on the following page. Take a point D on BC and a point F on AB such that $AF = CD$, as in figure (b) on the following page.

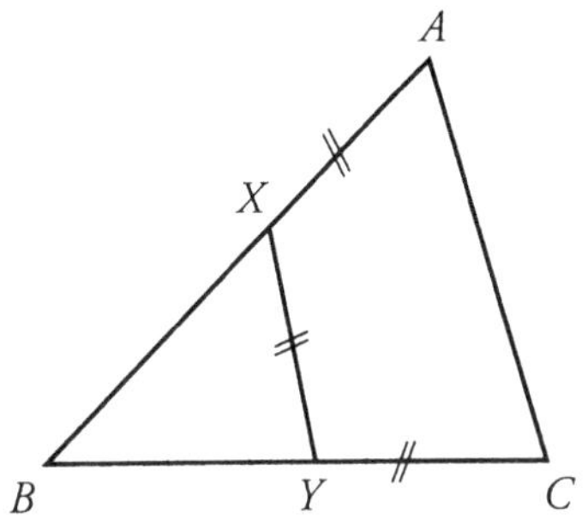

(a)

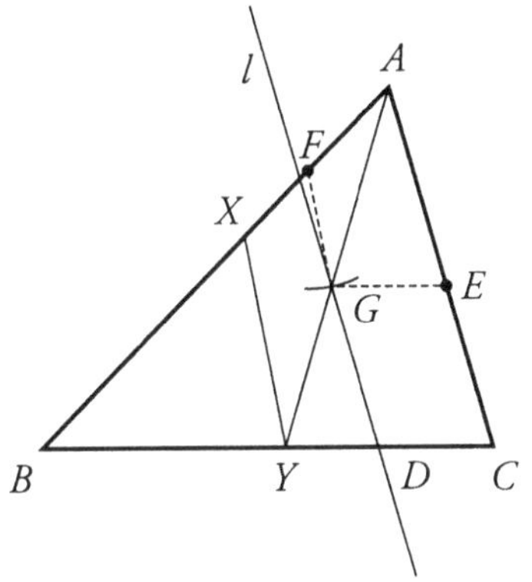

(b)

Draw a line l through D parallel to AC. Draw a circle with center F and radius FA, cutting l at a point G inside $\triangle ABC$. Complete the parallelogram $CDGE$, with E on AC. We have

$$AF = FG = GE.$$

Join A and G and extend the segment AG to cut BC at Y. Draw a line through Y parallel to FG, cutting AB at X. Then $AFGE$ and $AXYC$ are homothetic to each other with A being the center of homothety. Hence,

$$AX = XY = YC.$$

□

Example 11.3.3. *Given points P and X and line segments AB and CD and letting $k = CD/AB$, construct the point $\mathbf{H}_{P,k}(X)$.*

Solution. Construct a line m through P at an angle to PX and construct points D' and B' on m so that $PB' = AB$ and $PD' = CD$. Join B' to X and construct the line through D' parallel to $B'X$ meeting PX at Y. Therefore,

$$\frac{PY}{PX} = \frac{CD}{AB},$$

since triangles $PB'X$ and $PD'Y$ are similar. Thus, $Y = \mathbf{H}_{P,k}(X)$.

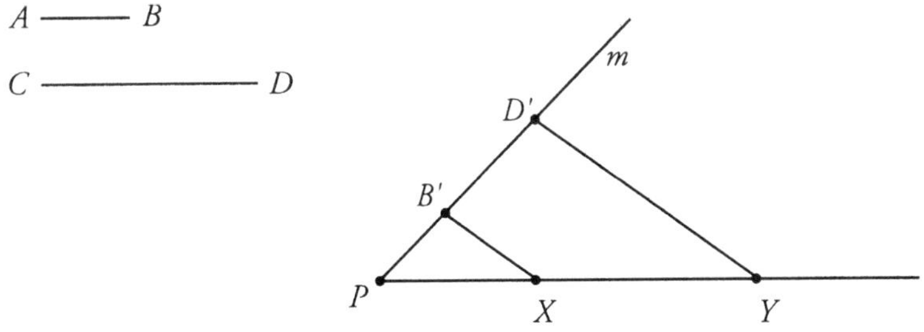

This shows that given any point X, we can construct the image of X under the homothety $\mathbf{H}_{P,k}$ where $k = CD/AB$.

□

Example 11.3.3 can be considered a basic construction. It shows us how to construct the homothetic images of all standard geometric figures.

For example, to construct the image of $\triangle ABC$ under $\mathbf{H}_{P,k}$, we can construct the images A', B', and C' of the vertices A, B, and C, respectively, under the homothety.

Also, to construct the image under $\mathbf{H}_{P,k}$ of a circle with center C and radius r, let X be a point on the circle and construct $\mathbf{H}_{P,k}(C)$ and $\mathbf{H}_{P,k}(X)$. These are, respectively, the center of the image circle and a point on the image circle, and we can now construct the image circle.

Example 11.3.4. *Given three points A, B, and C on a circle $\mathcal{C}$, explain how to find all chords through the point C that are bisected by AB.*

Solution. There are two ways to solve this problem.

The first way is to construct the image m of the line AB under $\mathbf{H}_{C,2}$, as in the figure on the right. Then for any point X' on the line m, the point

$$X = X'C \cap AB$$

is the midpoint of $X'C$. Thus, the points P and Q, where m meets the circle, will provide the chords PC and QC.

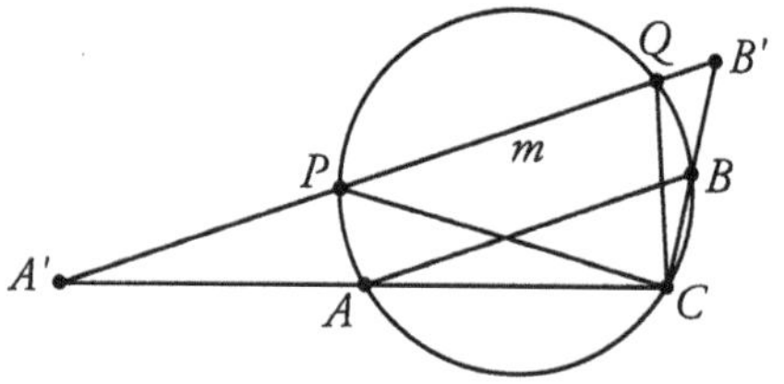

The second way is to apply $\mathbf{H}_{C,1/2}$ to the circle, obtaining another circle $\mathcal{D}$, as in the figure on the right. Let P' and Q' be the points where $\mathcal{D}$ intersects AB. Let P and Q be the points where CP' and CQ' meet $\mathcal{C}$; that is,

$$P = CP' \cap \mathcal{C} \text{ and } Q = CQ' \cap \mathcal{C},$$

so that, again, CP and CQ are the desired chords.

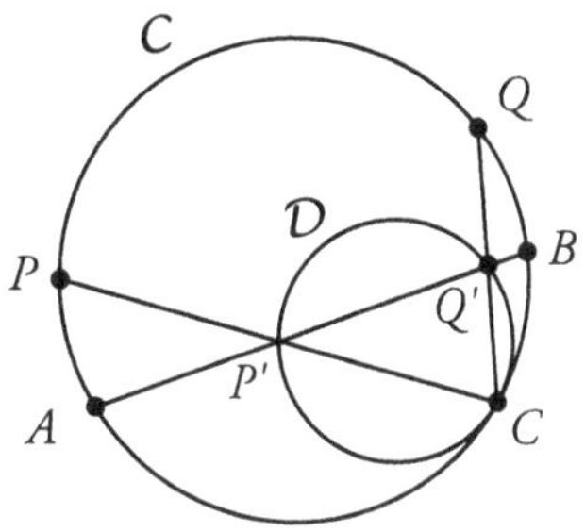

□

Example 11.3.5. *Given two lines l and m intersecting at Q and given a point P not on either line, construct a line through P cutting l at L and m at M such that $PL = 2PM$.*

Solution. There are two pairs of points L and M. To find the first pair L_1 and M_1, construct the line

$$l_1 = \mathbf{H}_{P,1/2}(l)$$

and let $M_1 = l_1 \cap m$, and then let $L_2 = PM_1 \cap l$.

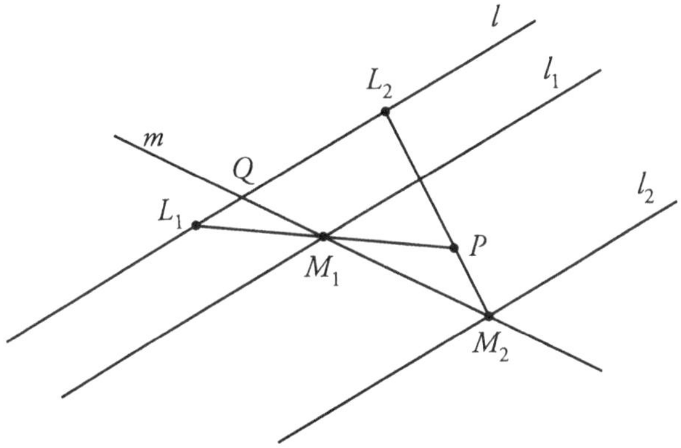

To find the second pair, construct the line

$$l_2 = \mathbf{H}_{P,-1/2}(l)$$

and let $M_2 = l_2 \cap m$, and then let $L_1 = PM_2 \cap l$.

□

Example 11.3.6. *Given $\angle ABC$ and a point P on the arm BC, construct a circle passing through P that is tangent to both arms of the angle.*

Solution. There are two possible solutions to this problem.

(1) Draw the angle bisector of ABC and let Q' be any point on it, as in the figure on the left. Drop the perpendicular $Q'P'$ to BC and construct the circle $\mathcal{C}'$ with center Q' and radius $Q'P'$. The circle $\mathcal{C}'$ is tangent to both arms of the angle. Thus, the desired circle $\mathcal{C}$ is obtained by constructing the image of $\mathcal{C}'$ under the homothety $\mathbf{H}_{B,k}$ where $k = PB/P'B$.

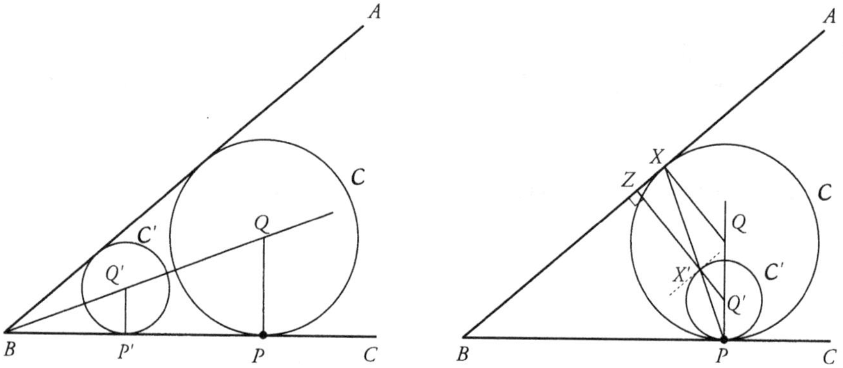

(2) Construct the line through P perpendicular to BC and let Q' be any point on it, as in the figure on the right above. Construct the circle $\mathcal{C}'$ with center Q' and radius $Q'P$, which is therefore tangent to BC. Drop the perpendicular from Q' to AB and let X' be the point where it intersects $\mathcal{C}'$. Note that the tangent to $\mathcal{C}'$ at X' is parallel to AB. Let X be the point $PX' \cap AB$, and the circle $\mathcal{C}$ is now obtained by constructing the image of $\mathcal{C}'$ under the homothety $\mathbf{H}_{P,k}$ where $k = PX/PX'$.

□

11.4 Using Homotheties in Proofs

Recall that the four major concurrency points of a triangle are:

1. the ***incenter***, the point where the angle bisectors meet,
2. the ***circumcenter***, the point where the right bisectors of the sides meet,
3. the ***centroid***, the point where the medians meet, and
4. the ***orthocenter***, the point where the altitudes meet.

The ***incircle*** is the circle centered at the incenter and is internally tangent to all three sides of the triangle.

The ***circumcircle*** is the circle centered at the circumcenter that passes through the three vertices of the triangle.

Despite their suggestive names, neither the centroid nor the orthocenter is the center of any significant circle associated with the triangle.

Theorem 11.4.1. *The incircle is the smallest circle that meets all three sides of a given triangle.*

Proof. Let C be a circle with center P that intersects all three sides of triangle ABC. If none of the sides are tangent to the circle C, contract it using a homothety centered at P, so that the image circle intersects all sides of ABC but is tangent to at least one side. Let T be the point of tangency.

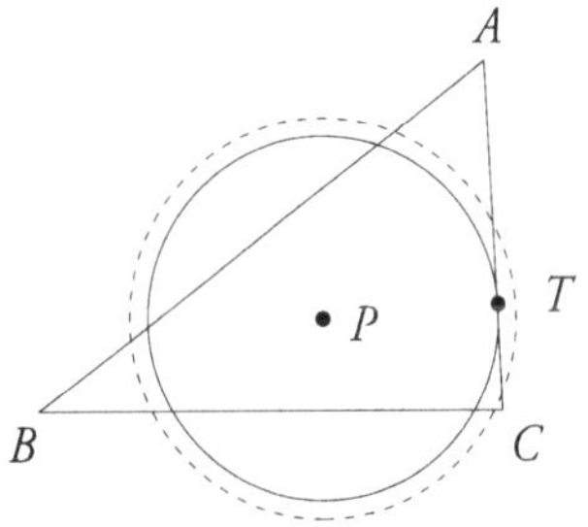

If the image circle is not tangent to two sides of triangle ABC, then contract it using a homothety centered at T, so that the new image circle still intersects all three sides of ABC and is tangent to at least two sides. Let the second point of tangency be denoted by S.

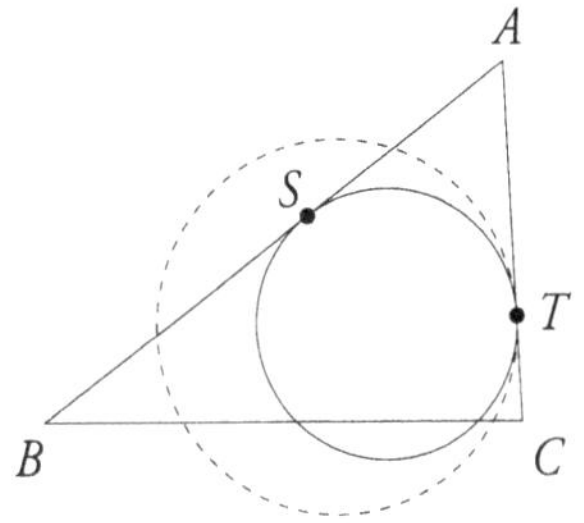

If this image circle is not tangent to all three sides of triangle ABC, then contract it again using a homothety centered at the vertex that is common to the two sides tangent to the circle.

This image circle is the incircle of triangle ABC. Since it was obtained using only *contractions*, it is smaller than the original circle C, unless C was already the incircle.

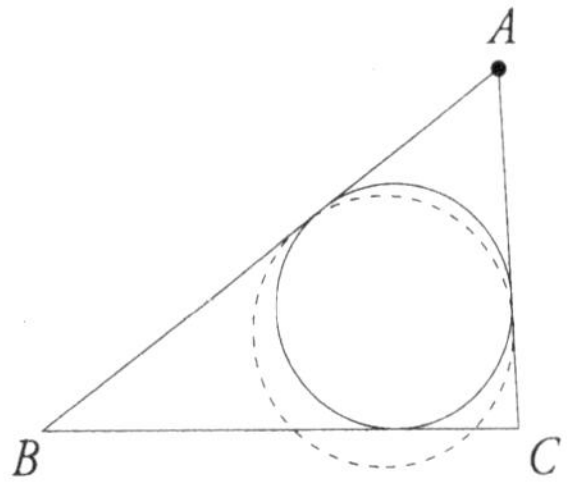

□

Example 11.4.2. *(Euler's Inequality)*
Prove that $R \geq 2r$, where R is the circumradius and r the inradius of a triangle. Equality holds if and only if the triangle is equilateral.

Solution. As in the figure below, let ABC be the triangle. Let D, E, and F be the midpoints of BC, CA, and AB, respectively. Let G be the centroid of $\triangle ABC$, and let $\mathcal{C}$ be the circumcircle.

Recalling that G is a trisection point of each median, we see that $\mathbf{H}_{G,-1/2}$ maps A, B, and C to D, E, and F, respectively. Hence, $\mathcal{C}'$, the image of $\mathcal{C}$ under $\mathbf{H}_{G,-1/2}$, is the circumcircle of triangle DEF, and the radius of $\mathcal{C}'$ is $R/2$.

Thus, $\mathcal{C}'$ is a circle that has a point in common with all three sides of ABC, and the smallest such circle is its incircle, so we can conclude that $R/2 \geq r$.

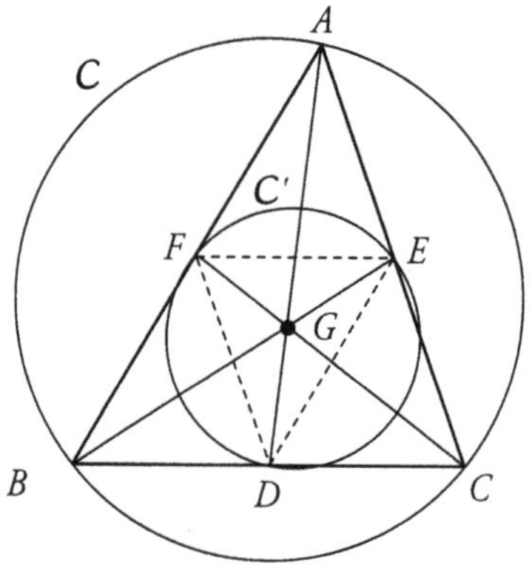

For equality to hold, $\mathcal{C}'$ must be the incircle of ABC, touching the sides at D, E, and F. Hence, the circumcenter coincides with the incenter of ABC, which must therefore be equilateral.

□

Example 11.4.3. *(The Euler Line)*
The centroid G, the circumcenter O, and the orthocenter H of a triangle are collinear. Moreover, G is between O and H and $\overline{GH} = 2\overline{GO}$.

Solution. This is also proved by applying the homothety $H_{G,-1/2}$. Given triangle ABC, let D, E, and F be the midpoints of BC, CA, and AB, respectively. As in the previous example, $\mathbf{H}_{G,-1/2}$ maps A, B, and C to D, E, and F, respectively.

The homothety maps the altitude AH of $\triangle ABC$ into a line DH' that is parallel to AH. In other words, the homothety maps the altitude from A into the right bisector of the opposite side. A similar thing happens to the other two altitudes, so the homothety maps the intersection of the altitudes to the intersection of the right bisectors. In other words, $\mathbf{H}_{G,-1/2}$ maps the orthocenter H of ABC into the circumcenter O of ABC.

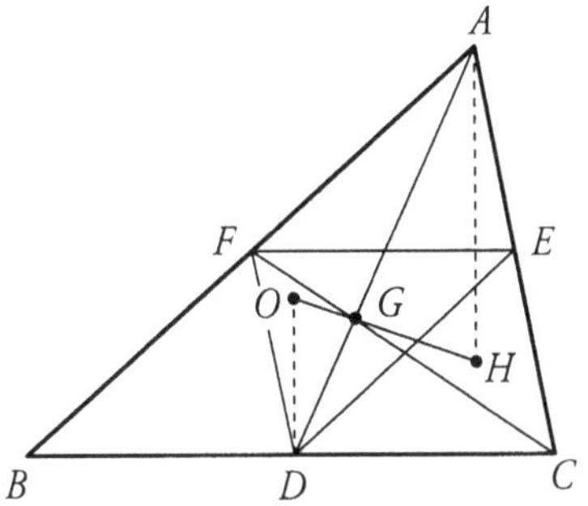

The definition of $\mathbf{H}_{G,-1/2}$ tells us that G, H, and O are collinear and also that

$$\overline{GH} = -2\overline{GO}.$$

□

Example 11.4.4. *$ABCD$ is a cyclic quadrilateral. Prove that the centroids of the triangles ABC, BCD, CDA, and DAB are cyclic, that is, that they lie on a circle.*

Solution. Let K and L be the midpoints of the diagonals AC and BD, and let M be the midpoint of KL. We will prove the result by showing that the quadrilateral formed by the centroids is the image $A'B'C'D'$ of $ABCD$ under $\mathbf{H}_{M,-1/3}$. Since $ABCD$ is a cyclic quadrilateral, it follows that the image is also a cyclic quadrilateral, since the circumcircle $\mathcal{C}$ of $ABCD$ will be mapped into the circumcircle of $A'B'C'D'$.

Let G be the centroid of triangle ABC. We will show that G is actually D', the image of D under $\mathbf{H}_{m,-1/3}$.

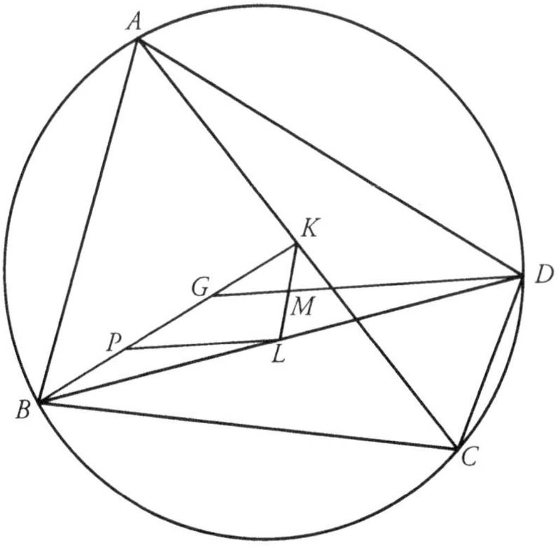

Let P be the midpoint of BG. Then P and G trisect BK. In triangle BGD, the points P and L are the midpoints of BG and BD, and the segments PL and GD are parallel, with $\overline{GD} = 2\overline{PL}$.

In triangle KPL, the points G and M are the midpoints of KP and KL, and the segments PL and GM are parallel, with $\overline{PL} = 2\overline{GM}$.

Since GD and GM are both parallel to PL, it follows that G, M, and D are collinear. Since $\overline{GD} = 2\overline{PL}$ and $\overline{PL} = 2\overline{GM}$, it follows that $\overline{GD} = 4\overline{GM}$ or, equivalently, that $\overline{MD} = -3\overline{MG}$.

This shows that $\mathbf{H}_{M,-1/3}(D) = G$, as claimed. Similarly, A', B', and C' are the centroids of BCD, CDA, and DAB, respectively. Hence, the four points A', B', C', and D' lie on the circle $\mathcal{C}'$, which is the image of $\mathcal{C}$ under $\mathbf{H}_{M,-1/3}$.

□

11.5 Dilatation

Any transformation of the plane that transforms a line segment into a parallel line segment is called a ***dilatation*** or a ***dilation***. The following theorem follows directly from the definition, and its proof is left to the reader.

Theorem 11.5.1. *The collection of all dilatations of the plane is a group.*

Examples of dilatations are translations, halfturns, homotheties, and the identity. It turns out that these are the only ones.

Theorem 11.5.2. *A dilatation that has two fixed points must be the identity.*

Proof. Suppose the dilatation T has A and B as fixed points. Let P be any point not on AB and let $T(P) = P'$.

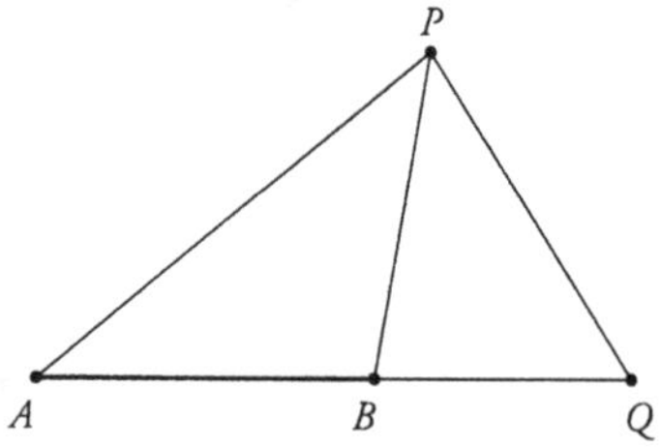

Then $P'A \parallel PA$ and $P'B \parallel PB$, so it follows that P' is on both PA and PB. Since PA and PB have only one point of intersection, it follows that $P' = P$.

Suppose now that Q is a point on AB other than A or B. Then Q is not on the line AP, and since T fixes A and P, the same proof shows that T maps Q onto itself.

This shows that T maps every point of the plane onto itself; that is, T is the identity.

□

Theorem 11.5.3. *A dilatation is completely determined by its action on any two given points.*

Proof. Suppose that the dilatations T and S map A and B to A' and B', respectively. Then T^{-1} maps A' to A and B' to B, and so

$$T^{-1}S(A) = A \qquad \text{and} \qquad T^{-1}S(B) = B.$$

By the previous theorem, $T^{-1}S = \mathbf{I}$, and multiplying by T shows that $S = T$.

□

Given parallel lines l and m, it is possible to compare directed segments on them so that if A and B are on l and C and D are on m, the meaning of an equation such as

$$\overline{AB} = -3\,\overline{CD}$$

is unambiguous (see the figure below).

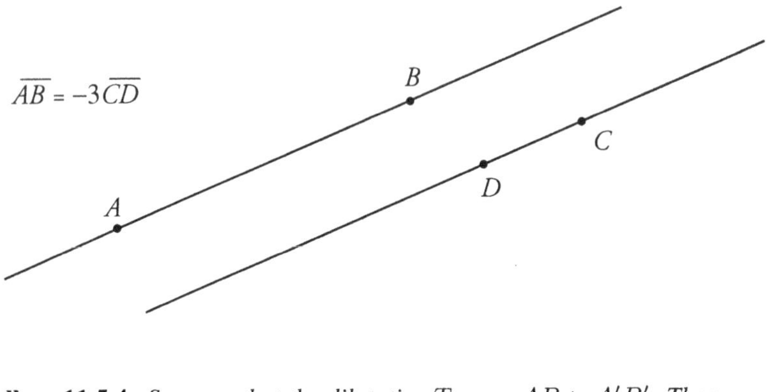

Corollary 11.5.4. *Suppose that the dilatation T maps AB to $A'B'$. Then:*

(1) $\overline{AB} = \overline{A'B'}$ *if and only if T is a translation (or the identity).*

(2) $\overline{AB} = -\,\overline{A'B'}$ *if and only if T is a halfturn.*

Theorem 11.5.5. *Any dilatation T that is not a translation or the identity is a homothety.*

Proof. Since T is not the identity, there is at least one point A that is not fixed. Let its image be A'. Now, let B be a point that is not on the line AA'. The segment $A'B'$ is parallel to AB, which is not parallel to AA', so B' cannot be on AA'.

The lines AA' and BB' cannot be parallel, for if they were, $AA'B'B$ would be a parallelogram; that is, $\overline{AB} = \overline{A'B'}$, which would contradict the fact that T is not a translation.

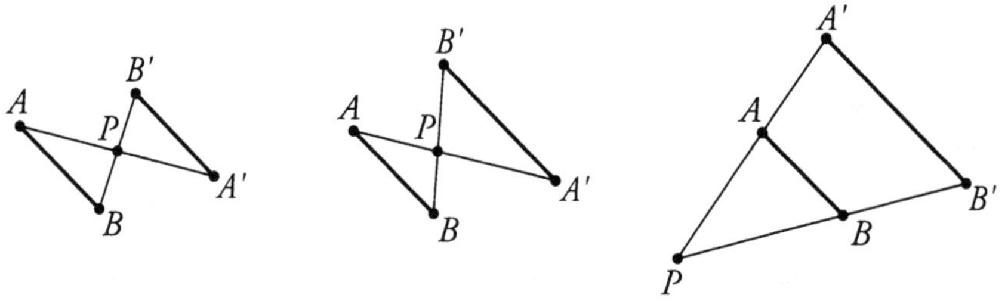

Thus, we may assume that the lines AA' and BB' intersect at a unique point P, as in one of the three situations depicted in the figure above. It follows that

$$\triangle PAB \sim \triangle PA'B',$$

and so

$$\frac{\overline{PA'}}{\overline{PB'}} = \frac{\overline{PA}}{\overline{PB}}.$$

Thus, the homothety $\mathbf{H}_{P,k}$, where $k = \overline{PA}/\overline{PB}$, maps A to A' and B to B', and therefore $T = \mathbf{H}_{P,k}$ by Theorem 11.5.3.

□

11.6 Problems

1. Find the two centers of homothety for the top and bottom of an isosceles trapezoid.

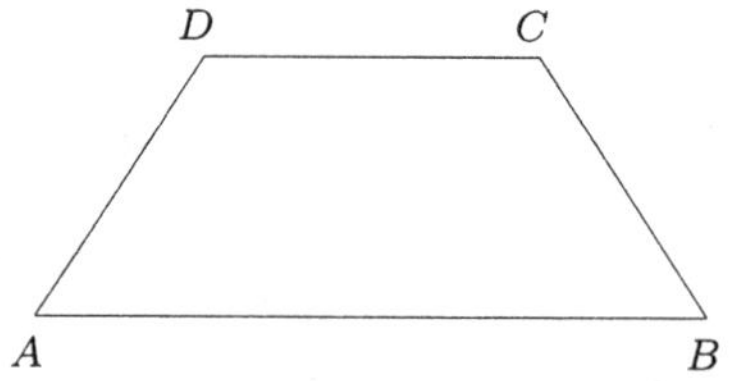

2. Show that two homotheties commute if and only if they have the same center or at least one of the ratios is $k = 1$.

3. Show that the product of two homotheties with the same center is a homothety and find its center and ratio.

4. Show that the product of two homotheties whose ratios are k and $1/k$ is a translation.

5. Show that if $\triangle ABC$ and $\triangle A'B'C'$ are similar, with AB parallel to $A'B'$, AC parallel to $A'C'$, and BC parallel to $B'C'$, then the lines joining corresponding vertices are concurrent, and there is a homothety $\mathbf{H}(O, k)$ such that $\triangle A'B'C'$ is the image of $\triangle ABC$ under $\mathbf{H}(O, k)$. Find the center O and the ratio k.

6. Show that a homothety preserves angles between lines.

7. Show that the inverse of the homothety $\mathbf{H}(O, k)$ is the homothety $\mathbf{H}(O, 1/k)$.

8. Show that the center of a homothety of ratio $k \neq 1$ is the only fixed point of the homothety and lines through the center are the only fixed lines.

9. Show that a product of three homotheties is a homothety or a translation.

10. Show that a similarity preserves angles.

11. $\overline{A_1B_1}$ and $\overline{A_2B_2}$ are two nonparallel segments such that $A_1B_1 = 2A_2B_2$, as in the figure below.

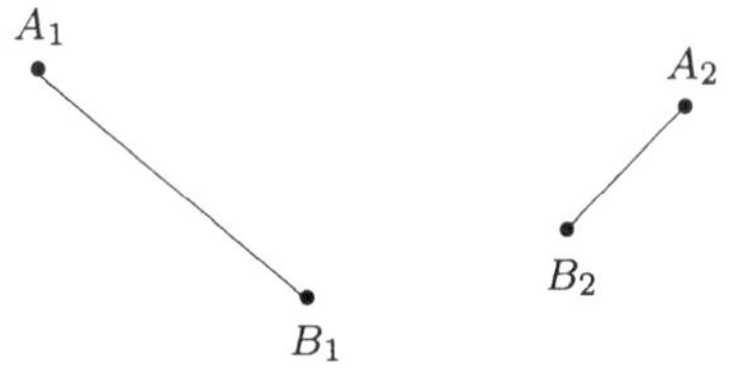

(a) Find a point O such that $\overline{A_2B_2}$ may be obtained from $\overline{A_1B_1}$ by means of a homothety centered at O with ratio $\frac{1}{2}$ followed by a rotation about O.

(b) Find a line ℓ and a point O on ℓ such that $\overline{A_2B_2}$ may be obtained from $\overline{A_1B_1}$ by a homothety centered at O with ratio $\frac{1}{2}$ followed by a reflection across ℓ.

12. Using homotheties, show that the figure formed by joining the midpoints of the sides of a square is a square having half the area of the original square.

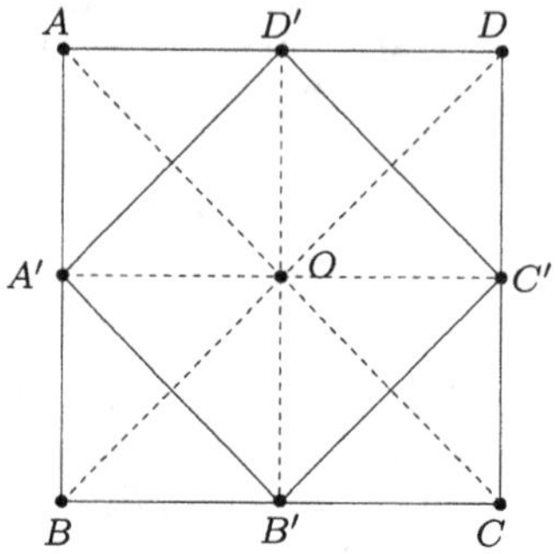

Hint: Show that the similarity composed of a 45° rotation and a homothety of ratio OA/OA', both with center O, maps the square $ABCD$ onto the quadrilateral $A'B'C'D'$.

13. Let P be a fixed point on a circle. Using homotheties, find the locus of midpoints of all chords PA.

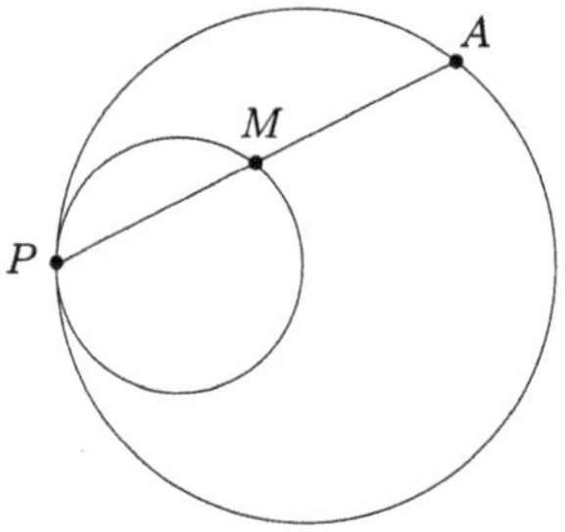

Hint: Consider the image of the given circle under the homothety $\mathbf{H}(P, 1/2)$.

14. If PT is a tangent and PAB a secant from an external point P to a circle $\mathcal{C}$, show that $\overline{PA}\,\overline{PB} = PT^2$.

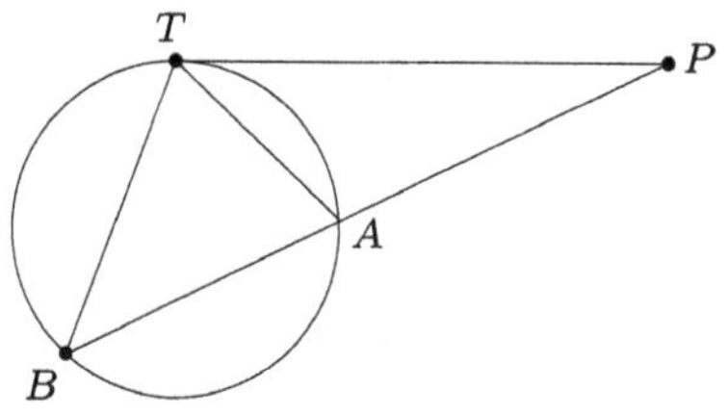

Hint: Reflect $\triangle PAT$ in the internal bisector of $\angle P$ and apply the homothety $H(P, PB/PT)$.

15. The point Q is a point inside $\angle AOB$, as in the figure.

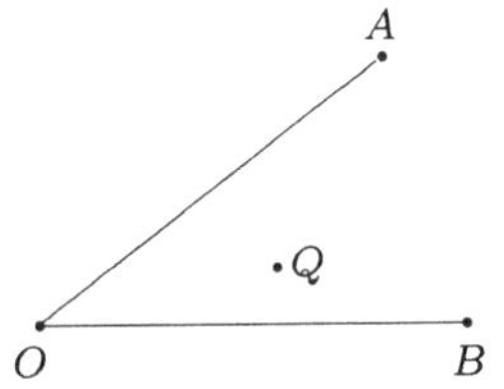

Construct a line through Q intersecting OA and OB at P and Q, respectively, such that $PQ = 2QR$.

16. Construct $\triangle ABC$ given its circumcenter O, its orthocenter H, and the vertex A, as in the figure.

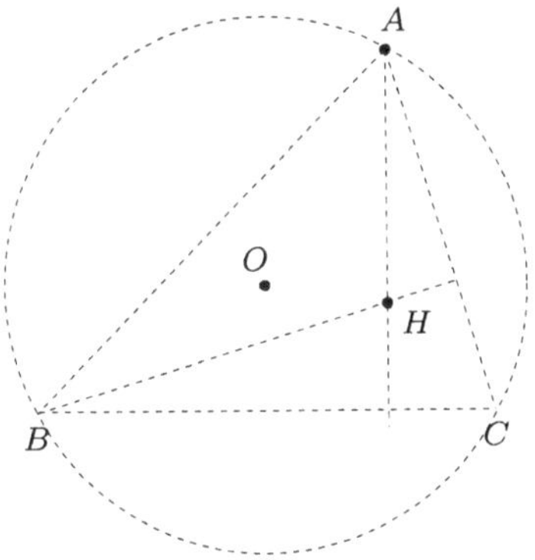

17. Construct a line intersecting three given concentric circles ω_1, ω_2, and ω_3 at A, B, and C, respectively, so that $AB = BC$.

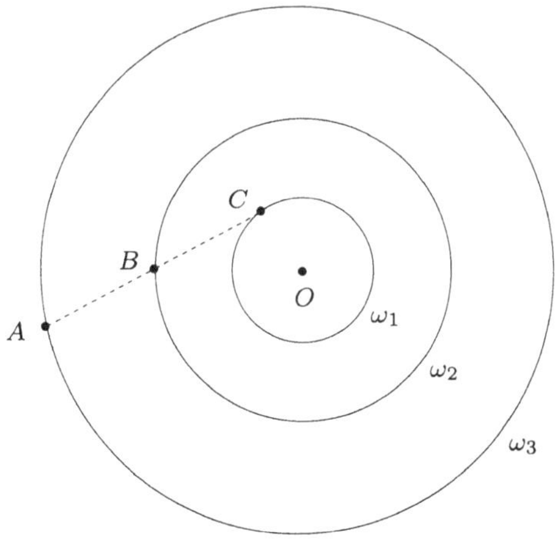

18. Let P be a point on side BC of equilateral triangle $\triangle ABC$ that is closer to C than to B. Construct a point Q on CA and a point R on AB so that $\angle RPQ = 90°$ and $PR = 2PQ$.

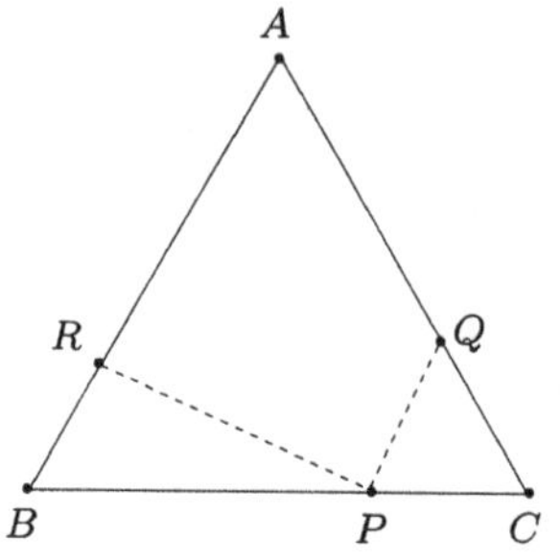

19. Two rods AD and BC are hinged at fixed points A and B on the ground. They are also connected to each other by means of a third rod CD, as in the figure, so that $ABCD$ is a parallelogram.

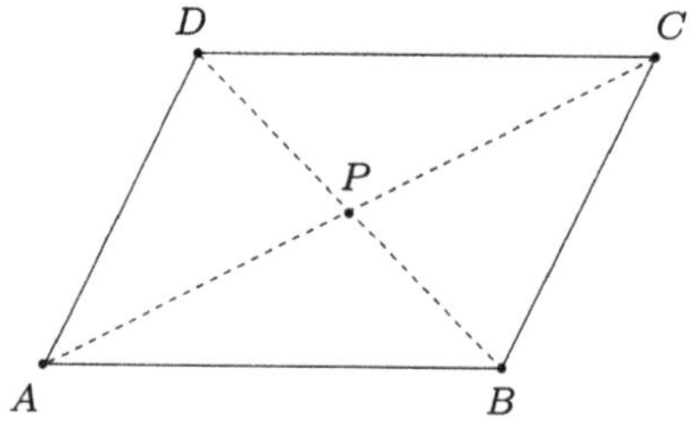

As the hinged rods move in a vertical plane, what is the locus of the point P of intersection of AC and BD?

20. The circles ω_1 and ω_2 intersect at M and N, and A_1 is a variable point on ω_1. A_2 is the point of intersection of the line $\overline{A_1M}$ with ω_2. B is the third vertex of an equilateral triangle A_1A_2B, with the vertices in counterclockwise order. Prove that the locus of B is a circle.

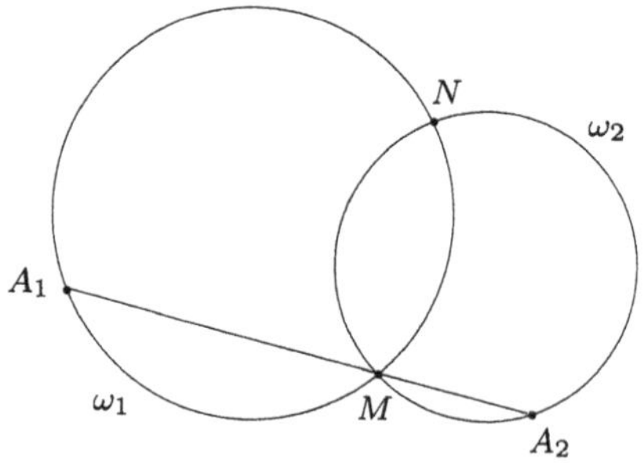

21. The circles ω_1 and ω_2 intersect at M and N, and A_1 is a variable point on ω_1. A_2 is the point of intersection of the line $\overline{A_1M}$ with ω_2, and L is the point of intersection of the tangent to ω_1 at A_1 and the tangent to ω_2 at A_2. Let O_1 and O_2 be the respective centers of ω_1 and ω_2. The line through O_1 parallel to LA_1 intersects the line through O_2 parallel to LA_2 at K, as in the figure below.

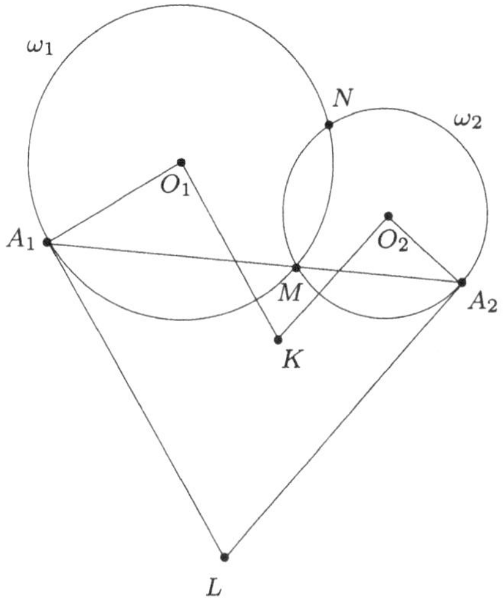

Prove the following:

(a) $\triangle A_1NA_2$ and $\triangle O_1NO_2$ are similar.

(b) A_1LA_2N is a cyclic quadrilateral.

(c) O_1KO_2N is a cyclic quadrilateral.

(d) K, L, and N are collinear.

22. Two circles ω_1 and ω_2 are tangent to each other at the point T. A line through T intersects ω_1 at A_1 and ω_2 at A_2. Prove that the tangent to ω_1 at A_1 is parallel to the tangent to ω_2 at A_2.

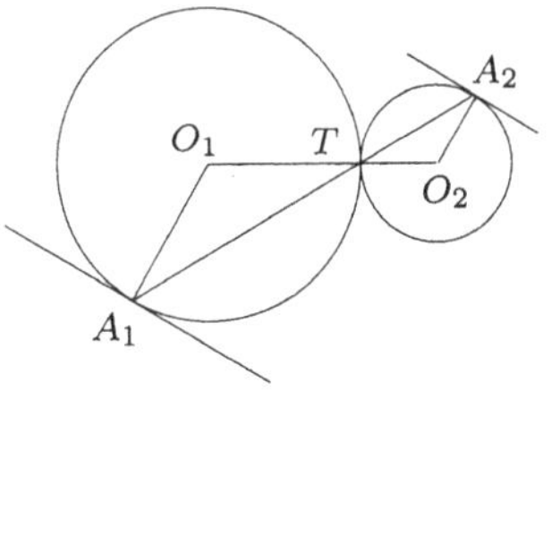

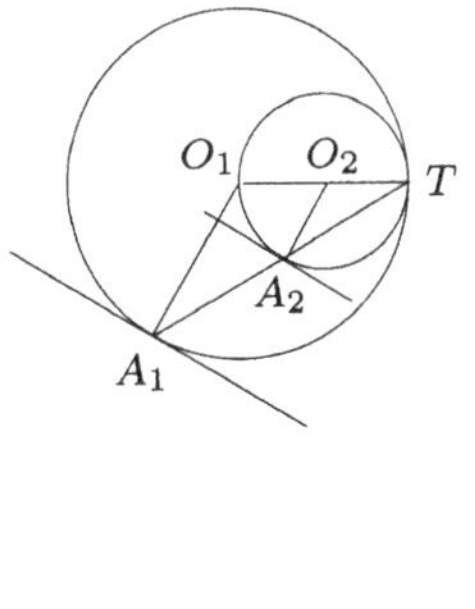

23. The incircle of triangle ABC touches BC at D. The excircle of triangle ABC opposite A touches BC at K. The line AK intersects the incircle at two points, and we let H be the one closer to A. Prove that DH is perpendicular to BC.

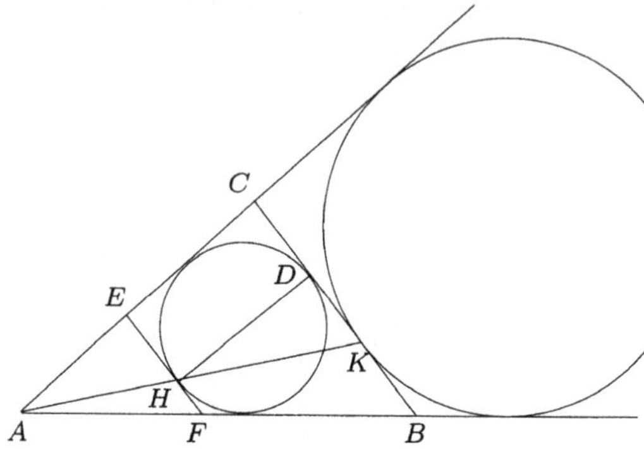

24. $ABCD$ is a quadrilateral with AB parallel to DC. The extensions of DA and CB intersect at P, and the diagonals AC and BD intersect at Q. Prove that PQ passes through the midpoints of AB and CD.

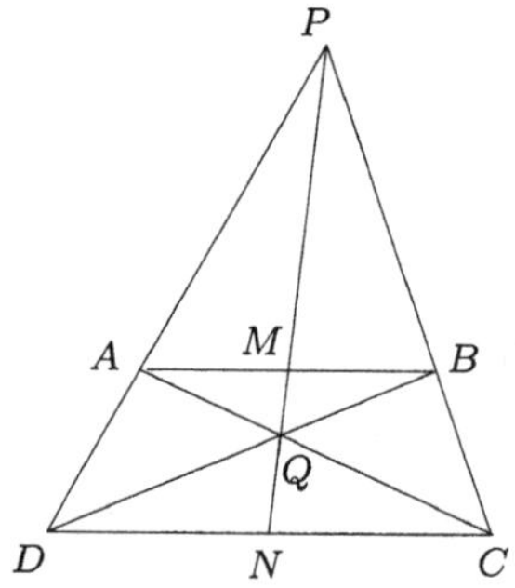

25. D, E, and F are the respective midpoints of the sides BC, CA, and AB of triangle ABC. P is a point inside ABC. K, L, and M are points such that D, E, and F are also the respective midpoints of PK, PL, and PM. Prove that AK, BL, and CM bisect one another at a common point.

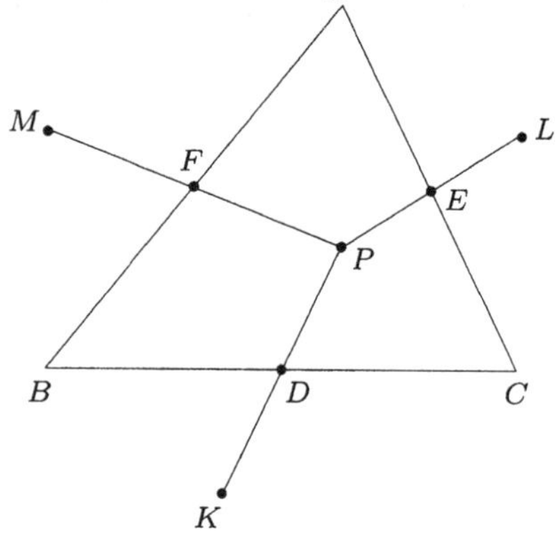

CHAPTER 12

TESSELLATIONS

12.1 Tilings

A ***tiling*** or ***tessellation*** of the plane is a division of the plane into regions

$$T_1,\ T_2,\ \dots\ ,$$

called ***tiles***, in such a manner that:

1. No region contains an interior point of another region.

2. Every point in the plane belongs to one of the regions.

In other words, the plane is completely covered by nonoverlapping tiles.

Classical Geometry: Euclidean, Transformational, Inversive and Projective, First Edition. By
I. E. Leonard, J. E. Lewis, A. C. F. Liu, G. W. Tokarsky
Copyright © 2014 John Wiley & Sons, Inc.

There is no requirement that the tiles be related in any way, but our interest is primarily in tilings where there are only a finite number of differently shaped tiles. A tessellation is of ***order-k***, or ***k-hedral***, if there is a finite set of k incongruent tiles S such that:

1. Every tile in the tessellation is congruent to some member of S.
2. Every member of S occurs at least once in the tessellation.

The members of S are called ***prototiles***, and we say that S ***tiles the plane***. The tiling in the figure below has a set of three prototiles, so it is an order-3 tiling.

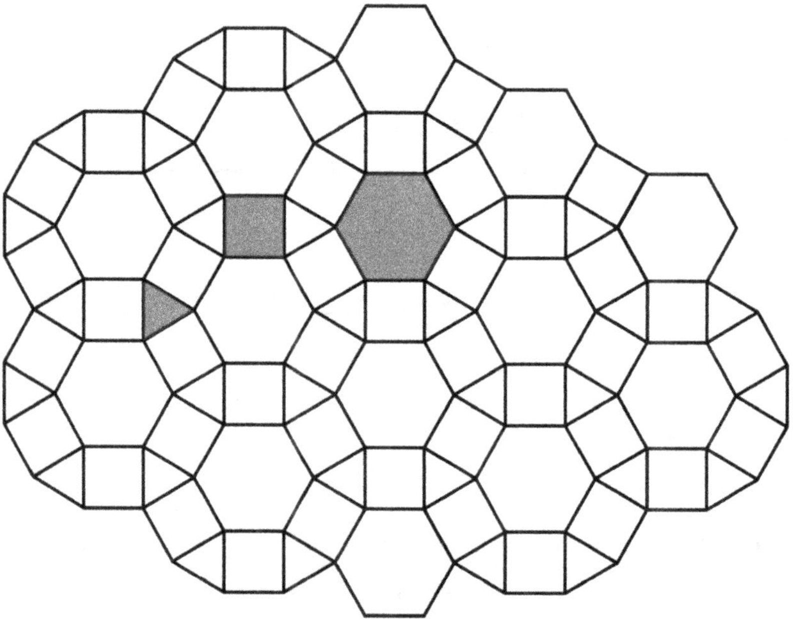

12.2 Monohedral Tilings

The most natural question is "Which polygons are monohedral prototiles?" The words ***monohedral***, ***dihedral***, and ***trihedral*** are commonly used as synonyms for ***1-hedral***, ***2-hedral***, and ***3-hedral***, respectively.

It is not difficult to see that every parallelogram tiles the plane, as in (a) below.

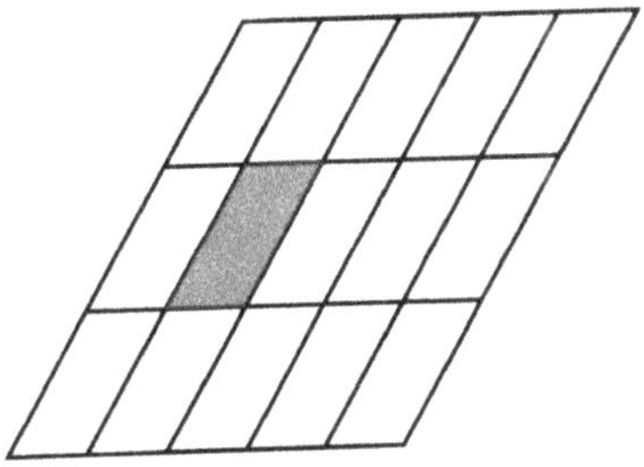

(a)

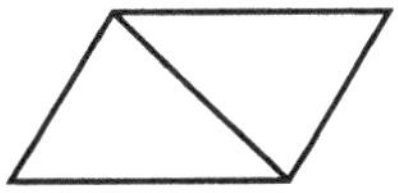

(b)

As a consequence, every triangle tiles the plane because two copies of the triangle can be combined to form a parallelogram, as in (b) above.

Every quadrilateral tiles the plane. The tiling can be obtained by successively rotating the quadrilateral around the midpoints of the sides, as in the figure below.

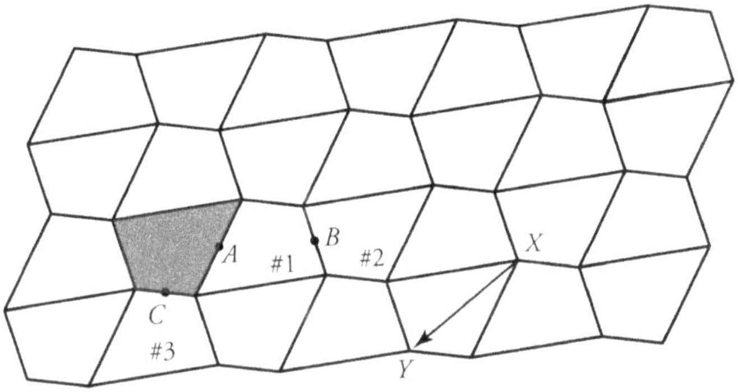

Tile #1 is obtained by applying $\mathbf{H}_A$ to the shaded tile, and tile #2 is obtained by applying $\mathbf{H}_B$ to tile #1. Continuing in this way, we can tile the entire horizontal strip containing tiles #1 and #2. The strip below this one can be obtained by similar rotations—for example, tile #3 may be obtained by applying $\mathbf{H}_C$ to the shaded tile. Alternatively, we can apply $\mathbf{T}_{XY}$ to the entire strip containing #1 and #2.

The regular pentagon will not tile the plane. One way to see this is to examine what happens when you try to tile around a vertex.

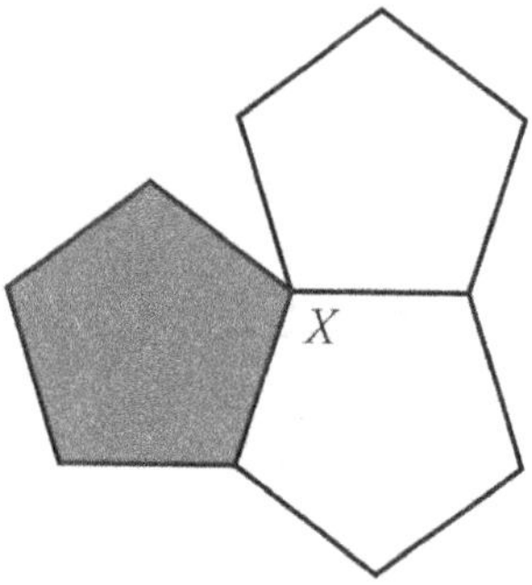

(a)

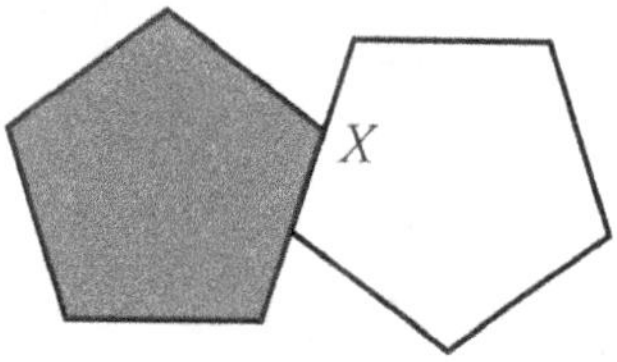

(b)

The figures above show what happens when you try to tile around vertex X of the shaded pentagon. You can either try a ***vertex-to-vertex tiling***, as in (a), or you can try an ***edge-to-vertex tiling***, as in (b). In (a) there is an angular gap of $36°$, and in (b) there is a $72°$ gap, neither of which can be covered without overlapping tiles.

Some hexagons tile the plane, but not all do. Everyone has seen the tiling in (a) below.

(a)

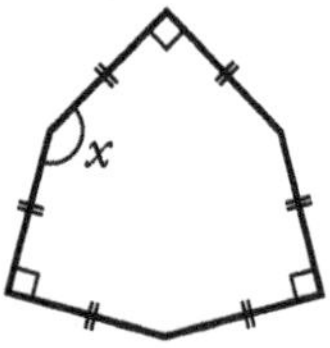

(b)

The hexagon in (b) above will not tile the plane, and, as before, this can be verified by trying to tile around the vertex of $x°$.

Although the regular pentagon does not tile the plane, there are some pentagons that do, and the figure below shows two of them.

The tiling in (a) is obtained by splitting the regular hexagonal tiling.

The tiling in (b), which also uses hexagons, is the beautiful "Cairo" tiling.

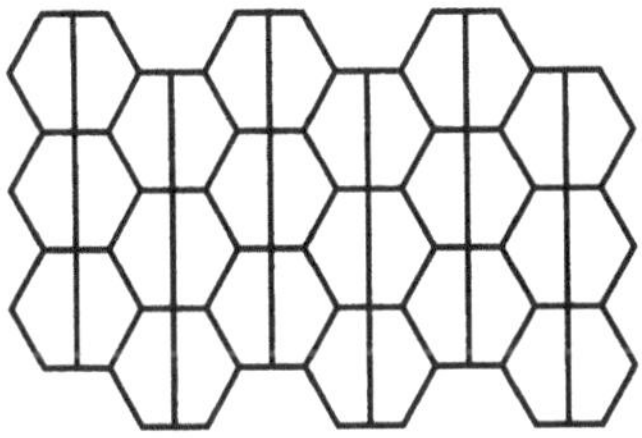

(a)

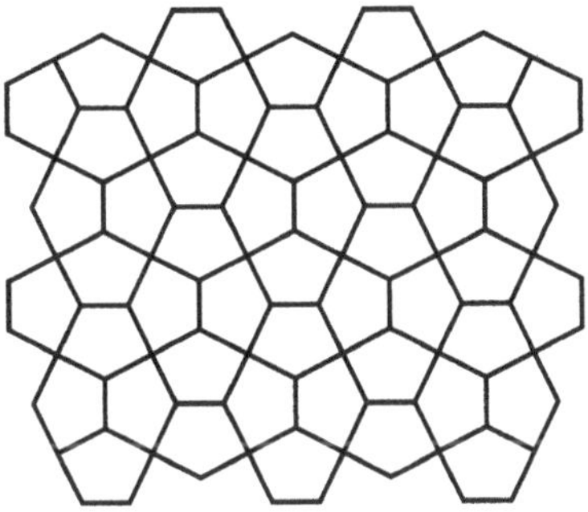

(b)

Convex Hexagons that Tile

A polygon is ***convex*** if all of its diagonals are interior to the polygon.

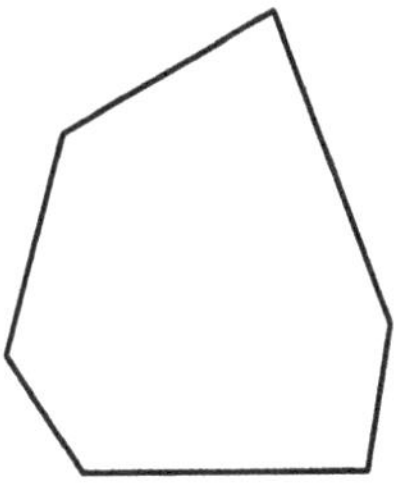

convex

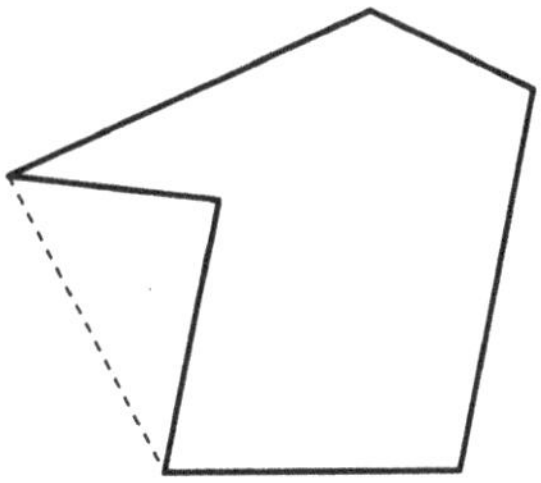

nonconvex

As shown earlier, some convex pentagons and some convex hexagons tile the plane. The situation for convex hexagons is completely understood—a convex polygon $ABCDEF$ will tile the plane if and only if it satisfies one of the following three criteria:

1. $\angle A + \angle B + \angle C = 360°$ and $AF = CD$.

2. $\angle A + \angle B + \angle D = 360°$, $AF = CD$, and $BC = DE$.

3. $\angle A = \angle C = \angle E = 120°$, $AF = AB$, $BC = CD$, and $DE = EF$.

The figures below give an example of each type.

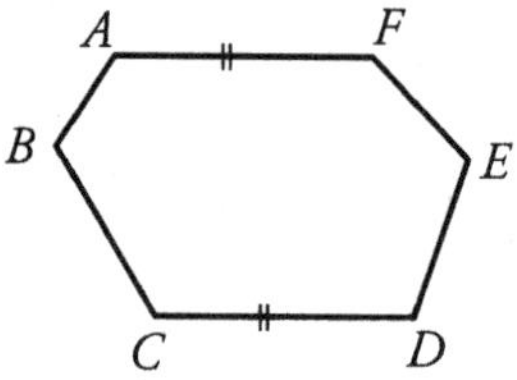

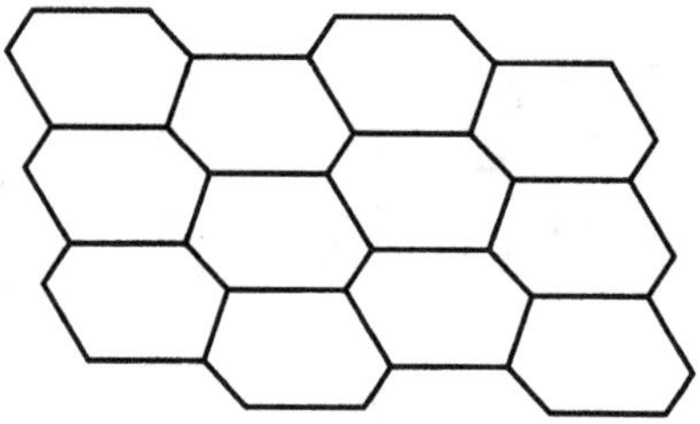

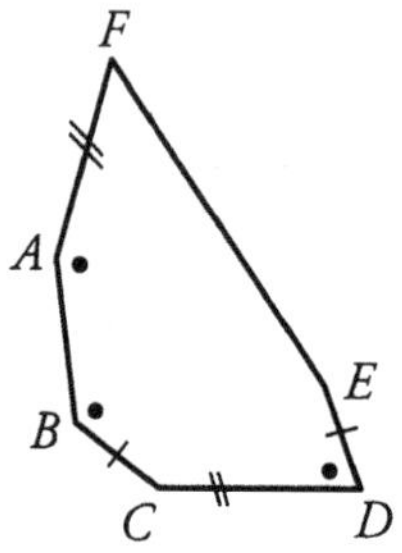

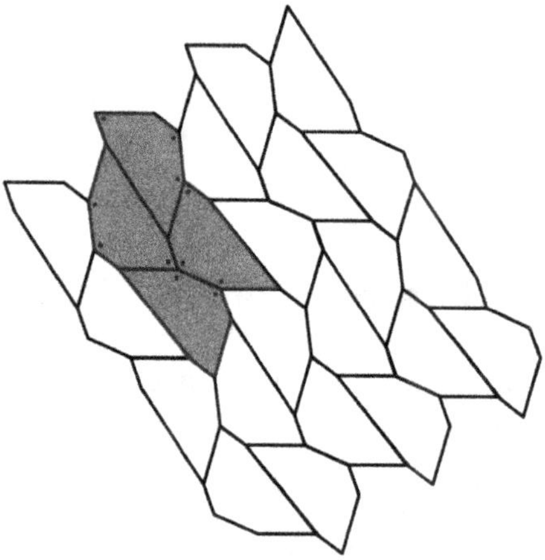

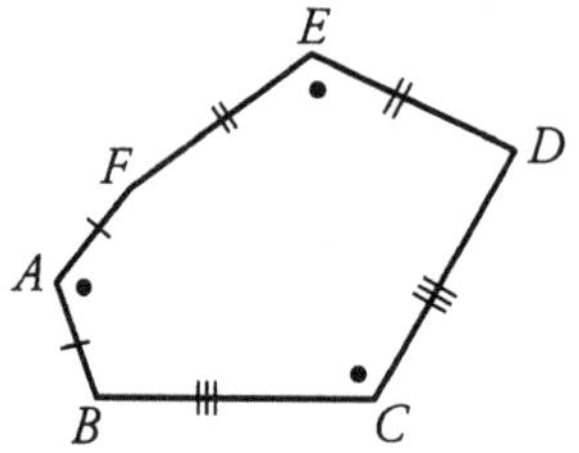

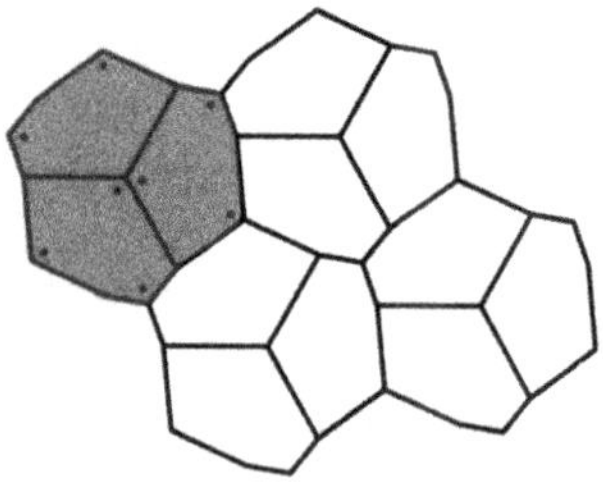

However, it is still not known which convex pentagons tile the plane. At the time of writing this text, we know that 15 different types of convex pentagons tile the plane. This problem has a fascinating history, and you can find much more information about the state of affairs of this problem on the web or in the following articles:

"Tiling with Convex Polygons," Chapter 13 of *Time Travels and Other Mathematical Bewilderments* (Martin Gardner; W. H. Freeman and Company, New York, 1988).

"In Praise of Amateurs," Doris Schattschneider, in *The Mathematical Gardner* (David A. Klarner, Editor; Prindle, Weber, and Schmidt, Boston, 1981).

12.3 Tiling with Regular Polygons

In this section, we will investigate tessellations or tilings with regular polygons whose sides are of unit length and such that two neighboring tiles share a complete edge. Such tessellations are said to be ***edge-to-edge tilings***. The diagram below shows three familiar examples of monohedral tessellations of this type.

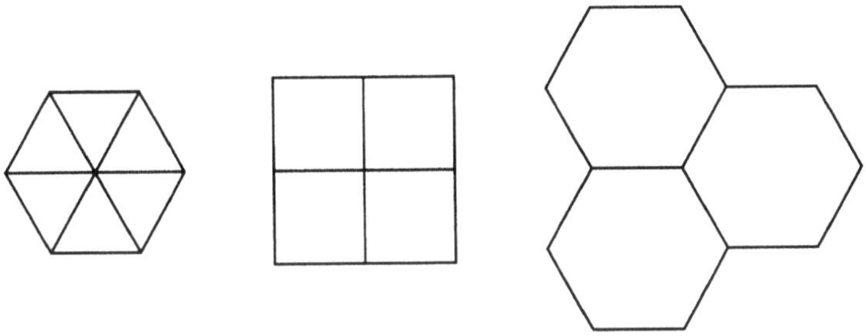

Recall that in order to form a tessellation, the tiles must be able to completely surround a vertex without overlapping.

The sequence of regular polygonal tiles that surround a point generates a ***vertex sequence of the point*** which gives the number of sides of each tile.

For example, in the first tessellation above, the vertex sequence of each vertex is (3,3,3,3,3,3). In the second and third tessellations, the vertex sequences are (4,4,4,4) and (6,6,6), respectively.

We will show that there are other combinations of regular polygonal tiles that can surround a point. In order to do this, we need to find a set of angles whose total measure is 360° and such that the measure of each angle is the measure of a vertex angle of an n-sided regular polygon, that is,

$$\frac{n-2}{n}\cdot 180^{\circ},$$

where n is an integer greater than or equal to 3.

Combinations of Three Polygons

We first consider combinations with three polygons. Let the vertex sequence be (a, b, c) with $a \leq b \leq c$. Then

$$\frac{a-2}{a}\cdot 180^{\circ} + \frac{b-2}{b}\cdot 180^{\circ} + \frac{c-2}{c}\cdot 180^{\circ} = 360^{\circ},$$

which simplifies to

$$\frac{1}{a} + \frac{1}{b} + \frac{1}{c} = \frac{1}{2}.$$

Case (i). $a = 3$.

In this case, the equation above becomes

$$\frac{1}{b} + \frac{1}{c} = \frac{1}{6}.$$

Solving for c in terms of b, we have

$$\frac{1}{c} = \frac{b-6}{6b}$$

so that

$$c = \frac{6b}{b-6},$$

and this implies that $b > 6$.

For $b = 7$, we have $c = 42$; for $b = 8$, we have $c = 24$; for $b = 9$, $c = 18$; for $b = 10$, $c = 15$; for $b = 11$, c is not an integer; and for $b = 12$, $c = 12$.

Since $b \leq c$, the only solutions in this case are $(3, 7, 42)$, $(3, 8, 24)$, $(3, 9, 18)$, $(3, 10, 15)$, and $(3, 12, 12)$.

Note. Vertex sequences that are not monohedral can have equivalent forms. For instance, the vertex sequence (3,7,42) has five other equivalent forms, namely,

$$(3, 42, 7), \quad (7, 3, 42), \quad (7, 42, 3), \quad (42, 3, 7), \quad \text{and} \quad (42, 7, 3).$$

They are listed in lexicographical order, that is, a vertex sequence with a smaller first term comes before one with a larger first term. Among those with equal first terms, a vertex sequence with a smaller second term comes before one with a larger second term, and so on.

Case (ii). $a = 4$.

In this case, the equation

$$\frac{1}{a} + \frac{1}{b} + \frac{1}{c} = \frac{1}{2}$$

becomes

$$\frac{1}{b} + \frac{1}{c} = \frac{1}{4}.$$

Solving for c in terms of b, we have

$$\frac{1}{c} = \frac{b-4}{4b}$$

so that

$$c = \frac{4b}{b-4},$$

and this implies that $b > 4$.

For $b = 5$, $c = 20$; for $b = 6$, $c = 12$; for $b = 7$, c is not an integer; and for $b = 8$, $c = 8$.

Since $b \leq c$, the only solutions in this case are

$$(4, 5, 20), \quad (4, 6, 12), \quad \text{and} \quad (4, 8, 8).$$

Case (iii). $a = 5$.

In this case, the equation

$$\frac{1}{a} + \frac{1}{b} + \frac{1}{c} = \frac{1}{2}$$

becomes

$$\frac{1}{b} + \frac{1}{c} = \frac{3}{10}.$$

Solving for c in terms of b, we have

$$\frac{1}{c} = \frac{3b-10}{10b}$$

so that

$$c = \frac{10b}{3b - 10},$$

and we have $b \geq a = 5$.

For $b = 5$, $c = 10$; for $b = 6$, 7, c is not an integer; and for $b = 8$, we have $b > c$. Thus, the only solution in this case is $(5, 5, 10)$.

Case (iv). $a = 6$.

In this case, if $c \geq 7$, then

$$\frac{1}{2} = \frac{1}{a} + \frac{1}{b} + \frac{1}{c} \leq \frac{2}{6} + \frac{1}{7} = \frac{20}{42},$$

which is a contradiction. Hence,

$$a = b = c = 6,$$

and the only solution in this case is (6,6,6).

Case (v). $a \geq 7$.

In this case,

$$\frac{1}{2} = \frac{1}{a} + \frac{1}{b} + \frac{1}{c} \leq \frac{3}{7},$$

which is a contradiction. Hence, there are no further solutions.

Note. In the case (3,7,42), neither the regular 7-gon nor the regular 42-gon has angles with integral measures. Thus, this combination is not easy to discover just by inspection.

Combinations of Four Polygons

Next we consider combinations with four polygons. Let the vertex sequence be a permutation of (a, b, c, d) with $a \leq b \leq c \leq d$. As before, we have

$$\frac{1}{a} + \frac{1}{b} + \frac{1}{c} + \frac{1}{d} = 1.$$

Case (i). $a = 3$.

In this case, the equation above becomes

$$\frac{1}{b} + \frac{1}{c} + \frac{1}{d} = \frac{2}{3}.$$

If $b = 3$, we have

$$\frac{1}{c} + \frac{1}{d} = \frac{1}{3},$$

and solving for d in terms of c, we have

$$\frac{1}{d} = \frac{c-3}{3c}$$

so that

$$d = \frac{3c}{c-3},$$

and therefore $c > 3$.

For $c = 4$, $d = 12$; for $c = 5$, d is not an integer; and for $c = 6$, $d = 6$.

Since $c \leq d$, the only solutions for $b = 3$ are (3,3,4,12) and (3,3,6,6).

If $b = 4$, we have

$$\frac{1}{c} + \frac{1}{d} = \frac{5}{12},$$

and solving for d in terms of c, we have

$$\frac{1}{d} = \frac{5c-12}{12c}$$

so that

$$d = \frac{12c}{5c-12},$$

and therefore $c \geq b = 4$.

For $c = 4$, $d = 6$; for $c = 5$, we already have $c > d$. Thus, the only solution when $b = 4$ is $(3, 4, 4, 6)$.

For $b \geq 5$,

$$\frac{2}{3} = \frac{1}{b} + \frac{1}{c} + \frac{1}{d} \leq \frac{3}{5},$$

which is a contradiction. Hence, there are no further solutions in this case.

Case (ii). $a = 4$.

In this case, for $d \geq 5$,

$$1 = \frac{1}{a} + \frac{1}{b} + \frac{1}{c} + \frac{1}{d} \leq \frac{3}{4} + \frac{1}{5} = \frac{19}{20},$$

which is a contradiction. Hence $a = b = c = d = 4$, and the only solution in this case is $(4, 4, 4, 4)$.

Case (iii). $a \geq 5$.

In this case,

$$1 = \frac{1}{a} + \frac{1}{b} + \frac{1}{c} + \frac{1}{d} \leq \frac{4}{5},$$

which is a contradiction. Hence there are no further solutions.

Combinations of Five Polygons

Now we consider combinations with five polygons. Let the vertex sequence be a permutation of (a, b, c, d, e) where $a \leq b \leq c \leq d \leq e$. As before, we have

$$\frac{1}{a} + \frac{1}{b} + \frac{1}{c} + \frac{1}{d} + \frac{1}{e} = \frac{3}{2}.$$

For $c \geq 4$,

$$\frac{3}{2} = \frac{1}{a} + \frac{1}{b} + \frac{1}{c} + \frac{1}{d} + \frac{1}{e} \leq \frac{2}{3} + \frac{3}{4} = \frac{17}{12},$$

which is a contradiction.

Hence, $a = b = c = 3$, so that

$$\frac{1}{d} + \frac{1}{e} = \frac{1}{2}.$$

As before, we have

$$(d, e) = (3, 6) \qquad \text{or} \qquad (4, 4),$$

so that the only solutions in this case are $(3, 3, 3, 3, 6)$ and $(3, 3, 3, 4, 4)$.

Combinations of Six Polygons

Finally, we consider combinations with six polygons. Let the vertex sequence be a permutation of (a, b, c, d, e, f) where $a \leq b \leq c \leq d \leq e \leq f$. As before, we have

$$\frac{1}{a} + \frac{1}{b} + \frac{1}{c} + \frac{1}{d} + \frac{1}{e} + \frac{1}{f} = 2.$$

For $f \geq 4$,

$$2 = \frac{1}{a} + \frac{1}{b} + \frac{1}{c} + \frac{1}{d} + \frac{1}{e} + \frac{1}{f} = \frac{5}{3} + \frac{1}{4} = \frac{23}{12},$$

which is a contradiction. Hence, $a = b = c = d = e = f = 3$, and the only solution in this case is (3,3,3,3,3,3).

Note. We cannot surround a point with seven or more regular polygons since the smallest of the angles at this point is at least $60°$, and the sum of these angles will exceed $6 \times 60° = 360°$.

Several of our solutions give rise to more than one vertex sequence. The combination (3,3,4,12) may be permuted as (3,4,3,12). The combination (3,3,6,6) may be permuted as (3,6,3,6). The combination (3,4,4,6) may be permuted as (3,4,6,4). Finally, the combination (3,3,3,4,4) may be permuted as (3,3,4,3,4). There are left-handed and right-handed versions of (3,3,3,3,6), but they are not considered to be different. This brings the total number of possible vertex sequences to 21.

12.4 Platonic and Archimedean Tilings

Now we consider tilings that are named after the Greek philosophers Plato and Archimedes.

For each vertex sequence, we wish to know if we can tile the entire plane with regular polygons such that every vertex has this vertex sequence. In this case, such a tessellation is said to be ***semiregular***.

Moreover, if all the terms in the vertex sequences are identical, the tessellation is said to be ***regular***. The regular tessellations of the plane are called ***Platonic*** tilings.

Semiregular tessellations that are not regular are called ***Archimedean*** tilings.

We divide the 21 possible vertex sequences into three groups.

• Group I.

$$(3,7,42),\quad (3,8,24),\quad (3,9,18),\quad (3,10,15),\quad (4,5,20),\quad (5,5,10).$$

These vertex sequences are all of the form (a,b,c) where a is odd and $b \neq c$ if we read (5,4,20) for (4,5,20).

If there is a tessellation in which every vertex has this vertex sequence, we must be able to surround the a-sided polygon. Thus, its neighbours must be the b-sided polygon and the c-sided polygon, alternately. However, this is impossible since a is odd. The figure below illustrates the case (5,5,10). Moreover, no combination of regular polygons can fill the void.

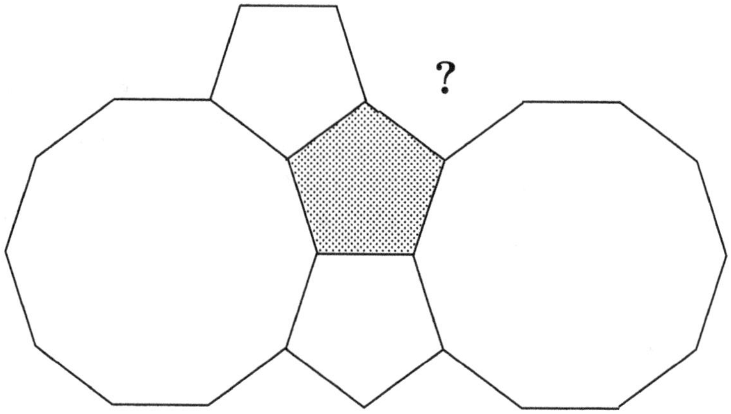

• Group II.

$$(3,3,4,12),\quad (3,3,6,6),\quad (3,4,3,12),\quad (3,4,4,6).$$

Each of these vertex sequences contains a triangular tile. We can surround two of its vertices properly, but the third vertex, indicated with a black dot in the diagram on the following page, requires a different vertex sequence.

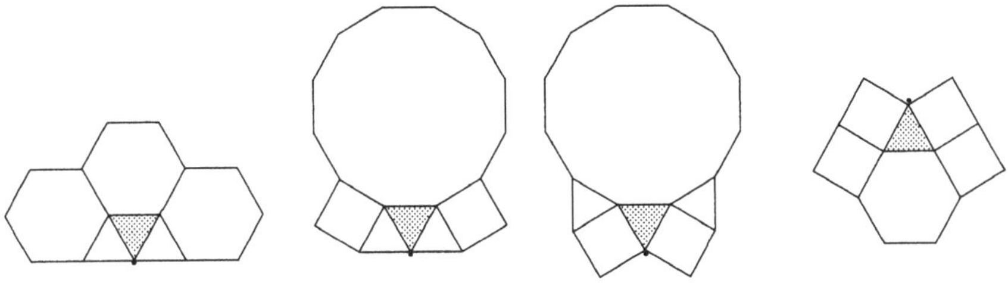

- Group III.

$$(3,12,12),\quad (4,6,12),\quad (4,8,8),\quad (6,6,6),$$
$$(3,4,6,4),\quad (3,6,3,6),\quad (4,4,4,4),$$
$$(3,3,3,3,6),\quad (3,3,3,4,4),\quad (3,3,4,3,4),$$
$$(3,3,3,3,3,3).$$

Here we have no local problem. It turns out that each of these vertex sequences leads to a semiregular tessellation. There are exactly three Platonic tessellations, namely,

$$(3,3,3,3,3,3),\quad (4,4,4,4),\quad \text{and}\quad (6,6,6),$$

which we saw at the beginning of Section 12.3.

The other eight are the only Archimedean tilings, and they will be discussed a later. However, we still have to prove that we have no global problem with any of these 11 vertex sequences. We shall use a direct approach and construct each of the 11 tessellations.

(a) The (3,3,3,3,3,3) and the (3,6,3,6) tessellations may be constructed with three infinite families of evenly spaced parallel lines forming 60° angles across the families, as shown in the diagram below.

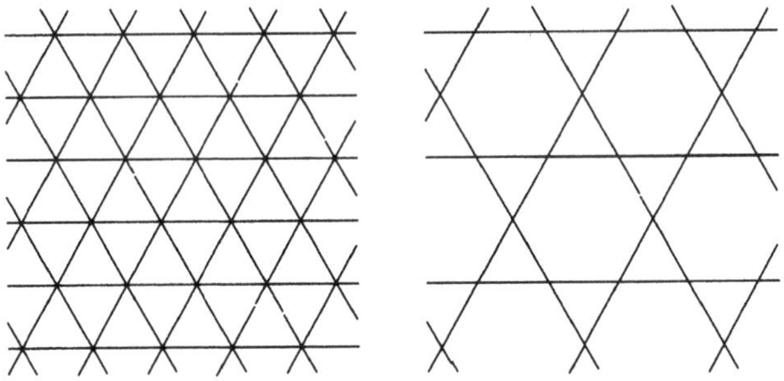

(b) The (4,4,4,4) tessellation may be constructed with two infinite familes of evenly spaced parallel lines forming 90° angles across families, as shown in the figure below on the left. This tessellation and the (3,3,3,3,3,3) and (3,6,3,6) tessellations are the three *basic* tessellations.

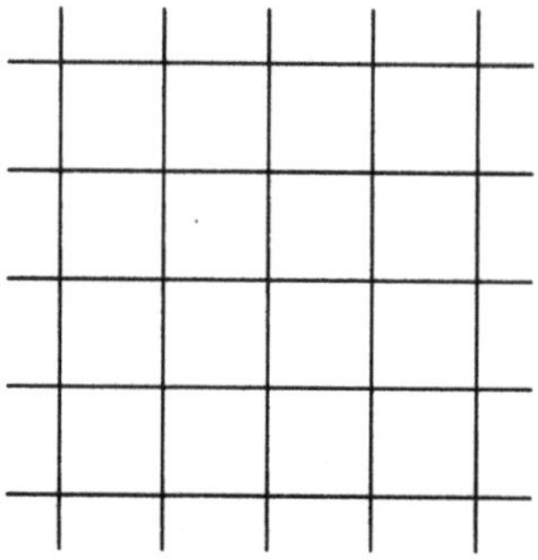
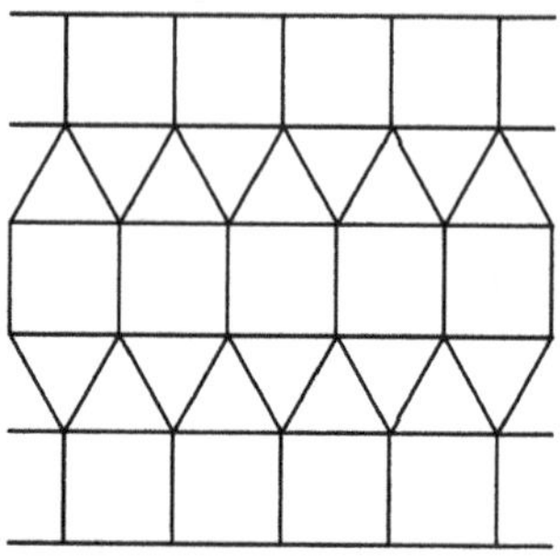

(c) The (3,3,3,4,4) tessellation is obtained by taking alternate strips from the basic (3,3,3,3,3,3) and (4,4,4,4,4) tessellations, as in the figure above on the right.

The remaining tessellations are obtained from others by the *cut-and-merge* method.

(d) From the basic (3,3,3,3,3,3) tessellation, we can merge a set of six equilateral triangles into a regular hexagon, as in the figure below on the left.

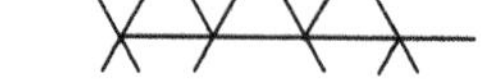

(e) The (6,6,6) and the (3,3,3,3,6) tessellations may be obtained from the basic (3,3,3,3,3,3) tessellations by merging various sets of six equilateral triangles, as shown in the figure above in the middle and on the right. No cutting is required in either case.

(f) The (3,4,6,4) tessellation can also be obtained from the basic (3,3,3,3,3,3) tessellation. As shown in the figure below on the left, we cut each triangular tile into seven pieces consisting of an equilateral triangle, three congruent half-squares, and three congruent kites with angles $120°$, $90°$, $60°$, and $90°$.

Let the edge length of the triangular tile be 1 and the length of the side of the equilateral triangle be x. Then the short sides of the kite have length $x/2$ and the long sides $\sqrt{3}x/2$. Since

$$\frac{\sqrt{3}x}{2} + x + \frac{\sqrt{3}x}{2} = 1,$$

we have

$$x = \frac{\sqrt{3}-1}{2} \approx 0.366.$$

When we merge the kites and half-squares across six triangular tiles, we obtain the $(3, 4, 6, 4)$ tessellations, as shown in the figure below on the right.

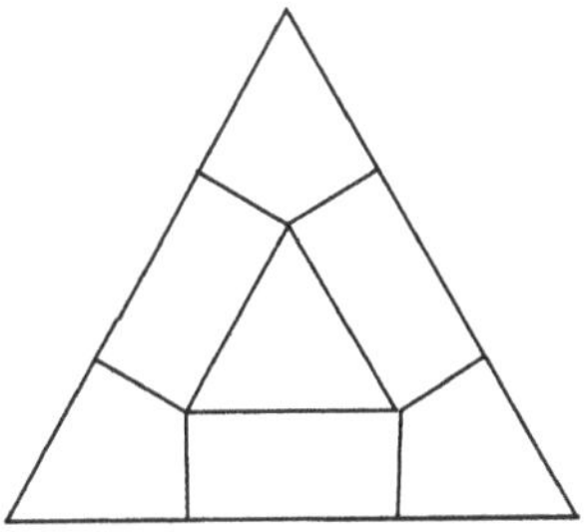

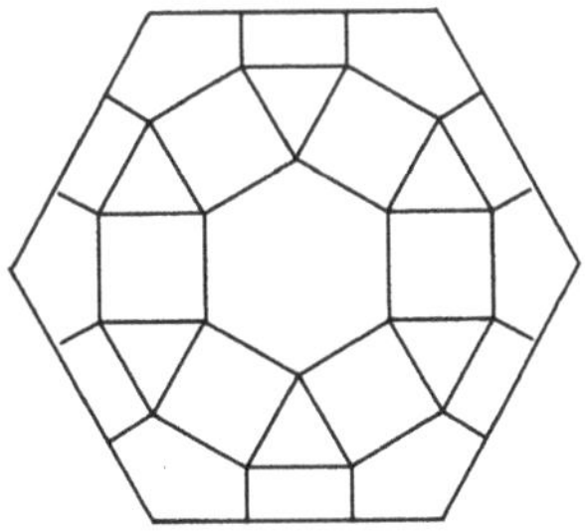

(g) The (4,6,12) tessellation can now be obtained from the (3,4,6,4) tessellation without cutting. Each dodecagon in the new tessellation is obtained by merging one regular hexagon, six squares and six equilateral triangles in the old tessellation, as shown in the figure on the following page.

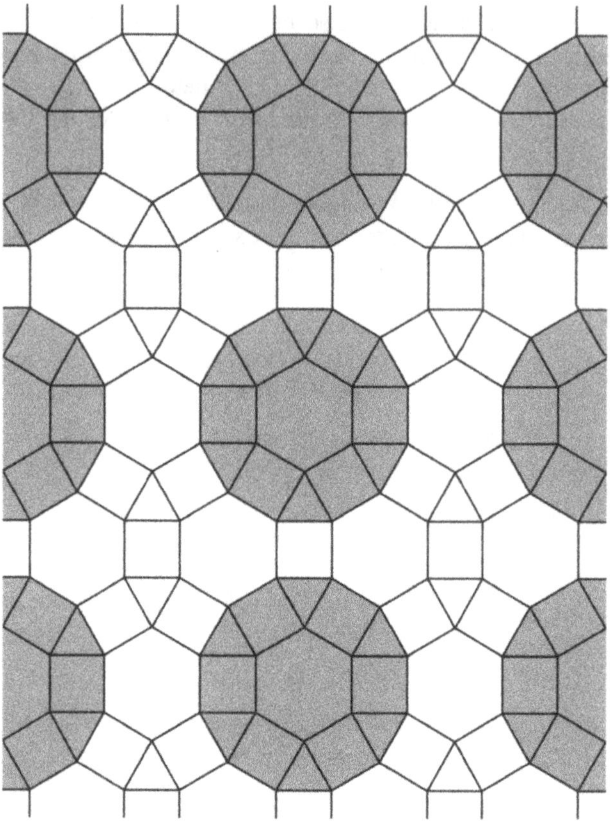

(h) The (3,12,12) tessellation may be constructed from the (6,6,6) tessellation. We cut each hexagonal tile into seven pieces consisting of a regular dodecagon and six congruent isosceles triangles with vertical angles $120°$, as in the figure on the following page on the left. Let the edge length of the hexagonal tile be 1 and the length of the base of the triangles be x. The equal sides of the triangles have length $x/\sqrt{3}$, and since

$$\frac{x}{\sqrt{3}} + x + \frac{x}{\sqrt{3}} = 1,$$

we have

$$x = 2\sqrt{3} - 3 \approx 0.464.$$

When we merge the triangles across three hexagonal tiles, we obtain the (3,12,12) tessellation, as in the figure on the following page on the right.

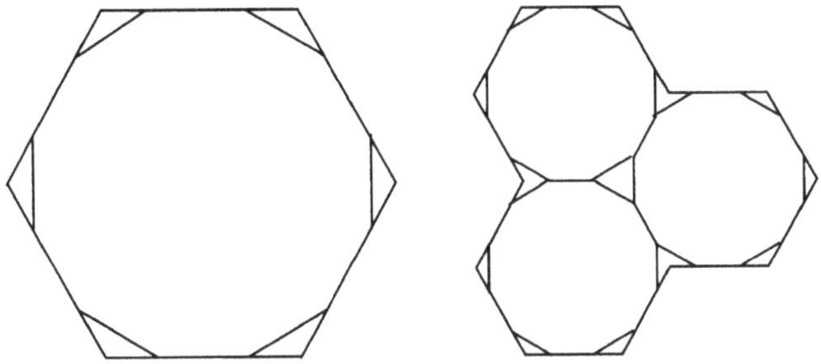

(i) Next, we construct the (4,8,8) tessellation from the basic (4,4,4,4) tessellation. We cut each square tile into five pieces consisting of a regular octagon and four congruent right isosceles triangles, as in the figure below on the left. Let the edge length of the square tile be 1 and the length of the hypotenuse of the triangles be x. Then the legs of the triangles have length $x/\sqrt{2}$. Since

$$\frac{x}{\sqrt{2}} + x + \frac{x}{\sqrt{2}} = 1,$$

we have

$$x = \sqrt{2} - 1 \approx 0.412.$$

When we merge the triangles across four square tiles, we obtain the (4,8,8) tessellation, as in the figure below on the right.

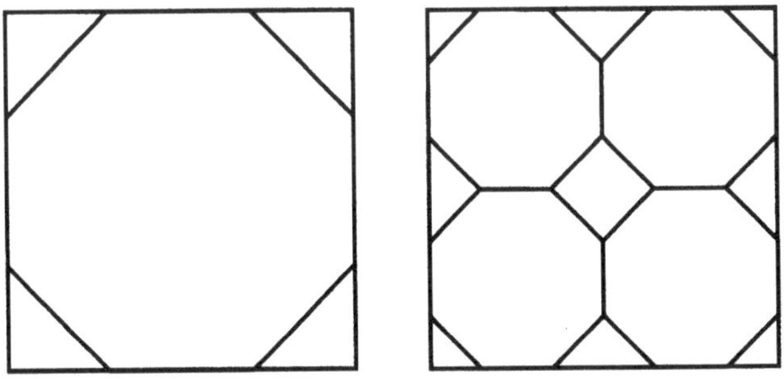

(j) The last tessellation, (3,3,4,3,4), is the most difficult to construct. It is obtained from the basic (4,4,4,4) tessellation with an intermediate step. We first modify the square tile as in the figure on the following page on the left. We cut out two isosceles triangles with vertical angles $150°$, based on two opposite sides of the square, and attach them to the other two sides. This modified tile can also tile the plane, as in the figure on the following page on the right.

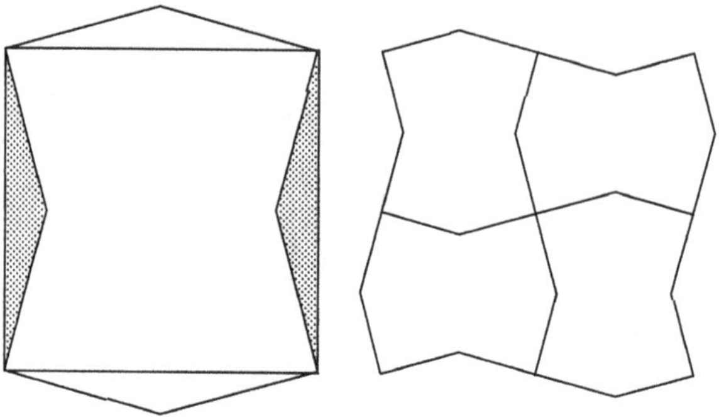

We now cut each modified tile into six pieces consisting of two congruent equilateral triangles and four congruent right isosceles triangles, as in the figure below on the left,

When we merge the right isosceles triangles across four modified tiles, we obtain the (3,3,4,3,4) tessellations, as in the figure below on the right.

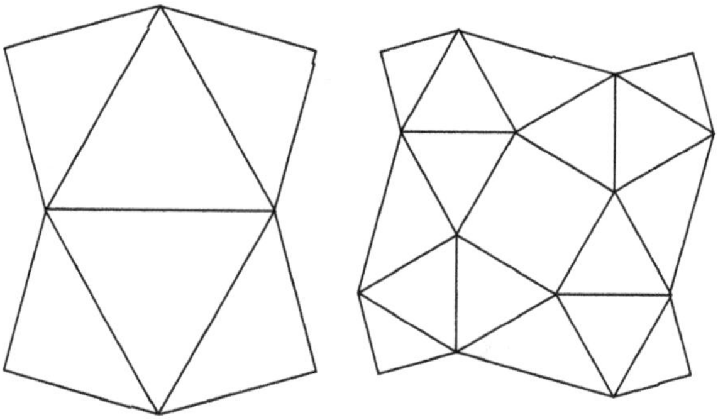

The eight Archimedean tilings are graphically illustrated on page 222 in the book *Sphere Packing, Lewis Carroll and Reversi* by Martin Gardner, published in 2009 by the Mathematical Association of America, Washington, DC.

12.5 Problems

1. There are 12 ways in which five unit squares can be joined edge to edge. The resulting figures are called ***pentominoes***, as shown in the figure on the following page.

They are given letter names, F, I, L, N, P, T, U, V, W, X, Y, and Z, respectively. Identify those that can tile a rectangle.

2. The figure below shows the tiling of a bent strip by a figure consisting of four unit squares joined edge to edge.

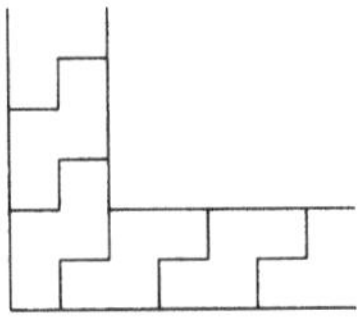

Obviously, a pentomino that can tile a rectangle can also tile some bent strip. Identify those that can tile a bent strip but not a rectangle.

3. Prove that a pentomino that can tile a bent strip can also tile some infinite strip.

4. Identify those pentominoes that can tile an infinite strip but cannot tile a bent strip.

5. Obviously, a pentomino that can tile an infinite strip can also tile the plane. Identify those that can tile the plane but cannot tile an infinite strip.

6. For each of the I-pentomino, L-pentomino, N-pentomino, U-pentomino, V-pentomino, and W-pentomino, dissect it into four pieces and reassemble them into a square.

7. (a) Show that the figure below can tile the plane.

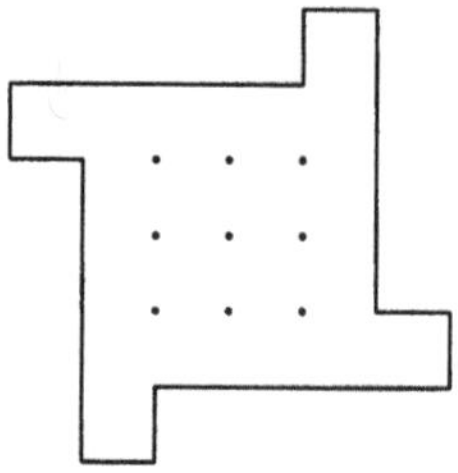

(b) For each of the P-pentomino, Y-pentomino, and Z-pentomino, dissect it into three pieces and reassemble them into a square.

8. (a) Show that the figure below can tile the plane.

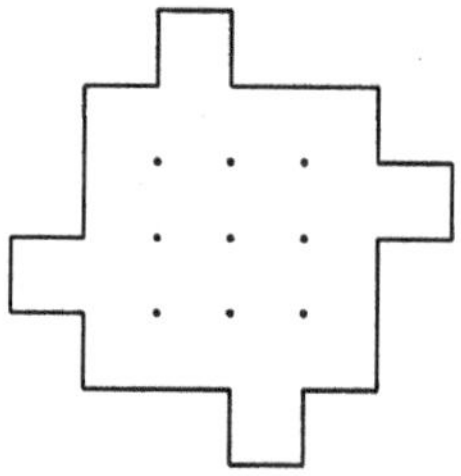

(b) For each of the F-pentomino, T-pentomino, and X-pentomino, dissect it into four pieces and reassemble them into a square.

9. Dissect the figure below into three pieces and reassemble them into a square.

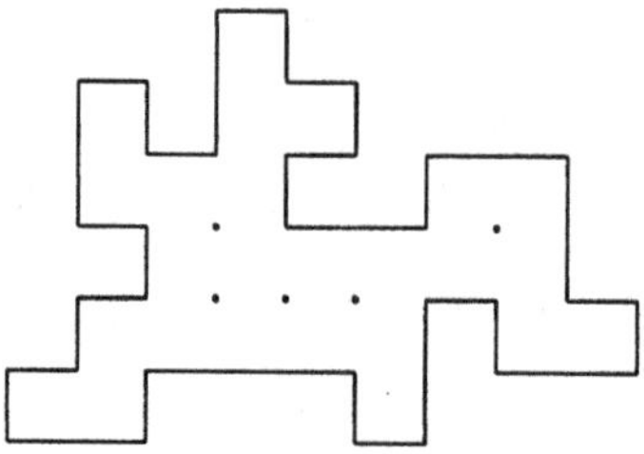

10. Dissect the figure below into three pieces and reassemble them into an equilateral triangle.

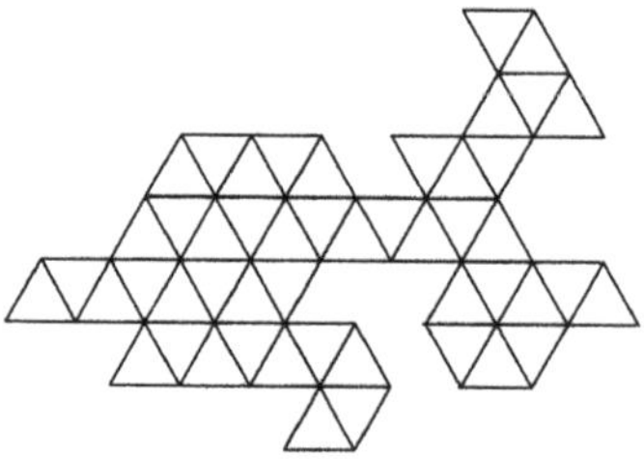

11. (a) Prove that the sum of measures of the exterior angles of a regular n-gon is $360°$ for any $n \geq 3$.

    (b) Use part (a) to prove that the sum of the measures of the central angles of a regular n-gon is given by $(n - 2)180°$.

12. For which values of n is the measure of the central angle of an n-sided regular polygon a positive integer?

13. Find three ways of obtaining the basic (3,6,3,6) tessellation from other tessellations.

14. Find another way of obtaining the (6,6,6) tessellation from the basic (3,3,3,3,3,3) tessellation.

15. Find a way of obtaining the (3,4,6,4) tessellation from the (6,6,6) tessellation.

16. Find a tessellation that has exactly two kinds of vertex sequences, one of which is (3,3,4,12).

17. Find a tessellation that has exactly two kinds of vertex sequences, one of which is (3,4,3,12).

18. Find three tessellations that have exactly two kinds of vertex sequences, one of which is (3,3,6,6).

19. Find three tessellations that have exactly two kinds of vertex sequences, one of which is (3,4,4,6).

PART III

INVERSIVE AND PROJECTIVE GEOMETRIES

CHAPTER 13

INTRODUCTION TO INVERSIVE GEOMETRY

13.1 Inversion in the Euclidean Plane

We introduce the concept of inversion with a simple example, that of constructing the midpoint of a line segment using only a compass.

Example 13.1.1. *Given a line through A and B, find the midpoint of the segment AB using only a compass.*

Solution. With center A and radius $r = AB$, draw the circle $\mathcal{C}(A, r)$ and locate the point P on the line AB so that B is the midpoint of AP.

With center P, draw the circle $\mathcal{C}(P, AP)$ intersecting the first circle at C, as in the figure on the following page.

Classical Geometry: Euclidean, Transformational, Inversive and Projective, First Edition. By I. E. Leonard, J. E. Lewis, A. C. F. Liu, G. W. Tokarsky
Copyright © 2014 John Wiley & Sons, Inc.

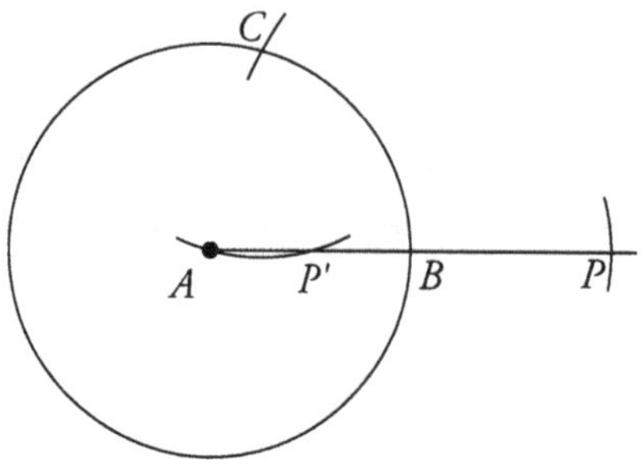

Finally, draw $\mathcal{C}(C, r)$ intersecting the line AB at P'. Then P' is the midpoint of AB.

To see that this is the case, note that the triangles $AP'C$ and ACP are similar isosceles triangles since they share a vertex angle at A, so that

$$\frac{AP'}{AC} = \frac{AC}{AP},$$

which implies that

$$\frac{AP'}{r} = \frac{r}{2r},$$

and this implies that

$$AP' = \frac{r}{2} = \frac{AB}{2}.$$

□

Note that with

$$AP = 2r \qquad \text{and} \qquad AP' = \tfrac{1}{2}r,$$

we have $AP \cdot AP' = r^2$. This relationship between P and P' is called an *inversion*. More generally, we have the following definition:

Definition. (*Inverse of a Point*)

Given a circle $\mathcal{C}(O, r)$ and a point P other than O, the point P' on the ray $\overrightarrow{OP}$ is the ***inverse*** of P if and only if $OP \cdot OP' = r^2$.

The circle $\mathcal{C}(O, r)$ is called the ***circle of inversion***, the point O is called the ***center of the inversion***, r is called the ***radius of inversion***, and r^2 is called the ***power of the inversion***.

Remark. Suppose P is a point other than the center of inversion. If P is outside the circle of inversion, then its inverse P' is interior to the circle of inversion. If P is on the circle, then it is its own inverse. If P is inside the circle, then its inverse P' is exterior to the circle of inversion.

Compass Method of Finding the Inverse

Note that given the ray $\overrightarrow{OP}$ and the circle $\mathcal{C}(O, r)$, the compass-only construction described above also works to find the inverse of P when P is outside the circle of inversion.

With center P and radius OP, draw an arc intersecting $\mathcal{C}(O, r)$ at Q. With center Q and radius OQ, draw an arc intersecting OP at P'. Then P' is the inverse of P.

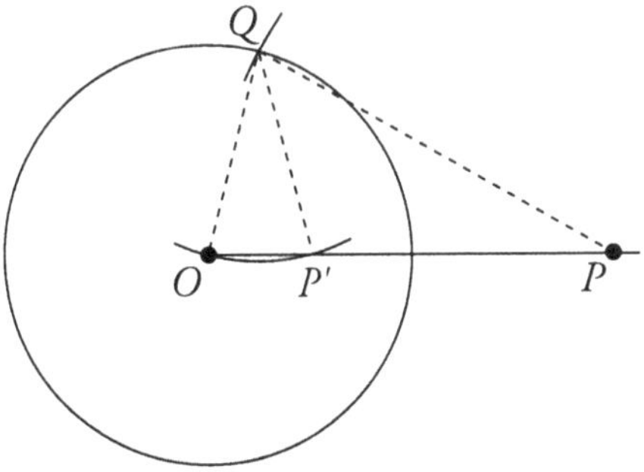

To see this, note that the isosceles triangles OQP and $OP'Q$ are similar by the **AA** similarity condition so that

$$\frac{OP}{OQ} = \frac{OQ}{OP'},$$

and therefore

$$OP \cdot OP' = OQ^2 = r^2.$$

The Tangent Method of Finding the Inverse

Another construction for finding the inverse using a compass and straightedge is as follows (several construction lines have been omitted).

Here we are given the circle $\mathcal{C}(O, r)$ and the point P outside the circle.

Draw the segment OP, and construct the tangents PS and PT to the circle with S and T being the points of tangency, as in the figure below.

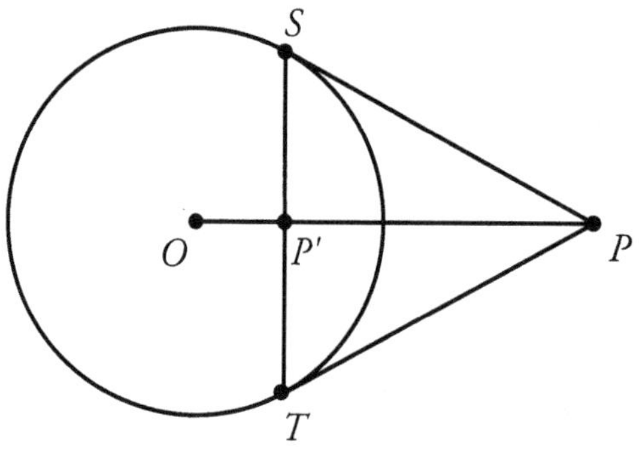

Let $P' = ST \cap OP$. Then P' is the inverse of P.

We leave the proof as an exercise.

Note that an easy modification works to find P' when P is inside the circle: draw the line through P perpendicular to OP intersecting the circle at S and T. Then draw the tangents at S and T meeting at P'.

The Perpendicular Diameter Method of Finding the Inverse

Another method that works when P is inside or outside the circle, as in the figure on the following page, is as follows:

1. Draw ST, the diameter perpendicular to OP.
2. Let Q be the point where the line SP meets the circle.
3. Let P' be the point where the line TQ meets OP.

Then P' is the inverse of P.

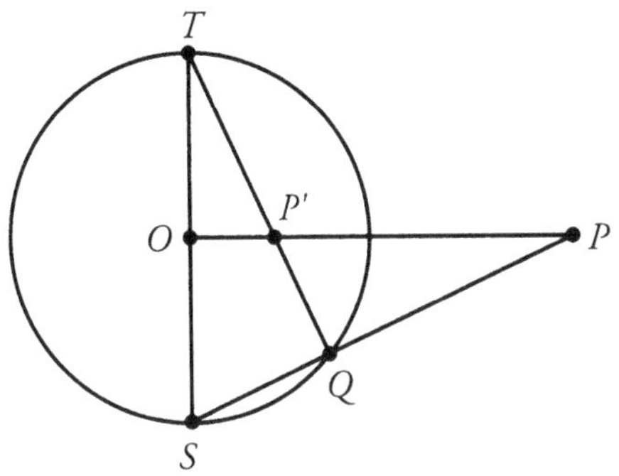

The proof is left as an exercise.

The Inversive Plane

Given a circle C, every point P in the plane has an inverse with respect to C *except* the center of the circle O. The point O has no inverse and is not itself the inverse of any point. As far as inversion is concerned, the point O may as well not exist.

In order to overcome this omission, we append a single ***point at infinity*** I to the plane so that the inversion maps O to I and vice-versa. The point I is also called the ***ideal point***, and it is considered to be on every line in the plane. The Euclidean plane together with this single ideal point is called the ***inversive plane***.

Note. There is only *one* ideal point in the inversive plane, in contrast to the extended Euclidean plane discussed in the earlier chapters, which has infinitely many ideal points that make up the ideal line.

When we want to exclude the ideal point from the discussion of the inversive plane, we refer to the nonideal points as ***ordinary points***.

In the inversive plane, all lines pass through the ideal point. Two lines that meet at a single point in the Euclidean plane meet at two points in the inversive plane. Two lines that are parallel in the Euclidean plane meet only at the ideal point in the inversive plane.

Technically speaking, there are no parallel lines in the inversive plane, although we continue to use the term "parallel lines" to mean that the lines meet only at the ideal point. As in the Euclidean plane, lines that coincide are also said to be parallel.

The following facts are immediate consequences of the definition of inversion in $\mathcal{C}(O, r)$:

Theorem 13.1.2.

(1) The point P' is the inverse of the point P if and only if P is the inverse of P'.

(2) If $OP = kr$, then $OP' = \frac{1}{k}r$.

(3) The inversion maps every point outside the circle to some point inside the circle and vice-versa.

(4) Each point on the circle of inversion is mapped onto itself.

Example 13.1.3. *Suppose that P and Q are points on the ray $\overrightarrow{OP}$. Let P' and Q' be the respective inverses. Show that if $OQ = k \cdot OP$, then $OP' = k \cdot OQ'$.*

Solution. Let r be the radius of inversion. Then

$$OP \cdot OP' = r^2 \qquad \text{and} \qquad OQ \cdot OQ' = r^2.$$

Multiplying both sides of the equation

$$OQ = k \cdot OP$$

by

$$OP' \cdot OQ',$$

we get

$$OP' \cdot OQ' \cdot OQ = OP' \cdot OQ' \cdot k \cdot OP,$$

which implies that

$$OP' \cdot r^2 = OQ' \cdot k \cdot r^2.$$

Thus,

$$OP' = k \cdot OQ'.$$

□

13.2 The Effect of Inversion on Euclidean Properties

A ***Euclidean property*** is one that is preserved by the *Euclidean transformations* (translations, rotations, reflections, and combinations thereof). Euclidean properties include distance, shape, and size. Inversion does *not* preserve Euclidean properties, but it does affect them in useful ways.

In this section, we denote the inversion in the circle $\mathcal{C}(O, r)$ by $I(O, r^2)$. Note that given a figure $\mathcal{G}$, its inverse $\mathcal{G}'$ is obtained by taking the inverse of each point of $\mathcal{G}$. Thus, it follows that $\mathcal{G}'$ is the inverse of $\mathcal{G}$ if and only if $\mathcal{G}$ is the inverse of $\mathcal{G}'$.

Lines and Circles

Before proving the result about how inversion affects lines and circles, we first prove a useful lemma.

Lemma 13.2.1. *Under the inversion $I(O, r^2)$, suppose that P and Q have inverses P' and Q', respectively. Then $\triangle OPQ \sim \triangle OQ'P'$.*

Proof. We have

$$OP \cdot OP' = r^2 = OQ \cdot OQ',$$

which implies that

$$\frac{OP}{OQ} = \frac{OQ'}{OP'}.$$

Since $\angle POQ \equiv \angle Q'OP'$, then $\triangle OPQ \sim \triangle OQ'P'$ by the **sAs** similarity criterion.

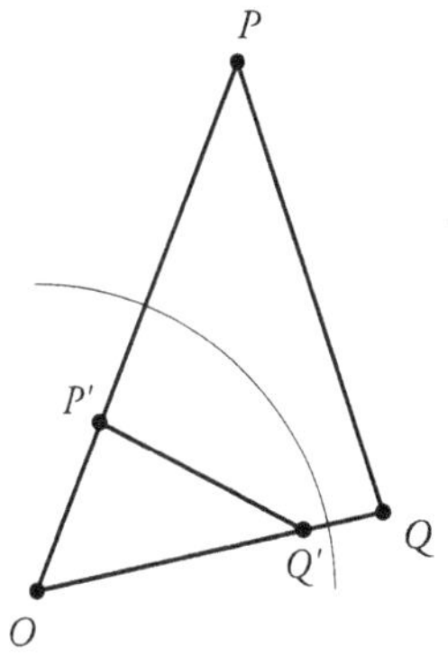

□

Remark. Note that in the similar triangles above we have

$$\angle OP'Q' = \angle OQP \qquad \text{and} \qquad \angle OQ'P' = \angle OPQ.$$

Note also that if we were discussing *signed* or *directed* angles, the direction of the angles would be reversed; that is,

$$m(\angle OP'Q') = -m(\angle OQP) \qquad \text{and} \qquad m(\angle OQ'P') = -m(\angle OPQ),$$

where $m(\angle ABC)$ denotes the measure of the angle $\angle ABC$, as defined in Part I, Section 1.1.

And now the theorem describing the effect of inversion on lines and circles.

Theorem 13.2.2. *The inversion operator $I(O, r^2)$ affects lines and circles as follows:*

(1) The inverse of a line through O is the same line through O.

(2) The inverse of a line not through O is a circle through O.

(3) The inverse of a circle through O is a line not through O.

(4) The inverse of a circle not through O is a circle not through O.

Proof.

(1) The first assertion is straightforward. However, note that a point of the line inverts to a different point of the line, except for the point where the line intersects the circle of inversion. The points I and O of the line are inverses of each other.

(2) Let Q be the foot of the perpendicular from O to the line, and let Q' be the inverse of Q. Let P be any other point on the line other than I, and let P' be the inverse of P, as in the figure on the following page. It follows from Lemma 13.2.1 that $\triangle OP'Q' \sim \triangle OQP$ and, therefore,

$$\angle OP'Q' = \angle OQP = 90^\circ.$$

Thus, P' is on the circle ω with diameter OQ' by the converse to Thales' Theorem. In a similar way, every point X on ω is the inverse of some point X' on the line.

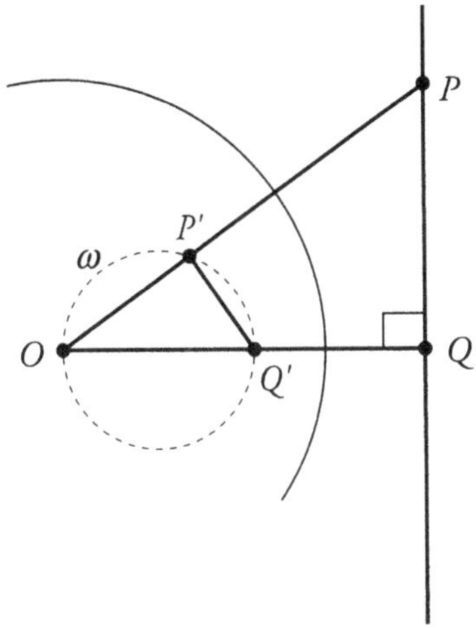

(3) This follows from the fact that since a circle through O is the inverse of a line not through O, then the line not through O is the inverse of a circle through O.

(4) Referring to the figure below, where α is the circle with diameter PQ to be inverted, Lemma 13.2.1 tells us that

$$\triangle OR'P' \sim \triangle OPR \qquad \text{and} \qquad \triangle OR'Q' \sim \triangle OQR.$$

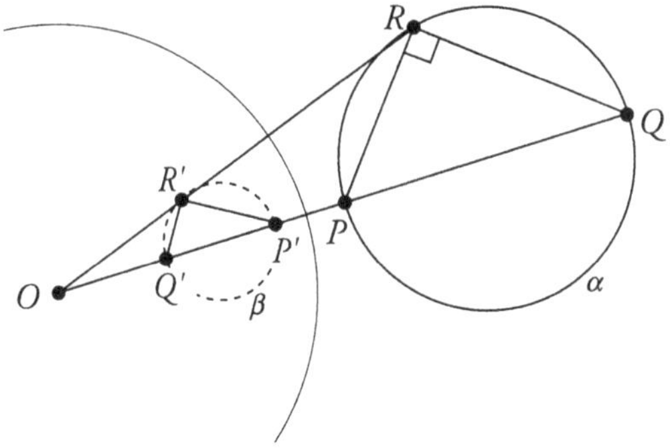

Thus,

$$\angle OR'P' = \angle OPR \qquad \text{and} \qquad \angle OR'Q' = \angle OQR.$$

Now,

$$\angle Q'R'P' = \angle OR'P' - \angle OR'Q',$$

so that

$$\angle Q'R'P' = \angle OPR - \angle OQR.$$

From the External Angle Theorem, we have

$$\angle OPR - \angle OQR = \angle PRQ,$$

while from Thales' Theorem, we have

$$\angle PRQ = 90^\circ.$$

Thus, $\angle QR'P' = 90^\circ$, and from the converse to Thales' Theorem we can conclude that R' is on β, the circle with diameter $Q'P'$. Similarly, any point on the circle β is the inverse of some point on the circle α.

□

Inversion and Distances

Theorem 13.2.3. *Let P and P' be inverse points with P outside the circle of inversion, and let B be the point where PP' meets the circle of inversion, as in the figure below. Then,*

$$BP' = \frac{BP}{1 + BP/r} \quad \textit{and} \quad BP = \frac{BP'}{1 - BP'/r}.$$

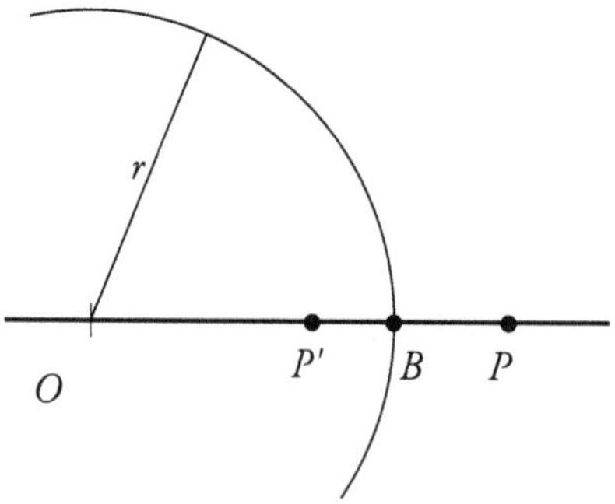

Proof. We have

$$\begin{aligned} BP' &= r - OP' \\ &= r - \frac{OP' \cdot OP}{OP} \\ &= r - \frac{r^2}{r + BP} \\ &= \frac{r \cdot BP}{r + BP} \\ &= \frac{BP}{1 + BP/r}. \end{aligned}$$

That is,

$$BP' = \frac{BP}{1 + BP/r}.$$

The proof that $BP = \dfrac{BP'}{1 - BP'/r}$ is similar.

□

Theorem 13.2.4. *If P' and Q' are inverse points for P and Q, respectively, under the inversion $I(O, r^2)$, then*

$$P'Q' = \frac{PQ}{OP \cdot OQ} r^2 .$$

Proof. We consider three distinct cases.

Case (i). P, Q, and O are not collinear.

By Lemma 13.2.1, $\triangle OP'Q' \sim \triangle OQP$, so that

$$\frac{P'Q'}{PQ} = \frac{OQ'}{OP},$$

which implies that

$$P'Q' = \frac{PQ \cdot OQ'}{OP}$$
$$= \frac{PQ \cdot OQ' \cdot OQ}{OP \cdot OQ}$$
$$= \frac{PQ}{OP \cdot OQ} r^2.$$

That is,

$$P'Q' = \frac{PQ}{OP \cdot OQ} r^2.$$

Case (ii). P, Q, and O are collinear, with O not between P and Q.

We may assume that $OP < OQ$, as in the figure below.

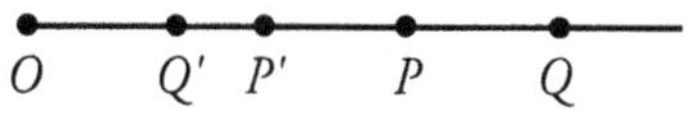

This implies that $OQ' < OP'$, since $OP \cdot OP' = OQ \cdot OQ'$, so that

$$P'Q' = OP' - OQ'$$
$$= \frac{r^2}{OP} - \frac{r^2}{OQ}$$
$$= \frac{OQ - OP}{OP \cdot OQ} r^2$$
$$= \frac{PQ}{OP \cdot OQ} r^2.$$

That is,

$$P'Q' = \frac{PQ}{OP \cdot OQ} r^2.$$

Case (iii). P, Q, and O are collinear, with O between P and Q.

The proof is almost the same as Case (ii) and is left as an exercise.

□

Theorem 13.2.5. *(Ptolemy's Theorem)*

If $ABCD$ is a convex cyclic quadrilateral, then

$$AC \cdot BD = BC \cdot AD + CD \cdot AB.$$

Proof. Consider the effect of $I(A, r^2)$ on the circumcircle of the cyclic quadrilateral. The circle inverts into a straight line, and the inverse points B', C', and D' are on this line, as shown in the figure below.

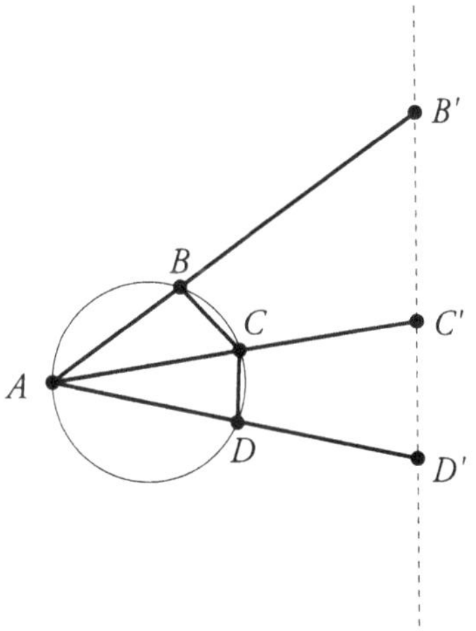

The convexity of $ABCD$ guarantees that C' is between B' and D', so that

$$B'D' = B'C' + C'D',$$

and thus,

$$\frac{BD}{AB \cdot AD} r^2 = \frac{BC}{AB \cdot AC} r^2 + \frac{CD}{AC \cdot AD} r^2,$$

from which the theorem follows.

□

Inversion and Angles

The angle between two curves at a point of intersection P is defined as the angle between the tangents to these curves at P.

If one or both of the curves fail to have a tangent at P, then the angle is not defined. We will be dealing only with curves that do have tangents.

Given two curves ω and ξ, there are two different magnitudes that are associated with the angle from ω to ξ: one measured counterclockwise from ω to ξ, shown as angle α in the figure below, and the other measured clockwise, as shown by β. In general, when we refer to the angle ***from*** ω ***to*** ξ without mentioning the direction, we mean the counterclockwise one.

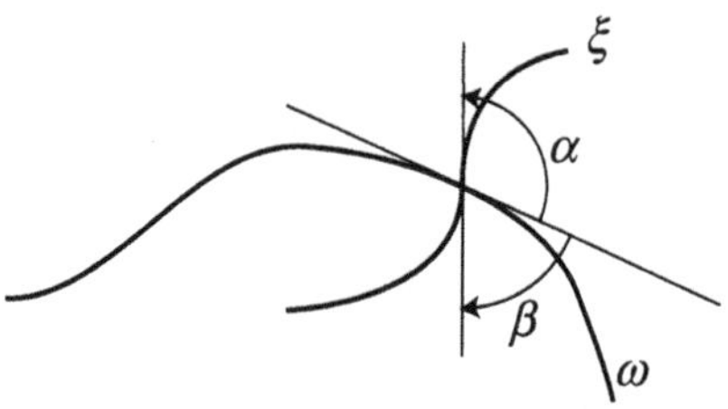

The main result concerning inversion and angles is:

Theorem 13.2.6. *Inversion preserves the magnitude of the angle between intersecting curves but reverses the direction.*

We omit the proof.

For us, the main consequence of Theorem 13.2.6 is that inversion preserves tangencies and orthogonality. For example, if two circles are tangent to each other, then their inverses are also tangent to each other. If two lines meet at 90°, then their inverses, which may be circles, also meet at 90°.

13.3 Orthogonal Circles

Theorem 13.3.1. *If two circles meet at P and Q, then the magnitude of the angles between the circles is the same at P and Q.*

Proof. Referring to the figure below, we have $\triangle APB \equiv \triangle AQB$ by the **SSS** congruency condition, so $\angle APB \equiv \angle AQB$. Since the tangents to the circles at P are perpendicular to the radii AP and BP, it follows that the angle between the tangents at P is equal in measure to $m(\angle APB)$. Likewise, the angle between the tangents at Q is equal in measure to $m(\angle AQB)$.

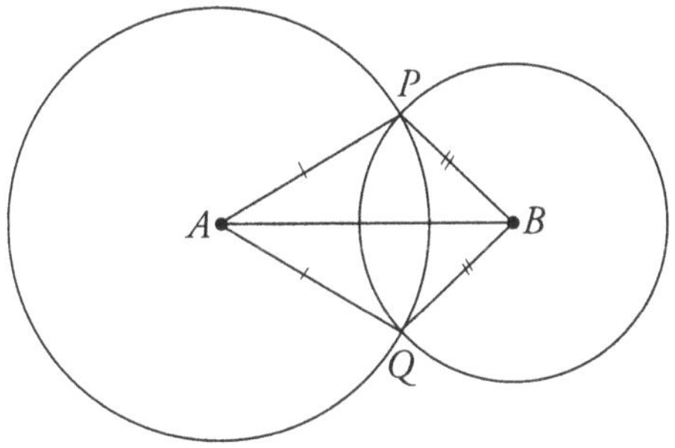

□

Remark. The previous theorem means that to determine the magnitude of the angle between two circles intersecting at P and Q, we only need to check one of the angles. Note, however, that the directions of the angles at P and Q are opposite.

Definition. Two intersecting circles α and β are said to be ***orthogonal*** if the angle between them is $90°$. We sometimes write $\alpha \perp \beta$ to indicate orthogonality.

Theorem 13.3.2. *If two circles α and β are orthogonal, then:*

(1) The tangents at each point of intersection pass through the centers of the other circle (figure (a) on the following page).

(2) Each circle is its own inverse with respect to the other.

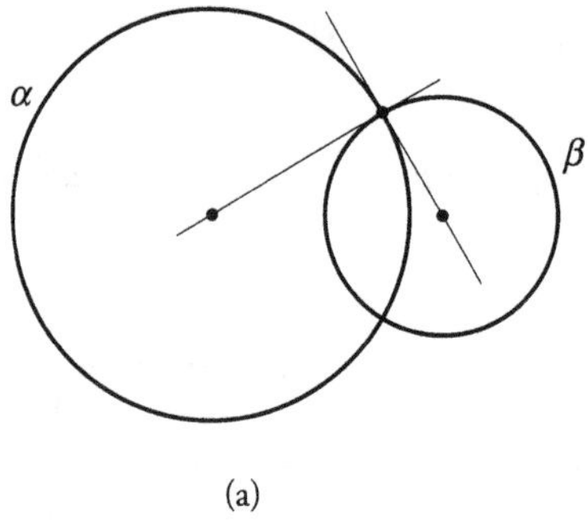

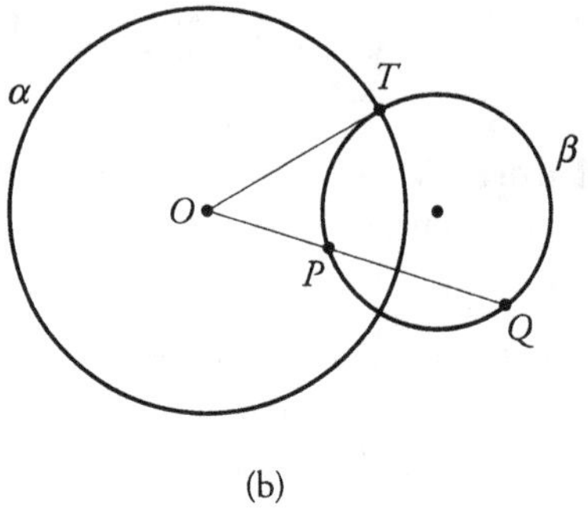

Proof.

(1) This follows because a line through the point of tangency perpendicular to the tangent must pass through the center of the circle.

(2) Let P be a point on the circle β. Join P to O, the center of α, and let r be the radius of α. Let Q be the other point where the ray $\overrightarrow{OP}$ meets β. Let T be the point of intersection of the two circles so that OT is the tangent to β by (1) above. By the power of the point O with respect to α, we have

$$OP \cdot OQ = OT^2 = r^2 ,$$

showing that the inverse of any point on β is another point on β.

□

The preceding theorem has the following converse:

Theorem 13.3.3. *Two intersecting circles α and β are orthogonal if any one of the following statements is true.*

(1) The tangent to one circle at one point of intersection passes through the center of the other circle (figure (a) below).

(2) One of the circles passes through two distinct points that are inverses with respect to the other circle.

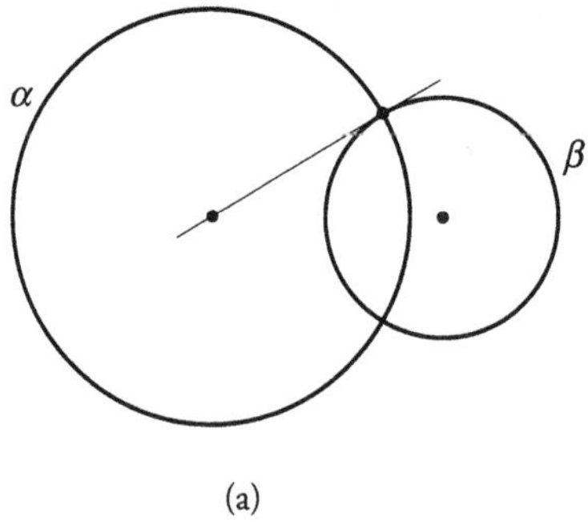

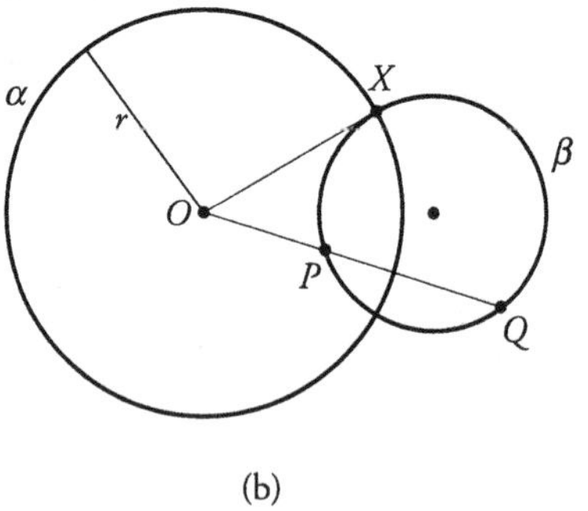

Proof.

(1) This implies that the two tangents at the point of intersection must be perpendicular.

(2) Suppose that the circle β passes through P and Q, which are inverses with respect to α. Let O be the center of α and let OX be tangent to β at X, as in figure (b) on the previous page. Then we have

$$OP \cdot OQ = OX^2$$

by the power of O with respect to β, and

$$OP \cdot OQ = r^2,$$

since P and Q are inverses.

This implies that $OX = r$, so X must be on α as well as on β; that is, X is a point of intersection of α and β, and the tangent to β at this point passes through the center of α. By (1), the circles must be orthogonal.

□

The Arbelos Theorem

Theorem 13.3.4. *(The Arbelos Theorem, a.k.a. Pappus' Ancient Theorem)*

Suppose that P, Q, and R are three collinear points with C, D, and K_0 being semicircles on PQ, PR, and QR, respectively. Let K_1, K_2, ..., be circles touching C and D, with K_1 touching K_0, K_2 touching K_1, and so on.

Let h_n be the distance of the center of K_n from PR and let r_n be the radius of K_n. Then $h_n = 2nr_n$.

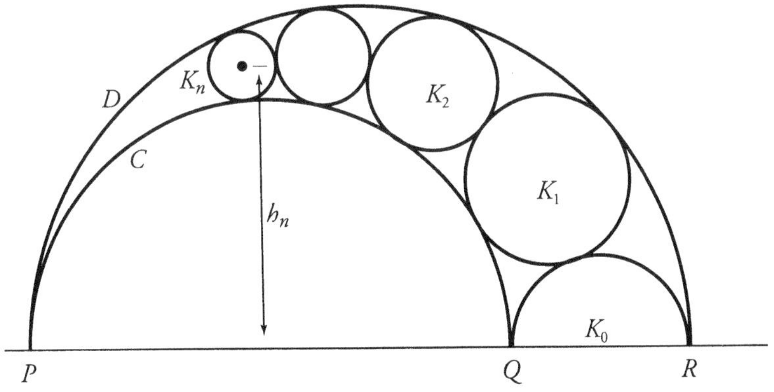

Proof. In the figure below, let t be the tangential distance from P to the circle K_n and apply $I(P, t^2)$.

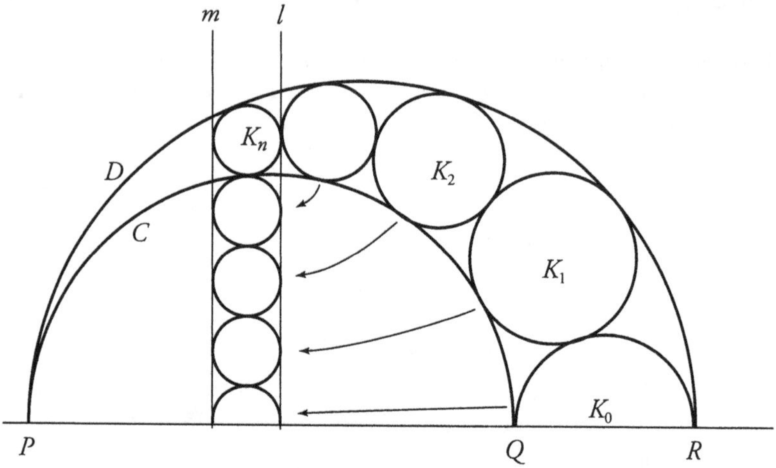

K_n is orthogonal to the circle of inversion, so it is its own inverse.

C inverts into a line l.

D inverts into a line m parallel to l.

K_0 inverts into a semicircle K'_0 tangent to l and m, since inversion preserves tangencies.

K_i inverts into a circle K'_i tangent to l and m.

Thus, all of the K'_is have the same radius, namely r_n, and the theorem follows.

□

Steiner's Porism

Given a point P outside a circle α, a point X of α is said to be ***visible*** from P if the segment PX meets α only at X.

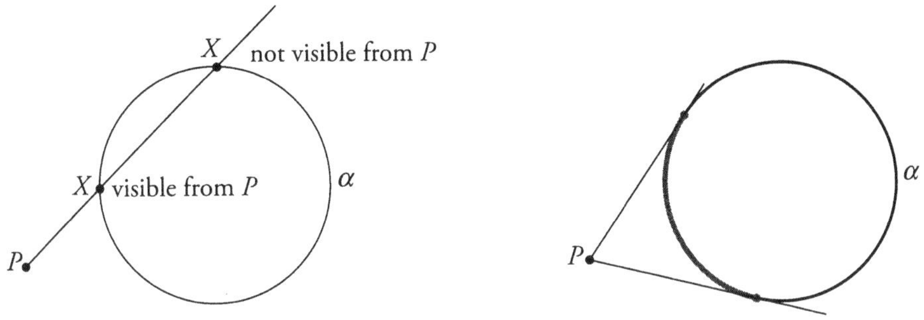

(a) (b)

Figure (b) above shows the set of points that are visible from P, namely, the two tangent points and the points on the arc between the tangent points. In other words, a point X of α is visible from P if and only if

- either PX is tangent to α
- or the tangent to α at X has α and P on opposite sides.

Note also that if a line m is tangent to α at X, and if P is on the same side of m as α but not on the line m, then X is not visible from P.

Lemma 13.3.5. *Suppose the line PQ misses the circle α. Then*

(1) there is a point X visible from both P and Q,

(2) there is a point Y visible from P but not from Q,

(3) there is a point Z visible from Q but not from P.

The figure below illustrates how to find points X and Y. Let m be a line parallel to PQ and tangent to α. X is the point of tangency of m with α. There are two lines from P tangent to α. Let l be the tangent line such that α and Q are both on the *same* side of l. Then Y is the point where l is tangent to α.

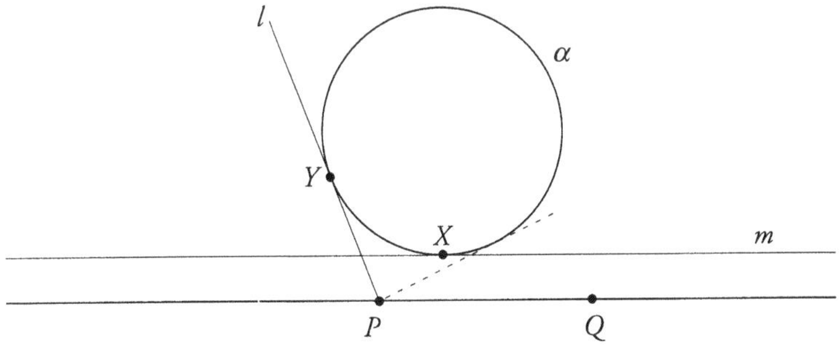

Lemma 13.3.6. *Given two circles ω and ξ and given a point P not on either circle, there is a circle through P orthogonal to both ω and ξ.*

Proof. Let Q be the inverse of P with respect to ω, and let S be the inverse of P with respect to ξ. Then there is a unique circle through P, Q, and S, and this circle must be orthogonal to both ω and ξ by Theorem 13.3.3.

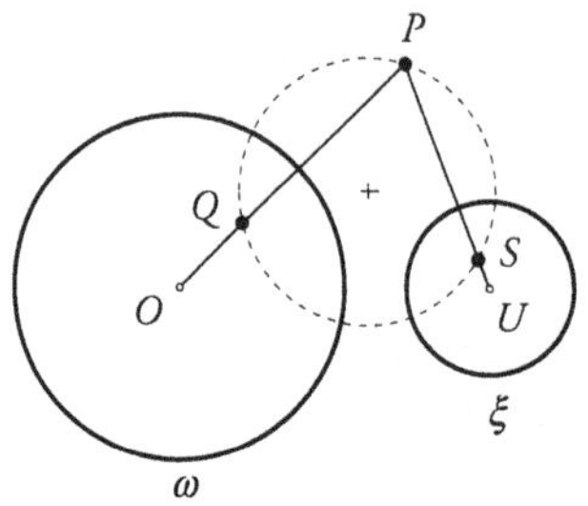

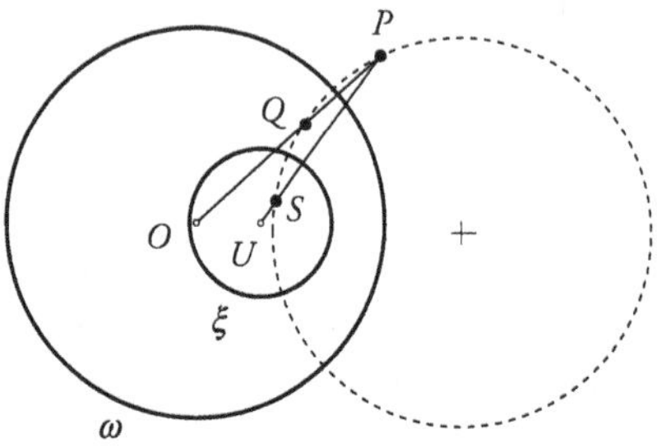

□

Note. If O, U, and P are collinear, then the orthogonal "circle" is a line. If O, U, and P are not collinear, then the orthogonal circle is a true circle.

Lemma 13.3.7. *Let ω and ξ be two nonintersecting circles with centers O and U, $O \neq U$. Then we can find points X and Y that are inverses to each other with respect to both ω and ξ.*

Proof. Let α be any circle other than a line that is orthogonal to both ω and ξ. We claim that the line OU intersects α in two points. In this case, the two points are X and Y and they are inverses to each other with respect to both ω and ξ.

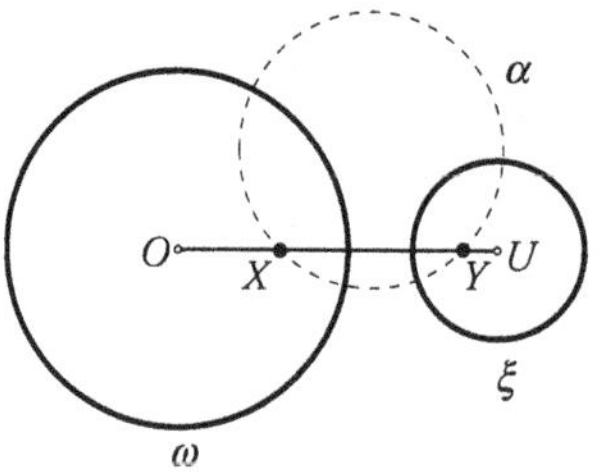

case (i)

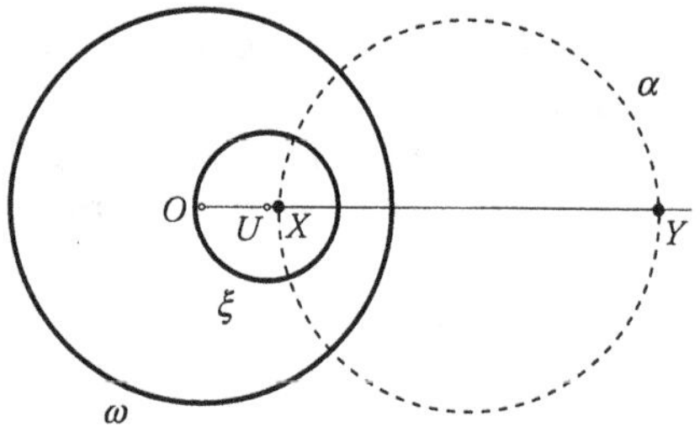

case (ii)

There are two cases to consider: (i) when the circles are exterior to each other and (ii) when one circle is inside the other.

(i) Suppose for a contradiction that OU misses α. Then there is a point Z on α that is visible from both O and U. Since Z is visible from O, it is inside or on ω. Since Z is visible from U, it is inside or on ξ. Then Z is inside or on both ω and ξ, which contradicts the fact that ω and ξ are exterior to each other. This proves case (i).

(ii) The proof of this case is left as an exercise.

□

Definition. Let ω and ξ be two nonintersecting circles.

Let α_1 be a circle tangent to both ω and ξ.

Let α_2 be a circle tangent to α_1, ω, and ξ.

Let α_3 be a circle tangent to α_2, ω, and ξ.

Continuing in this fashion, if at some point α_k is tangent to α_1, then we say that

$$\alpha_1, \quad \alpha_2, \quad \ldots, \quad \alpha_k$$

is a ***Steiner chain*** of k circles.

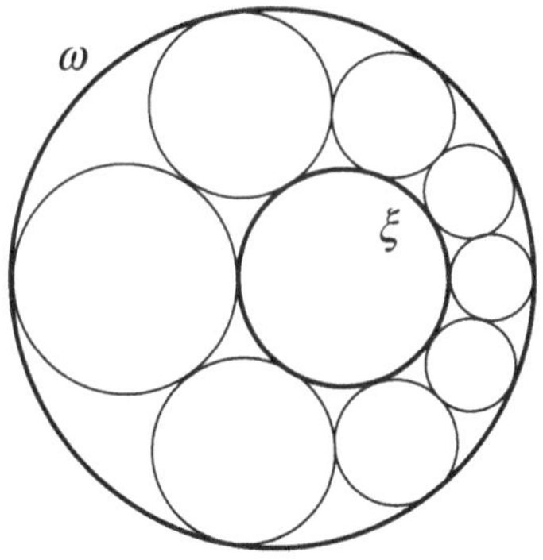

Remark. Given two circles ω and ξ, there is no guarantee that a Steiner chain exists for ω and ξ.

In order to prove the next theorem, we need the following lemma:

Lemma 13.3.8. *Given two nonintersecting circles ω and ξ that are not concentric, there is an inversion that transforms them into concentric circles.*

Proof. Using Lemma 13.3.6, we can find two circles α and β simultaneously orthogonal to both ω and ξ. These two circles intersect at the points X and Y referred to in Lemma 13.3.7.

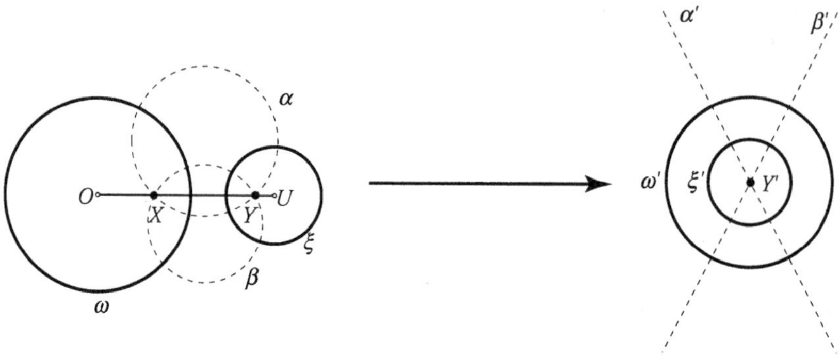

Perform the inversion $I(X, r^2)$ for some radius r. Then:

- α transforms to α', a straight line through Y' and not through X.
- β transforms to β', a straight line through Y' and not through X.
- ω transforms to a circle ω' and

$$\omega' \perp \alpha' \qquad \text{and} \qquad \omega' \perp \beta'$$

  since orthogonality is preserved.
- ξ transforms to a circle ξ' and

$$\xi' \perp \alpha' \qquad \text{and} \qquad \xi' \perp \beta'.$$

Since the circle ω' is orthogonal to the line α', then ω' must be centered at some point of α'. Similarly, ω' must be centered at some point of β'. Thus, ω' is centered at Y'. By the same argument, ξ' must also be centered at Y'.

□

Now we are able to prove the following:

Theorem 13.3.9. *(Steiner's Porism)*
Suppose that two nonintersecting circles ω and ξ have a Steiner chain of k circles. Then any circle tangent to ω and ξ is a member of some Steiner chain of k circles.

Proof. In the figure below, we invert α and β into concentric circles. The inversion preserves the Steiner chain of k circles.

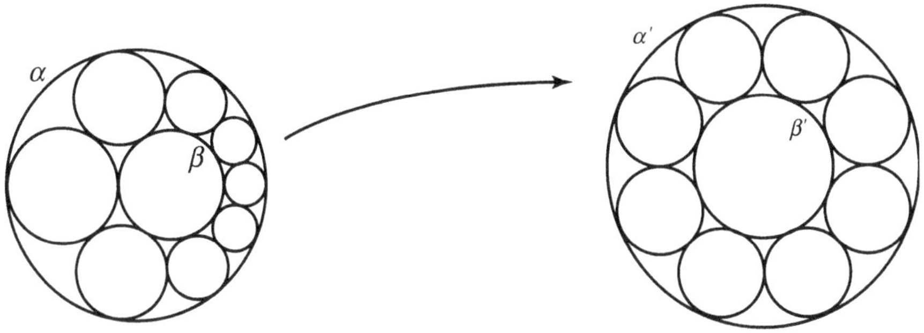

Using the same inversion, we transform ξ into a circle ξ', as in the figure below.

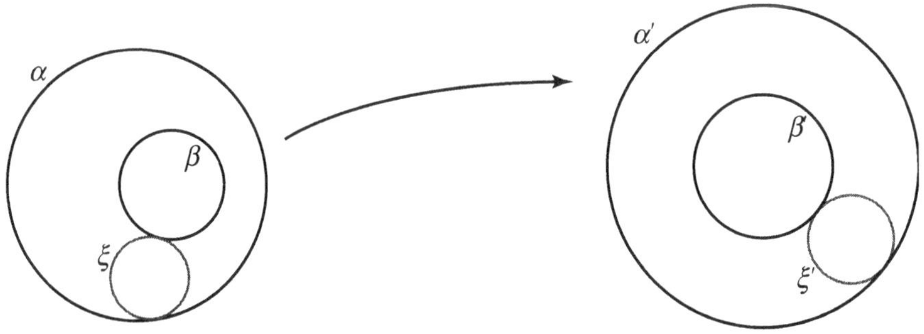

The circle ξ' is obviously part of a Steiner chain of k circles, so by the reverse inversion, ξ must also be part of a Steiner chain of k circles.

□

13.4 Compass-Only Constructions

We will use some special notation for this section only:

- $A(P)$ the circle with center A passing through the point P.
- $A(r)$ the circle with center A and radius r.
- $A(BC)$ the circle with center A and radius BC.

Note that when we say "draw $A(B)$," we will often draw only an arc of the circle rather than the entire circle. Also, *in this section only*, "construct" means "construct using only a compass."

The objective of this section is to show that, insofar as constructing points is concerned, a compass alone is just as powerful as the combination of compass and straightedge. To put it another way, even if a construction involves drawing straight lines, we can carry out the construction in such a manner that we postpone drawing the straight lines until the very end. Furthermore, at the end we only use the straightedge to draw lines between existing points—we never need to use the straightedge to perform any new construction. This does not mean, however, that using a compass alone will be as efficient as a compass and straightedge together.

We begin with some examples that have been discussed previously.

Example 13.4.1. *Given points A and B, construct the point C such that B is the midpoint of AC.*

Solution. We perform the following constructions, as shown in the figure on the following page:

1. construct $B(A)$,
2. construct $A(B)$, yielding point D,
3. construct $D(AB)$, yielding point E,
4. construct $E(AB)$, yielding point C.

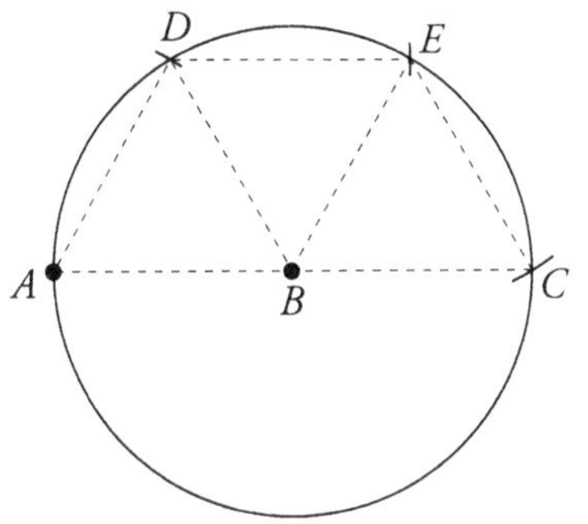

The justification is as follows. As in the figure, $\triangle ABD$, $\triangle DBE$, and $\triangle EBC$ are all equilateral triangles, so that $\angle ABD = \angle DBE = \angle EBC = 60°$. This means that ABC is a straight line, and so AC is a diameter of $B(A)$. Therefore, B is the midpoint of AC.

□

Example 13.4.2. *Given a point P outside a circle ω with center A, construct the inverse of P with respect to ω.*

Solution. We perform the following constructions, as in the figure below:

1. draw $P(A)$ meeting ω at B and C,
2. draw $B(A)$ and $C(A)$ meeting at Q.

Then Q is the inverse of P.

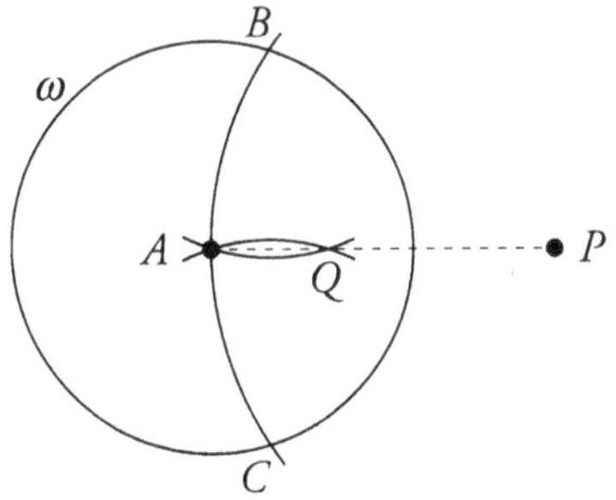

The justification is as follows. As in the figure, P' is on $B(A)$ and the line AP. Also, P' is on $C(A)$ and the line AP, so P' is the point $B(A) \cap C(A)$ other than A; that is, $P' = Q$.

□

Example 13.4.3. *Given points A and B, find the midpoint C of AB.*

Solution. Use Example 13.4.1 to construct the point D such that B is the midpoint of AD.

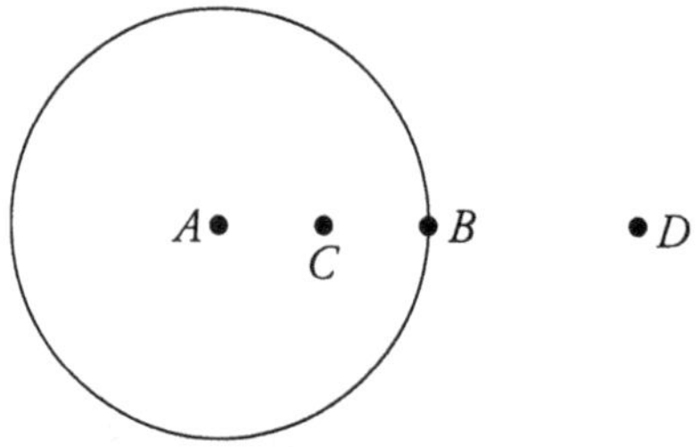

Now draw $A(B)$, and use Example 13.4.2 to find the inverse C of D with respect to $A(B)$. Then C is the midpoint of AB.

□

Example 13.4.4. *Given a circle ω with center A and radius r, and given a point P inside ω, construct the inverse of P with respect to ω.*

Solution. Repeatedly use Example 13.4.1 to construct points $P_1, P_2, \ldots, P_k$ so that P_k is outside ω, with

$$\begin{aligned} AP_1 &= 2AP, \\ AP_2 &= 2AP_1 = 4AP \\ AP_3 &= 2AP_2 = 8AP \\ &\vdots \\ AP_k &= 2AP_{k-1} = 2^k AP \end{aligned}$$

For example, with $k = 3$, we would have the following figure.

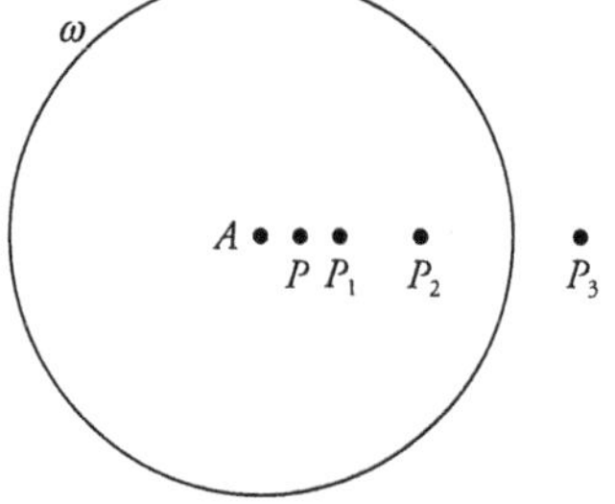

Use Example 13.4.2 to find the inverse S of P_k. Then $AS \cdot AP_k = r^2$.

Now use Example 13.4.1 to find points $S_1, S_2, \ldots, S_k$ so that

$$\begin{aligned} AS_1 &= 2AS, \\ AS_2 &= 2AS_1 = 4AS, \\ &\vdots \\ AS_k &= 2AS_{k-1} = 2^k AS. \end{aligned}$$

Then S_k is the inverse of P, since

$$AS_k \cdot AP = 2^k AS \cdot AP = AS \cdot 2^k AP = AS \cdot AP_k = r^2.$$

□

The following example is a famous problem due to Mohr.

Example 13.4.5. *Given a circle α with unknown center A, construct its center.*

Solution. We perform the following constructions, as shown in the figure below.

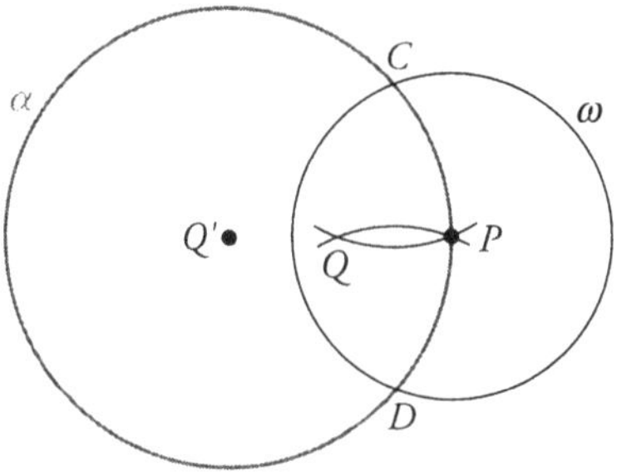

1. With any point P on α, construct a circle ω meeting α at C and D, with the radius of ω less than the radius of α.
2. Draw $C(P)$ and $D(P)$ meeting at Q.
3. Using Example 13.4.4, construct Q', the inverse of Q with respect to the circle ω. Then Q' is the center of α.

The justification is as follows. Let A be the center of α, and referring to the figure above, note that if we knew where A was, then Q would be the inverse of A by Example 13.4.2, so A must be Q'.

□

Example 13.4.6. *Given an arc BC with center A, construct the midpoint of the arc.*

Solution. It suffices to construct a point X of arc BC that is on the right bisector of chord BC, as in the analysis figure below.

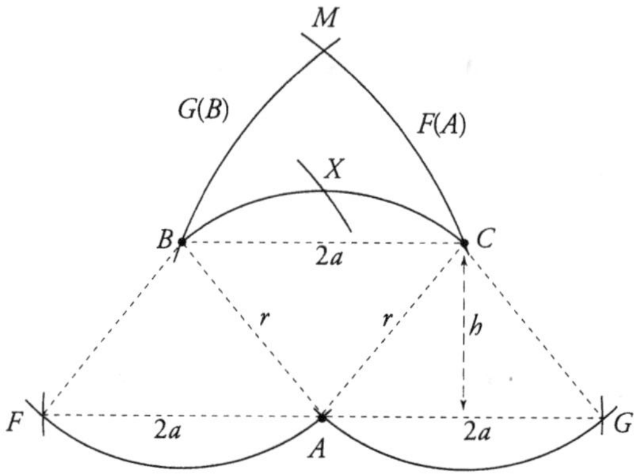

Construction:

1. Construct $A(BC)$, $B(A)$, yielding F, the fourth point of parallelogram $ACBF$.
2. Construct $A(BC)$, $C(A)$, yielding G, the fourth point of parallelogram $ABCG$. Then A is collinear with F and G.
3. Construct $F(C)$ and $G(B)$, yielding M. Note that AM is the right bisector of both BC and FG.
4. Construct $F(AM)$ intersecting the arc BC at X. Then X is the desired point.

Justification:

(a) Let BC have length $2a$. Then $FA = GA = BC = 2a$.

(b) Let h be the perpendicular distance from BC to FG. Then from Pythagoras' Theorem, we have

$$\begin{aligned} FX^2 = AM^2 &= FM^2 - 4a^2 \\ &= FC^2 - 4a^2 \\ &= \left[(3a)^2 + h^2\right] - 4a^2 \\ &= 9a^2 + r^2 - a^2 - 4a^2 \\ &= 4a^2 + r^2 \\ &= FA^2 + AX^2. \end{aligned}$$

Since $FX^2 = FA^2 + AX^2$, by the converse to Pythagoras' Theorem, we conclude that $\triangle FCX$ is a right triangle; that is, X is on the right bisector of FG and BC.

□

The Mohr-Mascheroni Theorem

A straightedge and compass construction allows only the following:

- Drawing a straight line through two different points.
- Drawing a circle centered at one point passing through another point.
- Drawing a circle centered at one point with a particular radius (typically specified by two other points).

All constructions are just a sequence of these basic operations. By using these operations we can construct new points and then use the new points to carry out more of the basic operations.

In the Euclidean plane, there are only three ways to construct new points:

(1) Construct a point as the intersection of two circles.

(2) Construct a point as the intersection of a line and a circle.

(3) Construct a point as the intersection of two lines.

The purpose of this section is to show that we can carry out all of these constructions with a compass alone. In a sense, this means that any construction we can perform with a straightedge and compass we can also perform with a compass alone. Thus, the objective here is to accomplish (2) and (3) using only a compass. We restate (2) as an example:

Example 13.4.7. *Given points A, B, C, and D, construct the intersection of $C(D)$ with the line AB.*

Solution. There are two cases to consider:

Case (i). A, B, C are not collinear. The analysis figure is shown below.

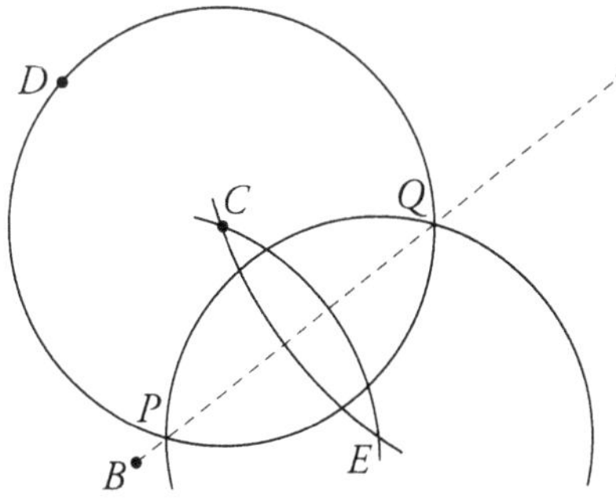

Construction:

1. Draw $C(D)$.
2. Draw $A(C)$, $B(C)$ intersecting at E. Then $ACBE$ is a kite, and hence AB is the right bisector of CE.
3. Draw $E(CD)$ intersecting $C(D)$ at P and Q. Then P and Q are the desired points.

Justification:

(a) $EPCQ$ is a rhombus, so PQ is the right bisector of CE.

(b) P and Q are on AB; that is, P and Q are the points where the line AB meets the circle $C(D)$.

Case (ii). A, B, C are collinear. The analysis figure is shown below.

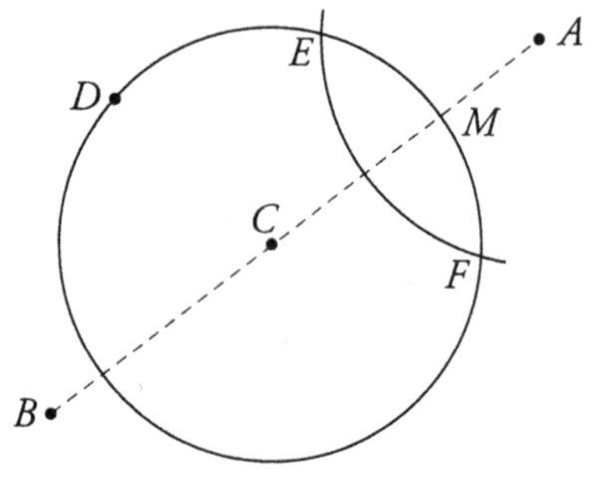

Construction:

1. Draw $C(D)$.
2. Draw $A(r)$ for some convenient r, intersecting $C(D)$ at E and F.
3. Use Example 13.4.6 to find the midpoint M of the arc EF. Then M is one of the desired points, and the other can be found in a similar way.

The justification is left as an exercise.

□

Accomplishing (3), that is, constructing the intersection of two lines, requires the construction of 12 different circles. We restate (3) as an example:

Example 13.4.8. *Given points A, B, C, and D, construct the intersection of the lines AB and CD.*

Solution. As in the figure below, we make the following constructions.

1. Draw $A(C)$, $B(C)$, yielding H. Then AB is the right bisector of CH because $ACBH$ is a kite.

2. Draw $B(D)$ and AD, yielding K. Then AB is the right bisector of KD because $AKBD$ is a kite. Note that this means that $HK = CD$ because of trapezoid $KCHD$.

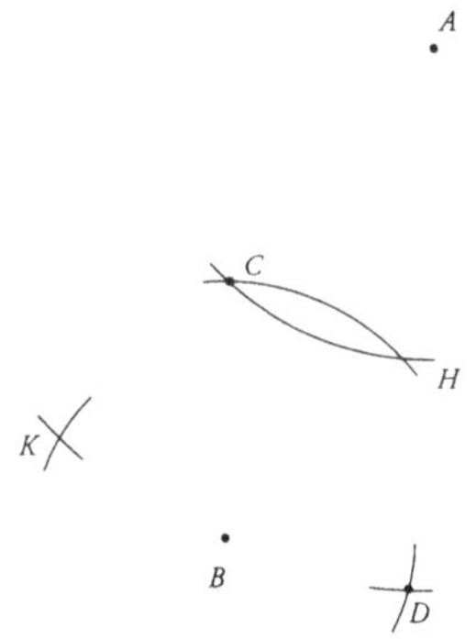

Next we perform the following constructions, as shown in the figure on the following page.

3. Draw $C(DK)$ and $K(CD)$, yielding point G. The point G is collinear with C and H since $KDCG$ is a parallelogram and $CH \parallel DK \parallel CG$.

4. Draw $H(G)$, $G(K)$, giving point E (one of two possible points). Note that

$$GE = GK = CD.$$

5. Draw $H(C)$, $G(CE)$ giving point F. Then F is collinear with H and E. This follows since $\triangle GHF \equiv \triangle EHC$ by the **SSS** congruency condition, and hence,

$$\angle GHF = \angle EHC = \angle EHG,$$

and the lines HE and HF coincide.

6. Draw $C(F)$ and $H(CF)$, yielding point X, which is the desired point.

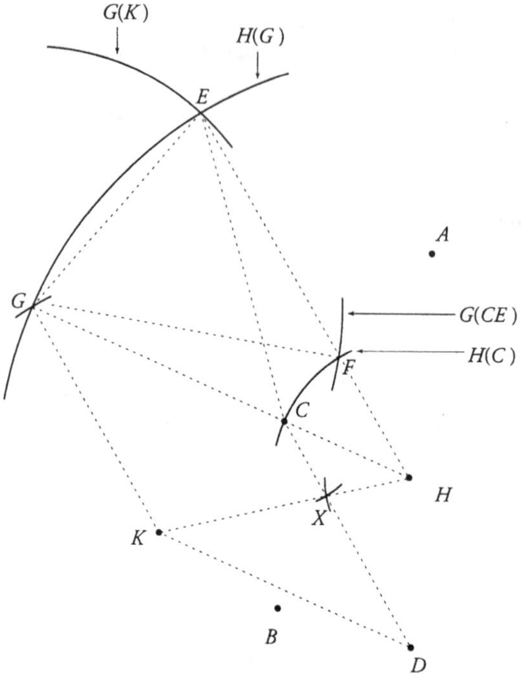

Now the justification. We need to show that $X = AB \cap CD$.

Since $ACXH$ is a kite with $XC = XH$, then X is on the right bisector of CH, but AB is the right bisector of CH by step 1.

To show that X is also on CD, we will show that $\angle HCX = \angle HCD$.

Consider the triangles HCX and HGK. We have

$$HX = CX = CF,$$

while from steps 2 and 4 we have

$$HK = KG = GE.$$

Thus,

$$\frac{HX}{HK} = \frac{CX}{KG} = \frac{CF}{GE}.$$

However, since $\triangle HCF$ and $\triangle HGE$ are isosceles with a common vertex angle at H, they are similar, and

$$\frac{CF}{GE} = \frac{HC}{HG}.$$

Therefore,

$$\frac{HX}{HK} = \frac{CX}{KG} = \frac{HC}{HG},$$

which shows that triangles HCX and HGK are similar by the **sss** similarity condition, so that $\angle HCX = \angle HGK$.

Hence $\angle HGK = \angle HCD$, and since $CD \parallel GK$, then $\angle HCX = \angle HCD$; that is, the points C, X, and D are collinear, which completes the proof.

□

The solutions to Example 13.4.7 and Example 13.4.8 prove:

Theorem 13.4.9. *(The Mohr-Mascheroni Construction Theorem)*

Any Euclidean construction, insofar as the given and required elements are points, may be completed with the compass alone.

13.5 Problems

1. Suppose that P and Q are inverse points with respect to a circle with center S, that $SP = m$, and that the radius of the circle of inversion is n. Find SQ.

2. Given that A and A' are two inverse points ($A \neq A'$) with respect to some circle ω, find:

   (a) the radius of ω and

   (b) the center O of ω.

3. Given points P and Q with $PQ = 8$, draw all circles ω of radius 3 such that P and Q are inverses with respect to ω.

4. In the figure below, the tangents from P to the circle $C(O, r)$ meet the circle at S and T. The point Q is the intersection of OP and ST. Prove that P and Q are inverses of each other with respect to the circle.

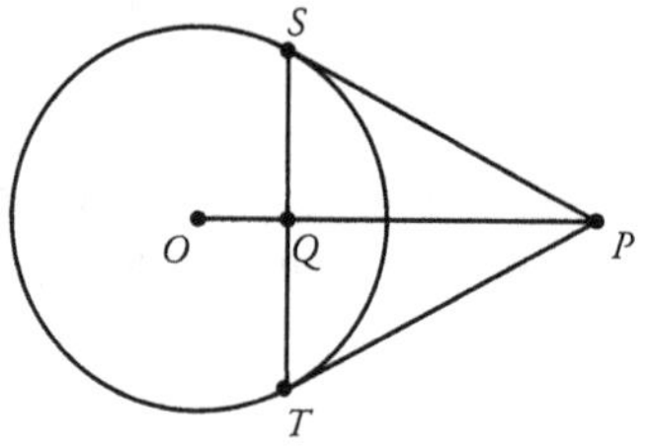

5. In the figure below, O is the center of the circle. The diameter ST is perpendicular to OP. PT intersects the circle at R, and SR intersects OP at Q. Prove that P and Q are inverses of each other with respect to the circle.

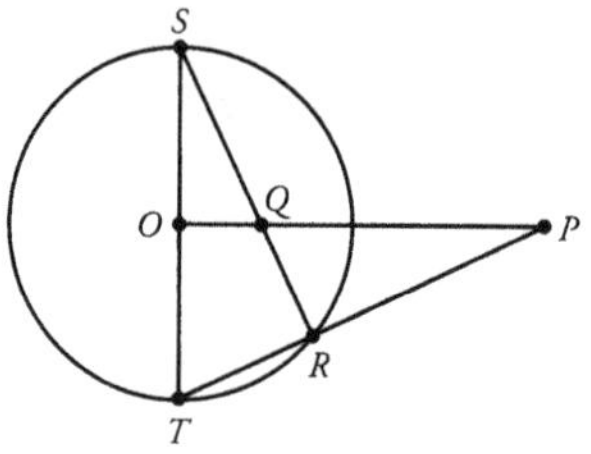

6. If the circle ξ passes through the center O of the circle ω, and if a diameter of ω meets the common chord of ω and ξ at P and meets the circle ξ at Q, show that P and Q are inverse points with respect to ω.

7. Draw the figure obtained by inverting a square with respect to its circumcircle.

8. What is the image under inversion $I(O, r^2)$ of the set of lines passing through P, where P is different from O and I? Include a sketch.

9. What is the inverse of a set of parallel lines?

10. Let P and P' be inverses under $I(O, r^2)$ with P outside the circle of inversion. Let B be the point where PP' meets the circle of inversion. Show that

$$BP = \frac{BP'}{1 - BP'/r}.$$

11. Let P and Q have inverses P' and Q', respectively, under $I(O, r^2)$, with O between P and Q. Show that

$$P'Q' = \frac{PQ}{OP \cdot OQ} r^2.$$

This is called the ***distortion teorem***.

12. If A and B are two distinct points inside some circle α, use inversion to show that there are exactly two circles through both A and B that are tangent to α.

13. A circle and an intersecting line (nontangential) can be inverses to each other in two different ways. Illustrate this by showing how to find two circles of inversion α and β such that the line and the given circle are inverses of each other.

14. If $ABCD$ is a convex noncyclic quadrilateral, show that

$$AC \cdot BD < AB \cdot CD + AD \cdot BC.$$

15. Given a triangle ABC with circumcenter O, let A', B', and C' be the images of the points A, B, and C under the inversion $I(O, r^2)$. Prove that

$$\triangle A'B'C' \sim \triangle ABC.$$

16. If a quadrilateral with sides of length a, b, c, and x is inscribed in a semicircle of diameter x, as shown, prove that

$$x^3 - (a^2 + b^2 + c^2)x - 2abc = 0.$$

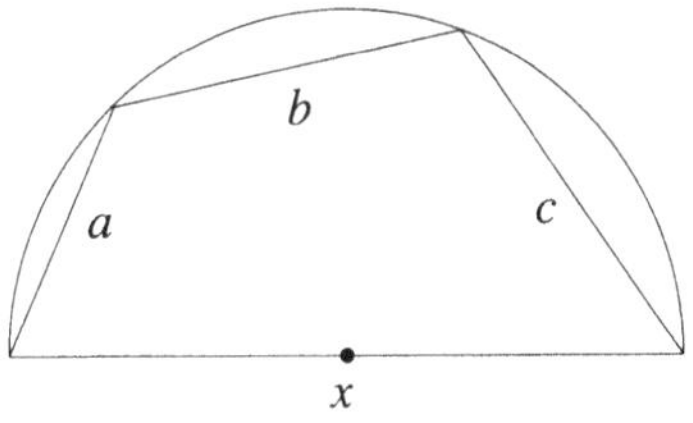

17. If PQ and RS are common tangents to two circles PAR and QAS, respectively, prove that the circles PAQ and RAS are tangent to each other.

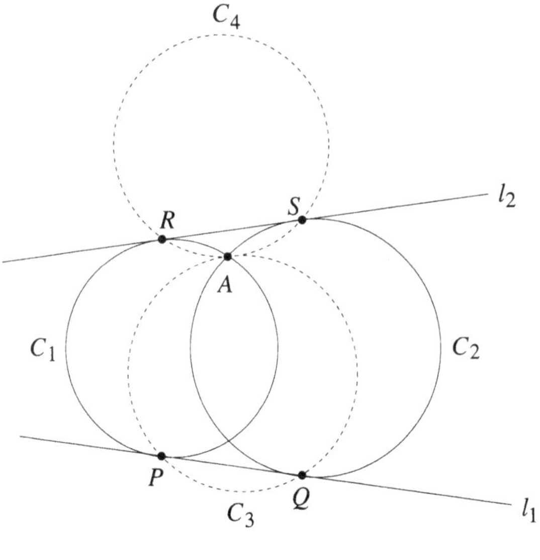

18. Given a circle ω with center O and a point A outside ω, construct the circle with center A orthogonal to ω.

19. Given a circle ω and two noninverse points P and Q inside ω, construct the circle through P and Q orthogonal to ω.

20. Two circles α and β intersect orthogonally at P. O is any point on a circle γ tangent to both former circles at Q and R. Prove that the circles ω and ξ through OPQ and OPR, respectively, intersect at an angle of $45°$.

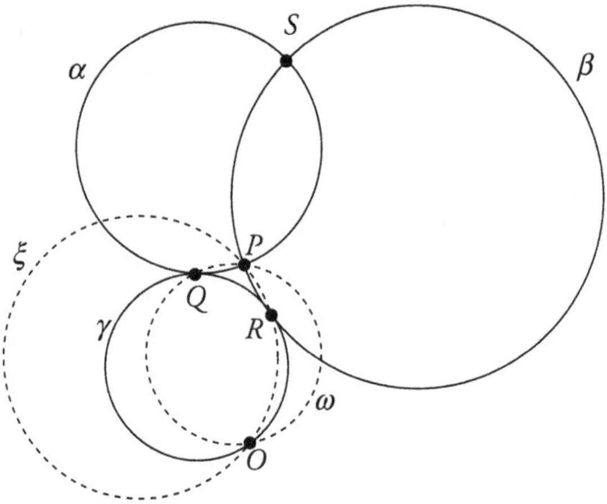

21. Construct (using a straightedge and compass) a circle orthogonal to a given circle having within it one-third of the circumference of the given circle.

22. Construct (using a straightedge and compass) a circle orthogonal to a given circle so that one-third of the circumference of the constructed circle lies within the given circle.

23. Let AC be a diameter of a given circle and chords AB and CD intersect (produced if necessary) in a point O. Prove that the circle OBD is orthogonal to the given circle.

24. Let ω and ξ be orthogonal circles intersecting at P and Q. Let AB be a straight line tangent to both circles at A and B. Show that one of $\angle APB$ and $\angle AQB$ is $45°$ and the other is $135°$.

25. In the Arbelos Theorem, show that the points of contact of K_i and K_{i+1}, $i = 0, 1, \ldots,$ lie on a circle.

CHAPTER 14

RECIPROCATION AND THE EXTENDED PLANE

14.1 Harmonic Conjugates

If A and B are two points on a line, any pair of points C and D on the line for which

$$\frac{AC}{CB} = \frac{AD}{DB}$$

is said to ***divide*** AB ***harmonically***. The points C and D are then said to be ***harmonic conjugates*** with respect to A and B.

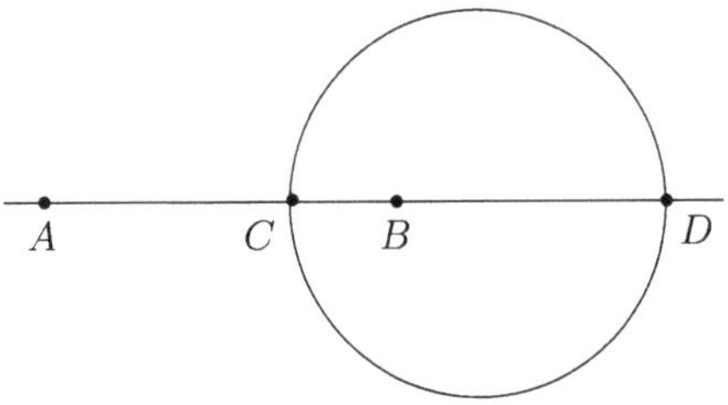

Classical Geometry: Euclidean, Transformational, Inversive and Projective, First Edition. By I. E. Leonard, J. E. Lewis, A. C. F. Liu, G. W. Tokarsky
Copyright © 2014 John Wiley & Sons, Inc.

Lemma 14.1.1. *Given ordinary points A and B, and given a positive integer k where $k \neq 1$, there are two ordinary points C and D such that*

$$\frac{AC}{CB} = \frac{AD}{DB} = k.$$

One of the points C and D is between A and B, while the other is exterior to the segment AB.

Proof. Choose a point C on the line AB such that

$$CB = \frac{AB}{1+k}.$$

Since $k > 0$, then $CB < AB$, and we may assume that C lies between A and B.

Now, we have

$$CB = \frac{AB}{1+k} = \frac{AC+CB}{1+k},$$

so that

$$CB + kCB = AC + CB;$$

that is,

$$\frac{AC}{CB} = k.$$

Now we find the point D, which will be exterior to the segment AB—beyond B if $k > 1$ and beyond A if $0 < k < 1$.

Assuming that $k > 1$, we set

$$k = \frac{AD}{DB} = \frac{AD}{AD - AB}$$

and solve for AD to get

$$AD = \frac{k}{k-1} \cdot AB.$$

Therefore, if k is a positive number such that $k > 1$ and C satisfies

$$\frac{AC}{CB} = k,$$

then there always exists a point $D \neq C$ such that

$$\frac{AD}{DB} = k.$$

We simply choose D such that

$$AD = \frac{k}{k-1} \cdot AB.$$

A similar argument will find the point D when $0 < k < 1$.

□

Note. The midpoint C of AB satisfies

$$\frac{AC}{CB} = 1,$$

and we will adopt the convention that

$$\frac{AI}{IB} = 1,$$

where I is the ideal point in the inversive plane.

Using this convention, given two ordinary points A and B, for every positive number k there are harmonic conjugates C and D with respect to A and B for which

$$\frac{AC}{CB} = \frac{AD}{DB} = k.$$

Recall that three positive numbers a, b, and c form a ***harmonic progression*** if and only if

$$\frac{1}{a}, \quad \frac{1}{b}, \quad \text{and} \quad \frac{1}{c}$$

form an arithmetic progression. A similar definition holds for an infinite sequence of positive numbers.

For example, the sequence

$$1, \quad \frac{1}{2}, \quad \frac{1}{3}, \quad \frac{1}{4}, \quad \ldots$$

forms a harmonic progression, since the sequence

$$1, \quad 2, \quad 3, \quad 4, \quad \ldots$$

forms an arithmetic progression.

Theorem 14.1.2. *Given four ordinary points A, B, C, and D, if AB is divided harmonically by C and D, then CD is divided harmonically by A and B.*

This terminology is explained by the following:

Theorem 14.1.3. *Suppose that P, Q, R, and S are consecutive ordinary points on a line and that Q and S divide PR harmonically. Then the sequence of distances PQ, PR, and PS forms a harmonic progression.*

Proof. The hypothesis says that

$$\frac{RQ}{QP} = \frac{RS}{SP}.$$

We want to show that

$$\frac{1}{PQ}, \quad \frac{1}{PR}, \quad \frac{1}{PS}$$

are in an arithmetic progression; that is, that

$$\frac{1}{PQ} - \frac{1}{PR} = \frac{1}{PR} - \frac{1}{PS}.$$

From the first equation, we have

$$\frac{RQ}{QP \cdot PR} = \frac{RS}{SP \cdot PR},$$

which implies that

$$\frac{PR - PQ}{PQ \cdot PR} = \frac{PS - PR}{PR \cdot PS},$$

which in turn implies that

$$\frac{1}{PQ} - \frac{1}{PR} = \frac{1}{PR} - \frac{1}{PS},$$

which is what we wanted to show.

□

The Circle of Apollonius

If we are given points A and B and a positive number $k \neq 1$, we can find precisely two ordinary points X on the line AB such that

$$\frac{AX}{XB} = k.$$

However, there are also points X not on the line AB for which

$$\frac{AX}{XB} = k,$$

and in fact, as we show in the following theorem, the set of all such points X lie on a circle.

Theorem 14.1.4. *(Circle of Apollonius)*

Given two ordinary points A and B, and a positive number $k \neq 1$, the set of all points X in the plane for which

$$\frac{AX}{XB} = k$$

forms a circle called the ***Circle of Apollonius for*** *A, B,* ***and*** *k.*

Proof. Let C and D be the two points on AB for which

$$\frac{AC}{CB} = \frac{AD}{DB} = k,$$

and let ξ be the circle with diameter CD.

We show first that every point X for which $AX/XB = k$ is on ξ.

Let X be a point such that $AX/XB = k$, and since $AC/CB = k$ and $AD/DB = k$, we know from the Angle Bisector Theorem that XC and XD are, respectively, internal and external bisectors of angle AXB. Referring to the figure below, we see that $\alpha + \beta = 90°$; that is, $\angle CXD$ is a right angle. Therefore, by the converse to Thales' Theorem, this means that X is on the circle ξ.

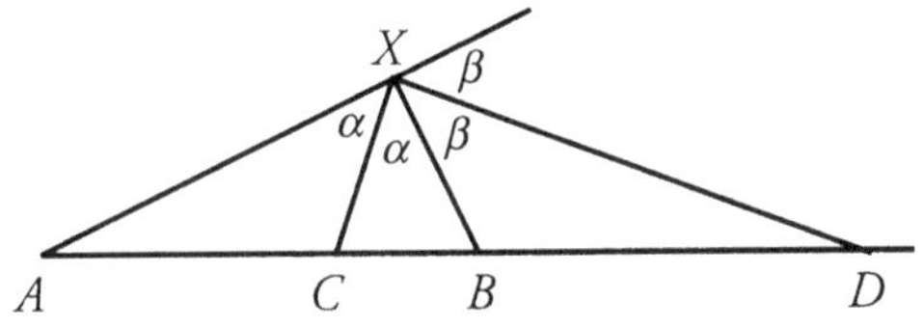

We show next that every point X on the circle ξ satisfies $AX/XB = k$.

Let X be a point on the circle ξ and draw $BP \parallel DX$ and $BQ \parallel CX$, as shown below.

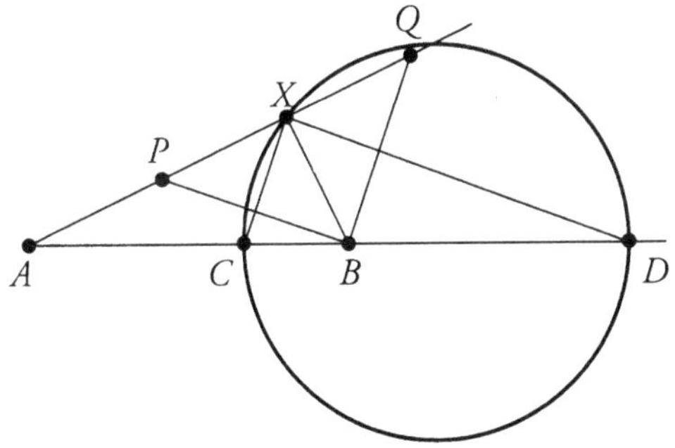

Since X is on the circle, then $\angle CXD = 90°$, and it follows that $\angle PBQ = 90°$.

Also, since

$$\triangle APB \sim \triangle AXD \quad \text{and} \quad \triangle AQB \sim \triangle AXC,$$

we have the following:

$$\frac{AX}{XP} = \frac{AD}{DB} \quad \text{and} \quad \frac{AX}{XQ} = \frac{AC}{CB}.$$

Since

$$\frac{AD}{DB} = \frac{AC}{CB} = k,$$

it follows that

$$\frac{AX}{XP} = \frac{AX}{XQ},$$

from which we get $XP = XQ$.

Now, $\angle PBQ$ is a right angle, and so the point B is on the circle centered at X with radius XP, by Thales' Theorem. Thus, $XB = XP$, so that

$$\frac{AX}{XB} = \frac{AX}{XP} = \frac{AD}{DB} = k.$$

□

Now we give some facts concerning the Circle of Apollonius.

Theorem 14.1.5. *Let O be the center and r the radius of the Circle of Apollonius for A, B, and k. Then:*

(1) O is on the line AB.

(2) The points A and B are to the same side of O.

(3) A and B are inverses with respect to the circle.

(4) If the circle meets AB at C and D, then C and D divide AB harmonically in the ratio k.

Proof. Statements (1) and (4) follow directly from Theorem 14.1.4.

(2) We may assume that the line AB is horizontal, that A is to the left of B, and that C is between A and B, but D is not.

Thus, D is located either to the left of A as in figure (a) below or to the right of B as in figure (b) below. We will show that statement (2) is true for case (a). The proof for case (b) is similar (see Problem 14.5 in this chapter).

(a)

(b)

For case (a), we have $CB < DB$, and since C and D are on the Circle of Apollonius, we also have

$$\frac{AC}{CB} = \frac{AD}{DB},$$

so that

$$\frac{AC}{CB} \cdot CB < \frac{AD}{DB} \cdot DB,$$

which implies that $AC < AD$. Thus, the midpoint O of CD is to the left of A and hence to the left of both A and B.

(3) Assuming that O is to the left of A, we have the following relationships, as in the figure below:

$$AC = r - OA, \qquad AD = r + OA, \qquad CB = OB - r, \qquad DB = OB + r.$$

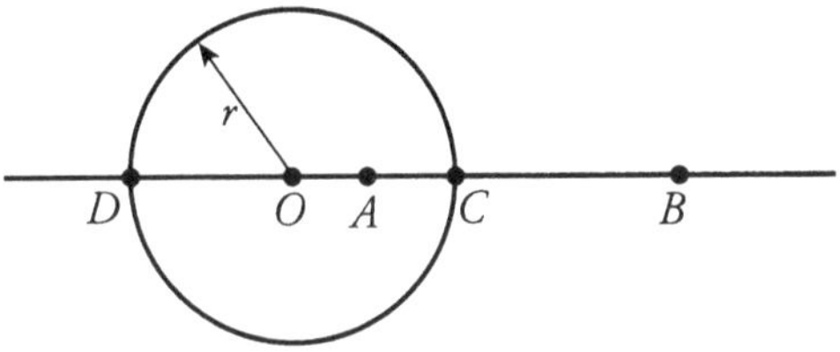

Since C and D are on the circle,

$$\frac{AC}{CB} = \frac{AD}{DB},$$

which implies that

$$\frac{r - OA}{OB - r} = \frac{r + OA}{OB + r},$$

and solving for r^2, we have $OA \cdot OB = r^2$.

□

Harmonic Conjugates and Inverses

Theorem 14.1.6. *A and B are harmonic conjugates with respect to C and D if and only if A and B are inverses with respect to the circle with diameter CD.*

Proof. Suppose that A and B are harmonic conjugates for CD. Then C and D are harmonic conjugates for AB; that is,

$$\frac{AC}{CB} = \frac{AD}{DB}.$$

Letting r be the radius of the circle ω with diameter CD, as in the figure below.

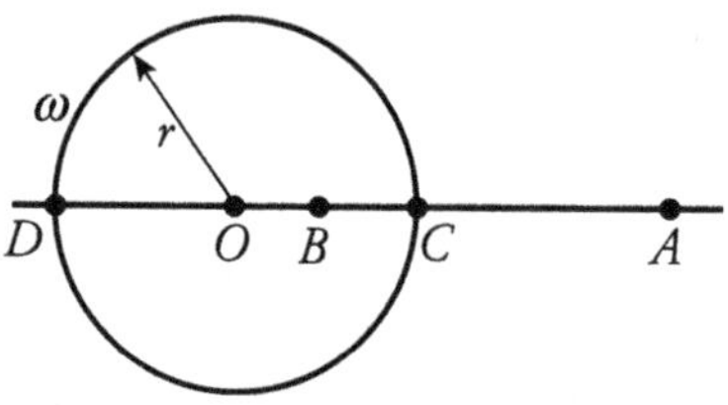

We want to show that $OA \cdot OB = r^2$, and the proof proceeds as in the proof of statement (3) of Theorem 14.1.5. We have

$$AC \cdot BD = AD \cdot BC,$$

that is,

$$(OA - r)(OB + r) = (OA + r)(r - OB),$$

which simplifies to $OA \cdot OB = r^2$.

Conversely, suppose that A and B are inverses with respect to the circle ω with diameter CD. Assuming that A is outside ω, as shown above, then to prove that A and B are harmonic conjugates for CD, it suffices to show that

$$\frac{CA/AD}{CB/BD} = 1.$$

Referring to the figure above, we have

$$\begin{aligned}\frac{CA/AD}{CB/BD} &= \frac{CA \cdot BD}{AD \cdot CB} = \frac{(OA - r) \cdot (OB + r)}{(OA + r) \cdot (r - OB)} \\ &= \frac{OA \cdot OB - r \cdot OB + r \cdot OA - r^2}{r \cdot OA + r^2 - OA \cdot OB - r \cdot OB},\end{aligned}$$

and since $OA \cdot OB = r^2$, we get

$$\frac{CA/AD}{CB/BD} = \frac{r \cdot OA - r \cdot OB}{r \cdot OA - r \cdot OB} = 1,$$

which completes the proof.

□

The relationship between harmonic conjugates and inverses allows us to show how a straightedge alone can be used to find the inverse of a point P that is outside the circle of inversion.

Example 14.1.7. *Given a point P outside a circle ω with center O, construct the inverse of P using only a straightedge.*

Solution. The analysis figure is shown below.

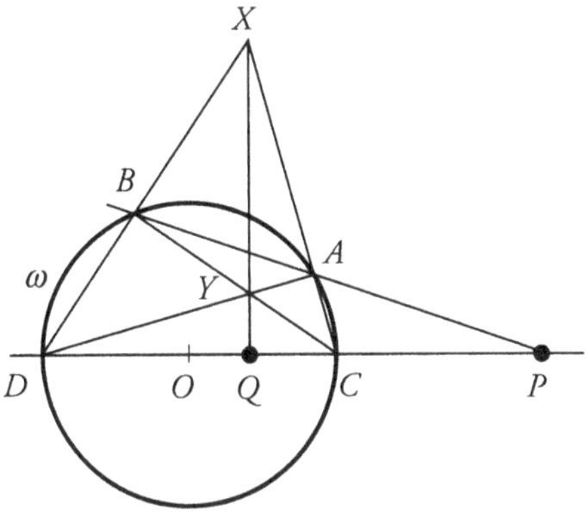

Construction:

(1) Draw the line OP intersecting ω at C and D.

(2) Draw a second line through P intersecting ω at A and B, as shown.

(3) Draw AC and BD intersecting at X.

(4) Draw AD and BC intersecting at Y.

(5) Draw the line through X and Y intersecting OP at Q.

Then Q is the inverse of P.

Justification:

(a) Apply Ceva's Theorem to $\triangle XCD$ and cevians XQ, CB, and DA. The cevians are concurrent at Y so that

$$\frac{XA}{AC} \cdot \frac{CQ}{QD} \cdot \frac{DB}{BX} = 1.$$

(b) Apply Menelaus' Theorem to $\triangle XCD$ with menelaus points P, A, and B. The points P, A, and B are collinear so that

$$\frac{XA}{AC} \cdot \frac{CP}{PD} \cdot \frac{DB}{BX} = 1.$$

(c) From (a) and (b), we get

$$\frac{CQ}{QD} = \frac{CP}{PD}.$$

(d) Thus, from (c), we see that P and Q are harmonic conjugates with respect to CD.

By Theorem 14.1.6, this means that P and Q are inverses with respect to ω.

□

Inversion and the Circle of Apollonius

We state here several theorems that are easy consequences of the results in the preceding sections.

Theorem 14.1.8. *If ω is the Circle of Apollonius for A, B, and k, then A and B are inverses with respect to ω.*

Theorem 14.1.9. *The Apollonian circle for A, B, and k is the same as the Apollonian circle for B, A, and $\frac{1}{k}$.*

Remark. Note the change in the order of the points A and B in the previous theorem.

Theorem 14.1.10. *If A and B are inverse points for a circle ω, then ω is the Circle of Apollonius for A, B, and some positive number k.*

Theorem 14.1.11. *If α and β are orthogonal circles, then whenever either circle intersects a diameter of the other, it divides that diameter harmonically.*

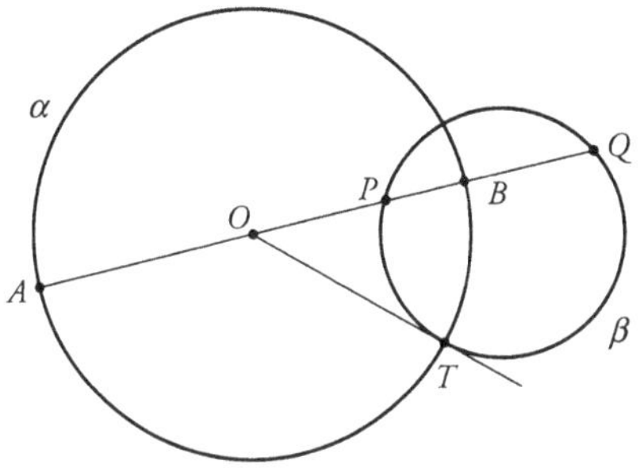

Proof. Referring to the figure, we know that if β cuts the diameter AB of α at P and Q, then P and Q are inverse points for α, since β is its own inverse by Theorem 13.3.2.

Thus, P and Q are harmonic conjugates with respect to A and B by Theorem 14.1.6.

□

The following is the converse of the previous theorem.

Theorem 14.1.12. *If α and β are two circles and β divides a diameter of α harmonically, then the two circles are orthogonal.*

14.2 The Projective Plane and Reciprocation

We augment the Euclidean plane with infinitely many ***ideal points***, or ***points at infinity***, in such a way that:

1. All lines parallel to a given line (including the given line) pass through the same ideal point.

2. Lines that are not parallel do not pass through the same ideal point.

Collectively, the ideal points form the ***ideal line*** or ***line at infinity***.

The resulting structure is called the ***extended plane*** or the ***projective plane***.

Nonideal points and nonideal lines are called ***ordinary points*** and ***ordinary lines***, respectively. The words "point" and "line" by themselves may refer to either an ordinary or ideal point and line.

Unlike the situation in inversive geometry, we can illustrate ideal points by using arrows or vectors to indicate the ideal point. Parallel vectors indicate the same ideal point. In the figure below, all three arrows indicate the same ideal point, and any line parallel to these vectors passes through that ideal point.

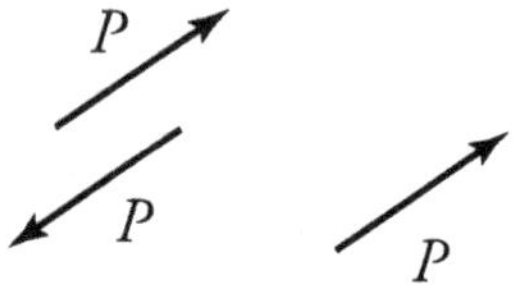

Immediate consequences of the definitions in the projective plane are as follows:

1. Every ordinary line contains exactly one ideal point.
2. Every two lines meet at exactly one point.
3. Every two points determine a unique line.

Reciprocation

Definition. Given a circle ω centered at an ordinary point O and given an ordinary point $P \neq O$, the ***polar*** of P is the line p that is perpendicular to OP passing through the inverse P' of P.

The circle ω is called the ***circle of reciprocation***. The center O is called the ***center of reciprocation***.

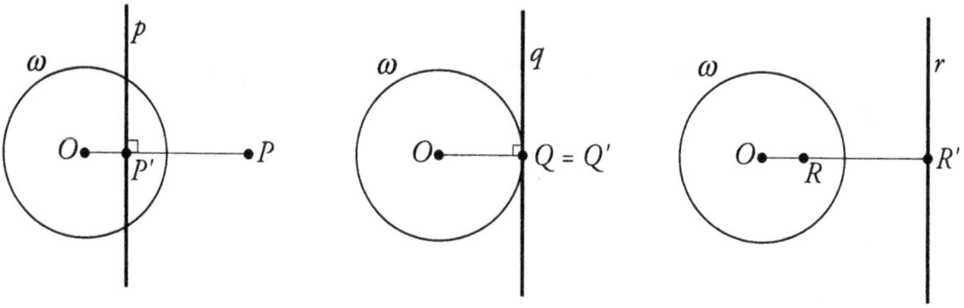

Note. We use the convention that uppercase letters denote points and corresponding lowercase letters denote the polars of those points.

- The polar of O is defined to be the ideal line.

- If P is an ideal point, then its polar is a line through O perpendicular to OP, that is, perpendicular to the arrow that points to P, as in the figure below.

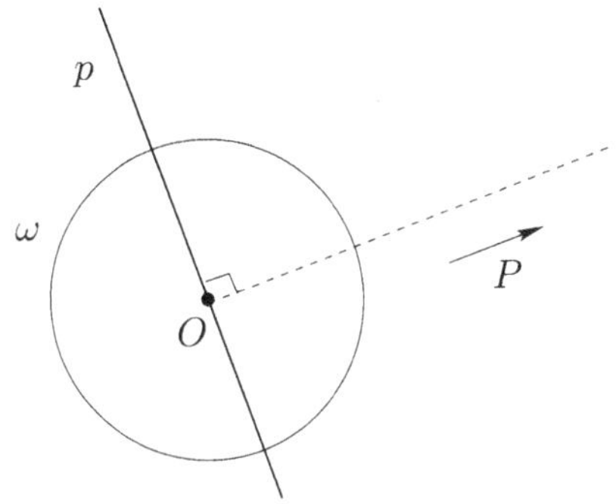

Definition. If m is a line, then the ***pole*** of m is the point M such that m is the polar of M, as in the figure below.

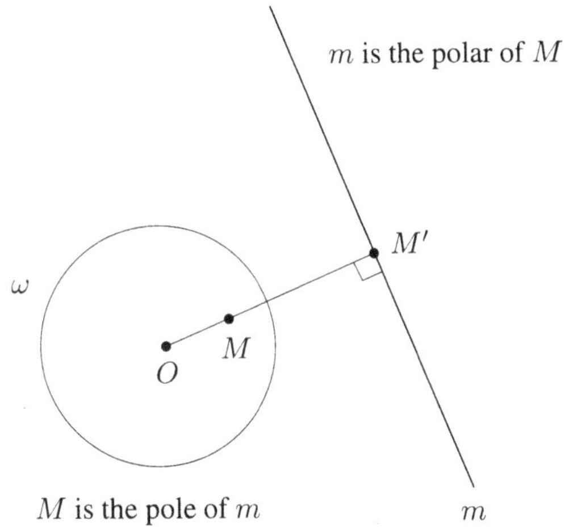

- The pole of the ideal line is the center O of the circle ω.

- If m is a line through the center O of the circle ω, then the pole of m is the ideal point M on any line perpendicular to m, as in the figure on the following page.

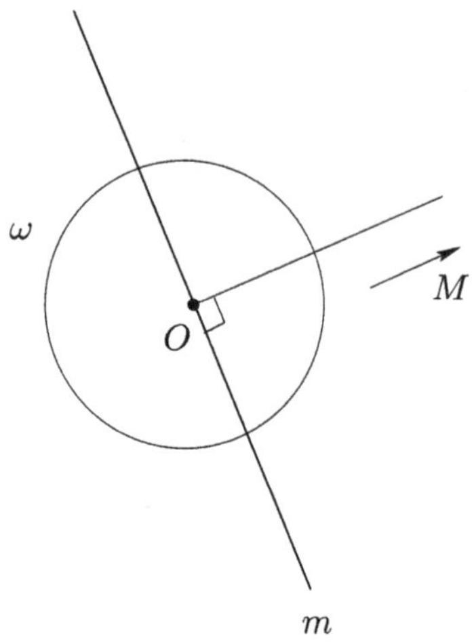

The most useful theorem about poles and polars follows:

Theorem 14.2.1. *(Reciprocation Theorem)*

P is on q if and only if Q is on p.

Proof. We will show that if P is on q, then Q is on p, and the converse can then be obtained by interchanging P and Q. We consider three cases.

Case (i). P is an ordinary point and q is an ordinary line.

Suppose that P is on q and let Q' be the inverse of Q.

On OP, let X be the foot of the perpendicular from Q, as in the figure below. Then from the **AA** similarity condition, we have

$$\triangle OXQ \sim \triangle OQ'P.$$

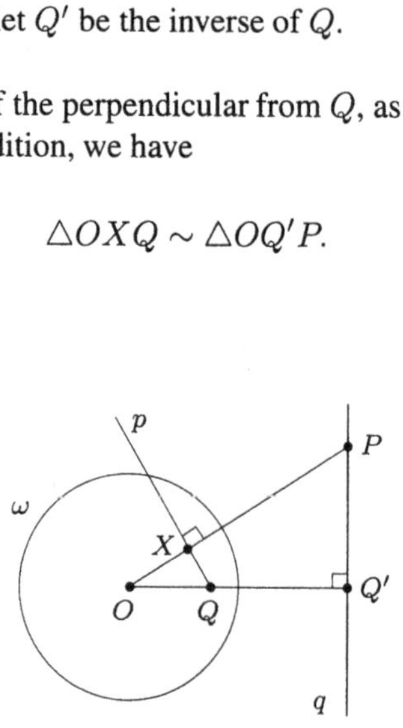

Therefore,

$$\frac{OX}{OQ} = \frac{OQ'}{OP},$$

which implies that

$$OX \cdot OP = OQ \cdot OQ' = r^2.$$

Thus, X is the inverse of P.

The definition of p now shows that p is the line QX, which means that if P is on q, then Q is on p.

Case (ii). P is an ideal point and q is the ideal line.

Since q is the ideal line, Q is the center of the reciprocating circle; that is, $Q = O$. Therefore, p is the line through O perpendicular to any line pointing to P, as in the figure below. Thus, Q is on p.

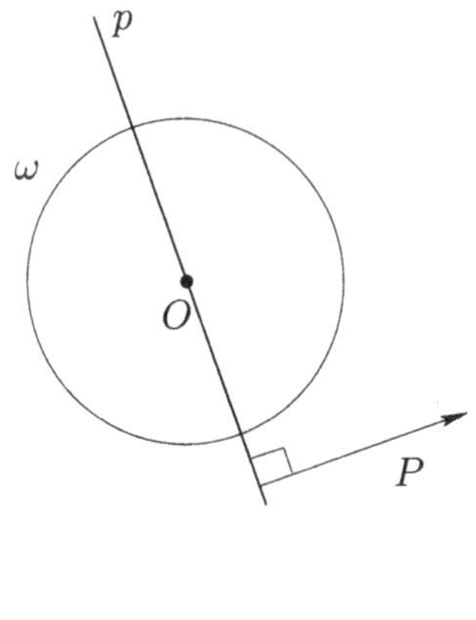

Case (iii). P is an ideal point and q is an ordinary line passing through P.

Let Q be the pole of q. Since q is an ordinary line, we can draw the line OQ so that OQ is perpendicular to q, as in the figure on the following page.

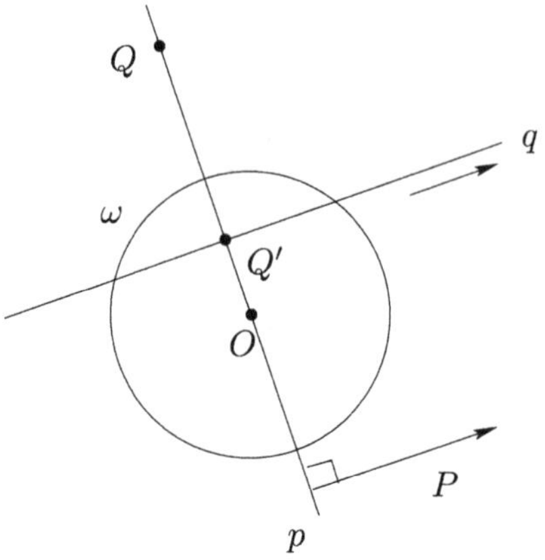

Thus, the line OQ is the polar of P; that is, $p = \overleftrightarrow{OQ}$ and Q is on p.

We leave as an exercise the situation where $P = O$, the center of the reciprocating circle.

□

Theorem 14.2.1 can be stated in different ways:

- P is on the polar of Q if and only if Q is on the polar of P.
- p is on the pole of q if and only if q is on the pole of p.

Definition. A ***range*** of points is a set of collinear points, and a ***pencil*** of lines is a set of concurrent lines.

An immediate consequence of Theorem 14.2.1 is the following:

Corollary 14.2.2. *The polars of a range of points on a line m are a pencil of lines concurrent at M, and vice versa; that is, the poles of a pencil of lines are a range of points.*

Example 14.2.3. *Given a circle ω with center O, suppose that P, Q, and R are a range of points that lie on the line m, and let M be the pole of m. Show that the polars p, q, and r are concurrent at M.*

Solution. We have the following:

1. P is on m, so that M is on p.
2. Q is on m, so that M is on q.
3. R is on m, so that M is on r.

Therefore, the polars p, q, and r of P, Q, and R, respectively, are concurrent at the point M; that is, polars p, q, and r form a pencil of lines through M, as in the figure below.

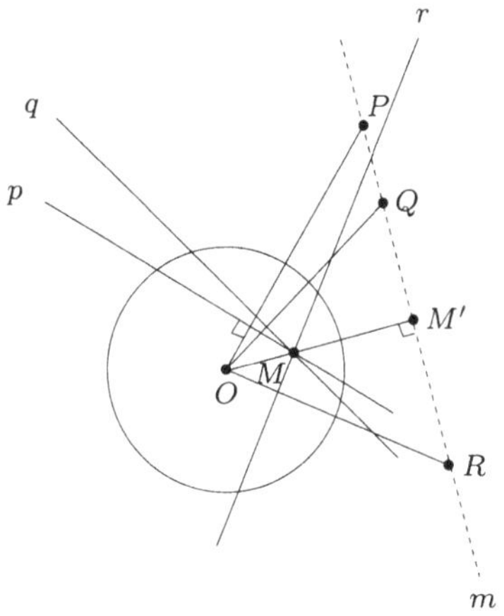

In the extreme situation where all the points are ideal points, we have a range of points on the ideal line, and their polars form a pencil of lines through the center of the circle of reciprocation, as in the figure below, where we are given ideal points P, Q, R, and S.

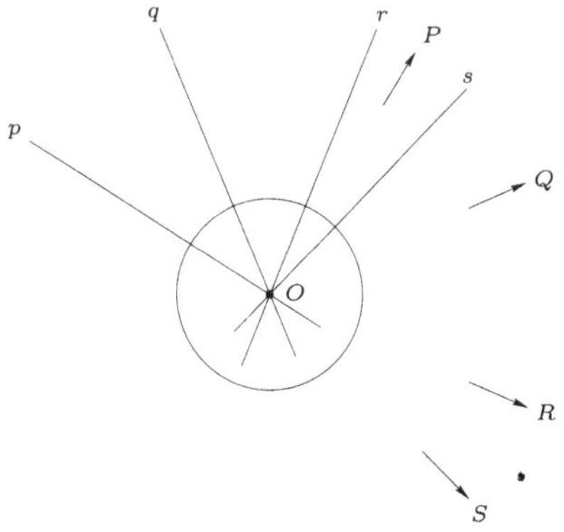

□

In the following, $\overleftrightarrow{AB}$ denotes the line through A and B.

Theorem 14.2.4. *Let ω be the circle of reciprocation. Then:*

(1) A is outside ω if and only if a cuts ω.

(2) A is on ω if and only if a is tangent to ω.

(3) A is inside ω if and only if a misses ω.

(4) The pole of $\overleftrightarrow{AB}$ is $a \cap b$.

(5) The polar of $a \cap b$ is $\overleftrightarrow{AB}$.

Proof. We will prove statements (4) and (5) and leave the others as exercises.

(4) Let $m = \overleftrightarrow{AB}$ and let the pole of m be M.

Since A is on m, then M is on a, and since B is on m, then M is on b. Therefore, M is on $a \cap b$; that is, $M = a \cap b$.

(5) We have $P = a \cap b$ if and only if P is on a and P is on b, and this is true if and only if A is on p and B is on p; that is, if and only if $p = \overleftrightarrow{AB}$.

□

Duality

Given a figure $\mathcal{F}$ that consists of lines (entire lines, not just segments) and points (which may or may not be on the lines), then the ***polar*** or ***dual*** of $\mathcal{F}$ is the figure that is obtained by taking the poles of the lines of $\mathcal{F}$ and the polars of the points of $\mathcal{F}$.

Example 14.2.5. *Draw the dual of the figure on the following page.*

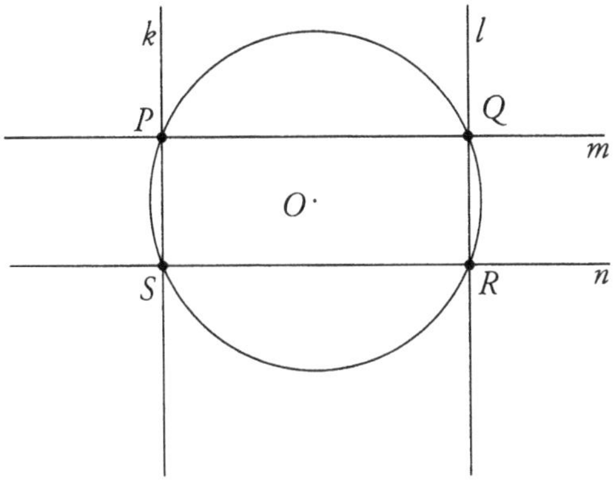

Solution. The dual is shown below.

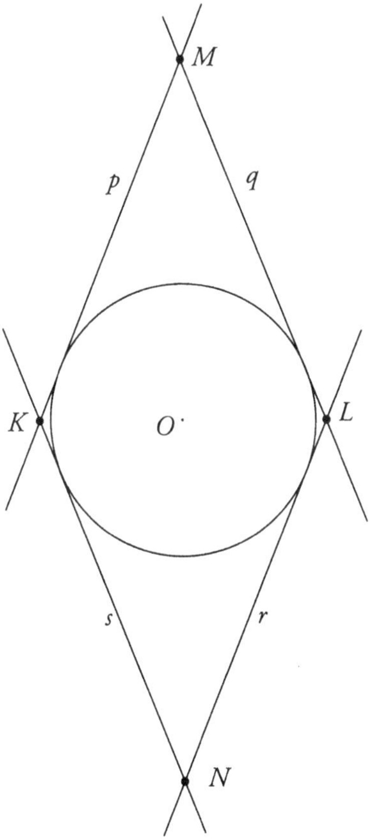

□

The following is a translation table for obtaining the dual of a figure or the dual of a statement.

- To obtain the dual, any word or phrase that appears in one column must be replaced by the corresponding word or phrase in the other column.
- The symbol ω is the circle of reciprocation.

point	line
lie on	pass through
ω	ω
concurrent	collinear
pole	polar
locus	envelope
point on a curve	line tangent to a curve
inscribed in ω	circumscribed about ω
hexagon	hexagon

The translation table is a direct consequence of Theorem 14.2.1.

Theorem 14.2.6. *(Principle of Duality)*

If a statement that involves only points, lines, and their incidence properties is true, then the dual statement is automatically true.

We will illustrate the use of the translation table by using it to obtain the dual of Pascal's Mystic Hexagon Theorem. The dual theorem is called Brianchon's Theorem, and it was discovered by Brianchon by taking the dual, as illustrated on the following page.

In the left column, we state Pascal's Theorem using only the incidence properties of points and lines. The right column is the translation obtained by using the table above.

Pascal's Theorem	**Brianchon's Theorem**
If A, B, C, D, E, and F	If a, b, c, d, c, and f
are the vertices of a hexagon	are the edges of a hexagon
inscribed in ω	circumscribed about ω
then the points $\overleftrightarrow{AB} \cap \overleftrightarrow{DE}$,	then the lines through $a \cap b$ and $d \cap e$,
$\overleftrightarrow{BC} \cap \overleftrightarrow{EF}$,	through $b \cap c$ and $e \cap f$,
and $\overleftrightarrow{CD} \cap \overleftrightarrow{FA}$	and through $c \cap d$ and $f \cap a$
are collinear.	are concurrent.

The points

$$a \cap b, \quad b \cap c, \quad c \cap d, \quad d \cap e, \quad e \cap f, \quad f \cap a$$

are the vertices of a hexagon, so in more familiar language, Brianchon's Theorem says:

Theorem 14.2.7. *(Brianchon's Theorem)*

If a hexagon is circumscribed about a circle, then the lines joining the opposite vertices are concurrent.

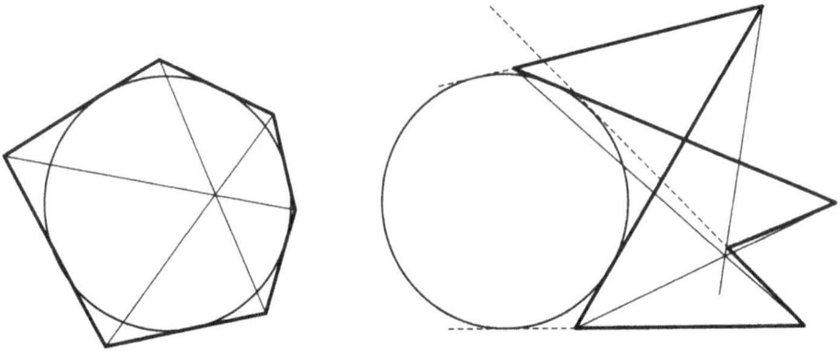

It is worth mentioning that in Brianchon's Theorem the hexagon is considered to circumscribe the circle if each edge, possibly extended, is tangent to the circle. In the figure above, the "hexagons" are shown as bold and the lines joining the opposite vertices are lighter.

If you try the same exercise to dualize Desargues' Theorem, you will find that you get nothing new—Desargues' Theorem is self-dual.

14.3 Conjugate Points and Lines

Let ω be a fixed circle with center O.

Two points A and B are said to be ***conjugate points*** with respect to ω if each lies on the polar of the other (that is, if A lies on b and B lies on a).

Two lines a and b are ***conjugate lines*** with respect to ω if each passes through the pole of the other.

A point or line which is conjugate to itself is said to be ***self-conjugate***.

Examples

The following properties concerning conjugate points and conjugate lines with respect to a circle ω are illustrated in the figure below and are easily proven.

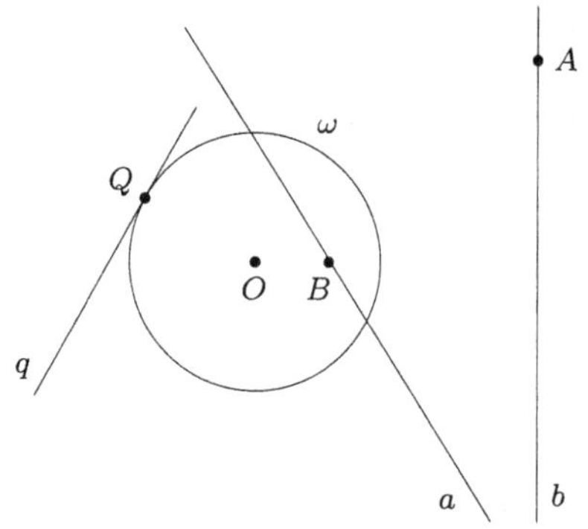

A and *B* are conjugate points

a and *b* are conjugate lines

Q and *q* are self-conjugate

1. A and B are conjugate points if and only if a and b are conjugate lines.
2. B is conjugate to A if and only if B lies on a.
3. b is conjugate to a if and only if b passes through A.
4. The set of lines conjugate to a is the pencil of lines through A.
5. The set of points conjugate to B is the range of points on b.
6. The following are equivalent:
   (a) A is self-conjugate.
   (b) A is on ω.
   (c) a is tangent to ω.

Example 14.3.1. *Each point on a line has a conjugate point on that line.*

Solution. Let A be on l and let $B = a \cap l$. Note that if l and a are parallel, then B is an ideal point.

If l and a are not parallel, then B is on a and, by the basic reciprocation theorem, A is on b.

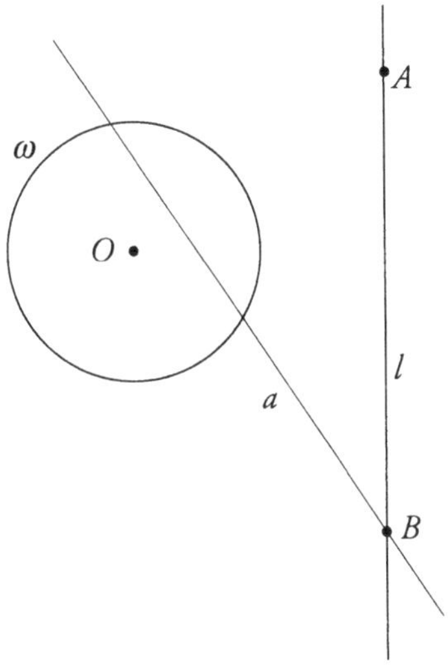

□

Example 14.3.2. *Each line through a point A has a conjugate line through A.*

Solution. This is the dual of the previous example.

A direct proof is as follows:

Let b be a line through A. Then $\overleftrightarrow{BA}$ is a line conjugate to b.

Recall that the pencil of lines at B is the set of all lines conjugate to b.

□

Example 14.3.3. *Of two distinct conjugate points on a line that cuts the circle of reciprocation, one point is inside or on the circle and the other point is outside the circle.*

Solution. Let ω be the circle of reciprocation and suppose that A and B are conjugate points on the line l that cuts ω.

If A is inside ω, then a misses ω, and since B is on a, B must be outside ω.

If A is on ω, then a is tangent to ω at A, and since B is on a and is different from A, then B must be outside ω.

If A is outside ω, then a cuts ω. Suppose for a contradiction that B is also outside ω, and let L be the pole of l. Since we are given that l cuts ω, L is outside ω. The situation is as shown in either of the two diagrams below.

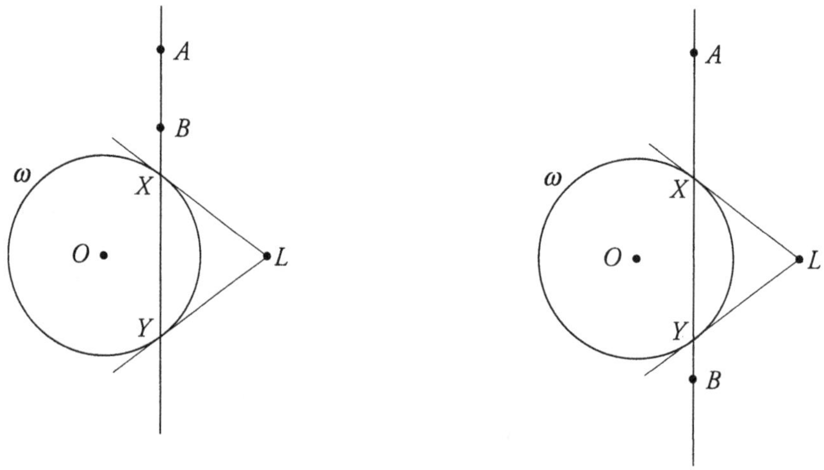

Now, since A is on l, L is on a, and since B is conjugate to a, B is on a. Together these imply that $a = \overleftrightarrow{LB}$.

Let X and Y be the points of tangency from L to ω. Then B is outside the segment XY and LB misses ω. However, this contradicts the fact that a cuts ω. Thus, we must conclude that B is on or inside ω. □

The following is the dual of the previous example and we leave the direct proof as an exercise.

Example 14.3.4. *Of two distinct conjugate lines that intersect outside a circle ω, one cuts the circle or is tangent to it, and the other misses the circle.*

Self-Polar Triangles

A triangle is ***self-polar*** if each vertex is the pole of the opposite side. Here, the sides are considered as lines.

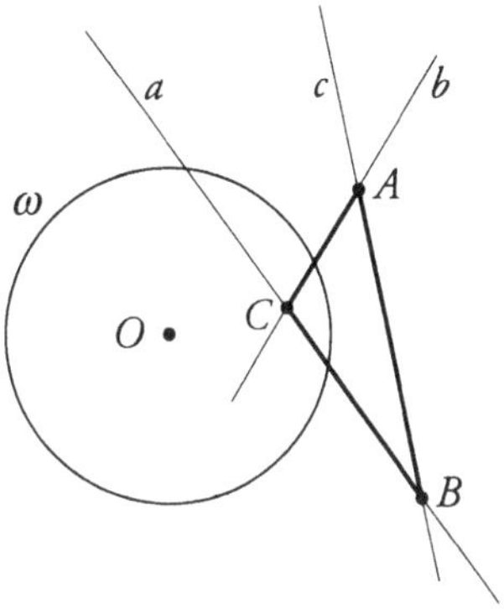

a, b, c are polars of *A, B, C*, respectively

Remark. For a self-polar triangle:

- Any two vertices are conjugate points.
- Any two sides are conjugate lines.
- Given any two conjugate points, $A \neq B$, they are the vertices of some self-polar triangle, and the third vertex is $C = a \cap b$; that is, $c = \overleftrightarrow{AB}$.

Theorem 14.3.5. *Every nondegenerate self-polar triangle is obtuse, with the obtuse angle inside the circle of reciprocation* ω.

Proof. Let ABC be self-polar. Then exactly one vertex must be inside ω.

Here are the reasons:

(1) Suppose one vertex (say, A) is inside ω. Then a misses ω, but both other vertices are on a, so B and C are outside ω. This shows that there is at most one vertex inside ω.

(2) It is impossible for A to be on ω, because B and C would have to be on a, in which case all three of A, B, and C would be on a; that is, ABC would be degenerate.

(3) Suppose A, B, and C are all outside ω. Then a cuts ω, and B and C are on a. Therefore, by Example 14.3.3, one of B or C is inside ω and the other is outside.

This proves that exactly one vertex is inside ω. Supposing that A is inside ω, it remains to show that $\angle BAC$ is obtuse.

In the figure below, A is on b. Join OB. Then b is the line through A perpendicular to OB. Note that C is on b and B is on c, since C and B are conjugates.

A is on c. Join OC. Then c is the line through A perpendicular to OC.

Referring to the figure, for angles α, β, and γ, we have

$$\beta = 180 - \alpha = 180 - (90 - \gamma) = 90 + \gamma,$$

showing that β is obtuse.

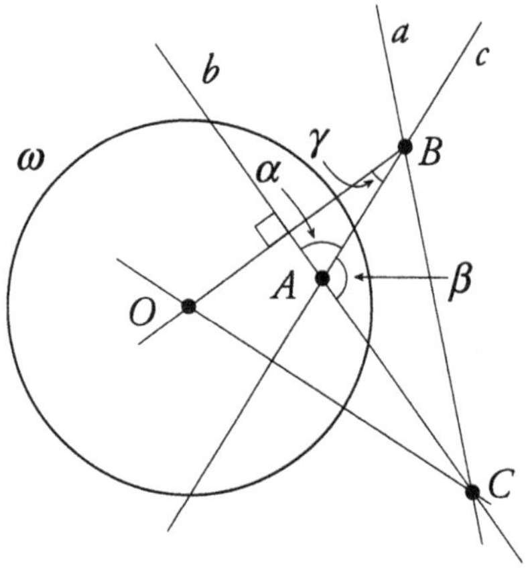

□

Theorem 14.3.6. *Every obtuse triangle ABC is self-polar with respect to a unique circle ω, which is called the* **polar circle** *for the triangle.*

Proof. In the figure on the following page, let O be the orthocenter of $\triangle ABC$.

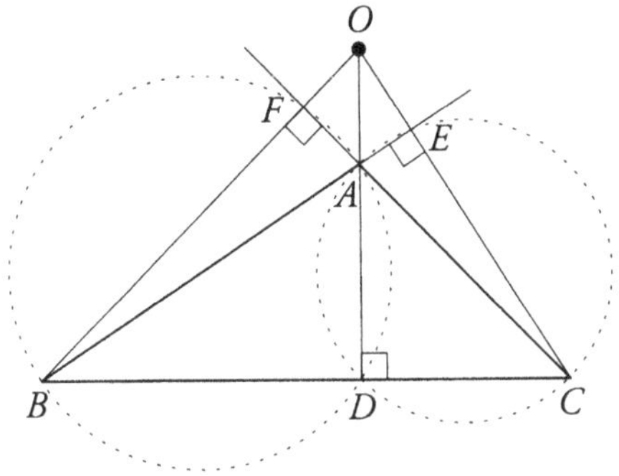

Referring to the figure above,

$$\angle AEC = 90^\circ = \angle ADC.$$

Therefore, $\square AECD$ is cyclic, and by the power of the point O with respect to the circumcircle of $\square AECD$, we have

$$OA \cdot OD = OC \cdot OE\,.$$

Similarly, $\square ADBF$ is cyclic, and by the power of the point O with respect to the circumcircle of $\square ADBF$, we have

$$OA \cdot OD = OF \cdot OB\,.$$

Let $OA \cdot OD = k^2$. Then

$$OC \cdot OE = k^2 \qquad \text{and} \qquad OF \cdot OB = k^2.$$

Now, let ω be the circle with center O and radius k. Then ω is the polar circle for triangle ABC.

Note that this works because:

- $OA \cdot OD = k^2$, which means that $D = A'$, so $BC = a$ (with respect to ω),
- $OC \cdot OE = k^2$, which means that $E = C'$, so $AB = c$, and
- $OF \cdot OB = k^2$, which means that $F = B'$, so $AC = b$.

Thus, each vertex is the pole of the opposite side, and the obtuse triangle ABC is self-polar with respect to ω.

□

Example 14.3.7. *Show that given an obtuse triangle, the circumcircle and the 9-point circle invert into each other with respect to the polar circle.*

Solution. Let ABC be an obtuse triangle with the obtuse angle at vertex B. From the previous theorem, $\triangle ABC$ is self-polar with respect to the polar circle, and the vertices A, B, and C invert into the points A', B', and C', respectively, as in the figure below.

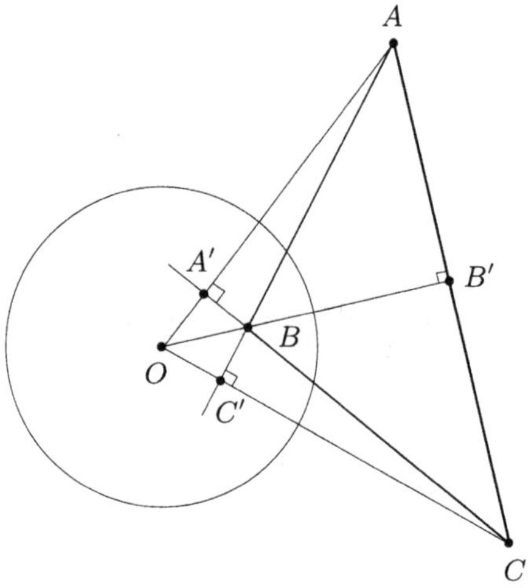

The circle through A, B, and C is the circumcircle of $\triangle ABC$. Since A', B', and C' are the feet of the altitudes, the inverse of $\triangle ABC$ with respect to the polar circle is just the 9-point circle of $\triangle ABC$.

□

14.4 Conics

Duality Revisited

A ***complete quadrangle*** consists of four points, A, B, C, and D, no three of which are collinear, and the six joins or lines determined by these points.

A ***complete quadrilateral*** consists of four lines, a, b, c, and d, no three of which are concurrent, and the six points determined by these lines.

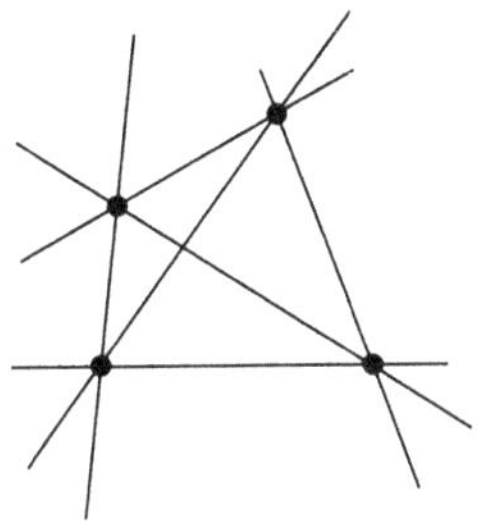

a complete quadrangle

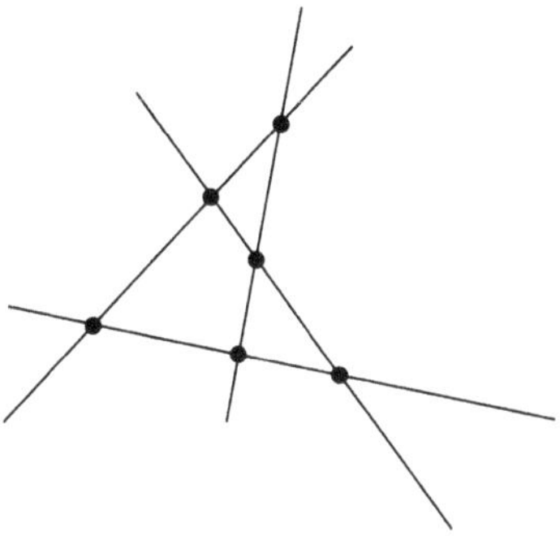

a complete quadrilateral

Exercise 14.4.1. *Show that the reciprocal of a complete quadrangle is a complete quadrilateral, and vice-versa.*

We can think of an n-gon as being composed of n points and the successive lines between these points, or, in the *opposite aspect*, as n lines and the successive points of intersections of these lines. In general, the dual or reciprocal of an n-gon is another n-gon of the opposite aspect.

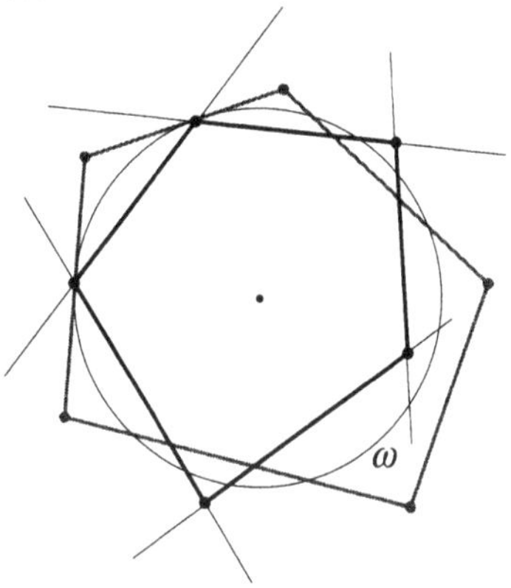

Reciprocals of Circles

The problem we are concerned with here is this: what is the image of a circle under reciprocation?

We can view a circle as a locus of points or as an envelope of lines, as in the figures on the following page, and in fact, every smooth curve can be viewed in these two aspects.

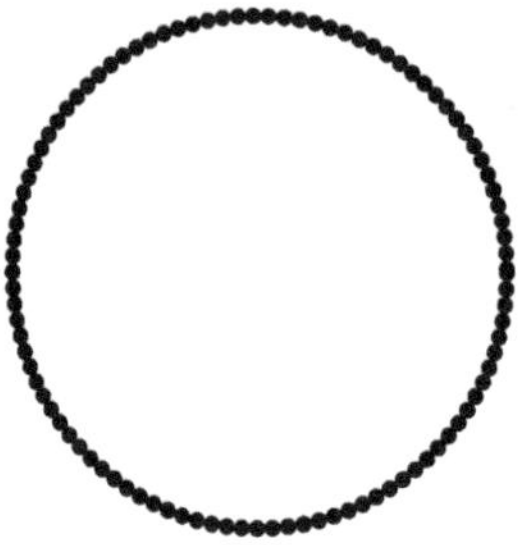

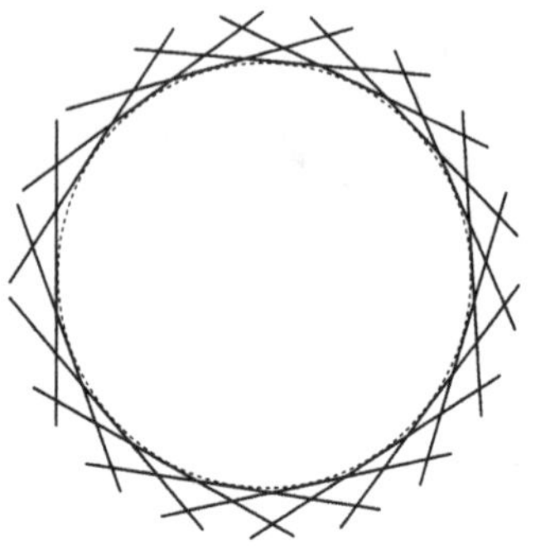

To find the reciprocal of the circle, view the circle in one aspect and see what curve is generated by the reciprocal aspect. We will use ω throughout to denote the reciprocating circle.

Special Cases

The reciprocal of the reciprocating circle ω is ω itself.

If the circle α is concentric with ω, and if the radius of α is s and the radius of ω is r, the reciprocal of α is another circle β concentric with ω with radius r^2/s.

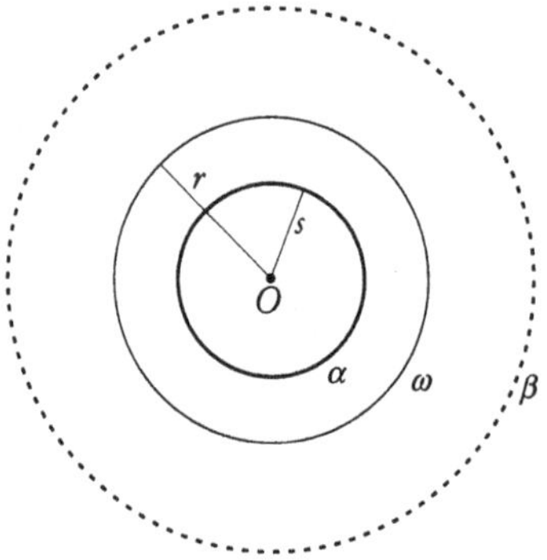

Theorem 14.4.2. *Let α and ω be two nonconcentric circles with centers A and O, respectively. The reciprocal of α with respect to ω is*

(1) an ellipse, if O is inside α;
(2) a parabola, if O is on α;
(3) a hyperbola, if O is outside α.

In each case, the focus of the conic section is O and the directrix is the polar of A. If the radius of α is s, the eccentricity ϵ of the conic is given by

$$\epsilon = OA/s\,.$$

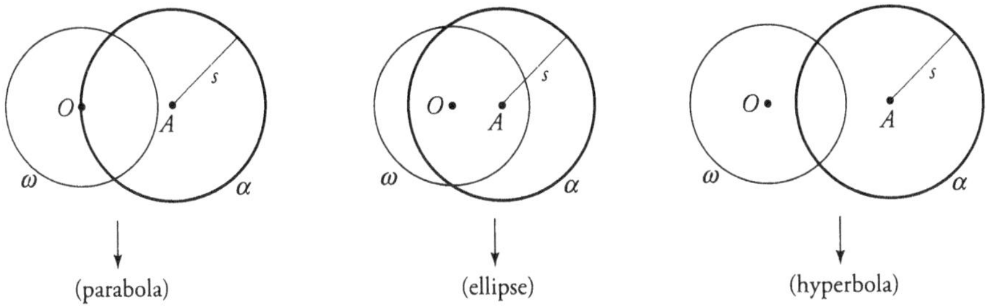

Proof. We will prove case (2). The proofs of (1) and (3) are similar.

Recall that a parabola with focus O and directrix a is the set of all points P such that $\text{dist}(P, a) = PO$. (See the next two sections for the connection between the focus-directrix definition and the more common Cartesian definition.)

As in the figure below, let p be tangent to α, let P be its pole, and let P' be the inverse of P.

Let $M = p \cap \overleftrightarrow{OA}$, let M' be the inverse of M, and let m be the polar of M. Note that P is on m since M is on p.

We will show that $PK/PO = 1$.

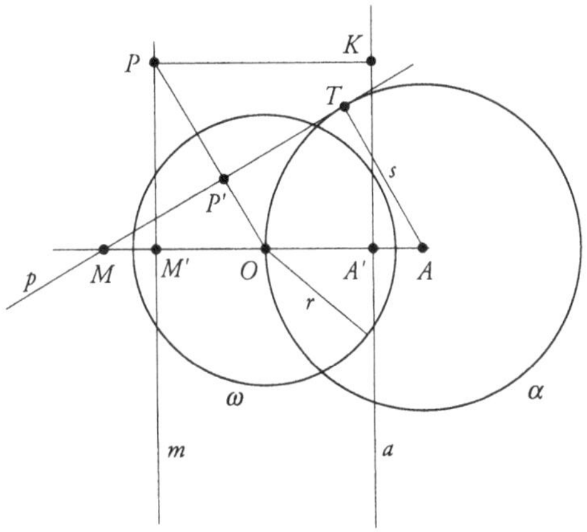

We have

$$\begin{aligned}\frac{PK}{PO} &= \frac{MO' + OA'}{PO} = \frac{r}{PO}\left(\frac{M'O}{r} + \frac{OA'}{r}\right)\\ &= \frac{P'O}{r}\left(\frac{r}{MO} + \frac{r}{OA}\right) = P'O\left(\frac{1}{MO} + \frac{1}{OA}\right).\end{aligned}$$

That is,

$$\frac{PK}{PO} = P'O\left(\frac{MO + OA}{MO \cdot OA}\right) = \frac{P'O}{MO} \cdot \frac{MA}{AO},$$

and since $\triangle MP'O \sim \triangle MTA$, we have

$$\begin{aligned}\frac{PK}{PO} &= \frac{TA}{MA} \cdot \frac{MA}{AO} \\ &= \frac{TA}{AO} = \frac{s}{s} = 1.\end{aligned}$$

□

Focus-Directrix Definition of a Conic

Let d be a fixed line, let F be a fixed point not on the line, and let $\text{dist}(X, d)$ denote the perpendicular distance from the point X to the line d. If ϵ is a fixed positive constant, then the set of all points X for which

$$\frac{FX}{\text{dist}(X, d)} = \epsilon$$

is a ***conic section***. The point F is the ***focus*** of the conic and the line d is the ***directrix*** of the conic. The positive constant ϵ is called the eccentricity of the conic, and:

- If $\epsilon = 1$, the conic is a parabola.
- If $0 < \epsilon < 1$, the conic is an ellipse.
- If $\epsilon > 1$, the conic is a hyperbola.

In the next section, we will show how to recover the Cartesian definitions of the conic sections from the focus-directrix definitions.

Cartesian Definitions from Focus-Directrix Definitions

Let d be the vertical line through $(-r, 0)$ and let F be the point $(0, 0)$, as in the figure on the following page.

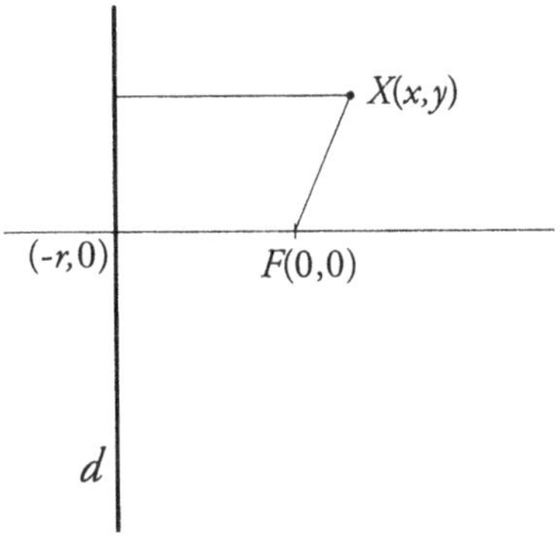

We have

$$\frac{FX}{\text{dist}(X, d)} = \epsilon,$$

which implies that

$$\frac{FX^2}{(\text{dist}(X, d)\ )^2} = \epsilon^2,$$

which in turn implies that

$$\frac{x^2 + y^2}{(x + r)^2} = \epsilon^2.$$

After rearranging, we get

$$(1 - \epsilon^2)x^2 - 2r\epsilon^2 x + y^2 = \epsilon^2 r^2.$$

If $\epsilon = 1$, the x^2 term disappears and the equation takes the form

$$y^2 - 2rx = r^2,$$

which we recognize as the Cartesian equation for a parabola.

If $\epsilon \neq 1$, then after some additional rearrangement we get

$$\left(x - \frac{r\epsilon^2}{1 - \epsilon^2}\right)^2 + \frac{y^2}{1 - \epsilon^2} = \left(\frac{r\epsilon}{1 - \epsilon^2}\right)^2.$$

This is of the form

$$(x - h)^2 + \frac{y^2}{1 - \epsilon^2} = a^2,$$

which is the Cartesian equation for an ellipse if $\epsilon < 1$ or a hyperbola if $\epsilon > 1$.

Pascal's Mystic Hexagon Theorem

Two facts that we will not prove in this text:

1. Given any conic in the plane, there is a reciprocating circle ω and a circle α such that the conic is the reciprocal of α with respect to ω.
2. The reciprocal of a conic is a conic.

Theorem 14.4.3. *Pascal's Mystic Hexagon Theorem holds for conics.*

Proof. Let the vertices of the hexagon inscribed in the conic be $A, B, C, D, E,$ and F, and let α and ω be circles such that the conic is the reciprocal of α with respect to ω. Then α is the reciprocal of the conic.

The points $A, B, C, D, E,$ and F on the conic have polars $a, b, c, d, e,$ and f that are tangent to the circle α. Brianchon's Theorem, which holds for the circle α, tells us that the joins of $a \cap b$ and $d \cap e$, $b \cap c$ and $e \cap f$, $c \cap d$ and $a \cap f$ are concurrent. Thus, taking reciprocals, the points $AB \cap DE$, $BC \cap EF$, and $CD \cap AF$ are collinear.

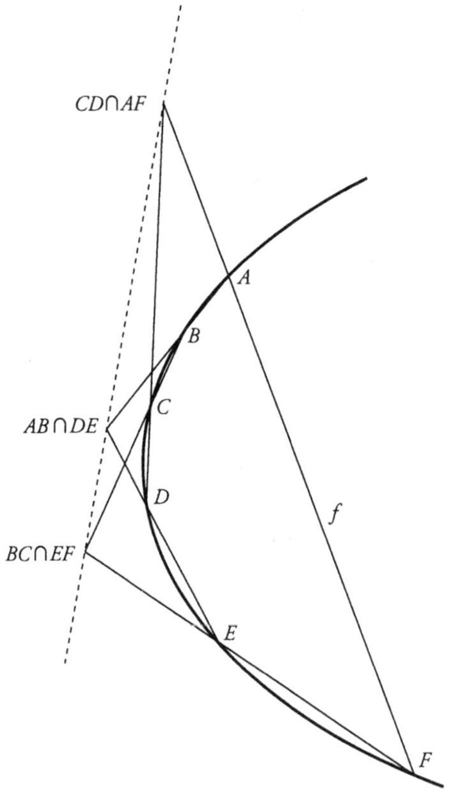

□

Using the same idea, the following can be seen to be true.

Theorem 14.4.4. *Brianchon's Theorem holds for conic sections.*

Another approach to showing that Pascal's Mystic Hexagon Theorem and Brianchon's Theorem are true for conics is to use the fact that all proper conics can be obtained via a central perspectivity of a circle. (See Theorem 16.6.1.)

14.5 Problems

1. Prove case (b) of statement (2) of Theorem 14.1.5.

2. Prove Theorem 14.1.9: The Apollonian circle for A, B, and k is the same as the Apollonian circle for B, A, and $1/k$.

3. Prove Theorem 14.1.10: If A and B are inverse points for a circle ω, then ω is the Circle of Apollonius for A, B, and some positive number k.

4. If PR is a diameter of circle α orthogonal to a circle β with center O, and if OP meets α in Q, prove that the line QR is the polar of P for β.

5. Show that one of the angles between the polars of A and B is equal to $\angle AOB$, where O is the center of the reciprocating circle.

6. Prove or disprove:

   (a) The reciprocal of a simple convex quadrilateral is a simple convex quadrilateral.

   (b) The reciprocal of a simple n-gon is a simple n-gon.

7. Use reciprocation to prove that given a triangle inscribed in a circle, then the points of intersection of the tangent lines at the vertices with the opposite sides are collinear.

8. If P and Q are conjugate points for a circle α, prove that the circle on PQ as diameter is orthogonal to α.

9. If two circles are orthogonal, prove that the extremities of any diameter of one are conjugate points for the other.

10. Sketch a circle α (with center A) and its polar with respect to a circle ω if the center O of ω is on α.

11. Given r and ϵ, $0 < \epsilon < 1$, then the equation

$$\frac{FX}{\text{dist}(X, d)} = \epsilon$$

in Cartesian coordinates becomes

$$\left(x - \frac{r\epsilon^2}{1-\epsilon^2}\right)^2 + \frac{y^2}{1-\epsilon^2} = \left(\frac{r\epsilon}{1-\epsilon^2}\right)^2.$$

Show that given positive numbers a and b, there are suitable values for r and ϵ (in terms of a and b) so that the Cartesian equation above becomes

$$\frac{(x-f)^2}{a^2} + \frac{y^2}{b^2} = 1.$$

12. Find the foci, directrices, and eccentricities of the following:

(a) $\dfrac{x^2}{25} + \dfrac{y^2}{9} = 1.$

(b) $\dfrac{x^2}{16} - \dfrac{y^2}{9} = 1.$

13. Find the Cartesian equations of the following conic sections:

(a) foci: $(\pm 8, 0)$, $e = 0.2$.

(b) foci: $(\pm 4, 0)$, directrix: $x = \dfrac{16}{3}$.

CHAPTER 15

CROSS RATIOS

15.1 Cross Ratios

Directed Distances

Recall that we denote directed distances, or signed distances, with a bar over the distance, as in $\overline{AB}$, and that

$$\overline{AB} = -\overline{BA}.$$

Also recall the following facts:

1. Given points A, B, and C on a line, if $\overline{AB} = \overline{AC}$, then $B = C$.
2. Given points A, B, C, and X on a line, if $\overline{AB}/\overline{BC} = \overline{AX}/\overline{XC}$, then $B = X$.

Classical Geometry: Euclidean, Transformational, Inversive and Projective, First Edition. By
I. E. Leonard, J. E. Lewis, A. C. F. Liu, G. W. Tokarsky
Copyright © 2014 John Wiley & Sons, Inc.

Directed Distances and Harmonic Conjugates

Recall that given points A, B, C, and D, then B and D are harmonic conjugates with respect to A and C if and only if

$$\frac{AB}{BC} = \frac{AD}{DC},$$

as in the figure below.

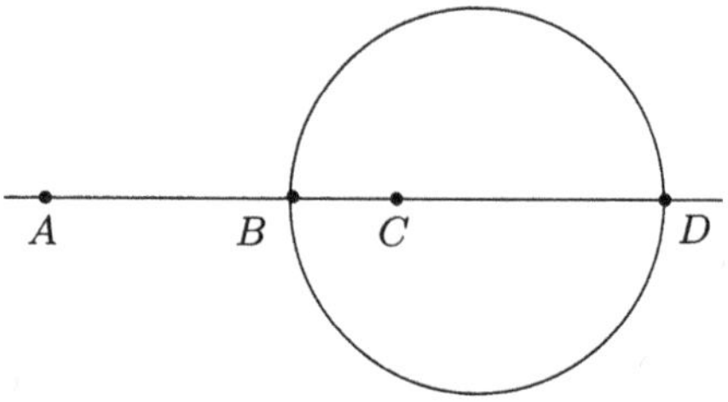

Here, we are using unsigned distances; for *signed distances*, B and D are harmonic conjugates with respect to A and C if and only if

$$\frac{\overline{AB}}{\overline{BC}} = -\frac{\overline{AD}}{\overline{DC}}.$$

Properties of Cross Ratios

Given a range of four points, A, B, C, and D, we define the quantity (AB, CD) by

$$(AB, CD) = \frac{\overline{AC}/\overline{CB}}{\overline{AD}/\overline{DB}}$$

and call it the ***cross ratio*** of the points A, B, C, and D, taken in that order.

Note.

1. The order refers to the order of the points in the notation, *not* the order of the points on the line.
2. If $(AB, CD) = -1$, then A and B are harmonic conjugates with respect to C and D.
3. The value of (AB, CD) is independent of the direction of the line on which the range of points lie.

Example 15.1.1. *Find (AB, CD), (AC, BD), and (BA, DC) where A, B, C, and D are collinear with coordinates along the line given by 0, 1, 2, and 3, respectively.*

Solution. Working directly from the definition of cross ratios, we get

$$(AB, CD) = \frac{\overline{AC}/\overline{CB}}{\overline{AD}/\overline{DB}} = \frac{2/(-1)}{3/(-2)} = \frac{4}{3}.$$

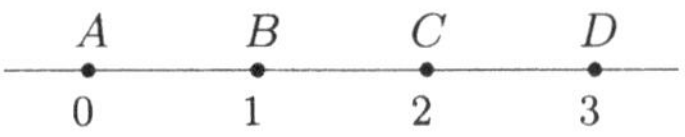

Similarly,

$$(AC, BD) = -\frac{1}{3},$$

and

$$(BA, DC) = \frac{4}{3}.$$

□

Theorem 15.1.2. *If $(AB, CD) = k$, then:*

(1) If we interchange any pair of points and also interchange the other pair of points, then the resulting cross ratio has the same value k. Thus, (AB, CD), (BA, DC), (CD, AB), and (DC, BA) all have the value k.

(2) Interchanging only the first pair or only the last pair of points results in a cross ratio with the value $1/k$. Thus, $(BA, CD) = (AB, DC) = 1/k$.

(3) Interchanging only the middle pair or only the outer pair of points results in a cross ratio with the value $1 - k$. Thus, $(AC, BD) = (DB, CA) = 1 - k$.

Proof. (1) and (2) follow directly from the definition of the cross ratio. To prove (3), we will use the fact that for three collinear points X, Y, and Z, the directed distances are related by $\overline{XZ} = \overline{XY} + \overline{YZ}$, whether Y is between X and Z or not.

Interchanging the middle pair, we have

$$\begin{aligned}(AC, BD) &= \frac{\overline{AB}/\overline{BC}}{\overline{AD}/\overline{DC}} = \frac{\overline{AB} \cdot \overline{CD}}{\overline{AD} \cdot \overline{CB}} \\ &= \frac{(\overline{AC} + \overline{CB})(\overline{CB} + \overline{BD})}{\overline{AD} \cdot \overline{CB}} \\ &= \frac{\overline{AC} \cdot \overline{BD}}{\overline{AD} \cdot \overline{CB}} + \frac{\overline{AC} \cdot \overline{CB} + \overline{CB}(\overline{CB} + \overline{BD})}{\overline{AD} \cdot \overline{CB}} \\ &= -\frac{\overline{AC} \cdot \overline{DB}}{\overline{AD} \cdot \overline{CB}} + \frac{\overline{CB}(\overline{AC} + \overline{CB} + \overline{BD})}{\overline{AD} \cdot \overline{CB}} \\ &= -k + 1.\end{aligned}$$

Interchanging the outer pair, from (1) we have

$$(DB, CA) = (CA, DB),$$

and interchanging the middle pair on the right-hand side, we have

$$(DB, CA) = 1 - (CD, AB),$$

and again by (1), we have

$$\begin{aligned}(DB, CA) &= 1 - (AB, CD) \\ &= 1 - k.\end{aligned}$$

□

Remark. Any permutation of the letters A, B, C, and D can be obtained by successively interchanging pairs using (1), (2), or (3) of Theorem 15.1.2.

Example 15.1.3. *Given* $(AB, CD) = k$, *find* (DA, CB).

Solution. From (3), we have

$$(DA, CB) = 1 - (BA, CD),$$

while from (2), we have

$$\begin{aligned}(DA, CB) &= 1 - \frac{1}{(AB, CD)} \\ &= 1 - \frac{1}{k}.\end{aligned}$$

Alternatively, we have

$$(AB, CD) = k,$$

which implies that

$$(BA, CD) = 1/k,$$

which in turn implies that

$$(DA, CB) = 1 - 1/k.$$

□

Ideal Points

Suppose that *one* of the four points A, B, C, or D is an ideal point I. We use the convention that

$$\frac{\overline{XI}}{\overline{IY}} = -1 \qquad \text{and} \qquad \frac{\overline{XI}}{\overline{YI}} = +1.$$

For example, if $B = I$, then

$$(AI, CD) = \frac{\overline{AC}/\overline{CI}}{\overline{AD}/\overline{DI}} = \frac{\overline{AC}}{\overline{CI}} \cdot \frac{\overline{DI}}{\overline{AD}} = \frac{\overline{AC}}{\overline{AD}} \cdot \frac{\overline{DI}}{\overline{CI}} = \frac{\overline{AC}}{\overline{AD}}.$$

Now suppose we interchange the two middle symbols in (AI, CD). Then by direct computation we get

$$(AC, ID) = \frac{\overline{AI}/\overline{IC}}{\overline{AD}/\overline{DC}} = \frac{-1}{\overline{AD}/\overline{DC}} = -\frac{\overline{DC}}{\overline{AD}}.$$

However,

$$1 - \left(-\frac{\overline{DC}}{\overline{AD}}\right) = \frac{\overline{AD} + \overline{DC}}{\overline{AD}} = \frac{\overline{AC}}{\overline{AD}},$$

which shows that the theorem on permutation of symbols remains true if one of the points is an ideal point.

There are 24 different arrangements of the symbols A, B, C, and D, giving rise to 24 different cross ratios: (AB, CD), (BA, CD), (BC, AD), and so on. However, there are only six different values for the 24 cross ratios, namely,

$$k, \quad \frac{1}{k}, \quad 1 - k, \quad \frac{1}{1-k}, \quad \frac{k-1}{k}, \quad \text{and} \quad \frac{k}{k-1}.$$

To see why this is so, note that all permutations of the symbols A, B, C, and D can be obtained via a sequence of interchanges of the types described in Theorem 15.1.2 and that only the operations of the second and third types produce different values. Thus, if the cross ratio (WX, YZ) has the value k, then the only new values we can obtain by operation of the second or third type are $1/k$ and $1 - k$. The value $1/k$ in turn can produce values of k or $1 - 1/k = (k - 1)/k$. If we were to continue in this manner, we would see that only the six different values listed above can be obtained.

Theorem 15.1.4. *If $(AB, CD) = (AB, CX)$, then D and X are the same point.*

Proof. We have

$$(AB, CD) = (AB, CX),$$

so that

$$\frac{\overline{AC}/\overline{CB}}{\overline{AD}/\overline{DB}} = \frac{\overline{AC}/\overline{CB}}{\overline{AX}/\overline{XB}},$$

which implies that

$$\overline{AD}/\overline{DB} = \overline{AX}/\overline{XB}.$$

From the remarks at the beginning of the chapter on the ratios of directed distances, this implies that $D = X$.

□

Corollary 15.1.5. *If $(AB, CD) = (AB, XD)$, then $C = X$.*

Proof. We have

$$(AB, CD) = (AB, XD),$$

so that

$$(AB, DC) = (AB, DX),$$

and from the previous theorem, $C = X$.

□

Cross Ratio of a Pencil of Lines

Let k, l, m, and n be a pencil of lines concurrent at a point P. This section will provide a definition of the cross ratio, denoted (kl, mn), of the pencil of lines.

Point of Concurrency is an Ordinary Point

Suppose k, l, m, and n are concurrent at an ordinary point P, and let A, B, C, and D be points other than P on k, l, m, and n, respectively, as in the figure below.

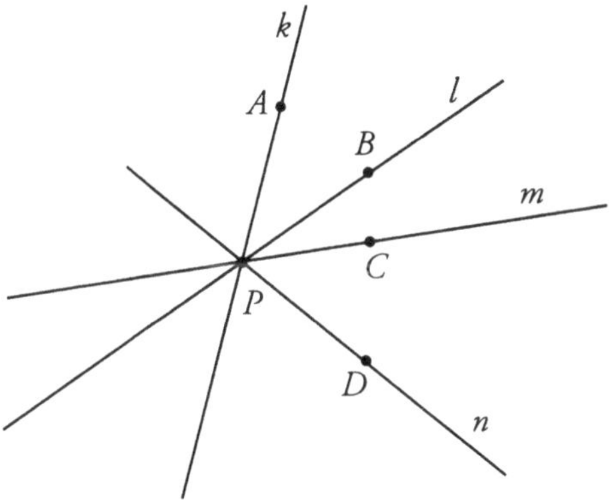

We define $P(AB, CD)$ as follows:

$$P(AB,CD) = \frac{\sin \overline{APC} / \sin \overline{CPB}}{\sin \overline{APD} / \sin \overline{DPB}},$$

where the directed angle $\overline{XPY}$ is the angle from the ray $\overrightarrow{PX}$ to the ray $\overrightarrow{PY}$ and whose magnitude is between $0°$ and $180°$.

Note that if A and A' are points on k on the opposite sides of P, then the directed angles $\overline{APC}$ and $\overline{A'PC}$ are different, as are the directed angles $\overline{APD}$ and $\overline{A'PD}$. In fact,

$$\overline{APC} = \overline{A'PC} - 180 \qquad \text{and} \qquad \overline{APD} = \overline{A'PD} - 180.$$

Since $\sin(x - 180) = -\sin x$, it follows that if A and A' are on opposite sides of P on the line k, then

$$\frac{\sin \overline{APC} / \sin \overline{CPB}}{\sin \overline{APD} / \sin \overline{DPB}} = \frac{-\sin \overline{A'PC} / \sin \overline{CPB}}{-\sin \overline{A'PD} / \sin \overline{DPB}} = \frac{\sin \overline{A'PC} / \sin \overline{CPB}}{\sin \overline{A'PD} / \sin \overline{DPB}},$$

or, in other words,

$$P(AB,CD) = P(A'B,CD).$$

Thus, we have the following result:

Theorem 15.1.6. *For a pencil of lines k, l, m, and n that are concurrent at the ordinary point P, the definition of $P(AB, CD)$ is independent of the choice of the points A, B, C, and D, as long as none of them are the point P.*

This allows us to make the following definition:

Definition. We define the ***cross ratio of a pencil of lines concurrent at an ordinary point*** P as

$$(kl, mn) = P(AB, CD),$$

where A, B, C, and D are points on k, l, m, and n other than P.

Point of Concurrency Is an Ideal Point

When the point of concurrency is an ideal point, the lines k, l, m, and n are parallel.

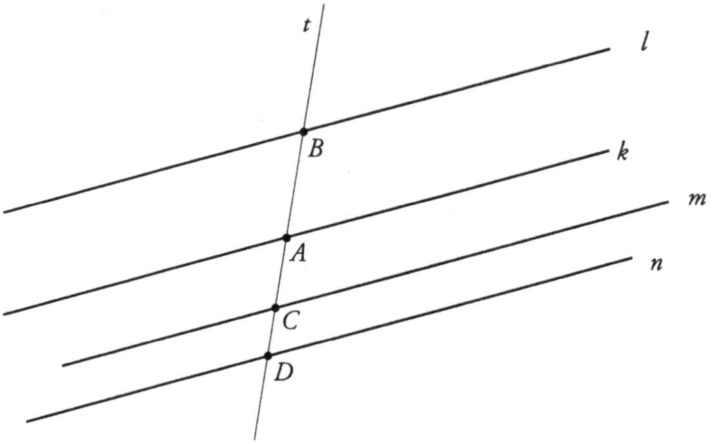

In this case, let t be any line intersecting k, l, m, and n at A, B, C, and D, respectively, and define (kl, mn) to be (AB, CD).

To check that this definition is independent of the choice of the line t, let t' be another line intersecting k, l, m, and n at A', B', C', and D'. There are two cases to consider:

Case (i). t and t' are parallel.

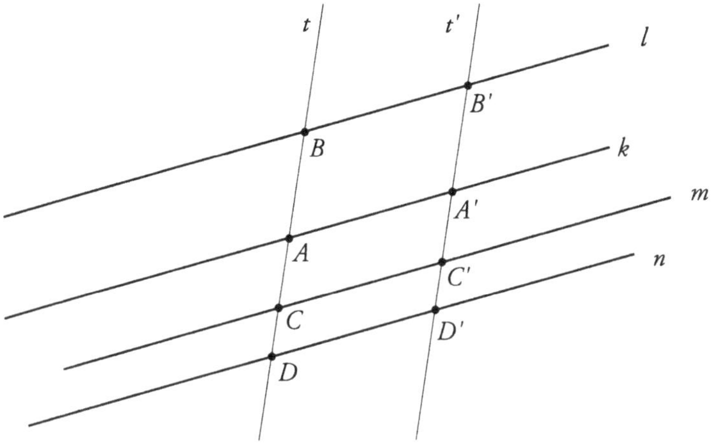

In this case, obviously $AC = A'C'$, $CB = C'B'$, etc., since they are opposite sides of a parallelogram, so that

$$(AB, CD) = (A'B', C'D').$$

Case (ii). t and t' meet at an ordinary point P.

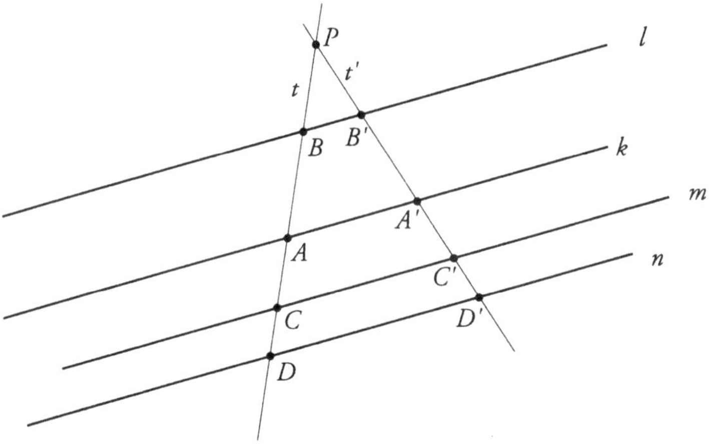

In this case, by similar triangles,

$$\frac{\overline{A'C'}}{\overline{C'B'}} = \frac{\overline{AC}}{\overline{CB}},$$

with similar results for the other ratios. From this it follows that

$$(AB, CD) = (A'B', C'D').$$

Theorem 15.1.7. *Suppose that k, l, m, and n form a pencil of lines concurrent at the ordinary point P. If a transversal cuts the lines k, l, m, and n at the points A, B, C, and D, respectively, then*

$$P(AB, CD) = (AB, CD) .$$

Proof. In order to prove the theorem, we will show two things:

(1) The signs (signum) of $P(AB, CD)$ and (AB, CD) are the same.

(2) The magnitudes of $P(AB, CD)$ and (AB, CD) are the same.

Proof of (1).

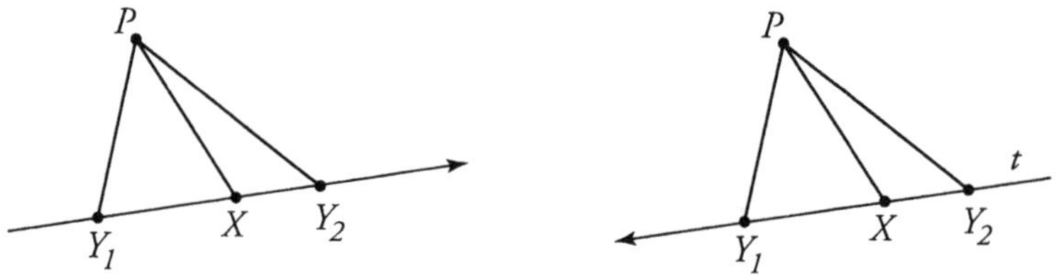

To check that (1) is true, note that given a directed line t and a point P not on t, then either

$$\operatorname{sgn}(\sin \overline{XPY}) = \operatorname{sgn}(\overline{XY})$$

for all pairs of points X and Y on t or else

$$\operatorname{sgn}(\sin \overline{XPY}) = -\operatorname{sgn}(\overline{XY})$$

for all pairs of points X and Y on t.

For points A, B, C, and D on t, the value of (AB, CD) is independent of the direction of t (see statement (3) in the note following the definition of (AB, CD) at the beginning of this chapter). In particular, we can choose the direction of t so that for all pairs of points X and Y,

$$\operatorname{sgn}(\sin \overline{XPY}) = \operatorname{sgn}(\overline{XY}).$$

Thus,

$$
\begin{aligned}
\operatorname{sgn}(P(AB,CD)) &= \operatorname{sgn}\left(\frac{\sin \overline{APC}/\sin \overline{CPB}}{\sin \overline{APD}/\sin \overline{DPB}}\right) \\
&= \frac{\operatorname{sgn}(\sin \overline{APC})/\operatorname{sgn}(\sin \overline{CPB})}{\operatorname{sgn}(\sin \overline{APD})/\operatorname{sgn}(\sin \overline{DPB})} \\
&= \frac{\operatorname{sgn}(\overline{AC}/\operatorname{sgn}(\overline{CB})}{\operatorname{sgn}(\overline{AD})/\operatorname{sgn}(\overline{DB})} \\
&= \operatorname{sgn}(AB,CD).
\end{aligned}
$$

Proof of (2).

The proof uses the fact that it is possible to compute the area of a triangle in two different ways:

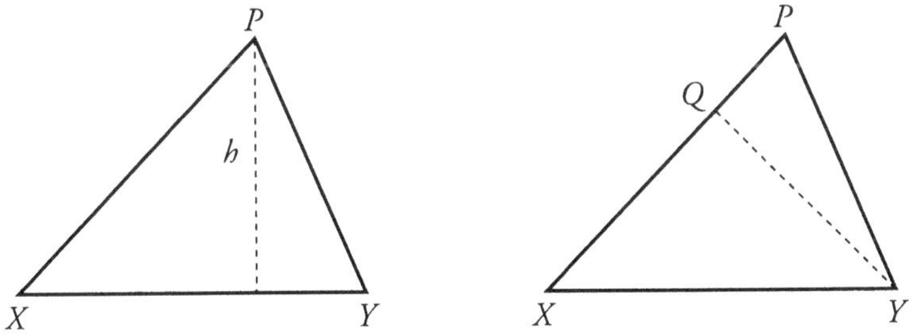

(a) (b)

In the figure above,

(a) area($\triangle XPY$) $= XY \cdot \frac{h}{2}$, and

(b) area($\triangle XPY$) $= \frac{1}{2}XP \cdot PY \cdot \sin(\angle XPY)$,

where (b) follows from the fact that $QY = PY \sin(\angle XPY)$.

Expanding $|(AB,CD)|$, we get

$$
\begin{aligned}
|(AB,CD)| &= \frac{|\overline{AC}|\,/\,|\overline{CB}|}{|\overline{AD}|\,/\,|\overline{DB}|} \\
&= \frac{(|\overline{AC}|\cdot\frac{h}{2})\ /\ (|\overline{CB}|\cdot\frac{h}{2})}{(|\overline{AD}|\cdot\frac{h}{2})\ /\ (|\overline{DB}|\cdot\frac{h}{2})} \\
&= \frac{\text{area}(\triangle APC)\,/\,\text{area}(\triangle CPB)}{\text{area}(\triangle APD)\,/\,\text{area}(\triangle DPB)}.
\end{aligned}
$$

Calculating the areas using (b) we get

$$\text{area}(\triangle APC) = \frac{1}{2}|AP| \cdot |PC| \cdot |\sin \overline{APC}|,$$

$$\text{area}(\triangle APD) = \frac{1}{2}|AP| \cdot |PD| \cdot |\sin \overline{APD}|,$$

$$\text{area}(\triangle CPB) = \frac{1}{2}|CP| \cdot |PB| \cdot |\sin \overline{CPB}|,$$

$$\text{area}(\triangle DPB) = \frac{1}{2}|DP| \cdot |PB| \cdot |\sin \overline{DPB}|.$$

Substituting these values into the expression for $|(AB, CD)|$ above, we get

$$|(AB, CD)| = \frac{|\sin \overline{APC}| \,/\, |\sin \overline{CPB}|}{|\sin \overline{APD}| \,/\, |\sin \overline{DPB}|} = |P(AB, CD)|\,.$$

This completes the proof of the theorem. □

15.2 Applications of Cross Ratios

Four Useful Lemmas

The following lemmas connect cross ratios with concurrency and collinearity.

Lemma 15.2.1. *In the figure below, P is an ordinary point and P, A, and A' are collinear; P, B, and B' are collinear; and P, C, and C' are collinear. Transversals cut the lines PA, PB, and PC at A, B, and C, respectively. The point D is on AC, and the point D' is on $A'C'$.*

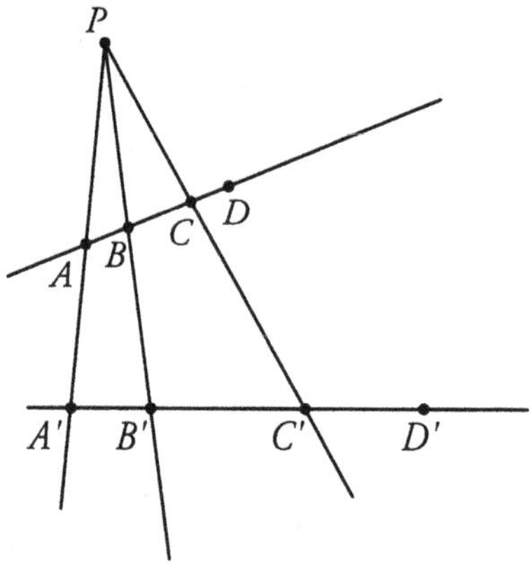

With this configuration, if $(AB, CD) = (A'B', C'D')$, then P is on DD'.

Proof. Suppose the line PD intersects $A'D'$ at some point E' (E' is not shown in the diagram). Then from Theorem 15.1.6, we have

$$P(AB, CD) = (A'B', C'E').$$

However,

$$P(AB, CD) = (AB, CD) = (A'B', C'D'),$$

so that

$$(A'B', C'E') = (A'B', C'D'),$$

which implies that $E' = D'$, and P is on DD'.

□

Lemma 15.2.2. *In the figure below, two lines intersect at the point A. The points B, C, and D are on one of the lines, while the points B', C', and D' are on the other line.*

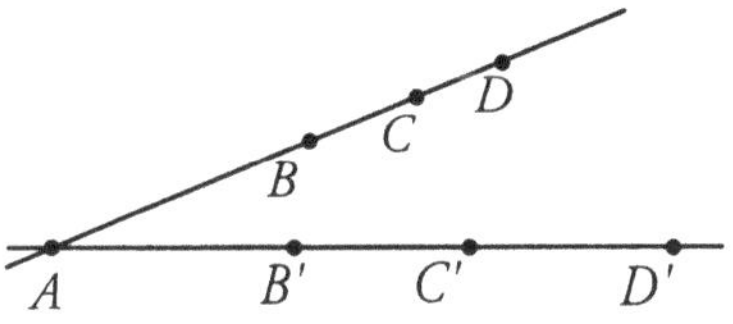

With this configuration, if $(AB, CD) = (AB', C'D')$, then BB', CC', and DD' are concurrent.

Proof. Let $P = BB' \cap CC'$ and let $E' = PD \cap AC'$, as shown below.

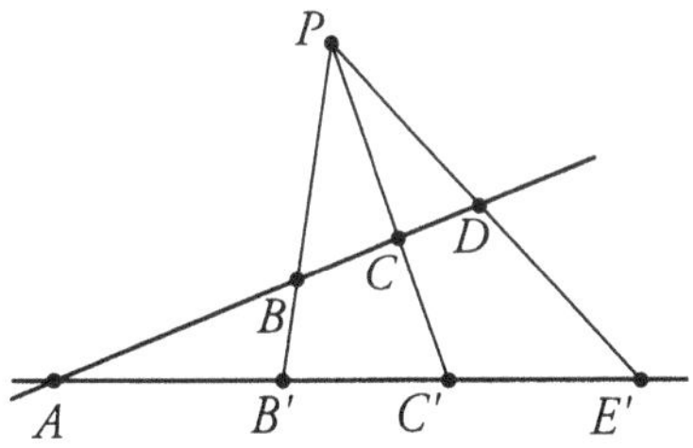

It suffices to show that $E' = D'$.

We have

$$P(AB, CD) = P(AB', C'E'),$$

which implies that

$$(AB, CD) = (AB', C'E'),$$

and since $(AB, CD) = (AB', C'D')$, it follows that

$$(AB', C'D') = (AB', C'E').$$

Therefore, $E' = D'$.

□

Lemma 15.2.3. *In the figure below, A, B, and C are points on a line m. P and Q are points not on m. D is a point other than A, B, C, P, or Q.*

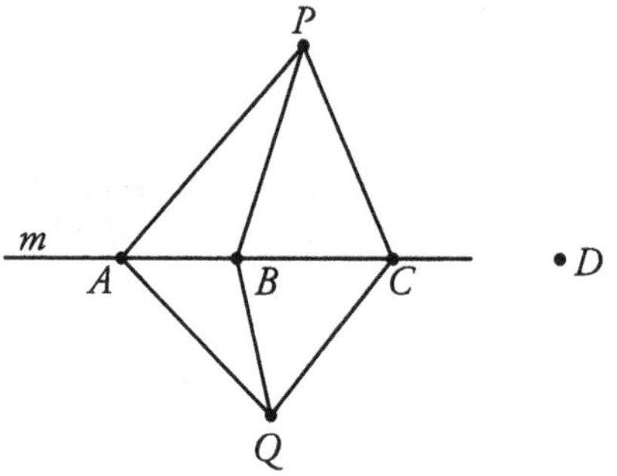

With this configuration, if $P(AB, CD) = Q(AB, CD)$, then D is on m.

Proof. If D is not on m, let PD intersect m at X and let QD intersect m at Y, as shown below.

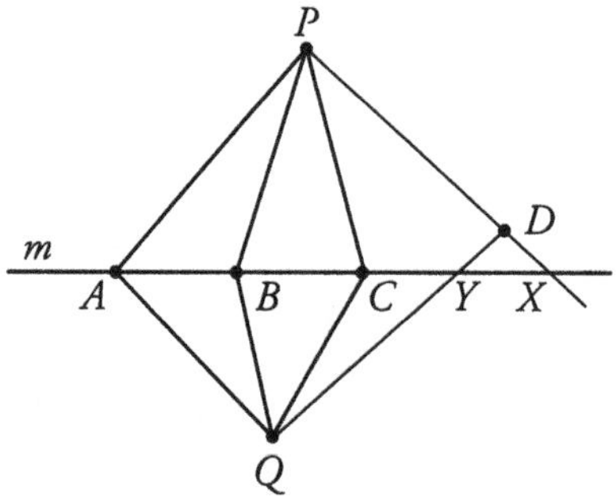

We have

$$P(AB, CD) = P(AB, CX) = (AB, CX)$$

and

$$Q(AB, CD) = Q(AB, CY) = (AB, CY),$$

and since $P(AB, CD) = Q(AB, CD)$, we must have $(AB, CX) = (AB, CY)$.

However, this can only happen if $PD \cap QD = X = Y = D$; that is, D is on m.

□

Lemma 15.2.4. *In the figure below, A, B, C, and D are points other than P and Q, and the point A is on PQ.*

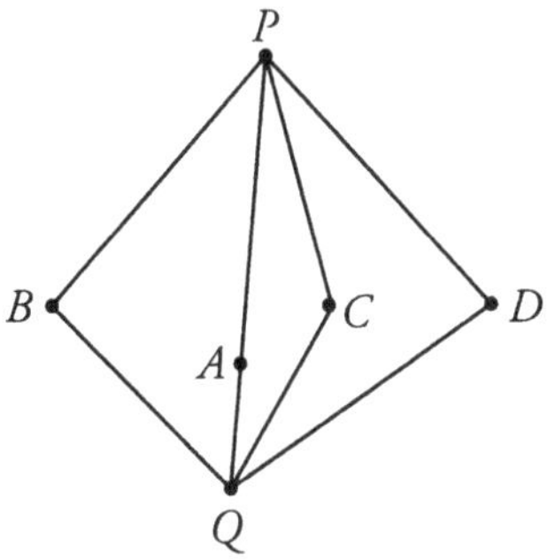

With this configuration, if

$$P(AB,CD) = Q(AB,CD),$$

then B, C, and D are collinear.

Proof. Let A' be the point $PQ \cap BC$ and let m be the line BC. Then

$$P(AB,CD) = P(A'B,CD),$$
$$Q(AB,CD) = Q(A'B,CD),$$

which implies that

$$P(A'B,CD) = Q(A'B,CD),$$

which implies that D is on m, by the previous lemma.

□

Theorems of Desargues, Pascal, and Pappus

Desargues' Two Triangle Theorem (Theorem 4.4.3) was proven in Part I using Ceva's Theorem and Menelaus' Theorem. We restate it here and prove it using cross ratios.

Theorem 15.2.5. *Copolar triangles are coaxial and conversely.*

Proof. Let the copolar triangles be ABC and $A'B'C'$. Then, as in the figure on the following page,

$$AA' \cap BB' \cap CC' = O.$$

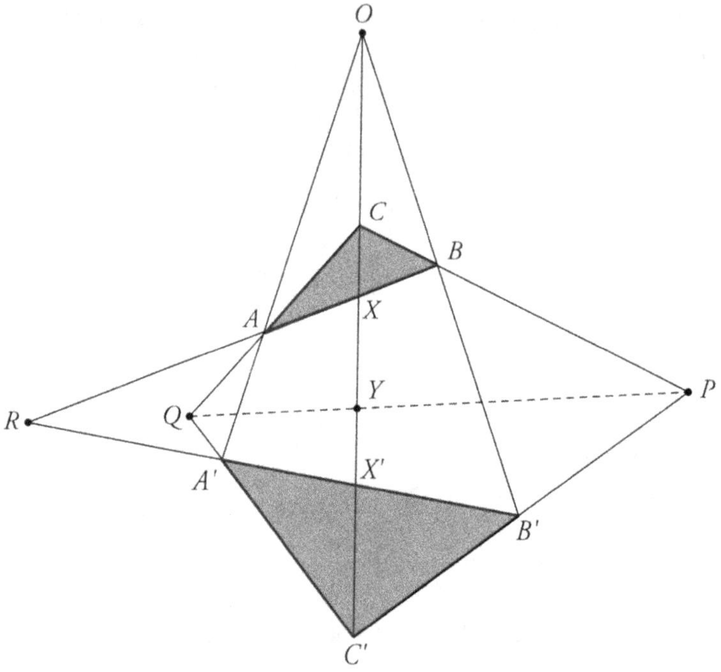

Let P, Q, and R be the points of intersection of the corresponding sides of the triangle. In order to show that P, Q, and R are collinear, we will show that R is on the line PQ.

Let X, X', and Y be as shown; that is, let $X = OC' \cap AB$, $X' = OC' \cap A'B'$, and $Y = OC' \cap PQ$.

There are three pencils of lines: one concurrent at C, one concurrent at O, and one concurrent at C'.

Since BA is a transversal for the pencil at C, we have

$$C(PY, QR) = (BX, AR),$$

and since $B'A'$ is a transversal for the pencil at O, we have

$$(BX, AR) = O(BX, AR) = (B'X', A'R) = C'(B'X', A'R),$$

and replacing B' by P, X' by Y, and A' by Q in the last expression, we have

$$C'(B'X', A'R) = C'(PY, QR).$$

Therefore, $C(PY, QR) = C'(PY, QR)$, and by Lemma 15.2.4, P, Q, and R are collinear.

□

Pascal's Mystic Hexagon Theorem from Part I (Theorem 4.4.4) says the following:

Theorem 15.2.6. *If a hexagon is inscribed in a circle, the points of intersection of the opposite sides are collinear.*

There are many possible configurations, three of which are shown in the following figure.

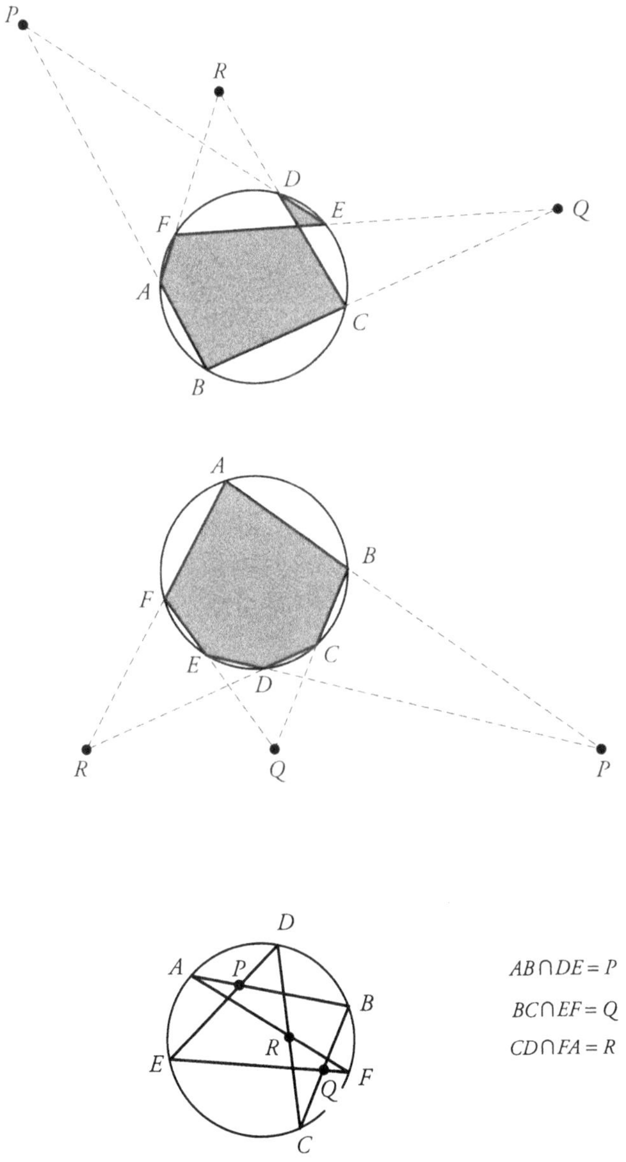

In order to prove the theorem using cross ratios we first need the following lemma:

Lemma 15.2.7. *If A, B, C, D, and P, Q are distinct points on a circle, then* $P(AB,CD) = Q(AB,CD)$.

Proof. There are two cases to consider, as illustrated in the following figures.

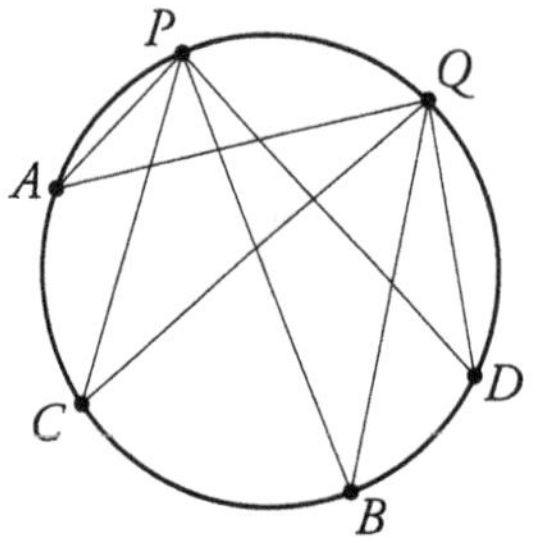

(1)

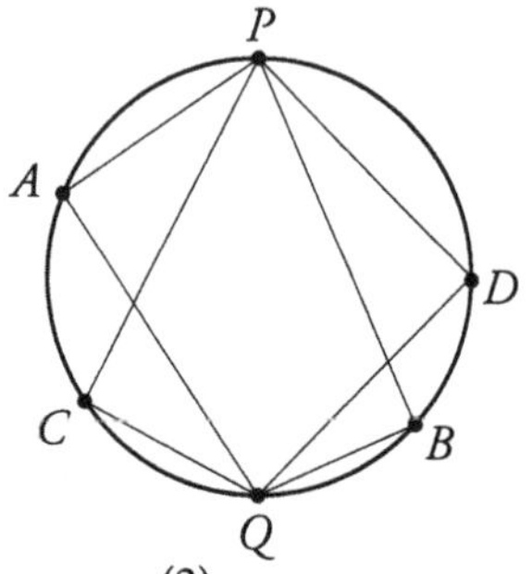

(2)

Case (1). P and Q are not separated by any of the points A, B, C, or D.

In this case, Thales' Theorem implies that

$$\begin{aligned}\angle\overline{APC} &= \angle\overline{AQC},\\ \angle\overline{CPB} &= \angle\overline{CQB},\\ \angle\overline{APD} &= \angle\overline{AQD},\\ \angle\overline{DPB} &= \angle\overline{DQB}.\end{aligned}$$

Thus,

$$P(AB,CD) = \frac{\sin\overline{APC}/\sin\overline{CPB}}{\sin\overline{APD}/\sin\overline{DPB}} = \frac{\sin\overline{AQC}/\sin\overline{CQB}}{\sin\overline{AQD}/\sin\overline{DQB}} = Q(AB,CD).$$

Case (2). P and Q are separated by some of the points A, B, C, or D.

The proof here is similar to that for Case (1), but now we have

$$\begin{aligned}\angle\overline{APC} &= \angle\overline{AQC},\\ \angle\overline{CPB} &= 180 + \angle\overline{CQB},\\ \angle\overline{APD} &= 180 + \angle\overline{AQD},\\ \angle\overline{DPB} &= \angle\overline{DQB}.\end{aligned}$$

The positive signs in the second and third equations arise since the signed angles are in opposite directions. From the second equation, we get

$$\sin\overline{CPB} = \sin(180 + \overline{CQB}) = -\sin\overline{CQB}.$$

Similarly, from the third equation, we get

$$\sin \overline{APD} = \sin(180 + \overline{AQD}) = -\sin \overline{AQD}\,.$$

The lemma now follows from the definitions of $P(AB, CD)$ and $Q(AB, CD)$ as in Case (1).

□

For convenience, we restate Pascal's Theorem.

Theorem 15.2.8. *If $ABCDEF$ is a hexagon inscribed in a circle, then the points of intersection of the opposite sides are collinear.*

Proof. The proof works for any configuration, and for clarity we use the one in the figure below.

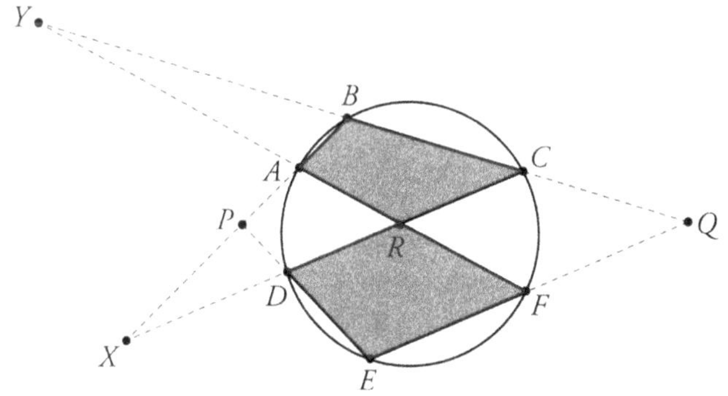

Let $X = AB \cap CD$ and $Y = BC \cap AF$.

Consider the pencils at D and F. Since P is on DE and X is on DC, then

$$D(AE, CB) = (AP, XB),$$

and from Lemma 15.2.7, we have

$$D(AE, CB) = F(AE, CB).$$

Since Y is on FA and Q is on FE, then

$$F(AE, CB) = (YQ, CB),$$

and, therefore,

$$(AP, XB) = (YQ, CB)\,.$$

Note that we have the following configuration:

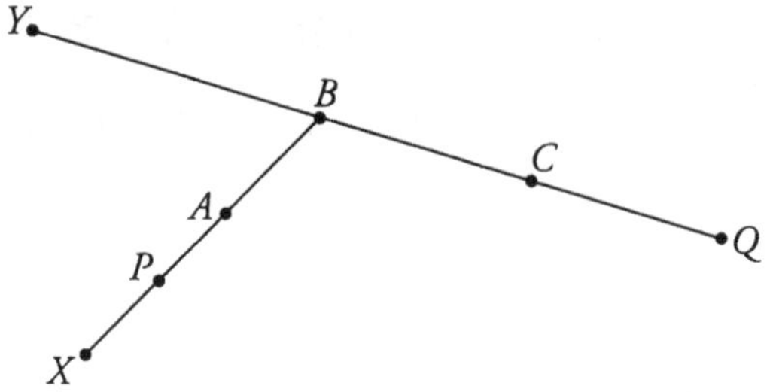

Since $(AP, XB) = (YQ, CB)$, then from Lemma 15.2.2 it follows that AY, PQ, and XC are concurrent. However, $AY \cap XC = R$, so PQ passes through R.

□

As a final example, we prove Pappus' Theorem (Theorem 4.4.5) using cross ratios:

Theorem 15.2.9. *Given points A, B, and C on a line l and points A′, B′, and C′ on a line l′, then the points of intersection*

$$P = AB' \cap A'B, \quad Q = AC' \cap A'C, \quad R = BC' \cap B'C$$

are collinear.

Proof. Introduce points $X = AC' \cap A'B$ and $Y = CA' \cap C'B$, as shown in the figure below.

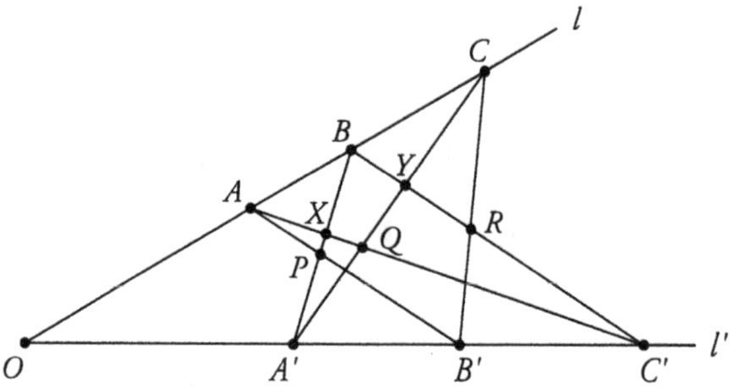

Using the pencil through A, replace B by O, X by C', and P by B'. Then

$$A(BX, PA') = A(OC', B'A'),$$

and since these are transversals for the pencil through A, we have

$$(BX, PA') = (OC', B'A').$$

Using the pencil through C, replace O by B, B' by R, and A' by Y. Then

$$C(OC', B'A') = C(BC', RY),$$

and since these are transversals for the pencil through C, we have

$$(OC', B'A') = (BC', RY).$$

Thus,

$$(BX, PA') = (OC', B'A') = (BC', RY).$$

It follows from Lemma 15.2.2 that XC', PR, and $A'Y$ are concurrent, and since $XC' \cap A'Y = Q$, then Q is on PR; that is, P, Q, and R are collinear.

□

15.3 Problems

1. Given $(AB, CD) = k$, find (BC, AD) and (BD, CA).

2. Given three points A, B, and C, as shown below,

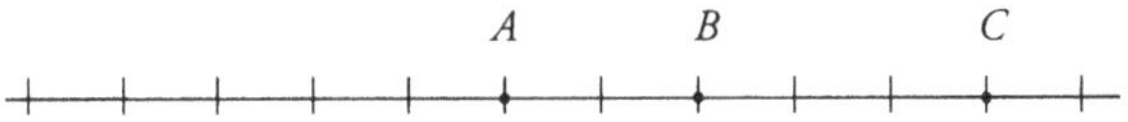

   find points D_i, $i = 1, 2, 3, 4$ such that

   (a) $(AB, CD_1) = 5/6$,
   (b) $(AB, CD_2) = -5/3$,
   (c) $(AB, CD_3) = 10/3$,
   (d) $(AB, CD_4) = 5/3$.

3. Using the definition of the cross ratio, show that $(AB, CD) = (CD, AB)$.

4. Show that for collinear points A, B, C, D, and E we have

   (a) $(AB, CE) \cdot (AB, ED) = (AB, CD)$,
   (b) $(AE, CD) \cdot (EB, CD) = (AB, CD)$.

5. Find x if $(AB, CD) = (BA, CD) = x$.

6. Let L, M, and N be the respective midpoints of the sides BC, CA, and AB of $\triangle ABC$. Prove that

$$L(MN, AB) = -1.$$

7. If P, Q, and R are the respective feet of the altitudes on the sides of BC, CA, and AB of $\triangle ABC$, show that

$$P(QR, AB) = -1.$$

8. Given $\mathcal{C}(O, r)$ and ordinary points A, B, C, and D on a ray through O, inverting into A', B', C', and D', respectively, show that $(AB, CD) = (A'B', C'D')$; that is, that the cross ratio is invariant under inversion.

Hint: Use the distance formula $A'B' = \dfrac{r^2}{OA \cdot OB} AB.$

9. Prove the following: If PA, PB, PC, PD and QA, QB, QC, QD are two pencils of lines, and if $P(AB, CD) = Q(AB, CD)$ and A is on PQ, then B, C, and D are collinear.

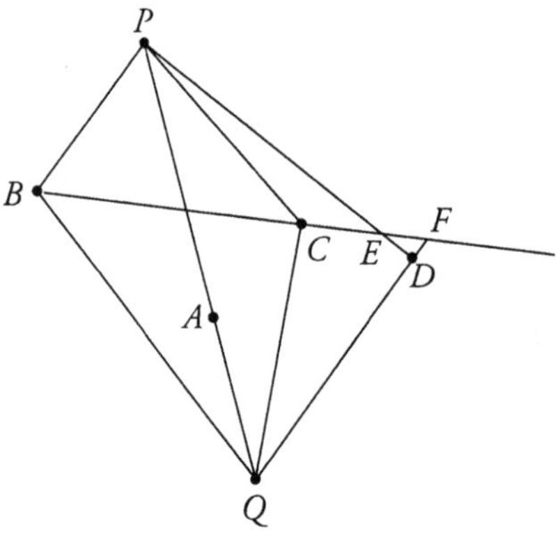

10. In $\triangle ABC$ on the following page we have

$$(BC, PP') = (CA, QQ') = (AB, RR') = -1.$$

Show that AP', BQ', and CR' are concurrent if and only if P, Q, and R are collinear.

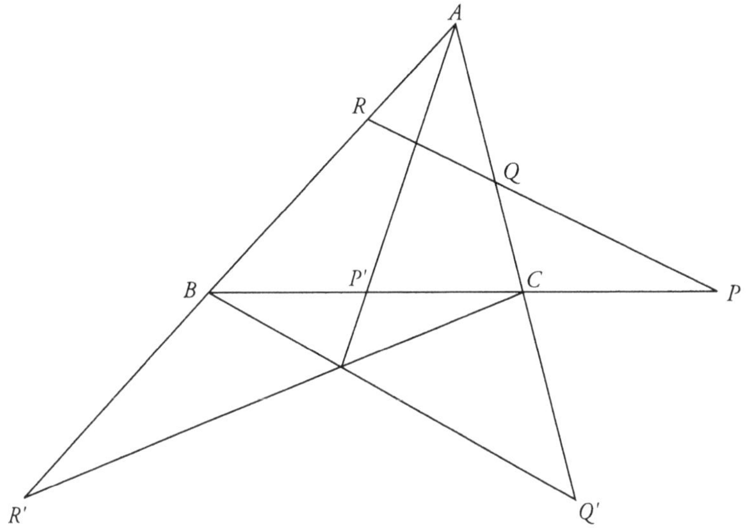

11. Given a variable triangle $\triangle ABC$ whose sides BC, CA, and AB pass through fixed points P, Q, and R, respectively, then if the vertices B and C move along given lines through a point O collinear with Q and R, find the locus of the vertex A.

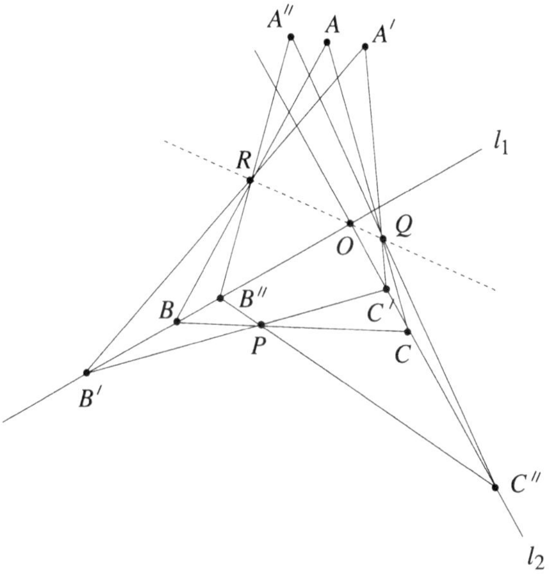

12. If $V(AB, CD) = -1$ and if VC is perpendicular to VD, show that VC and VD are the internal bisector and external bisector of $\angle AVB$.

13. The bisector of angle A of $\triangle ABC$ intersects the opposite side at the point T. The points U and V are the feet of the perpendiculars from B and C, respectively, to the line AT. Show that U and V divide AT harmonically; that is, that $(AT, UV) = -1$.

14. A line through the midpoint A' of side BC of $\triangle ABC$ meets the side AB at the point F, side AC at the point G, and the parallel through A to side CB at the point E. Show that the points A' and E divide FG harmonically; that is, that $(FG, A'E) = -1$.

15. Prove the second part of Desargues' Theorem using cross ratios; that is, show that coaxial triangles are copolar.

CHAPTER 16

INTRODUCTION TO PROJECTIVE GEOMETRY

16.1 Straightedge Constructions

We saw earlier that a compass alone is as "powerful" as a compass combined with a straightedge. We begin this section by indicating why a straightedge alone is not as powerful as a straightedge and compass or a compass alone. There are only a few admissible operations that can be done with a straightedge by itself.

Admissible Straightedge Operations

1. Draw an arbitrary line.

2. Draw a line through a given or previously constructed point.

3. Draw a line through two given or previously constructed points.

4. Construct a point as the intersection of two different lines.

Classical Geometry: Euclidean, Transformational, Inversive and Projective, First Edition. By I. E. Leonard, J. E. Lewis, A. C. F. Liu, G. W. Tokarsky
Copyright © 2014 John Wiley & Sons, Inc.

A ***straightedge construction*** is a finite sequence of the above operations.

We will give informal proofs that certain well-known constructions with straightedge and compass are not possible with a straightedge alone.

One of the standard straightedge and compass constructions is bisecting a given line segment.

Theorem 16.1.1. *Using only a straightedge, we cannot construct the midpoint of a given segment.*

Proof. The idea behind the proof is that a straightedge construction is projectively invariant. Here we give an intuitive justification of the theorem.

Suppose that there is a finite sequence of the possible straightedge operations that yield the midpoint of a segment AB. In other words, there is a sequence of instructions that, when followed, produces the midpoint of AB. For example, the first few instructions might be:

(1) Draw a line l through endpoint A.

(2) Draw a line m through endpoint B.

(3) Let C be the point of intersection of l and m.

$\vdots$

In the plane $\mathcal{P}$, carry out the instructions that yield the midpoint M of the segment AB. Now let $\mathcal{P}'$ be a plane that is not parallel to $\mathcal{P}$, as shown in the figure below.

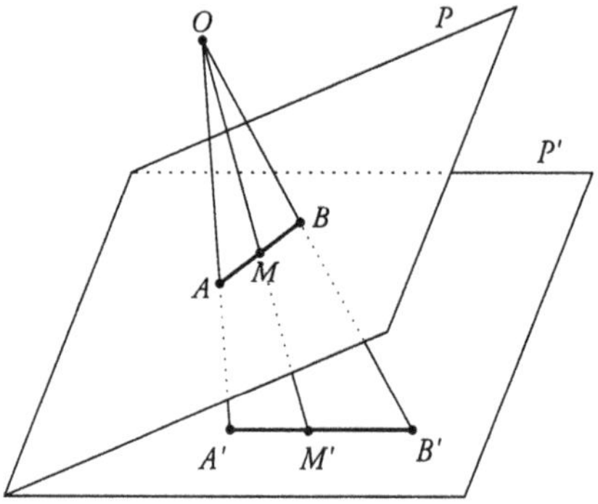

Let O be a point not on $\mathcal{P}$ or $\mathcal{P}'$ and "project" AB onto $\mathcal{P}'$ from O.

Points in $\mathcal{P}$ are projected to points in $\mathcal{P}'$. Straight lines in $\mathcal{P}$ are projected to straight lines in $\mathcal{P}'$. The segment AB in P is projected into a segment $A'B'$ in $\mathcal{P}'$, and each point M of AB projects to a point M' in $A'B'$.

Each of the four permissible operations in $\mathcal{P}$ projects into exactly the same operation in $\mathcal{P}'$. Thus, if a finite sequence of these operations yields the point M of AB, the same finite sequence of instructions carried out in $\mathcal{P}'$ would yield the projected point M' of $A'B'$. However, projection from the point O does not preserve midpoints, so it seems that on the one hand the sequence of instructions does yield a midpoint while on the other hand it does not yield a midpoint.

□

Thus, we see that this construction is possible with a straightedge and compass, but not with a straightedge alone. However, if we have a line parallel to the segment, then we can construct the midpoint of the segment using a straightedge alone, as in the following example.

Example 16.1.2. *Given two parallel lines and points A and B on one of them, construct the midpoint of the segment $\overline{AB}$ using a straightedge alone.*

Solution. Using the straightedge, construct a point C such that the sides AC and BC of $\triangle ABC$ are cut by the other parallel at E and D, respectively, as in the figure. Let O be the point of intersection of AD and BE, and draw the line through C and O hitting AB at F. Let CF meet m at X. Since l and m are parallel, then $\triangle CFA \sim \triangle CXE$, which implies that

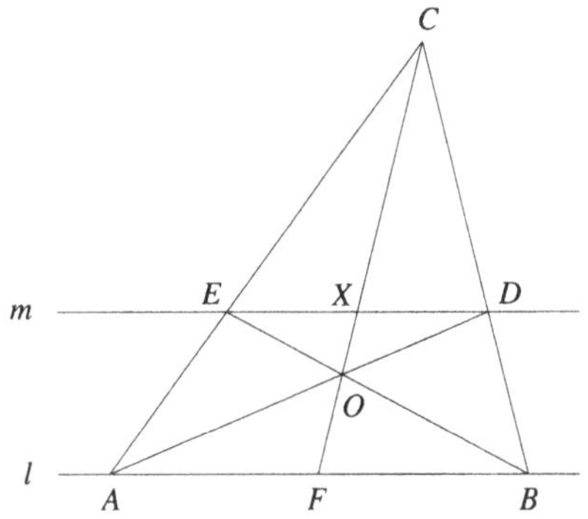

$$\frac{CA}{CE} = \frac{CF}{CX},$$

so that

$$\frac{CE + EA}{CE} = \frac{CX + XF}{CX},$$

that is, that

$$\frac{EA}{CE} = \frac{XF}{CX}.$$

Also, since $\triangle CFB \sim \triangle CXD$, a similar argument shows that

$$\frac{BD}{DC} = \frac{XF}{CX},$$

so that

$$\frac{BD}{DC} = \frac{EA}{CE}.$$

Now, since the three cevians AD, BE, and CF are concurrent at O, and all divisions are internal, from Ceva's Theorem we have

$$1 = \frac{\overline{AF}}{\overline{FB}} \cdot \frac{\overline{BD}}{\overline{DC}} \cdot \frac{\overline{CE}}{\overline{EA}} = \frac{AF}{FB} \cdot \frac{BD}{DC} \cdot \frac{CE}{EA} = \frac{AF}{FB} \cdot \frac{EA}{CE} \cdot \frac{CE}{EA} = \frac{AF}{FB},$$

so that $AF = FB$ and F is the midpoint of AB.

□

Harmonic Conjugates and Complete Quadrilaterals

What can be constructed using only a straightedge? Problem 6.10 in Section 6.10 illustrates that there are some constructions that can be done.

We will show how to construct the harmonic conjugate of a point using only a straightedge. It is convenient at this point to introduce some terminology.

The figure on the left below consists of four lines a, b, c, and d, no three of which are concurrent, along with the six points formed by the intersections of each pair of lines. Such a figure is called a ***complete quadrilateral***. In the projective plane, some lines may be parallel, in which case some points of intersection may be ideal points. Also, one of the lines may be the ideal line.

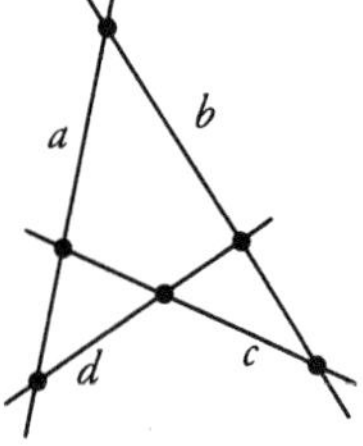

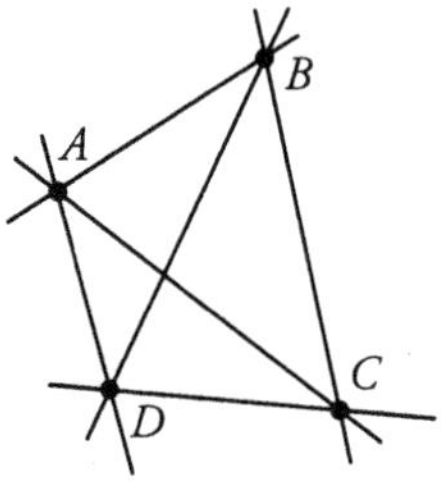

The figure on the right illustrates the dual notion, which is a configuration consisting of four points, no three of which are collinear, together with the six lines connecting each pair of points. This configuration is called a ***complete quadrangle***. In the projective plane, one or two of the points may be ideal points, and one of the lines may be the ideal line.

Before we give the construction, we prove a lemma that we alluded to at the beginning of Chapter 15 (Note 2 in Section 15.1).

Lemma 16.1.3. *Given four collinear points A, B, C, and D, the points C and D divide AB harmonically if and only if $(AB, CD) = -1$ or, equivalently, if and only if $(CD, AB) = -1$.*

Proof. Suppose C and D divide AB harmonically. Then

$$\frac{AC}{CB} = \frac{AD}{DB},$$

which implies that

$$\frac{AC/CB}{AD/DB} = 1.$$

However, one of C or D is between A and B while the other is not.

If C is between A and B, we have $\overline{AC}/\overline{CB} = 1$; if not, we have $\overline{AC}/\overline{CB} = -1$. Thus, one of $\overline{AC}/\overline{CB}$ and $\overline{AD}/\overline{DB}$ is positive and the other is negative, which implies that

$$\frac{\overline{AC}/\overline{CB}}{\overline{AD}/\overline{DB}} = -1.$$

Conversely, suppose that $(AB, CD) = -1$. Then

$$\frac{\overline{AC}/\overline{CB}}{\overline{AD}/\overline{DB}} = -1,$$

and taking absolute values, this implies that

$$\frac{AC}{CB} = \frac{AD}{DB}.$$

□

Theorem 16.1.4. *Given a complete quadrilateral, the two points of intersections of one given diagonal with the other two diagonals divide the vertices of the given diagonal harmonically.*

Proof. The following figure illustrates the situation, where the six points of intersection of the sides are A, B, C, D, E, and F, and the three possible diagonals are shown as dashed lines.

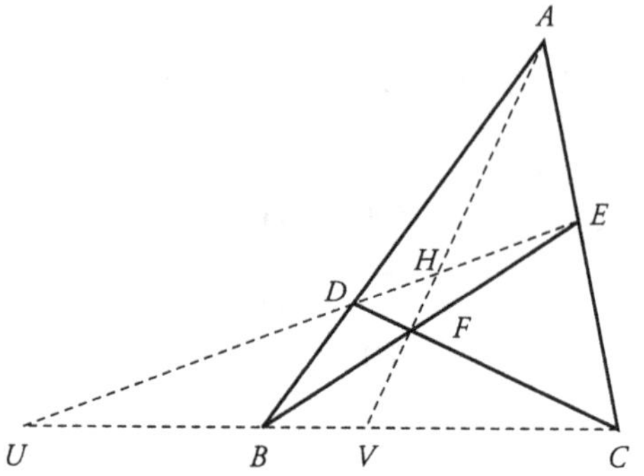

Here we wish to show that the points where diagonals ED and AF intersect BC divide BC harmonically; that is, we want to show that $(UV, BC) = -1$.

Consider the pencils at A and F. We have

$$(UV, BC) = A(UV, BC) = (UH, DE) = F(UV, CB) = (UV, CB).$$

Now recall that if $(UV, BC) = k$, then $(UV, CB) = 1/k$. Therefore,

$$(UV, BC) = 1/(UV, BC)$$

so that $(UV, BC)^2 = 1$; that is, (UV, BC) is either 1 or -1. Since U and V separate B and C, we must have $(UV, BC) = -1$.

□

Example 16.1.5. *Given points A, B, and C on a line l, construct the harmonic conjugate D of C using only a straightedge.*

Solution. The analysis figure is shown below.

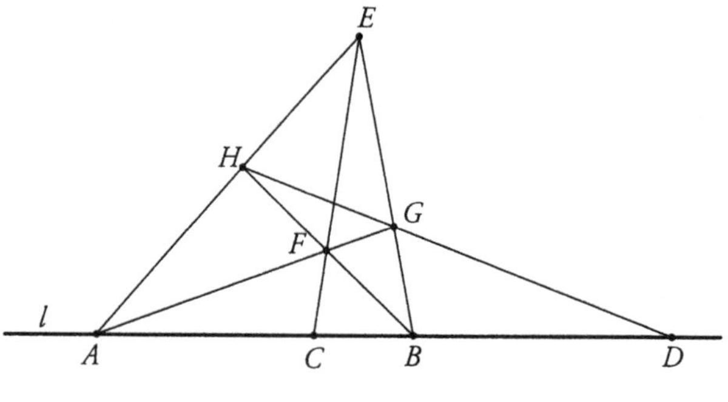

Construction:

1. Choose point E not on l.
2. Construct the lines EA, EC, and EB.
3. Choose a point F on EC.
4. Construct the lines FA and FB meeting EA and EB at G and H.
5. Construct the line HG and let D be the intersection of HG and l.

Justification:

The fact that D is the harmonic conjugate of C follows from Theorem 16.1.4.

□

Theorem 16.1.6. *Given a circle with unknown center, we cannot construct its center using a straightedge alone.*

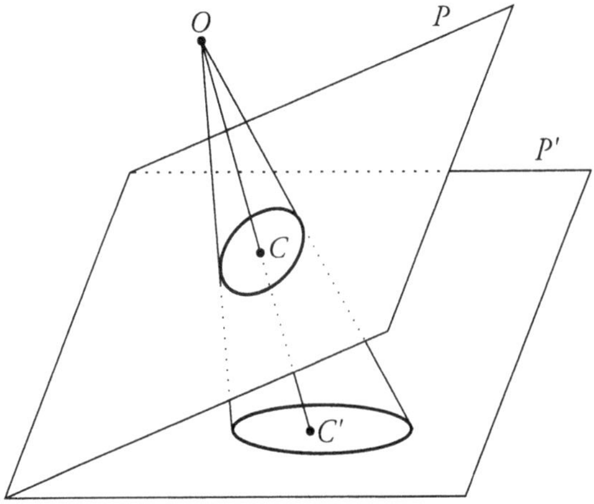

Proof. The proof is essentially the same as the proof of Theorem 16.1.1.

Using O as the center of projection, project a circle in plane $\mathcal{P}$ into an ellipse in plane $\mathcal{P}'$. The center C of the circle in plane $\mathcal{P}$ projects to a point C' in plane $\mathcal{P}'$, but C' is not the center of the ellipse in $\mathcal{P}'$.

Let M be the center of the ellipse. Then there is a point C'' in $\mathcal{P}'$ (not shown in the figure) such that M is the midpoint of $C'C''$. However, a sequence of instructions for the straightedge construction of C in $\mathcal{P}$ would "project" into the same sequence of instructions for the straightedge construction of C' in $\mathcal{P}'$.

However, with regard to the ellipse in $\mathcal{P}'$, why would the sequence of instructions always yield C' and never C''? As far as the operations in $\mathcal{P}'$ are concerned, there is no distinction between C' and C''.

□

Remark. When it was realized that a straightedge alone could not be used to solve all the construction problems that could be done with a straightedge and compass, the question arose as to what sort of minimal "equipment" was needed in addition to a straightedge. In 1822, Victor Poncelet asserted that all that was needed was a single circle with its center, and in 1833, Jacob Steiner gave a systematic proof of this fact. Such constructions are called ***Poncelet-Steiner constructions***.

Theorem 16.1.7. *If A and B are conjugate points on a line that cuts the circle β at C and D, then A and B are divided harmonically by C and D.*

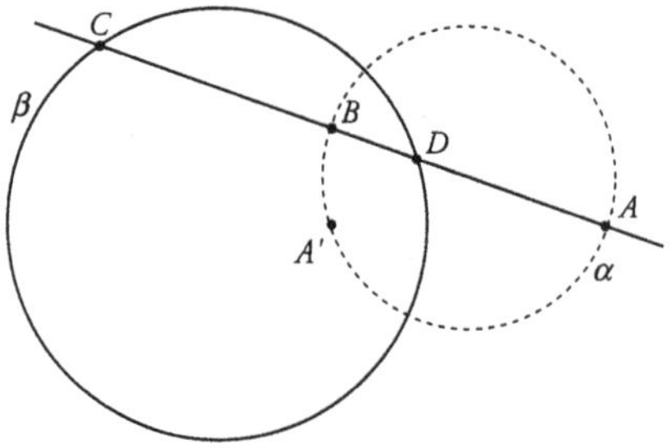

Proof. If B is the inverse of A, then this is part of Theorem 14.1.6.

Supposing that B is not the inverse of A, assume that A is outside β. Then B is inside since AB cuts β and B is on the polar of A by assumption. Let A' be the inverse of A and let α be the circle through A, A', and B. Recalling that two intersecting circles are orthogonal if one of the circles passes through two distinct points that are inverses with respect to the other circle (see Theorem 13.3.3), this means that α is orthogonal to β.

Since B is on the polar of A, the line $A'B$ is the polar of A, so $\angle AA'B$ is a right angle. From the converse of Thales' Theorem, AB is a diameter of α, so that β intersects the diameter of α at C and D. By Theorem 14.1.11, β divides the diameter AB of α harmonically; that is, C and D divide A and B harmonically.

□

The converse to the previous theorem is also true.

Theorem 16.1.8. *If a line AB cuts a circle β at C and D, and if C and D divide A and B harmonically, then A and B are conjugate points.*

Proof. Assume that A is outside β, and assume that E is a point on the line AB such that A and E are conjugate points. Then C and D divide A and E harmonically by the previous theorem; that is, $(AE, CD) = -1$.

However, by hypothesis, $(AB, CD) = -1$, so that

$$(AB, CD) = (AE, CD),$$

and it follows that $B = E$; that is, A and B are conjugate points.

□

The preceding theorems and examples lead to the straightedge construction of the polar of a point with respect to a given circle without its center.

Theorem 16.1.9. *Given a circle ω without its center, and given a point P outside ω, it is possible to construct the polar of P using only a straightedge.*

Proof. Through P, draw two lines intersecting the circle at B, C and D, E as shown in the figure below.

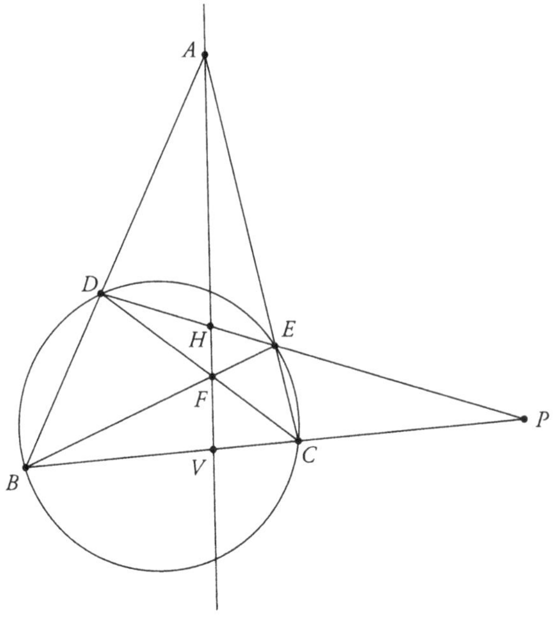

Let $A = BD \cap CE$ and $F = CD \cap BE$. Then the lines AB, AE, CD, and BE are the sides of a complete quadrilateral with diagonals AF, DE, and BC. From Theorem 16.1.4, AF and DE divide BC harmonically; that is, V is the harmonic conjugate of P, while from Theorem 16.1.8, P and V are conjugate points; that is, V is on the polar of P.

Similarly, H, the intersection of the diagonals AF and DE, is on the polar of P. Thus, VH is the polar of P, and the construction is complete.

□

Theorem 16.1.10. *Let $ABCD$ be a complete quadrangle inscribed in a circle. Let P, Q, and R be the points of intersection of the opposite sides. Then PQR is a self-polar triangle; that is, P is the pole of QR, Q is the pole of PR, and R is the pole of PQ.*

Proof. Referring to the figure below, note that the lines RD, RC, AC, and BD form a complete *quadrilateral*.

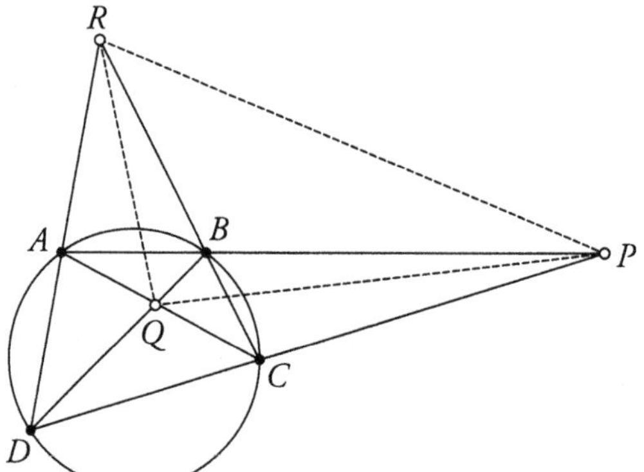

Thus, by the proof of the previous theorem, it follows that $RQ = p$, the polar of P, and similarly, $PQ = r$. Since Q is on both p and r, then by the reciprocation theorem, P and R are both on q; that is, $PR = q$.

□

Another approach to this is given at the end of Section 16.6.

16.2 Perspectivities and Projectivities

Given two planes $\mathcal{P}$ and $\mathcal{P}'$ and a point O not on either plane, a ***perspectivity*** or ***perspective transformation*** from $\mathcal{P}$ onto $\mathcal{P}'$ is a one-one correspondence between points X of $\mathcal{P}$ and points X' of $\mathcal{P}'$ such that if X is transformed into X', then the line XX' passes through O. The point O is called the ***center of perspectivity***.

The three-dimensional analogue of the projective plane is ***projective 3-space***, which is obtained by appending a "plane at infinity" to Euclidean 3-space. Note that the appended plane is a projective plane.

The same definitions of "perspectivity" and "center of perspectivity" apply to projective 3-space.

Note.

1. The transformation is one-one and onto, so every point in $\mathcal{P}$ is sent to a point in $\mathcal{P}'$, and every point in $\mathcal{P}'$ is the image of some point in $\mathcal{P}$.
2. If the center of perspectivity is an ideal point, the transformation is called a ***parallel perspectivity***.

Questions.

1. What gets mapped to the ideal points on $\mathcal{P}'$?
2. Let $\mathcal{P}''$ be a plane parallel to $\mathcal{P}$. What gets mapped to the points X' on $\mathcal{P}' \cap \mathcal{P}''$?
3. Where are the ideal points on $\mathcal{P}$ mapped to?

Given two lines l and l' in a plane and a point O not on either line, a ***perspectivity*** or ***perspective transformation*** from l onto l' is a one-one correspondence between points X of l and points X' of l' such that if X is transformed into X', then the line XX' passes through O. The point O is called the ***center of perspectivity***.

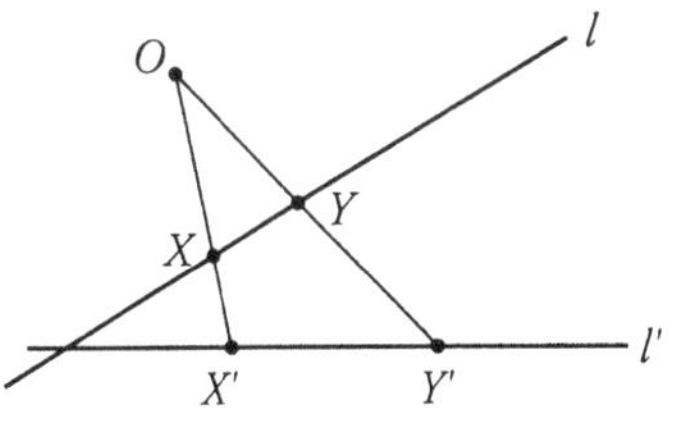

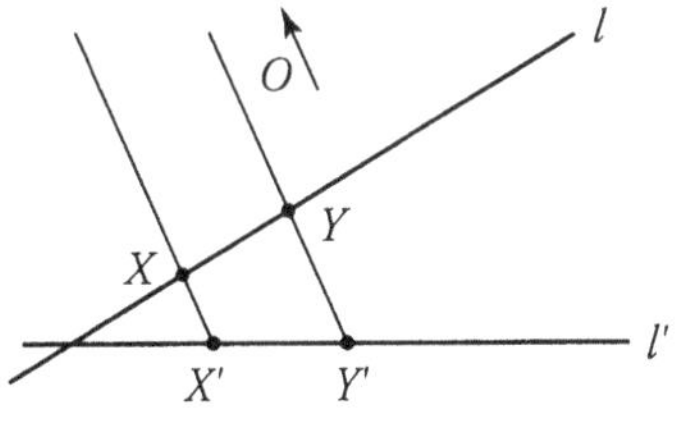

If O is an ideal point, as illustrated in the figure on the right above, the result is called a ***parallel perspectivity***. As before, the transformation is one-one and onto.

A finite sequence of perspectivities in a plane is called a ***projectivity***. Note that a perspectivity is always a projectivity, but a projectivity is not necessarily a perspectivity.

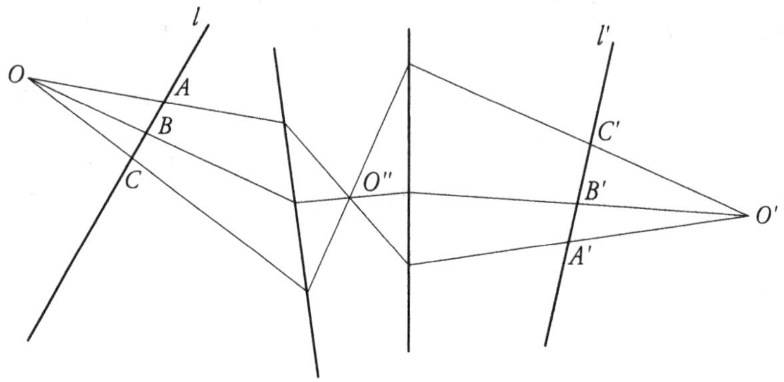

Perspectivities and projectivities from one plane to another in 3-space are defined similarly.

We will use lowercase Greek letters to denote perspectivities and projectivities. Two projectivities π_1 and π_2 from l to l' are ***equal*** if $\pi_1(X) = \pi_2(X)$ for each X in l. Note that for four perspectivities $\sigma_1 \neq \sigma_1'$ and $\sigma_2 \neq \sigma_2'$, it is possible that the projectivities $\sigma_2 \circ \sigma_1$ and $\sigma_2' \circ \sigma_1'$ are equal.

Effect on Euclidean Properties

In the following table, the properties on the left are preserved by projectivities, and the properties on the right are not preserved by projectivities.

Preserved	Not Preserved
• points	• midpoints
• straight lines	• distance
• collinearity	• angle size
• concurrency	• circularity
• incidence	• order
• triangles	
• cross ratios	

Note that the table on the previous page indicates that ***cross ratios are preserved by perspectivities***. This fact is an immediate consequence of Theorem 15.1.7, and it plays a crucial role in projective geometry.

The following figure illustrates that order is not preserved.

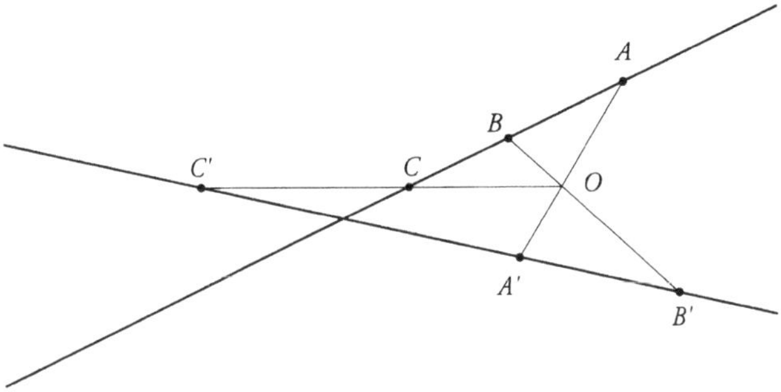

Projectivities in 2-Space

You may recall from linear algebra that a linear transformation from $\mathbb{R}^n$ to $\mathbb{R}^n$ is completely determined by its action on n linearly independent points; that is, if the points

$$a_1, \quad a_2, \quad \ldots, \quad a_n$$

are linearly independent and are mapped respectively to

$$a_1', \quad a_2', \quad \ldots, \quad a_n',$$

then this enables us to determine what every other point is mapped to.

There is an analogous situation regarding perspectivities and projectivities that we explore in this section. In particular, the following questions arise:

1. How many points are required to completely determine a perspectivity?
2. How many points are required to completely determine a projectivity?

Before addressing these questions, we remind ourselves that two projectivities are equal if and only if they have the same effect on every point. The two projectivities could be the composition of different perspectivities, or even different numbers of perspectivities, but that is immaterial to the definition of equality.

Theorem 16.2.1. *Given two distinct points A and B on a line l and two distinct points A' and B' on a line l', with $l \neq l'$ and none of the four points being $l \cap l'$, there is a unique perspectivity that takes A to A' and B to B'.*

Proof. Let $O = AA' \cap BB'$. Then the perspectivity centered at O takes A to A' and B to B'.

Call this perspectivity π_1, and suppose that π_2 is another perspectivity that takes A to A' and B to B'. The center of π_2 is $AA' \cap BB' = O$, and since a perspectivity is completely determined by its center, π_1 and π_2 are identical.

□

The proof of the corresponding theorem for projectivities is a bit more delicate.

Theorem 16.2.2. *Suppose l and l' are two distinct lines, that A, B, and C are three distinct points on l, and that A', B', and C' are three distinct points on l'. Then there is a unique projectivity from l to l' that takes A to A', B to B', and C to C'.*

Proof. First we prove existence. We may assume that $A \neq A'$ because at least two of A, B, and C differ from $l \cap l'$.

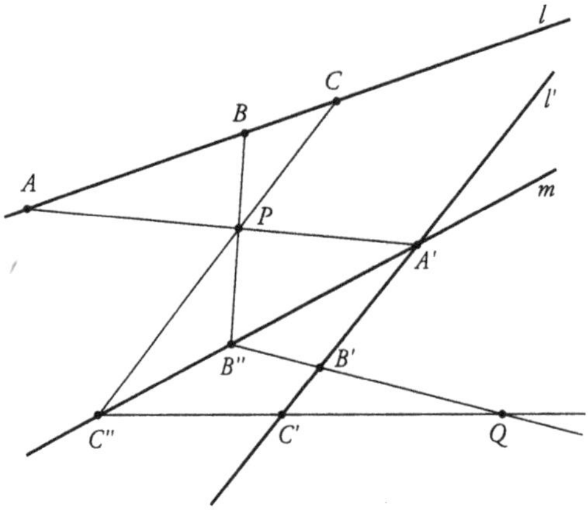

Draw a line m through A' that does not coincide with l' and that misses A. Pick a point P on AA' other than A or A'. Using P as the center of perspectivity, map l onto m and denote this perspectivity by σ_1. This takes B to B'' and C to C''.

Let $Q = B'B'' \cap C'C''$. Using Q as the center of perspectivity, map m onto l' and denote this perspectivity by σ_2. Let $\pi = \sigma_2 \circ \sigma_1$. The projectivity π maps A to A', B to B', and C to C'.

Next we prove uniqueness. We have to show that if π_1 and π_2 are projectivities that both take A, B, and C to A', B', and C', respectively, then

$$\pi_1(X) = \pi_2(X)$$

for every X in l.

Suppose that for some X we have

$$\pi_1(X) = X_1,$$
$$\pi_2(X) = X_2.$$

Since projectivities preserve cross ratios, we have

$$(AB, CX) = (A'B', C'X_1),$$
$$(AB, CX) = (A'B', C'X_2),$$

so that

$$(A'B', C'X_1) = (A'B', C'X_2),$$

which implies that

$$X_1 = X_2.$$

□

Theorem 16.2.3. *(The Fundamental Theorem of Projective Geometry)*

A projectivity in the plane from a line l to a line l' is completely determined by its action on three distinct points.

Proof. There are two cases to consider: where l and l' are different and where $l = l'$. The first case is the previous theorem.

For the case where $l = l'$, we will show that given three distinct points A, B, and C on l and three more distinct points A', B', and C' also on l, there is a unique projection that takes A to A', B to B', and C to C'. To establish the existence of such a projection, draw a line m different than l, and let σ be a perspectivity that maps l onto m. Let A'', B'', and C'' be the images of A, B, and C under σ. By the previous theorem, there is a projectivity π that maps m to l and which takes A'', B'', and C'' to A', B', and C', respectively. The composition $\pi \circ \sigma$ is a projectivity from l to l that takes A, B, and C to A', B', and C', respectively. This establishes the existence of a projectivity, and its uniqueness follows as before via cross ratios.

□

Corollary 16.2.4. *A projectivity from a line l to a different line l' can always be expressed as a sequence of two or fewer perspectivities or a sequence of three or fewer perspectivities if $l = l'$.*

16.3 Line Perspectivities and Line Projectivities

The perspectivities and projectivities described so far are sometimes referred to as being ***central perspectivities*** and ***central projectivities***. These two notions can be dualized. Two pencils L and L' of lines are said to be ***perspective from a line*** p if there is a mapping from L to L' that takes each line x of L to a line x' of L' in such a way that p, x, and x' are concurrent. The line p is called the ***axis of perspectivity***. Such a perspectivity is called a ***line perspectivity***, and a finite sequence of line perspectivities is called a ***line projectivity***.

It is important to realize that any theorem about central projectivities has a dual theorem about line projectivities that we can obtain free of charge. For example, here is the dual of the Fundamental Theorem:

Theorem 16.3.1. *A line projectivity from a pencil L to a pencil L' is completely determined by its action on three distinct lines a, b, and c of L.*

When the words *projectivity* and *perspectivity* are used without modifiers, they are understood to mean *central projectivity* and *central perspectivity* unless the context makes it perfectly obvious that the dual meaning should be used.

16.4 Projective Geometry and Fixed Points

When Is a Projectivity a Perspectivity?

Given a transformation π, any point A for which

$$\pi(A) = A$$

is called a ***fixed point*** of π.

Note that if π is a transformation from a line l to a line m, then for $l \neq m$ the only possible fixed point of π is $l \cap m$.

Theorem 16.4.1. *Suppose l and m are distinct lines. A projectivity π from l to m is a perspectivity if and only if π has a fixed point.*

Proof. If π is a perspectivity, then $l \cap m$ is a fixed point.

On the other hand, suppose that $A = l \cap m$ is a fixed point of π (the only point that could possibly be a fixed point). Let B and C be two other points on l, and let $B' = \pi(B)$ and $C' = \pi(C)$.

Now, by Theorem 16.2.1, there is a unique perspectivity σ from l to m that takes B and C on l to B' and C' on m, respectively.

However, σ also maps A to $A' = A$. Consequently, both π and σ map A, B, and C to A', B', and C', respectively, and so $\pi = \sigma$ by the Fundamental Theorem of Projective Geometry.

□

Theorem 16.4.2. *A projectivity π from a line to itself is a perspectivity if and only if it has at least three fixed points.*

We leave the proof as an exercise.

Alternate Characterizations

Theorem 16.4.3. *Suppose that*

$$\pi : k \to l \qquad \textit{and} \qquad \sigma : l \to m$$

are two perspectivities with different centers. Then $\sigma \circ \pi$ is a perspectivity if and only if k, l, and m are concurrent.

Proof. Let A and A' be the centers of perspectivities for π and σ.

There are two things to prove:

(1) If $\sigma \circ \pi$ is a perspectivity, then k, l, and m are concurrent.

(2) If k, l, and m are concurrent, then $\sigma \circ \pi$ is a perspectivity.

We will prove (1) and leave (2) as an exercise.

Assuming that $\sigma \circ \pi$ is a perspectivity from k to m, let P be the center of the perspectivity.

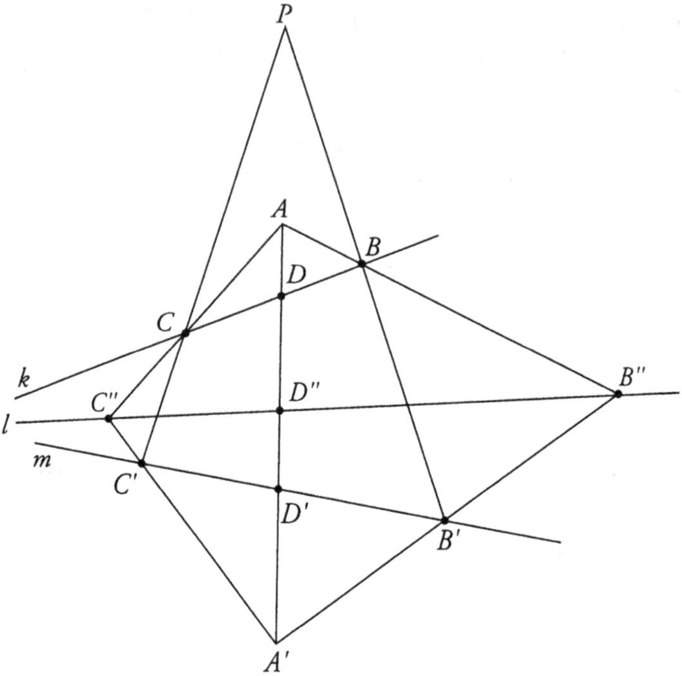

Let B and C be distinct points on k, and let

$$B'' = \pi(B) \qquad \text{and} \qquad B' = \sigma(B'')$$

and

$$C'' = \pi(C) \qquad \text{and} \qquad C' = \sigma(C'').$$

Then $P = BB' \cap CC'$.

Now draw the line AA' and let D, D'', and D' be defined as follows:

$$D = AA' \cap k,$$
$$D'' = AA' \cap l,$$
$$D' = AA' \cap m.$$

Then π maps D to D'' and σ maps D'' to D'.

Consequently, $\sigma \circ \pi$ maps D to D', and since $\sigma \cap \pi$ is really a perspectivity with center P, the line DD' must pass through P. Also, since $\overleftrightarrow{DD'} = \overleftrightarrow{AA'}$, this means that $\overleftrightarrow{AA'}$ passes through P.

Therefore, $\triangle ABC$ and $\triangle A'B'C'$ are perspective from P; that is, they are *copolar*. Hence, the triangles are perspective from a line; that is, they are *coaxial* by Desargues'

Two Triangle Theorem. However,

$$AB \cap A'B' = B'' \qquad \text{and} \qquad AC \cap A'C' = C'',$$

and so the axial line is $B''C''$, namely, l. Since $BC = k$ and $B'C' = m$, the point $k \cap m$ is on l, showing that k, l, and m are concurrent.

□

16.5 Projecting a Line to Infinity

Straight lines and incidence properties are preserved by projections, and this is sometimes advantageous. In the figure below, the quadrilateral $ABCD$ lies in a plane $\mathcal{P}$. Another plane $\mathcal{P}'$ intersects $\mathcal{P}$ in a line parallel to the diagonal BD of the quadrilateral. The point O is the center of a perspectivity from $\mathcal{P}$ to $\mathcal{P}'$, and O is positioned so that the plane defined by the points O, B, and D is parallel to $\mathcal{P}'$. With this perspectivity, points B and D are projected to ideal points B' and D' since the lines OB and OD are parallel to $\mathcal{P}'$. Points A and C are projected to ordinary points A' and C'. In this particular case, $C' = C$, since the point C happens to be on the line $\mathcal{P} \cap \mathcal{P}'$.

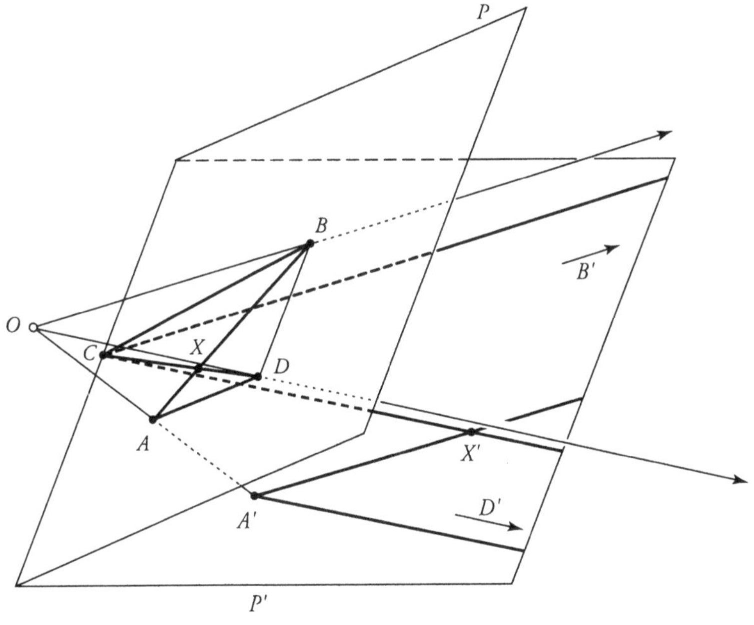

In the plane $\mathcal{P}'$, the line $B'D'$ is the ideal line, that is, the line at infinity. Consequently, the process above is called *sending the line BD to infinity*.

The projected lines $A'B'$ and $C'B'$ are necessarily parallel, as are the lines $A'D'$ and $C'D'$. This is all that we need to know in order to depict the figure with the line sent to infinity. In figure (a) below we have quadrilateral $ABCD$, with its projected image $A'B'C'D'$ shown in figure (b), and there is really no need to draw the planes $\mathcal{P}$ and $\mathcal{P}'$.

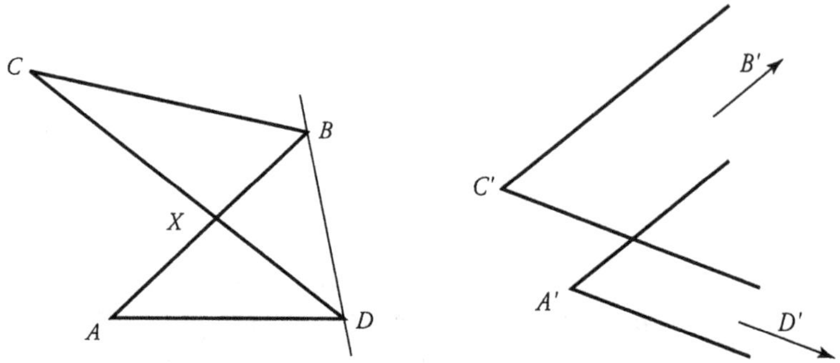

(a) (b)

We illustrate the use of this technique to prove the following version of Pappus' Theorem.

Theorem 16.5.1. *If the three lines APP', UQQ', and URR' meet two lines OX and OX' at P, Q, R and P', Q', R', respectively, then the points*

$$PQ' \cap P'Q, \qquad PR' \cap P'R, \qquad \text{and} \qquad QR' \cap Q'R$$

are collinear.

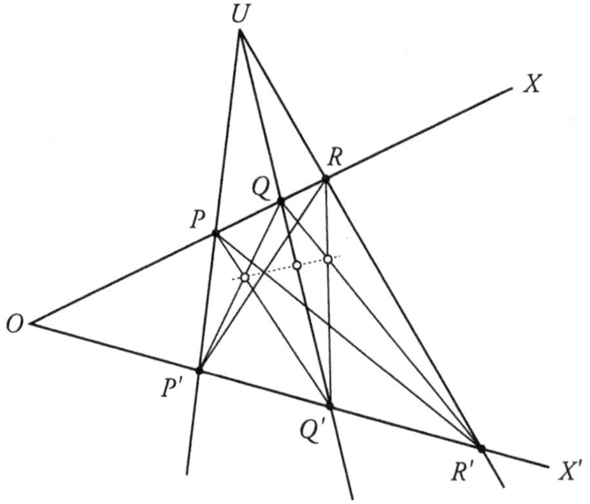

Proof. Send the line OU to infinity. In the diagram below, we use the same letter before and after the projection.

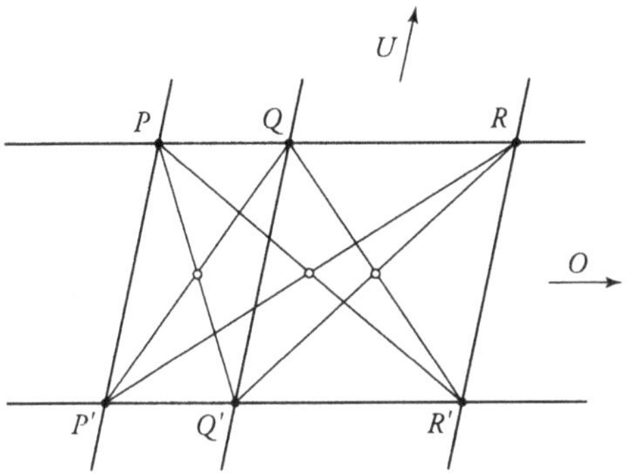

The lines PP', QQ', and RR' are parallel, as are PR and $P'R'$. Then $PQ' \cap P'Q$ is the intersection of the diagonals of the parallelogram $PQQ'P'$. Similarly, $PR' \cap P'R$ and $QR' \cap Q'R$ are the intersections of the diagonals of parallelograms, so all three points lie on a line midway between PR and $P'R'$.

□

As another example, we use the same technique to prove part of Desargues' Theorem.

Example 16.5.2. *Show that copolar triangles are coaxial.*

Solution. In the figure below, triangles ABC and $A'B'C'$ are perspective from the point O. We want to show that the intersections of the corresponding sides, namely, P, Q, and R, are collinear.

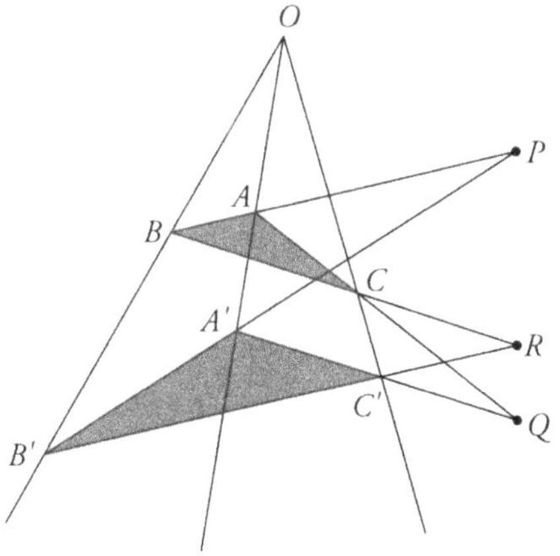

A solution may be obtained by sending the line QR to infinity and then showing that the projection also sends point P to the ideal line. The reverse projection then maps the ideal line back to the original line QR, and since it preserves incidence, the point P must lie on the line QR.

The projection of the line QR to infinity is shown below. As before, we use the same letters before and after the projection.

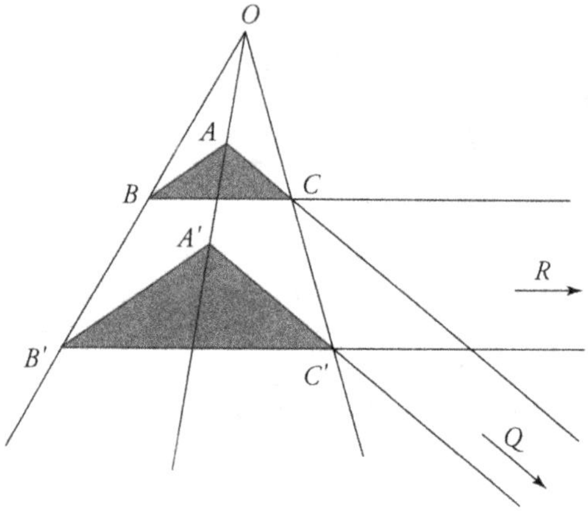

For triangles OBC and $OB'C'$, since BC and $B'C'$ are parallel, we have

$$\frac{OB}{BB'} = \frac{OC}{CC'}.$$

For triangles OAC and $OA'C'$, since AC and $A'C'$ are parallel, we have

$$\frac{OA}{AA'} = \frac{OC}{CC'}.$$

Consequently,

$$\frac{OB}{BB'} = \frac{OA}{AA'},$$

which implies that AB is parallel to $A'B'$ and, thus, the projected point

$$P = AB \cap A'B'$$

is on the ideal line.

□

16.6 The Apollonian Definition of a Conic

Originally, the ancient Greeks defined conic sections as cross sections of particular types of right circular cones. Apollonius recognized that all of the conic sections can be obtained from any given circular cone. He defined a circular cone as follows:

> Let K be a point and ω a circle whose plane does not contain K. If a straight line, passing always through K, be made to move around the circumference of ω, the moving straight line will trace out the surface of a ***circular cone***.

The line through K that passes through the center of ω is called the ***axis*** of the cone. If the axis of the cone is perpendicular to the plane of the circle, the cone is called a ***right circular cone***, otherwise the cone is called an ***oblique circular cone***. Note that the cone has two components, so they are often called ***two-napped cones***.

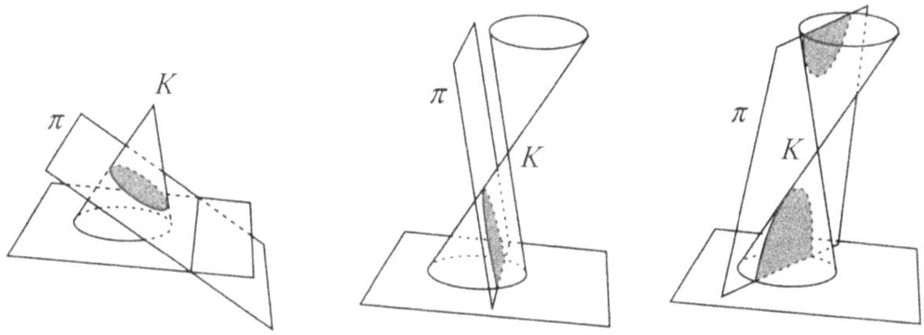

Ellipse Parabola Hyperbola

Apollonius defined a (proper) ***conic section*** as a curve that is formed by intersecting a plane with a circular cone in such a way that the plane does not contain the vertex K of the cone. The figure above shows what happens as the tilt of the plane is increased. When the plane is tilted so that it intersects only one nappe of the cone and is not parallel to any of the generating lines of the cone, the result is an ellipse. If the plane is parallel to one of the lines generating the cone, but it still only intersects one nappe, the result is a parabola. When it intersects both nappes of the cone, the result is a hyperbola.

It is not too difficult to derive the Cartesian equation of the conics from Apollonius' description. Here is how it is done for the ellipse. In the figure on the following page, the plane ξ, and any plane parallel to it, cuts the oblique circular cone in a circle. The plane π is another plane that cuts the cone, forming the curve α. The planes ξ and π meet in a straight line m.

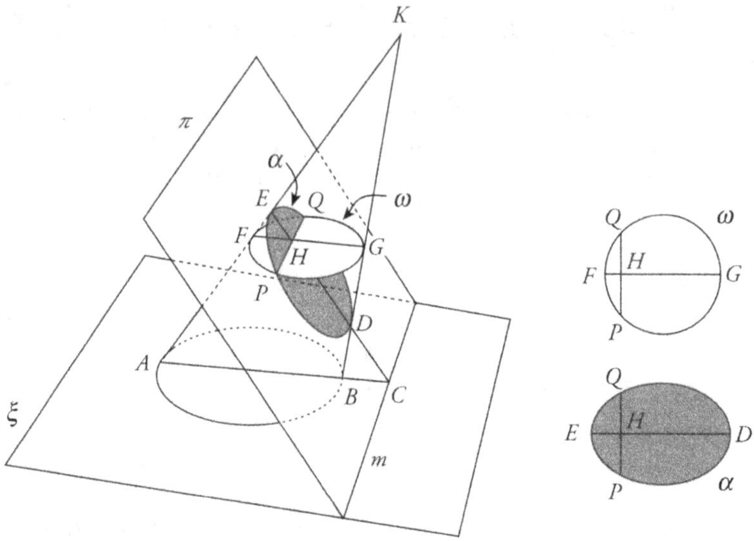

(a) (b)

In the plane ξ, the line AB is a diameter of the circle and is perpendicular to m and $C = AB \cap m$. The plane of the triangle ABK cuts the plane π in the line EDC. To obtain a Cartesian equation for α, let P be an arbitrary point on α, and through P pass a plane ϕ (not shown in the diagram) parallel to ξ cutting the cone in a circle ω, as shown in figure (a) above.

The intersection of ϕ and π is a line parallel to m cutting ω at P and Q. The plane ϕ cuts AK at F and BK at G, so that FG is a diameter of the circle ω. The lines ED, FG, and PQ are concurrent at H.

Let E be the point $(0, 0)$ in π, let x be the distance EH, and let y be the distance PH so that P is the point (x, y). Since $PQ \perp FG$, by the power of the point H with respect to ω (see figure (b) above), we have $PH^2 = FH \cdot HG$. This gives us

$$y^2 = FH \cdot HG. \tag{1}$$

From similar triangles EFH and EAC, we have $FH/AC = EH/EC$, so that

$$FH = \frac{AC}{EC}x. \tag{2}$$

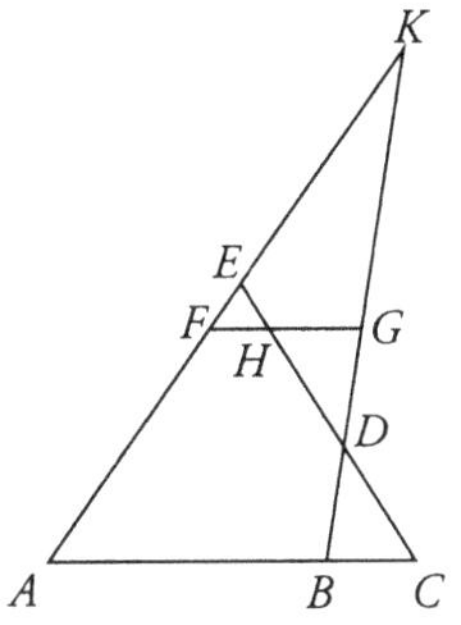

From similar triangles HGD and CBD, we have $HG/CB = HD/CD$. Since $HD = ED - x$, we get

$$HG = \frac{CB}{CD}(ED - x). \tag{3}$$

Letting $ED = 2a$, and substituting (2) and (3) into (1), we obtain

$$y^2 = \frac{AC \cdot CB}{EC \cdot CD} x(2a - x).$$

Denoting the positive quantity $(AC \cdot CB)/(EC \cdot CD)$ by k^2, we have

$$y^2 = k^2 x(2a - x),$$

and so

$$\frac{(x-a)^2}{a^2} + \frac{y^2}{a^2k^2} = 1.$$

This shows how the Cartesian equation for the ellipse arises from the Apollonian definition of the conic.

Poles and Polars of Conics

The Apollonian definition of conics can be described by saying that every conic can be interpreted as being the image of a circle under a central perspectivity. Thus, we can state without proof:

Theorem 16.6.1. *Every theorem about the incidence properties of straight lines and circles remains true when the word "circle" is replaced by the word "conic."*

For example, both Pascal's Mystic Hexagon Theorem and Brianchon's Theorem are true for parabolas, ellipses, and hyperbolas.

One of the important consequences of Theorem 16.6.1 is that in order to prove something about the incidence properties of straight lines and conics, we can reduce it to proving the same assertion about circles. There is a caveat, however, in that the properties must be described purely in terms of incidence properties.

For example, the notions of pole and polar for a circle are useful concepts that were not initially described in terms of incidence properties. If P is a point outside a circle ω, the polar of P can be described using incidence properties: let A and B be the points where the tangents from P meet the circle. Then $\overleftrightarrow{AB}$ is the polar of P. If P is a point inside the circle, a description of the polar in terms of the incidence is not immediately evident.

Theorem 16.6.2. *Let P be a point not on the circle ω, and let l and m be lines through P, with $l \cap \omega = \{A, B\}$ and $m \cap \omega = \{C, D\}$. Let a, b, c, and d be the tangent lines through A, B, C, and D. Then the line through $a \cap b$ and $c \cap d$ is the polar of P.*

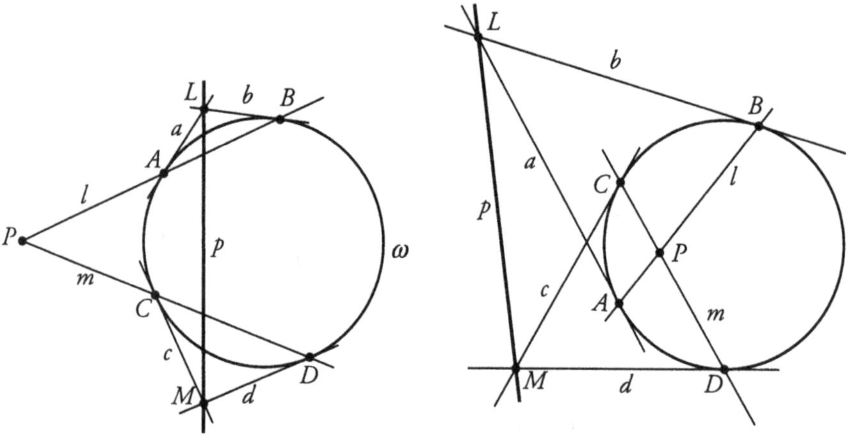

Proof. Let $L = a \cap b$ and $M = c \cap d$. Then L is the pole of l, and M is the pole of m. Since P is on l and m, the Reciprocation Theorem implies that L and M are on p. □

We indicated earlier that the ***polar*** of a point P on a conic is the line p that is tangent to the conic at P. The lemma on the following page allows us to describe the polar of a point P not on the conic entirely in terms of incidence properties.

Definition. Suppose ω is a conic. If P is not on the conic, then the ***polar*** of P is the line p described in the preceding theorem with the word "circle" replaced by the word "conic".

Theorems 16.1.8 and 16.1.9 illustrate how to construct the polar using only a straightedge and without necessarily drawing tangents. The next few lemmas provide an alternate approach that illustrates the connection with Pascal's Mystic Hexagon Theorem. In the lemmas, we follow the convention that lowercase letters refer to the polars of points designated by the corresponding uppercase letters.

Lemma 16.6.3. *Suppose that A, E, B, and D are four distinct points on a circle ω. Let $P = AE \cap DB$ and $Q = AB \cap DE$. Let $S = e \cap b$ and $T = a \cap d$. Then P, Q, S, and T are collinear.*

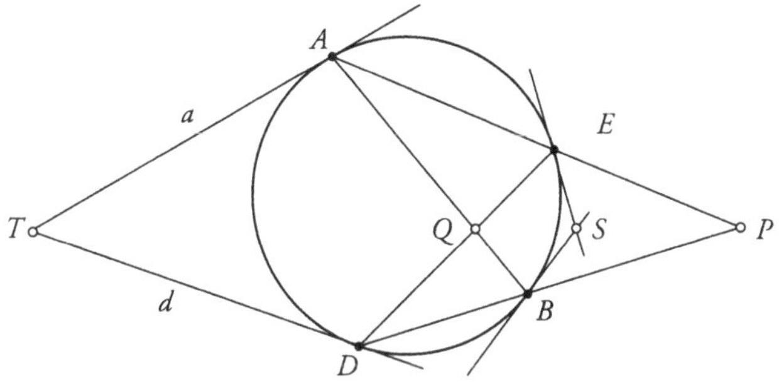

Proof. Introduce points F and C on the circle, as shown in the figure below, so that $ABCDEF$ is an inscribed hexagon. By Pascal's Mystic Hexagon Theorem, the opposite edges meet in three collinear points Q, M, and N, as depicted in the figure.

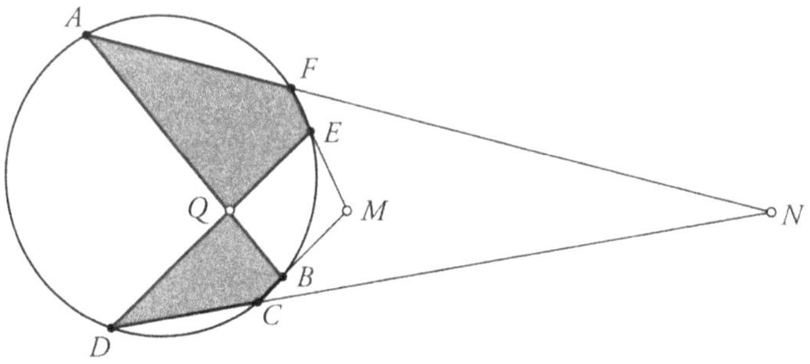

Keep the points A, E, B and D fixed, and let $F \to E$ and $C \to B$ along the circle. As this happens, M and N move but Q, M, and N remain collinear. The limiting points of M and N are S and P, respectively, so Q, S, and P are collinear.

Now apply a similar procedure to A and D, as shown on the following page. Here, the points M, Q, and N are collinear.

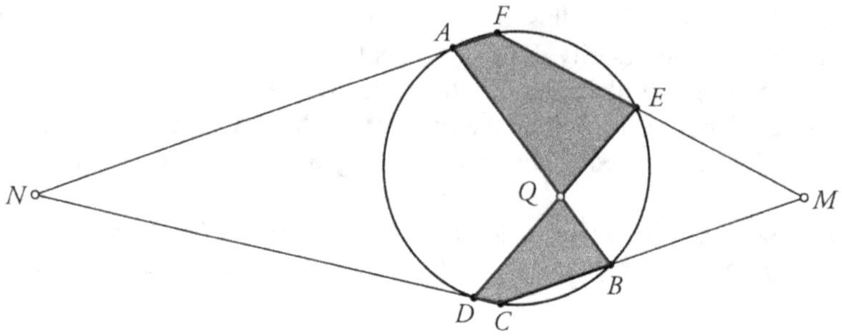

Letting $F \to A$ and $C \to D$ along the circle, the points M and N converge to P and T, respectively, and so P, Q, and T are collinear.

Consequently P, Q, S, and T are collinear.

□

The following is proven in a similar way:

Lemma 16.6.4. *Suppose that A, E, B, and D are distinct points on a circle, as in the previous lemma. Let $R = AD \cap EB$ and let $Q = AB \cap DE$, as in the previous lemma, and let $U = a \cap e$ and $V = d \cap b$. Then R, Q, U, and V are collinear.*

Theorem 16.6.5. *Suppose that A, E, B, and D are distinct points on a circle. Let $P = AE \cap BD$, $Q = AB \cap DE$, and $R = AD \cap BE$. Then RQ is the polar of P, PQ is the polar of R, and PR is the polar of Q.*

Proof. In the proof we continue to employ the convention that lowercase letters refer to the polars of points designated by the corresponding uppercase letters.

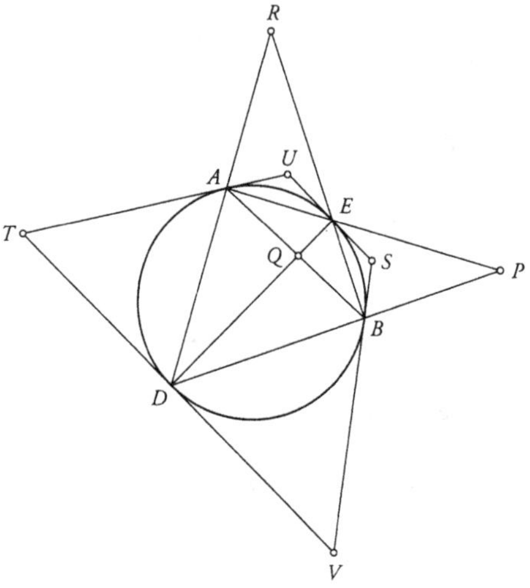

Introduce the polars a, b, c, and d, and let $U = a \cap e$, $S = e \cap b$, $V = b \cap d$, and $T = a \cap d$.

By the previous lemmas, the points P, Q, S, and T are collinear, as are R, U, Q, and V. By Theorem 16.6.2, UV is the polar of P, which implies that RQ is the polar of P, as claimed. Similarly, ST is the polar of R, which implies that PQ is the polar of R.

Since Q is on r, R must be on q by the Reciprocation Theorem. Similarly, since Q is on p, P must be on q. That is, R and P are on q, so $RP = q$.

□

16.7 Problems

1. Prove that the construction of the midpoint in Example 16.1.2 works by using the fact that the lines l, m, and CO are the diagonals of a complete quadrilateral.

2. (a) In the complete quadrilateral the sides intersect at points A, B, C, D, E, and F, as shown below, and the dashed lines are the three diagonals with intersection points H, U, and V. Prove that if V is the midpoint of the segment BC, then ED and BC are parallel.

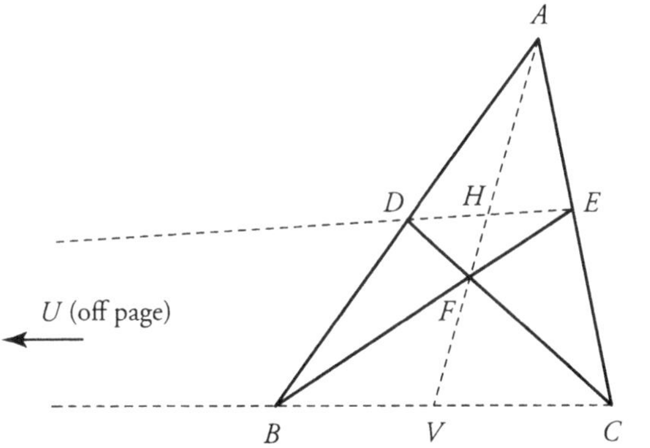

   (b) Given a line segment AB with midpoint M and a point P not on the line AB, explain how to construct a line through P parallel to AB using only a straightedge.

3. Using a straightedge alone, is it possible to construct a right angle? Explain.

4. Using only a straightedge, inscribe a square in a given circle whose center is also given.

5. Given a circle ω without its center, and given a point P outside ω, construct the tangents to ω from P.

6. Given a circle ω without its center, and given a point P inside ω, construct the polar of P with respect to ω.

7. Given a circle ω without its center, and given a point P on ω, construct the tangent to ω at P.

8. Prove that if $\{C, D\}$ divides AB harmonically on the line l and $\{C', D'\}$ divides $A'B'$ harmonically on the line l', then there exists a unique projectivity that maps A, B, C, and D to A', B', C', and D', respectively.

9. Given that π is a central perspectivity from l to l', show that the information that π maps A, B to A, B', respectively, is ***not*** sufficient to determine π uniquely.

10. In the figure below, there is a unique projectivity that takes A, B, and C to A', B', and C', respectively. Using only a straightedge construct the image of X under the projectivity.

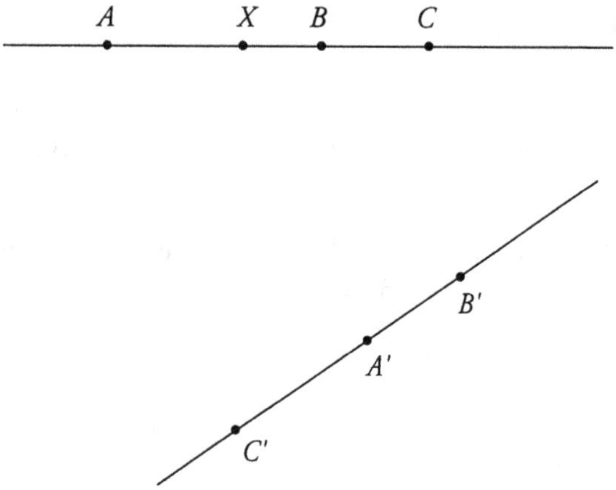

11. In the figure below, there is a line perspectivity from the pencil at A to the pencil at B that takes l to l' and m to m'. Using only a straightedge, construct the image of n under this perspectivity.

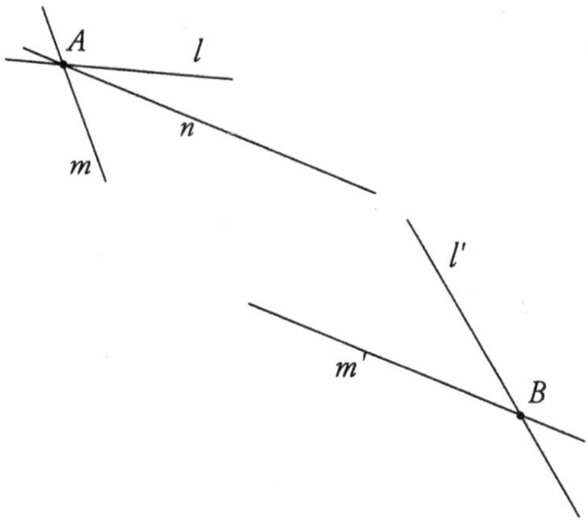

12. Theorem 16.4.3 states: suppose that $\pi : k \to l$ and $\sigma : l \to m$ are two perspectivities with different centers. Then $\sigma \circ \pi$ is a perspectivity if and only if k, l, and m are concurrent. State the dual of this theorem.

13. Give an example of a projectivity from m onto m, the same line, with two distinct fixed points, but which is not a perspectivity.

14. Show that coaxial triangles are copolar by projecting the polar axis to infinity; that is, given triangles ABC and $A'B'C'$ such that the points

$$P = AB \cap A'B', \qquad Q = AC \cap A'C', \qquad R = BC \cap B'C'$$

are collinear, show that AA', BB', and CC' are concurrent by projecting the line PQR to infinity.

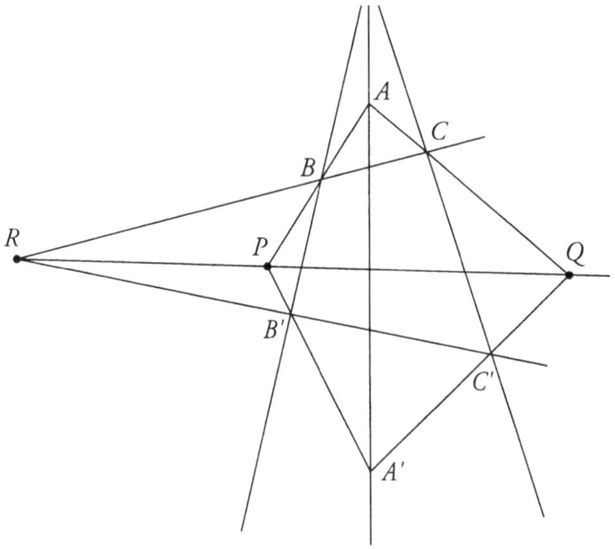

15. Let $PQRS$ be a complete quadrangle, and let

$$A = PQ \cap RS, \quad B = PR \cap SQ, \quad C = PS \cap QR, \quad D = AB \cap PS.$$

Show that $(PS, DC) = -1$ by projecting the line AC to infinity.

16. Is it possible to plant 10 trees in 10 straight rows, with 3 trees in each row? *Hint*: Desargues says it is!

17. The points D, E, and F are collinear and lie on the sides BC, CA, and AB, respectively, of triangle ABC. The line BE cuts CF at X, the line CF cuts AD at Y, and the line AD cuts BE at Z. Prove that AX, BY, and CZ are concurrent by projecting the line BY to infinity.

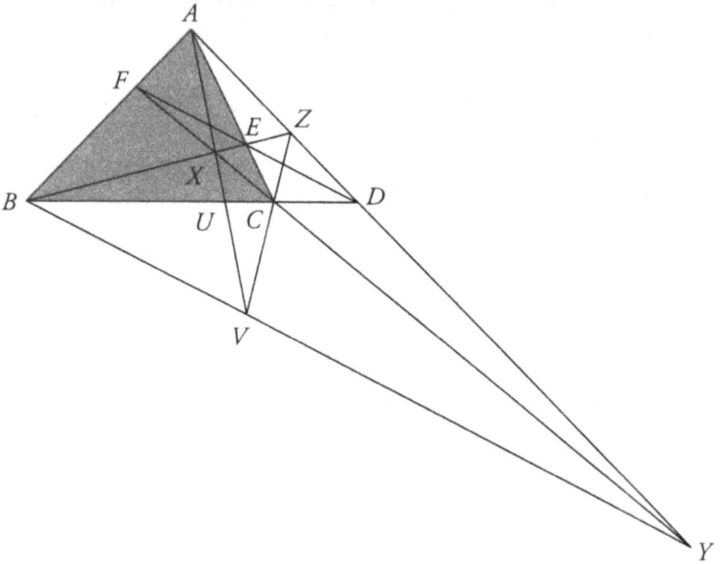

18. Find two coaxial triangles at nonideal points P, Q, and R whose projected image is as shown in the figure below and where P', Q', and R' are ideal points.

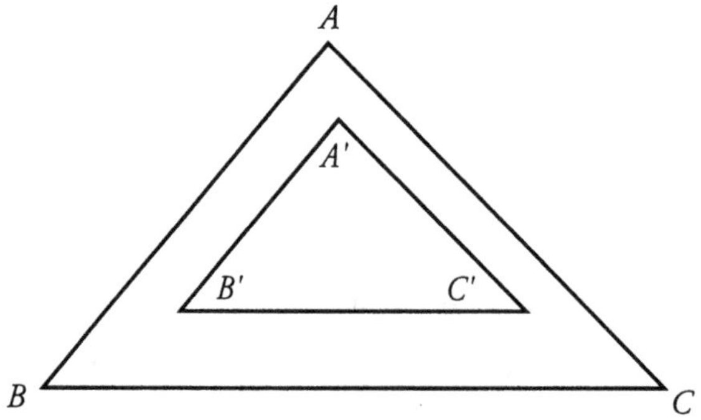

Hint: $AB \parallel A'B'$, $AC \parallel A'C'$, and $BC \parallel B'C'$.

19. Let π be a plane that cuts an oblique circular cone in such a way that one of the generating lines of the cone is parallel to π. Prove that the intersection of π with the cone can be described by the Cartesian equation $y = kx^2$.

BIBLIOGRAPHY

1. Agricola, I., and T. Friedrich, *Elementary Geometry*, American Mathematical Society, Providence, RI, 2008.

2. Altshiller-Court, N., *College Geometry*, Dover Publications, Inc., New York, NY, 2007.

3. Aref, M.N., and W. Wernick, *Problems and Solutions in Euclidean Geometry*, Dover Publications, Inc., New York, NY, 1968.

4. Baragar, A., *A Survey of Classical and Modern Geometries*, Prentice-Hall, Inc., Upper Saddle River, NJ, 2001.

5. Berger, M., *Geometry I*, Springer-Verlag, Berlin, Germany, 1987.

6. Berger, M., *Geometry II*, Springer-Verlag, Berlin, Germany, 1987.

7. Birkhoff, G.D., and R. Beatley, *Basic Geometry*, 3rd ed., Chelsea Publishing Company, New York, NY, 1959.

8. Boltyanski, V., and A. Soifer, *Geometric Etudes in Combinatorial Mathematics*, Center for Excellence in Mathematical Education, Colorado Springs, CO, 1991.

9. Brannan, D.A., Esplen, M.F., and J.J. Gray, *Geometry*, Cambridge University Press, Cambridge, England, 1999.

10. Burn, R.P., *Groups, A Path to Geometry*, Cambridge University Press, Cambridge, England, 1988.

Classical Geometry: Euclidean, Transformational, Inversive and Projective, First Edition. By I. E. Leonard, J. E. Lewis, A. C. F. Liu, G. W. Tokarsky

Copyright © 2014 John Wiley & Sons, Inc.

11. Byer, O., Lazebnik, F., and D.L. Smeltzer, *Methods for Euclidean Geometry*, Mathematical Association of America, Washington, DC, 2010.

12. Carslaw, H.S., *Non-Euclidean Plane Geometry and Trigonometry*, Chelsea Publishing Company, New York, NY, 1969.

13. Choquet, G., *Geometry in a Modern Setting*, Houghton Mifflin Company, Boston, MA, 1969.

14. Clark, D.M., *A Guided Inquiry Approach*, American Mathematical Society, Providence, RI, 2012.

15. Coxeter, H.S.M., *Introduction to Geometry*, 2nd ed., John Wiley & Sons, Inc., New York, NY, 1969.

16. Coxeter, H.S.M., *Projective Geometry*, 2nd ed., University of Toronto Press, Toronto, Canada, 1974.

17. Coxeter, H.S.M., and S.L. Greitzer, *Geometry Revisited*, Mathematical Association of America, Washington, DC, 1967.

18. Dieudonné, J., *Linear Algebra and Geometry*, Houghton Mifflin Company, Boston, MA, 1969.

19. Dodge, C.W., *Euclidean Geometry and Transformations*, Addison-Wesley Publishing Company, Inc., Reading, MA, 1972.

20. Eves, H., *A Survey of Geometry*, vols. 1 and 2, Allyn and Bacon, Boston, 1964.

21. Eves, H., *Great Moments in Mathematics before 1650*, Mathematical Association of America, Washington, DC, 1983.

22. Eves, H., *Great Moments in Mathematics after 1650*, Mathematical Association of America, Washington, DC, 1983.

23. Ewald, G., *Geometry: An Introduction*, Wadsworth Publishing Company, Inc., Belmont, CA, 1971.

24. Faber, R.L., *Foundations of Euclidean and Non-Euclidean Geometry*, Marcel Dekker, Inc., New York, NY, 1983.

25. Fishback, W.T., *Projective and Euclidean Geometry*, 2nd ed., John Wiley & Sons, Inc., New York, NY, 1969.

26. Fomin, D., Genkin. S., and I. Itenburg, *Mathematical Circles: Russian Experience*, American Mathematical Society, Providence, RI, 1996.

27. Frederickson, G.N., *Dissections: Plane and Fancy*, Cambridge University Press, Cambridge, England, 1997.

28. Frederickson, G.N., *Hinged Dissections: Swinging & Twisting*, Cambridge University Press, Cambridge, England, 2002.

29. Frederickson, G.N., *Piano-Hinged Dissections*, A. K. Peters, Natick, MA, 2006.

30. Golos, E.B., *Foundations of Euclidean and Non-Euclidean Geometry*, Holt, Rinehart, and Winston, Inc., New York, NY, 1968.

31. Greenberg, M.J., *Euclidean and Non-Euclidean Geometries, Development and History*, W. H. Freeman and Company, San Francisco, CA, 1974.

32. Grunbaum, B., and G.C. Shephard, *Tilings and Patterns*, W. H. Freeman, New York, NY, 1987.

33. Hadlock, C.R., *Field Theory and Its Classical Problems*, The Carus Mathematical Monographs, Mathematical Association of America, Washington, DC, 1978.

34. Hausner, M., *A Vector Approach to Geometry*, Dover Publications, Inc., Mineola, NY, 1998.

35. Heath, T.L., *Euclid: The Thirteen Books of the Elements*, vols. 1, 2, and 3, 2nd ed., Dover Publications, Inc., Mineola, NY, 1956.

36. Henle, M., *Modern Geometries*, Prentice-Hall, Inc., Upper Saddle River, NJ, 1996.

37. Hidetoshi, T., and T. Rothman, *Sacred Mathematics, Japanese Temple Geometry*, Princeton University Press, Princeton, NJ, 2008.

38. Hilbert, D., and S. Cohn-Vossen, *Geometry and the Imagination*, American Mathematical Society, Providence, RI, 1983.

39. Isaacs, I.M., *Geometry for College Students*, Brooks/Cole Publishing Company, Pacific Grove, CA, 2001.

40. Jacobs, H., *Geometry*, 2nd ed., W. H. Freeman, New York, NY, 1987.

41. Johnson, R.A., *Advanced Euclidean Geometry*, Dover Publications, Inc., Mineola, NY, 2007.

42. Kazarinoff, N.D., *Geometric Inequalities*, Mathematical Association of America, Washington, DC, 1961.

43. Kay, D.C., *College Geometry*. Holt, Rinehart and Winston, Inc., New York, NY, 1969.

44. Levi, H., *Foundations of Geometry and Trigonometry*, Robert E. Krieger Publishing Company, Huntington, NY, 1975.

45. Levi, H., *Topics in Geometry*, Robert E. Krieger Publishing Company, Huntington, NY, 1975.

46. King, J., and D. Schattschneider, *Geometry Turned On*, Mathematical Association of America, Washington, DC, 1997.

47. Klein, F., *Famous Problems of Elementary Geometry*, Dover Publications, Inc., New York, NY, 1956.

48. Libeskind, S., *Euclidean and Transformational Geometry, A Deductive Inquiry*, Jones and Bartlett Publishers, Sudbury, MA, 2008.

49. Liu, A., and P.J. Taylor, *International Mathematics Tournament of the Towns: 2002–2007*, Australia Mathematics Trust, Canberra, Australia, 2009.

50. Martin, G.E., *Geometric Constructions*, Springer-Verlag, New York, NY, 1997.

51. Maxwell, E.A., *Fallacies in Mathematics*, Cambridge University Press, Cambridge, England, 1969.

52. Melzak, Z.A., *Invitation to Geometry*, John Wiley & Sons, New York, NY, 1983.

53. Melzak, Z.A., *Companion to Concrete Mathematics*, vols. I and II, Dover Publications, Inc., Mineola, NY, 2007.

54. Moise, E.E., *Elementary Geometry from an Advanced Standpoint*, Addison-Wesley Publishing Company, Inc., Reading, MA, 1963.

55. Ogilvy, C.S., *Excursions in Geometry*, Dover Publications, Inc., Mineola, NY, 1969.

56. Pedoe, D., *A Course of Geometry for Colleges and Universities*, Cambridge University Press, Cambridge, England, 1970.

57. Petersen, J., *Methods and Theories for the Solution of Problems of Geometrical Construction*, Chelsea Publishing Company, New York, NY, 1969.

58. Posamentier, A.S., and C.T. Salkind, *Challenging Problems in Geometry*, Dover Publications, Inc., Mineola, NY, 1996.

59. Posamentier, A.S., *Advanced Euclidean Geometry*, Key College Press, Emeryville, CA, 2002.

60. Ringenberg, L.A., *College Geometry*, Robert E. Krieger Publishing Company, Huntington, NY, 1977.

61. Robinson, G. de B., *Vector Geometry*, Allyn and Bacon, Inc., Boston, MA, 1962.

62. Sally, J.D., and P.J. Sally, *Geometry: A Guide for Teachers*. American Mathematical Society, Providence, RI, 2011.

63. Sibley, T.Q., *The Geometric Viewpoint, A Survey of Geometries*, Addison-Wesley Longman, Reading, MA, 1998.

64. Smart, J., *Modern Geometries*, 4th ed., Brooks/Cole Publishing Company, Pacific Grove, CA, 1993.

65. Spain, B., *Analytical Geometry*, Pergamon Press Ltd., London, England, 1963.

66. Stehney, A.K., Milnor, T.K., D'Atri, J.E., and T.F. Banchoff, *Selected Papers on Geometry, The Raymond Brink Selected Mathematical Papers*, vol. 4, Mathematical Association of America, Washington, DC, 1979.

67. Stillwell, J., *Numbers and Geometry*, Springer-Verlag, Inc., New York, NY, 1998.

68. Storozhev, A.M., *International Mathematics Tournament of the Towns: 1997–2002*, Australia Mathematics Trust, Canberra, Australia, 2006.

69. Storozhev, A.M., and P.J. Taylor, *International Mathematics Tournament of the Towns: 1993–1997*, Australia Mathematics Trust, Canberra, Australia, 2003.

70. Taylor, P.J., *International Mathematics Tournament of the Towns: 1980–1984*, Australia Mathematics Trust, Canberra, Australia, 1993.

71. Taylor, P.J., *International Mathematics Tournament of the Towns: 1984–1989*, Australia Mathematics Trust, Canberra, Australia, 1992.

72. Taylor, P.J., *International Mathematics Tournament of the Towns: 1989–1993*, Australia Mathematics Trust, Canberra, Australia, 1994.

73. Tondeur, P., *Vectors and Transformations in Plane Geometry*, Publish or Perish, Inc., Houston, TX.

74. Veblen, O., and J.W. Young, *Projective Geometry*, vol. I, Blaisdell Publishing Company, New York, NY, 1938.

75. Wells, D., *The Penguin Dictionary of Curious and Interesting Geometry*, Penguin, London, England, 1991.

76. Yaglom, I.M., *Geometric Transformations I*, Mathematical Association of America, Washington, DC, 1962.

77. Yaglom, I.M., *Geometric Transformations II*, Mathematical Association of America, Washington, DC, 1968.

78. Yaglom, I.M., *Geometric Transformations III*, Mathematical Association of America, Washington, DC, 1973.

79. Yaglom, I.M., *Geometric Transformations IV*, Mathematical Association of America, Washington, DC, 2009.

80. Zwikker, C., *The Advanced Geometry of Plane Curves and Their Applications*, Dover Publications, Inc., New York, NY, 1963.

INDEX

Classical Geometry: Euclidean, Transformational, Inversive and Projective, First Edition. By I. E. Leonard, J. E. Lewis, A. C. F. Liu, G. W. Tokarsky
Copyright © 2014 John Wiley & Sons, Inc.